Nanotechnology in the Life Sciences

Series Editor

Ram Prasad
Department of Botany
Mahatma Gandhi Central University
Motihari, Bihar, India

D1806714

Nano and biotechnology are two of the 21st century's most promising technologies. Nanotechnology is demarcated as the design, development, and application of materials and devices whose least functional make up is on a nanometer scale (1 to 100 nm). Meanwhile, biotechnology deals with metabolic and other physiological developments of biological subjects including microorganisms. These microbial processes have opened up new opportunities to explore novel applications, for example, the biosynthesis of metal nanomaterials, with the implication that these two technologies (i.e., thus nanobiotechnology) can play a vital role in developing and executing many valuable tools in the study of life. Nanotechnology is very diverse, ranging from extensions of conventional device physics to completely new approaches based upon molecular self-assembly, from developing new materials with dimensions on the nanoscale, to investigating whether we can directly control matters on/in the atomic scale level. This idea entails its application to diverse fields of science such as plant biology, organic chemistry, agriculture, the food industry, and more.

Nanobiotechnology offers a wide range of uses in medicine, agriculture, and the environment. Many diseases that do not have cures today may be cured by nanotechnology in the future. Use of nanotechnology in medical therapeutics needs adequate evaluation of its risk and safety factors. Scientists who are against the use of nanotechnology also agree that advancement in nanotechnology should continue because this field promises great benefits, but testing should be carried out to ensure its safety in people. It is possible that nanomedicine in the future will play a crucial role in the treatment of human and plant diseases, and also in the enhancement of normal human physiology and plant systems, respectively. If everything proceeds as expected, nanobiotechnology will, one day, become an inevitable part of our everyday life and will help save many lives.

More information about this series at http://www.springer.com/series/15921

Ram Prasad • Busi Siddhardha
Madhu Dyavaiah
Editors

Nanostructures
for Antimicrobial
and Antibiofilm Applications

Springer

Editors
Ram Prasad
Department of Botany
Mahatma Gandhi Central University
Motihari, Bihar, India

Busi Siddhardha
Department of Microbiology
School of Life Sciences
Pondicherry University
Pondicherry, India

Madhu Dyavaiah
Department of Biochemistry & Molecular
Biology
School of Life Sciences
Pondicherry University
Pondicherry, India

ISSN 2523-8027   ISSN 2523-8035   (electronic)
Nanotechnology in the Life Sciences
ISBN 978-3-030-40339-3   ISBN 978-3-030-40337-9   (eBook)
https://doi.org/10.1007/978-3-030-40337-9

This Springer imprint is published by the registered company Springer Nature Switzerland AG
The registered company address is: Gewerbestrasse 11, 6330 Cham, Switzerland

Preface

In the quest of technological advancement in the field of microbial nanobiotechnology and pharmaceutical industries in the last few decades to counteract health-related issues, microbial infections remain a major cause of morbidity and mortality. The ability of microbial pathogens to form biofilms further agglomerates the situation by showing resistance to conventional antibiotics. To overcome the serious issue, bioactive metabolites and other natural products were exploited to combat bacterial infections and biofilm-related health consequences. Though the natural products exhibited promising results in vitro, their efficacy in in vivo conditions remains obscured due to their low solubility, bioavailability, and biocompatibility issues. In this scenario, nanobiotechnological interventions bring a multifaceted platform for targeted delivery of bioactive compounds by slow and sustained release of drug-like compounds. The unique physicochemical assets, biocompatibility, and eco-friendly nature of the bioinspired nanostructures have revolutionized the biomedical field to eradicate microbial infections and biofilm-related complications. The green nanotechnology-based metal, metal oxide, and polymeric nanoparticles have been regularly employed for antimicrobial and antibiofilm applications without causing damage to the host tissues. The recent development in carbon-based nanomaterials and silica-based nanomaterials such as mesoporous silica nanoparticles has also been exploited to target dreadful healthcare conditions such as cancer, immunomodulatory diseases, and microbial and biofilm-related infections. Recently, novel physical approaches such as photothermal therapy and antimicrobial photodynamic therapy also revolutionized the conventional strategies and are engaged in eradicating microbial biofilm-related infections and related health consequences. These promising advancements in the development of novel strategies to treat microbial infections and biofilm-related multidrug resistance phenomenon could possibly provide new avenues and a helping hand to conventional antimicrobial therapeutics.

This book should be immensely useful to nanoscience scholars, especially microbiologists, nanotechnologists, researchers, technocrats, and biomedical scientists of microbial nanobiotechnology. We are honored that the leading scientists who have extensive, in-depth experience and expertise in microbial nanobiotechnology took the time and effort to develop the outstanding chapters.

We wish to thank Mr. Eric Stannard, Senior Editor, Springer, Mr. Anthony Dunlap, Mr. Rahul Sharma, Project Coordinator, Springer Nature, and Ms. Kala Palanisamy, Project Manager, SPi Global, for generous assistance, constant support, and patience in initializing the volume. Ram Prasad particularly is very thankful to Honorable Vice Chancellor Professor Dr. Sanjeev Kumar Sharma, Mahatma Gandhi Central University, Bihar, for constant encouragement. The authors are also grateful to their esteemed friends and well-wishers and all faculty colleagues of the School of Life Sciences, Mahatma Gandhi Central University, Motihari, Bihar, and Pondicherry University, Puducherry, India.

Motihari, Bihar, India   Ram Prasad
Puducherry, India   Busi Siddhardha
Puducherry, India   Madhu Dyavaiah

Contents

Contributors

Anna V. Aleshukina Laboratory of Virology, Microbiology and Molecular Biological Research Methods, Rostov Research Institute of Microbiology and Parasitology, Rostov-on-Don, Russia

Iraida S. Aleshukina Laboratory of Virology, Microbiology and Molecular Biological Research Methods, Rostov Research Institute of Microbiology and Parasitology, Rostov-on-Don, Russia

V. T. Anju Department of Biochemistry and Molecular Biology, School of Life Sciences, Pondicherry University, Pondicherry, India

Bianca Pizzorno Backx Universidade Federal do Rio de Janeiro—Campus Duque de Caxias, Rio de Janeiro, Brazil

Matej Baláž Department of Mechanochemistry, Institute of Geotechnics, Slovak Academy of Sciences, Košice, Slovakia

Ľudmila Balážová Department of Pharmacognosy and Botany, University of Veterinary Medicine and Pharmacy, Košice, Slovakia

Abhijit Banik Department of Microbiology, Vidyasagar University, Midnapore, West Bengal, India

Department of Human Physiology with Community Health, Vidyasagar University, Midnapore, West Bengal, India

Zdenka Bedlovičová Department of Chemistry, Biochemistry and Biophysics, Institute of Pharmaceutical Chemistry, University of Veterinary Medicine and Pharmacy, Košice, Slovakia

Birru Bhaskar Brien Holden Eye Research Centre, L V Prasad Eye Institute, Banjara Hills, Hyderabad, India

Samuel Ignatious Bolleddu Morsani College of Medicine, University of South Florida, Tampa, FL, USA

Utpal Bora Department of Biosciences and Bioengineering, Indian Institute of Technology Guwahati, Guwahati, Assam, India

Papori Buragohain Department of Biosciences and Bioengineering, Indian Institute of Technology Guwahati, Guwahati, Assam, India

Bojjibabu Chidipi Morsani College of Medicine, University of South Florida, Tampa, FL, USA

Abhimanyu Dev Department of Pharmaceutical Sciences and Technology, Birla Institute of Technology, Mesra, Ranchi, India

Jintu Dutta Center for the Environment, Indian Institute of Technology Guwahati, Guwahati, Assam, India

Madhu Dyavaiah Department of Biochemistry and Molecular Biology, School of Life Sciences, Pondicherry University, Pondicherry, India

Chandradipa Ghosh Department of Human Physiology with Community Health, Vidyasagar University, Midnapore, West Bengal, India

Elena V. Goloshva Laboratory of Virology, Microbiology and Molecular Biological Research Methods, Rostov Research Institute of Microbiology and Parasitology, Rostov-on-Don, Russia

Andrey V. Gorovtsov Southern Federal University, Rostov-on-Don, Russia

Archita Gupta Department of Bio-Engineering, Birla Institute of Technology, Mesra, Ranchi, India

Suman Kumar Halder Department of Microbiology, Vidyasagar University, Midnapore, West Bengal, India

Achanta Jagadeesh College of Pharmacy, Seoul National University, Seoul, South Korea

Mani Jayaprakashvel Department of Marine Biotechnology, AMET Deemed to be University, Chennai, Tamil Nadu, India

Deepika Jothinathan Department of Life Sciences, Central University of Tamil Nadu, Thiruvarur, Tamil Nadu, India

Jonjyoti Kalita Department of Biosciences and Bioengineering, Indian Institute of Technology Guwahati, Guwahati, Assam, India

Paramita Karfa Department of Chemistry, Indian Institute of Technology (Indian School of Mines), Dhanbad, Jharkhand, India

Zinaida M. Kostoeva Southern Federal University, Rostov-on-Don, Russia

Mária Kováčová Department of Mechanochemistry, Institute of Geotechnics, Slovak Academy of Sciences, Košice, Slovakia

Pethakamsetty Lakshmi Department of Microbiology, AUCST, Andhra University, Visakhapatnam, Andhra Pradesh, India

J. Lakshmipraba Post Graduate and Research Department of Chemistry, Bishop Heber College, Tiruchirappalli, Tamil Nadu, India

Rashmi Madhuri Department of Chemistry, Indian Institute of Technology (Indian School of Mines), Dhanbad, Jharkhand, India

Kartick Chandra Majhi Department of Chemistry, Indian Institute of Technology (Indian School of Mines), Dhanbad, Jharkhand, India

Sarangam Majumdar Dipartimento di Ingegneria Scienze Informatiche e Matematica, Università degli Studi di L' Aquila, L' Aquila, Italy

Devanabanda Mallaiah Department of Biotechnology and Bioinformatics, Yogi Vemana University, Kadapa, Andhra Pradesh, India

Alalvala Mattareddy Department of Zoology, School of Life and Health Sciences, Adikavi Nannaya University, Rajahmundry, Andhra Pradesh, India

Keshab Chandra Mondal Department of Microbiology, Vidyasagar University, Midnapore, West Bengal, India

Ponnala Vimal Mosahari Center for the Environment, Indian Institute of Technology Guwahati, Guwahati, Assam, India

Prabhakaran Mylsamy Post Graduate and Research Department of Botany, Pachaiyappa's College, Chennai, Tamil Nadu, India

Vinod Kumar Nigam Department of Bio-Engineering, Birla Institute of Technology, Mesra, Ranchi, India

Kiyoshi Omine Geo-Tech Laboratory, Department of Civil Engineering, Graduate School of Engineering, Nagasaki University, Nagasaki, Japan

Padmini Padmanabhan Department of Bio-Engineering, Birla Institute of Technology, Mesra, Ranchi, India

Parasuraman Paramanantham Department of Microbiology, School of Life Sciences, Pondicherry University, Puducherry, India

Sudhakar Pola Department of Biotechnology, AUCST, Andhra University, Visakhapatnam, Andhra Pradesh, India

Rupesh N. Prabhu Post Graduate and Research Department of Chemistry, Bishop Heber College, Tiruchirappalli, Tamil Nadu, India

Lavanyasri Rathinavel Post Graduate and Research Department of Botany, Pachaiyappa's College, Chennai, Tamil Nadu, India

Aneta Salayová Department of Chemistry, Biochemistry and Biophysics, Institute of Pharmaceutical Chemistry, University of Veterinary Medicine and Pharmacy, Košice, Slovakia

Mnif Sami Laboratory of Molecular and Cellular Screening Processes, Centre of Biotechnology of Sfax, Sfax, Tunisia

Ramganesh Selvarajan Department of Environmental Sciences, College of Agriculture and Environmental Sciences, University of South Africa-Science Campus, Florida, South Africa

Shalini Department of Chemical Engineering, National Institute of Technology Warangal, Warangal, Telangana, India

Busi Siddhardha Department of Microbiology, School of Life Sciences, Pondicherry University, Pondicherry, India

Sneha Singh Department of Bio-Engineering, Birla Institute of Technology, Mesra, Ranchi, India

V. Sivasankar Post Graduate and Research Department of Chemistry, Pachaiyappa's College, Chennai, Tamil Nadu, India

Department of Civil Engineering, School of Engineering, Nagasaki University, Nagasaki, Japan

Pattnaik Subhaswaraj Department of Microbiology, School of Life Sciences, Pondicherry University, Pondicherry, India

Ramesh Subramani School of Biological and Chemical Sciences, The University of the South Pacific, Suva, Republic of Fiji

Elena A. Vereshak Southern Federal University, Rostov-on-Don, Russia

About the Editors

Ram Prasad, PhD is associated with Department of Botany, Mahatma Gandhi Central University, Motihari, Bihar, India. His research interest includes applied and environmental microbiology, plant-microbe-interactions, sustainable agriculture, and nanobiotechnology. Dr Prasad has more than 150 publications to his credit, including research papers, review articles and book chapters, and five patents issued or pending, and edited or authored several books. Dr. Prasad has 12 years of teaching experience and has been awarded the Young Scientist Award & Prof. J.S. Datta Munshi Gold Medal by the International Society for Ecological Communications; FSAB fellowship by the Society for Applied Biotechnology; the American Cancer Society UICC International Fellowship for Beginning Investigators, the USA; Outstanding Scientist Award in the field of Microbiology by Venus International Foundation; BRICPL Science Investigator Award and Research Excellence Award, etc. He has been serving as editorial board member for Frontiers in Microbiology, Frontiers in Nutrition, Archives of Phytopathology and Plant Protection, Phyton-International Journal of Experimental Botany; Academia Journal of Biotechnology including Series editor of Nanotechnology in the Life Sciences, Springer Nature, the USA. Previously, Dr. Prasad served as Assistant Professor Amity University, Uttar, Pradesh, India; Visiting Assistant Professor, Whiting School of Engineering, Department of Mechanical Engineering at Johns Hopkins University, Baltimore, the USA; and Research Associate Professor at School of Environmental Science and Engineering, Sun Yat-sen University, Guangzhou, China.

Busi Siddhardha, PhD is a full-time Assistant Professor at the Department of Microbiology, Pondicherry University (a central university) Puducherry, India, from 2011. Prior to joining the Pondicherry University, he was teaching at the Gurukul Kangri University, Haridwar. In brief, he has got more than 12 years of

research and 7 years of teaching experience at the university level. Currently, there are five Ph.D. students under his guidance. He has published more than 60 research articles/book chapters in the leading national and international journals. Dr. Siddhardha has successfully completed two research projects on "quorum sensing attenuation and antibiofilm activity of natural products" funded by DST-SERB and UGC, India. Currently, he is a principle investigator of a project on "attenuation of quorum sensing and biofilm mediated virulence of *Pseudomonas aeruginosa* by selected phytochemicals and understanding the mechanism of action" funded by DST-SERB (2018–2021). His current areas of research interests include antimicrobial drug discovery, nanobiotechnology, antivirulence therapy targeting bacterial quorum sensing and biofilms, and antimicrobial photodynamic therapy. He pursued his Ph.D. in Biology Division from CSIR-IICT, Hyderabad, India. He has been awarded Junior Research Fellowship by Council for Scientific and Industrial Research (CSIR), India, in the year 2005. He is editorial board member of several reputed journals. He is a member of many national and international scientific societies.

Madhu Dyavaiah, PhD is an Assistant Professor in the Department of Biochemistry and Molecular Biology; Pondicherry University, Pondicherry. He has served as a Research Scientist in the Gen∗NY∗sis Center for Excellence in Cancer Genomics, University at Albany, USA (2006–2012), Postdoctoral fellow at Wadsworth Center, New York Department of Health, USA (2003–2006) and IISc, Bangalore, India (2002–2003). His research interests include molecular pathogenesis, DNA damage response, tRNA modification and translational regulation, and aging biology. He has research experience in working with microbial pathogens, including *Cryptococcus*, *Candida,* and *Fusarium* species. Currently, he is working with different models systems such as *S. cerevisiae* and mice model to study the effect of natural compounds on age-related diseases, including neurodegenerative diseases and cancer. He has been conferred with various prestigious awards in the USA and India. He has served as a referee for a number of International journals, including Food Sciences and Toxicology Research. He has more than 15 years of research and 8 years of teaching experience in Genomics, Proteomics, Molecular Biology, Clinical Biochemistry Biology, and Drug Discovery. He has also published more than 20 research articles and a review in the peer-reviewed international journal, and authored three book chapters, which include a chapter, "Yeasts: Candida and Cryptococcus" in the book entitled *Bacterial and Mycotic Infections in Immunocompromised Hosts: Clinical and Microbiological Aspects*. He is a member of many scientific societies and organizations.

Chapter 1
Nanomaterials: Therapeutic Agent for Antimicrobial Therapy

Kartick Chandra Majhi, Paramita Karfa, and Rashmi Madhuri

Contents

K. C. Majhi (✉) · P. Karfa · R. Madhuri
Department of Chemistry, Indian Institute of Technology (Indian School of Mines),
Dhanbad 826004, Jharkhand, India

© Springer Nature Switzerland AG 2020
R. Prasad et al. (eds.), *Nanostructures for Antimicrobial and Antibiofilm
Applications*, Nanotechnology in the Life Sciences,
https://doi.org/10.1007/978-3-030-40337-9_1

Abstract In recent time, nanomaterials have been developed as the most auspicious therapeutic remedy toward the infectious microbes, which cannot be healed through traditional treatments. The ancient age treatments via antibiotic drugs are now failed toward microbes, owing to their heavy and unnecessary high dose consumption by the common people. Now, the microbes have become resistant to these antibiotic medicines, and therefore, the nanomaterials came in light to tackle these rising problems related to microbe infections.

Among the various nanomaterials, the carbonaceous nanomaterials (including carbon nanotubes, fullerene, graphene oxide, reduced graphene oxide) and heavy metals (gold, silver) and their oxides (silver oxide, titanium dioxide, zinc oxide, copper oxide) are more commonly employed as antimicrobial agent. In this chapter, we have discussed the antimicrobial activity of these nanomaterials and their mode of action/mechanism. Their unique small size, high surface/volume ratio, large inner volume, and unique chemical and physical properties resulted in efficient antimicrobial activity or antibiofilm activity. Generally, antimicrobial activity/property of the nanomaterials is mainly dependent on the composition, surface modification, and intrinsic properties of the nanomaterials as well as type of microorganism. In this book chapter, the future aspect and challenges faced by nanomaterials toward efficient and effective bactericidal effect are also discussed.

Keywords Nanomaterials · Metal and metal oxides · Carbonaceous materials · Bacterial cell

1.1 Introduction

Nowadays, bacterial infection and their adhesion and rapid growth have become a serious concern for everyone, either they are common person or industry peoples associated with textile, food packing, water treatment, and marine transport-based industries (Diez-Pascual 2018). In recent years, due to the antibiotic resistant behavior of bacterial cells, their infections and related problems have been hiked enormously. According to the World Health Organization (WHO), approximately 25,000 deaths have been occurring by bacteria per year in Europe (Cambau et al. 2018). Researchers and scientists are working very hard to find an effective and helpful solution for not only killing the bacterial cells but restrict their growth in very initial phase also.

In general, an antimicrobial agent is substance/compound/material, which can either kill or prevent the growth of microorganism like fungi, bacteria, protozoans, and other microorganisms. In the nineteenth century, two microbiologists Louis Pasteur and Jules Francois Joubert had their first observation of microorganisms and first discovered one type of bacteria that can prevent the growth of another type of bacteria. From that time till date, several antimicrobial agents have been discovered to prevent or kill the microorganisms' growth. Mainly model antimicrobial agents must have the following criteria to work effectively against microorganisms:

1. The model antimicrobial agents should have capacity to selectively inhibit the growth of microorganism.
2. Destroy the targeted organism while keeping other organisms safe and intact.
3. Their solubility as well as stability should be higher in body fluids.
4. Model antimicrobial agents must have easy and cost-effective availability in the market.
5. They should have high activity for longer period of time.

The most commonly used antibacterial agents can be either pure natural products like aminoglycosides or chemically modified natural compounds (like penicillins, cephalosporins etc.) or pure synthetic antibiotics like sulfonamides also (Hajipour et al. 2012). In general, the antibacterial agents can be classified in two classes: (1) bactericidal and (2) bacteriostatic. Bactericidal agents are class of compounds used to kill the bacterial cell, while bacteriostatic agents retard the growth of bacteria. Antibacterial agents, in the form of antibiotic, are usually provided to fight against the infectious diseases and kill the bacterial cells, but due to their high-dose use and frequent intake, now, bacterial cells have created a resistant for the ancient antibiotic drugs. Now, the antibiotics or antibacterial agents failed to kill the bacterial cell, and this has become the major problem all over the world.

Therefore, newer, essential, and effective antibacterial agents are required to be developed to overcome the problem of antibacterial resistance. We need new, cost-effective antibacterial agents which should be highly effective toward bacteria but exhibit almost negligible side effects. In the last few decades, nanomaterials have been well researched, developed, and used in several fields, and therefore researchers hope that they will solve this problem also. Therefore, nanomaterials like gold, silver, zinc oxide, and carbonaceous materials are nowadays very popularly synthesized and used as antibacterial agents. Now comes the very important question why nanomaterials have improved properties as antibacterial agent. The most common answer to this question would be the large surface to volume ratio of nanomaterials, which results in their various outstanding properties and features. In this chapter, we have focused on the use of nanomaterials as an effective antibacterial agent and also discussed the true reasons for their high success rate.

1.1.1 Basic Discussion About Bacterial Cells

Before discussing about the role of nanomaterial as antibacterial agent, let's discuss some basic things about the bacterial cells and their structure. The bacteria cell wall is prepared in such way that it can protect themselves from mechanical stress and osmotic rupture. Two types of bacteria cell wall are commonly studied, i.e., Gram-positive and Gram-negative cell wall. In the Gram-positive bacterial cell, a thin layer (20–50 nm) of peptidoglycan is present at the outer surface along with teichoic

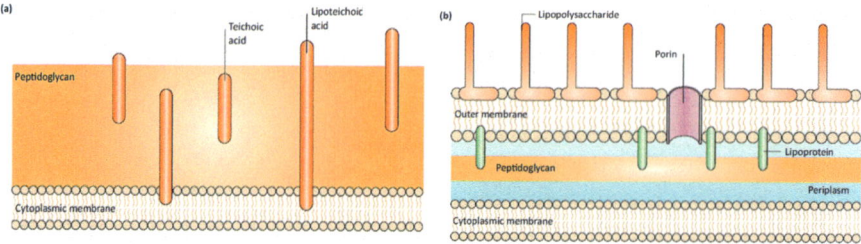

Fig.1.1 Cell wall structures of Gram positive (**a**) and Gram negative (**b**) bacteria (Reproduced with permission from Hajipour et al. 2012)

acid. Structure of Gram-positive bacteria cell is not very complex, but structure of Gram-negative bacteria cell wall is quite complex. Gram-negative bacteria contain outer membrane and peptidoglycan layer, which is connected by the lipoproteins. The outer membrane of Gram-negative bacteria plays crucial role to protect them from any kind of external effect and impact (Hajipour et al. 2012). The cell wall of the bacteria plays important function to efficiently resist from antimicrobial agents. The cell wall structures of Gram-positive and Gram-negative bacteria are shown in Fig. 1.1 (Hajipour et al. 2012).

1.1.2 How Nanomaterials or Antibacterial Agent Interact with Bacteria: Probable Mechanism

The exact mechanism of nanomaterial action toward the bacterial cells is still on debate. However, various literatures have been reported where some of mechanisms have been proposed and discussed (Fernando et al. 2018; Aziz et al. 2014, 2015, 2016). According to the literature, some of the common mechanisms are given below (Deus et al. 2013):

1. Nanomaterials that interact with cell membrane bind directly or indirectly through it and finally kill the cell.
2. Release of metal ions/compounds from the nanomaterials that can bind the sulfur containing proteins of the cell membrane inhibit the cell penetrability and destroy the cell membrane.
3. Generation of reactive oxygen species (ROS) can destroy the cell membrane.
4. Protein oxidation, obstacle the electron transport, and membrane potential breakdown because of nanomaterials contact with the cell membrane.

The probable mechanism discussed above does not operate individually, more than one mechanism are operated at a time. In the next section, we have discussed the details of these mechanisms.

1.1.2.1 Interaction of Nanomaterials with Bacterial Cell Membrane

It has been found that nanomaterials or antibacterial agents firstly interact with the bacterial cell membrane and then resulted in damage of cell membrane (Santos et al. 2013). For example, polymyxin is an antibiotic, which destroys the protective covering of bacterial cell membrane. Polymixin antibiotics consist of hydrophobic tail part with the positive charge containing cyclic peptide; these special arrangements can interact/bind the cell membrane and lead to destroy the Gram-negative bacterial membranes. When the nanomaterials come in contact with the cell membrane; it leads to somehow change in their cell permeability and destroy the cell. Leroueil et al. nicely explained the cell membrane damage; when the nanomaterials come in contact to the cell membrane, it leads to "hole" or "pore" formation in the cell membrane (Aruguete et al. 2013). However, the extent of cell membrane damage totally depends on the nanoparticle size, charge, and morphology (Aruguete et al. 2013). Sometimes, nanomaterials bind with bacterial cell membrane through specific proteins. For example, silver nanoparticles (AgNPs) can only bind the sulfur-containing protein. In some of the literatures, it was mentioned that when nanomaterials bind with cell membrane either directly or indirectly, cell damage occurred by the formation of reactive oxygen species (ROS). The details of ROS mediated cell damage are discussed in the next section.

1.1.2.2 Release of Compounds/Metal Toxic to the Bacterial Cell

In different way, when nanomaterials interact with bacterial cell, the metal ions or some compounds have been released from nanomaterials, which can bind the various functional groups of proteins present in bacteria. For example, silver nanomaterials release the silver ion (Ag^+), which binds the functional group of protein in bacteria. The antimicrobial activity of silver nanomaterials is highly particular for Gram-negative bacteria like *E. coli*, where silver ion inhibits the replication of bacterial deoxyribonucleic acid (DNA) and leads to their damage and cell death (Aruguete et al. 2013). Sometimes, the released Ag^+ get precipitated in the form of silver chloride (AgCl), after reacting with chloride ion, as a consequence stops the cell respiration in cytoplasm in the cell membrane, which also leads to cell death. Similarly, Cd^{2+} and Zn^{2+} binds with sulfur-containing protein in cell membrane leads to their damage or destruction.

1.1.2.3 Role of Reactive Oxygen Species (ROS) in Cell Damage

ROS are chemically reactive oxygen-based species of different type like superoxide, peroxide, hydroxyl radical, singlet oxygen, and triplet oxygen. Reduction of molecular oxygen (O_2) produces superoxide ($^{\bullet}O_2^-$), which acts as a precursor for generation of most of the other reactive oxygen species. Oxygen is the potent oxidizing agent and it is the best electron acceptor during respiration (Aruguete et al. 2013). The ground state of oxygen molecule is the triplet oxygen (or $^3O_2$) and triplet oxygen can be exceptionally toxic for cell membranes. Singlet oxygen ($^1O_2$) is also toxic for bacteria but less toxic than triplet state oxygen. According to the literature, a wide range of nanomaterials can generate reactive oxygen species (ROS), which leads to the catastrophic failure of many biological systems like different types of protein, deoxyribonucleic acid (DNA), and ribonucleic acid (RNA). For example, TiO_2, SiO_2, ZnO, Cu, and silver nanomaterials can catalyze the generation of ROS in the presence of light (Santos et al. 2013; Din et al. 2017). Nature also generates low concentrations of ROS to defend themselves against the formation or growth of bacteria on them (Santos et al. 2013). In reverse, to fight against these ROS, bacteria generate some enzymes like superoxide dismutase, which may be able to neutralize the oxidative stress or effect of ROS. However, based on the literature, generation of ROS through nanomaterials is the best possible mechanism to kill bacterial cell.

1.1.2.4   Obstacle in Electron Transport and Protein Oxidation

In general, the metal ions delivered from nanomaterials have positive charge, while the bacterial cell membrane carries negative charge. Till now clear mechanism is not known, but from the literature, it can be concluded that metal ions can induce the membrane-bound respiratory enzymes as well as induce the efflux bombs ions leading to cell death (Chiriac et al. 2016). The pictorial representation of nanomaterial and bacterial cell interaction is also shown in Fig. 1.2.

In this section, we have discussed the probable mechanism on how the antibacterial agent actually works to stop the growth of bacteria or kill them. Now comes the next important point; how to measure the antibacterial activity of any agent. In the next section, we have discussed the most commonly employed assay/methods/techniques reported in the literature to measure the ability of any antibacterial/antimicrobial agent.

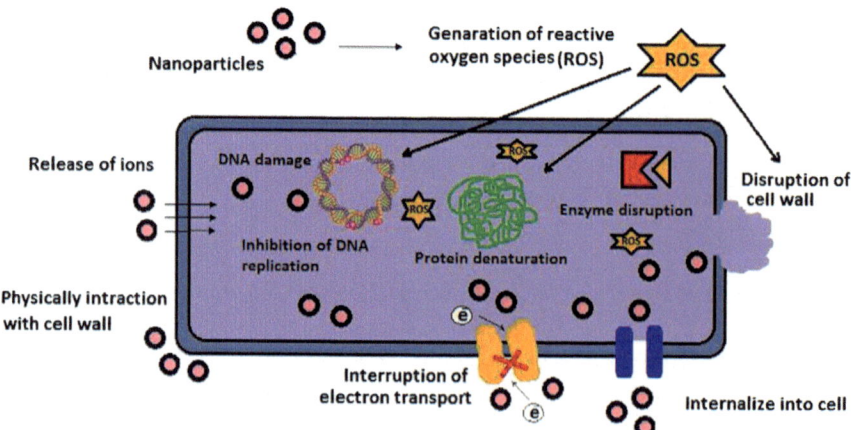

Fig. 1.2 Pictorial representation showing antimicrobial activity of nanomaterials (Chiriac et al. 2016)

1.2   Essay for Measuring the Antimicrobial Activity of Nanomaterials

The evaluation of activity of nanomaterials as an antimicrobial agent needs some experimental methods, which will determine the bacteria sustainability, after the nanomaterials have been subjected to them. Several methods have been discovered to measure the antimicrobial activity of nanomaterials like optical density or OD measurement study, cell counting method, crystal violet staining, etc. In general, the techniques were applied against Gram-negative and Gram-positive bacteria (Seil and Webster 2012). In this section, we have discussed the general and most popular methods used to determine the antimicrobial activity of nanomaterials.

1.2.1   Susceptibility of Nanomaterials Toward Microorganisms

As a foremost step, susceptibility assays are performed to access the responsiveness or sensitivity of nanomaterials toward microorganisms. In this regard, determination of minimum inhibitory concentration (MIC) and minimum bactericidal concentration (MBC) are the most popularly used methods to detect the susceptibility of nanomaterials toward microorganisms. MIC is actually the lowest concentration of antibacterial agent (expressed in μg/L or mg/L) required to inhibit the growth of bacteria in given time interval (usually overnight). However, MBC is the lowest concentration of antibacterial agent required to kill 99.9% of bacteria in a given time period. The graphical representation showing the difference between these two terms is shown in Fig. 1.3. Commonly, to evaluate the MIC and MBC, dilution method is used. However, susceptibility can be studied through disc diffusion method also, which is another popular method to access the effect of nanomaterial on the microorganisms.

1.2.1.1   Dilution Method

In order to determine the efficiency of nanomaterials toward bacterial cell, broth dilution method is the most popular one. In this method, different concentrations of nanomaterials were subjected to the bacterial cell and kept for incubation at 37 °C for 24 h (Balouiri et al. 2016; Karaman et al. 2017). The bacterial cell in broth medium is taken as positive control (i.e., without nanomaterials), while negative control has only broth medium. As we know, MIC is the lowest concentration of nanomaterial required to visually inhibit almost 99% growth of bacterial cell. Here, MIC is determined by observing the visual turbidity appearance in the vial containing cells and nanomaterials.

Similarly, for MBC determination, a small amount (around 50–100 μL) of suspension was taken from each tube (used for MIC determination), which does not

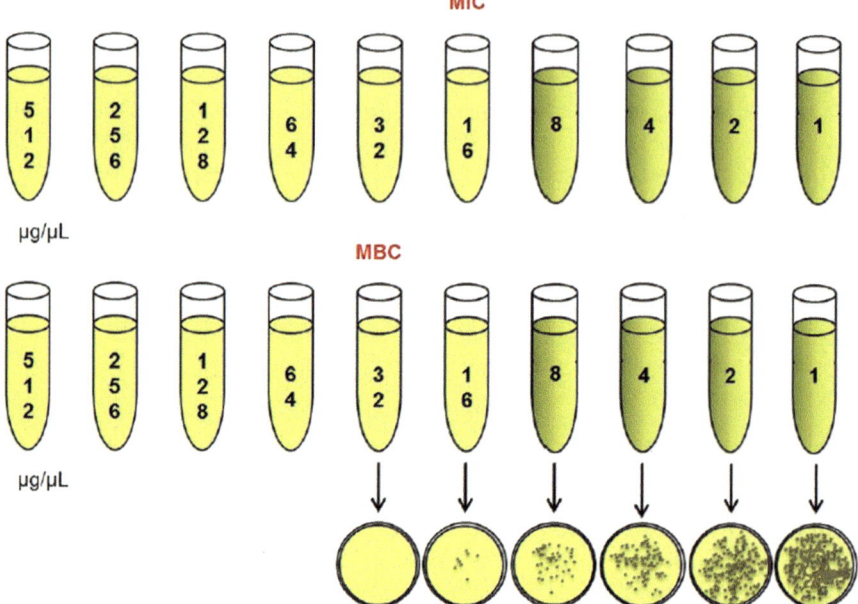

Fig. 1.3 Schematic representation showing minimum inhibitory concentration (MIC) versus minimum bactericidal concentration (MBC) (Karaman et al. 2017)

showed any visible bacterial growth and seeded in agar plates and kept for incubation at 37 °C for additional 24 h. The MBC was determined by observing the presence or absence of bacterial growth in each agar plate (Balouiri et al. 2016).

1.2.1.2 Disc-Diffusion Method

Starting from 1940, disc-diffusion method has become an official method in clinical laboratories to study the antimicrobial susceptibility (Balouiri et al. 2016). In this method, firstly, on the Petri dish, agar coated bacterial strain are homogeneously spread by the help of cotton bud or spatula. After that, a disc of filter paper (about 4–5 mm diameter) containing nanomaterials of different concentrations are place in the Petri dish. The dishes were closed and sealed to keep it for incubation at suitable condition. From filter paper, nanomaterials diffuse to the bacterial cell and kill them, depending on their efficiency and concentrations. The spherical region around the filter paper has been measured as zone of inhibition, which provides the efficacy of nanomaterials toward bacterial cell.

1.2.2   Methods for Quantification of Antibacterial Activity

The method discussed in the previous sections is devoted to study the visual changes caused by the nanomaterials. However, to quantify the percentage of cell killing, other methods are deployed.

1.2.2.1   Optical Density (OD) Measurement

Optical density or OD measurement technique is the most common method to determine the bacterial cell concentration. Here, the bacterial cell concentration was monitored by using UV-visible spectrophotometry. From this method, the degree of proliferation can be predicted by determining the cell concentration of bacteria at different time intervals. In this technique, the light is passed in the bacteria cell or sample and then spectrophotometer record the corresponding optical density. For comparison, the same experiment has been done with standard sample, whose bacteria cell concentration is known to us. If the sample penetrated less light, it means bacteria cell concentration is high in the sample, with respect to the standard sample or vice versa. Similarly, using the standard calibration curve and regression equation, the bacteria cell concentration can also be determined. This technique is very simple in use and therefore widely applied to explore the effect of nanomaterials on the bacteria. Some of the advantages and disadvantages of this technique are summarized in Table 1.1 (Seil and Webster 2012).

Table 1.1 Some advantage and disadvantage of commonly used methods for quantification of antibacterial activity of nanomaterials [reproduced from reference (Seil and Webster 2012)]

Name of the method	Advantages	Disadvantages
Optical density (OD) measurement	Fast, easy to handle, no chemical required	Spectrophotometer instrument is required, low accuracy
Cell counting method	Accuracy is high	Costly instrument required
Spread-plate colony counts	Accuracy is high	Can't counts cell size of CFU, time taking process, required disposal materials
Crystal violet staining	Identify and counts biofilm formation	Spectrophotometer instrument is required and not applicable for planktonic bacteria cell growth
Live/dead cell staining and imaging	Offers imagining of sample surface	Cost effect due to reagents required, fluorescent plate reader is required
Tetrazolium salt reduction	Counts cell viability on the surfaces and also in solution	Spectrophotometer instrument is required, costly reagents required

1.2.2.2 Cell Counting Method

For counting the cell number in a liquid suspension of bacteria, the conductivity of the solution is also measured sometimes (Seil and Webster 2012). From this method, the presence of cell density and their size distribution can be exactly determined.

1.2.2.3 Spread-Plate Colony Counts

In another method, colony-forming units (CFUs) can be calculated by the help of microscopes. In this method, the bacteria cells were monitored on agar plate. From this method, a comparison can be made between the bacterial cell incubated agar plates, with and without nanomaterials. This technique also has some advantages and disadvantages, which are portrayed in Table 1.1 (Seil and Webster 2012).

1.2.2.4 Crystal Violet Staining

For the estimation of formed bacteria colony biofilm, hexamethyl pararosaniline chloride (which is crystal violet color) can be used as a staining agent. Crystal violet is a cost-effective and popular dye, which is used to measure the effect of nanomaterials on the total biomass of bacterial film. This dye can bind to Gram-negative bacteria as well as extracellular polymeric substance like polysaccharides (Seil and Webster 2012). After staining was performed, the adsorbed dye will be eluted in different types of solvents like ethanol, methanol, etc. The dye eluted in the solvent can be measured by any spectrophotometric methods and will be directly proportional to the biomass of the bacterial film.

1.2.2.5 Live/Dead Cell Staining and Imaging

Confocal fluorescence imaging of bacterial cell is nowadays very popular to measure the concentration of dead and live bacteria after the treatment of nanomaterials. For this, Syto® 9 green fluorescent nucleic acid stain with excitation 480 nm and emission 500 nm can be used for detection of both dead and live cells, while propidium iodide with excitation 490 nm and emission 635 nm can be used for the identification of only dead cells (Seil and Webster 2012). Sometimes the combination of these two is also used to the individual cell strains, to observe the dead and living cells.

1.2.2.6 Tetrazolium Salt Reduction

From this technique, different tetrazolium salts are used to measured cell viability using spectrophotometry. Different tetrazolium salts such as MTS (3-[4,5-dimethyl thiazol-2-yl]-5-[3-carboxymethoxyphenyl]-2-[4-sulfophenyl]-2Htetrazolium),

XTT (2,3-bis-[2-methoxy-4-nitro-5-sulfophenyl]-2H-tetrazolium-5-carboxanilide), and MTT (3-[4,5-Dimethylthiazol-2-yl]-2,5-diphenyltetrazolium bromide) are used for determination of cell viability in the presence or absence of nanomaterials. On the high popularity of MTT use in this method, the detection method is popularly called as MTT assay. In this process, the tetrazolium salts get reduced to different color formazan derivatives, when came in contact to the dehydrogenase and reductase enzymes present in the living cells. The formation/conversion of formazan derivatives directly depend on the presence of viable cells, which offered the cell concentration. The common advantages and disadvantages of this method are portrayed in Table 1.1 (Seil and Webster 2012).

Till the point, we have discussed how the effect of nanomaterials on the bacterial cell can be observed or detected. We have also discussed the possible and probable mechanism for bacterial cell killing. Now, in the next sections, we have focused on the nanomaterials used or reported in the literature and showed high antibacterial properties.

1.3   Role of Nanomaterials as Antimicrobial Agent

1.3.1   The Ancient Era

It is not like nanomaterials are the only one class of compounds and have ability to kill the bacterial cell or retard their cell growth. Prior to that, when nanomaterials were not in picture, a wide number of organic compounds have been used as an antimicrobial agent. For example, Janeczko et al. have synthesized naphthalene-1,4-dione derivatives and checked their antimicrobial activity against eight bacterial cells such as *Proteus, Enterobacter, Staphylococcus, Pseudomonas, Escherichia, Klebsiella, Salmonella,* and *Enterococcus* (Janeczko et al. 2016). All the synthesized derivatives of naphthalene-1, 4-dione has shown pronounced antimicrobial activity with minimum inhibition concentration (MIC) in the range of 5.8–500 µg/mL. Raimondi et al. have prepared various types of pyrrolomycin compounds by microwave-assisted methods and tested their antimicrobial properties against both Gram-positive and Gram-negative bacteria (Raimondi et al. 2019). Recently, thiouracil derivatives having triazolo-thiadiazole compounds exhibited promising antimicrobial properties toward the Gram-negative and Gram-positive bacteria, as reported by Cui et al. (2017). Sekhar and co-workers have prepared some heterocyclic compounds such as 1,3,4-thiadiazoles, pyrimidinyl 1,3,4-oxadiazoles, and 1,2,4-triazoles and tested their antimicrobial and antifungal activity (Sekhar et al. 2018). All the prepared compounds have shown very good antimicrobial and antifungal activity, due to the presence of 1,3,4-thiadiazole backbone in their structures. Although various organic compounds exhibited promising antimicrobial activity against large number of bacteria, the major disadvantage behind these compounds is their large particle size and complicated/tedious synthesis approach. This leads to low surface area to volume ratio, more cost, and difficulty in their separation from their pure form. Other than this aspect, more importantly, the bacterial cells have developed resistance toward these antimicrobial agents, and to overcome this effect, their high dose is required to cure the disease, which leads to hostile side effects.

1.3.2   Why Nanomaterials Have Replaced the Ancient Antimicrobial Agents?

To overcome drawbacks of conventional antimicrobial agent, nanomaterials came in light and well explored as antimicrobial agent, owing to their easy synthesis and purification, affordable cost, eco-friendly nature, small size, and high surface to volume ratio (which is the most prominent factor). In the last few decades, nano-technology has been developed very rapidly as well as successfully in the field of antimicrobial treatment. Nanoscale structured (particle size 1–100 nm) materials possess high surface area to volume ratio, that is why it has unique physical, chemical, electrical, mechanical, magnetic, electro-optical, magneto-optical, and optical properties entirely different from their corresponding bulk materials (Whitesides 2005). Having large surface to volume ratio and possessing high reactivity, when nanomaterials came in contact with bacterial cell, then large amount of cell get in contact with these nanomaterials, leads to easy destruction of cells and finally death. Because of their unique features and properties, nanomaterials are also called as wonder of modern medicine. Different viruses, bacteria, and fungi can be stopped or killed within a few minutes, after treatment with nanomaterials. This credit can be given to their small size, due to which nanomaterials can easily enter/penetrate into the cell membrane and start their action within small time span (Haghi et al. 2012; Prasad et al. 2018). And this can be the reason for popularity of nanomaterials as antimicrobial agents.

1.4   Different Class of Nanomaterials Used as an Antimicrobial Agent

In this section we have summarized some of the nanomaterials which are used as an antimicrobial agent and their probable working mechanism. The most popular nanomaterials as antimicrobial agents come from the metal nanoparticles. The nanomaterials/nanoparticles of Ag, Pt, Au, and Cu and metal oxides like Ag_2O, SiO_2, TiO_2, MgO, ZnO, CaO, CuO, and Fe_2O_3 have shown their antimicrobial activities very successfully in various literatures (Lemire et al. 2013). Some of the recent ones are discussed in the next section.

1.4.1 Antimicrobial Properties of Silver-Based Nanomaterials

From the very old times, silver is used as antimicrobial agent, while from the last few decades, their nanomaterials came into picture and were very popularly used in recent times. Koduru et al. have nicely explained the function of silver nanoparticles (AgNPs) in microorganisms or bacteria (Patil and Kim 2017). According to them, AgNPs or silver-based nanomaterials are able to inhibit the growth of microorganisms. Although exact mechanism is still under debate, possible mechanism involves two stages. At first the AgNPs or silver-based nanomaterials were fabricated using different active organic or bioorganic molecules. The use of organic and bioorganic molecules in their synthesis can accelerate the antimicrobial activity of resulted nanomaterials, which inhibit the cell growth. In second step, silver nanomaterials enter into the bacterial cell, which totally depends on the size, shape, and surface chemistry of the entering agent. Silver nanomaterials give silver ion (Ag^+) by oxidative dissolution of silver nanomaterials (present Ag^0). These silver ions have affinity to strongly bind with marpapto groups of enzymes, organic functional groups, proteins, and DNA of bacteria cell via formation of covalent bond between silver ions and sulfur (Ag-S), which have bond energy 65 kcal/mol. The strong binding between Ag^+ and bacterial cell destroys the bacteria DNA to inhibit the bacterial replication and finally kill the cell (Li et al. 2010).

Chernousova and Epple have reported that with changing morphology or shape of the particles, the interaction between the nanomaterials (silver nanomaterials) and bacterial protein can be changed (Chernousova and Epple 2013). The silver nanomaterials were also able to generate the reactive oxygen species (ROS), which can destroy the various biochemical systems in bacteria (Chernousova and Epple 2013). It is already evidenced that the high-level ROS have been detected by applying silver nanomaterials in the bacterial cell, which leads to high oxidative stress and damages the bacterial protein (Patil and Kim 2017). Sharma et al. have investigated the antimicrobial activity of silver nanoparticles with the nanomolar and micromolar concentration, whereas micromolar concentration showed better result, i.e., antimicrobial activity than nanomolar concentration (Sharma et al. 2009). This group also reports the role of silver ions toward antimicrobial applications. According to the literature, firstly, silver ions generated from silver nanomaterials get chemisorbed on the surface of silver nanomaterials. Therefore, surface concentration of silver ions in silver nanomaterials largely reflects the antimicrobial activity. After that, silver nanomaterials adsorbed on the cell membrane wall lead to prevent the cell growth and proliferation. Then, silver nanomaterials enter/penetrate into the cell membrane and as a result destroy or inhibit the function of the proteins, DNA damage and death the cell. The credit behind the cell death can be attributed to silver ions, which bind the proteins and destroy the cell membrane wall (Prasad and Swamy 2013; Aziz et al. 2019).

According to other literature, firstly, silver nanomaterials come in contact with the cell wall and after that gradually release the silver ions from the silver nanomaterials and accordingly produced ROS by several biochemical processes (Patil and Kim 2017). Finally, destroy or inhibit the function of proteins, enzymes in

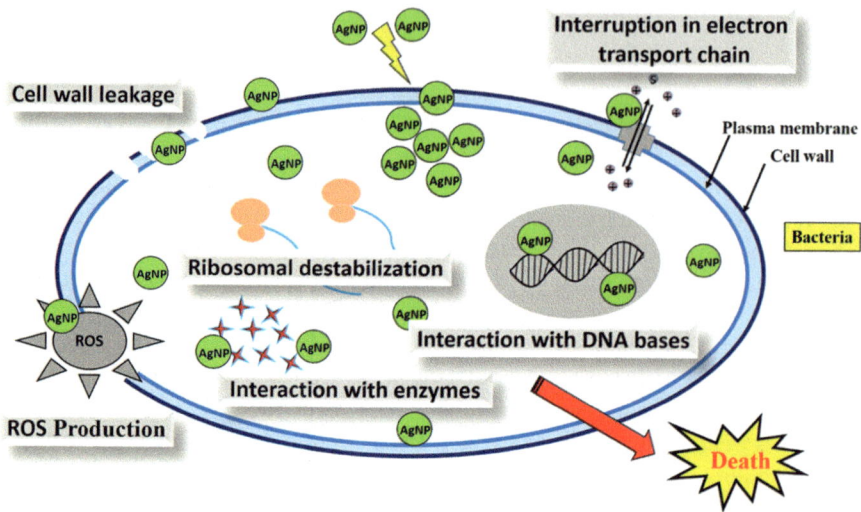

Fig. 1.4 Schematic representation showing antimicrobial activity of AgNMs (Patil and Kim 2017)

cytoplasma and lead to cell death. The role of silver nanomaterials in bacteria cell is shown in Fig. 1.4 (Patil and Kim 2017). Some of the synthesized silver nanomaterials from different precursors, its shape, size, and antimicrobial activity toward particular bacterial cell are summarized in Table 1.2 (Ajitha et al. 2016; Ibrahim 2015; Dhand et al. 2016; Hassanien and Khatoon 2019; Alfuraydi et al. 2019; Varghese et al. 2017; Sulaiman et al. 2013; Padalia et al. 2015; Palanisamy et al. 2017; Rao et al. 2016; Verma and Mehata 2016; Ravichandran et al. 2016; Nzekwe et al. 2016; Jena et al. 2015).

1.4.2   *Antimicrobial Activity of Zinc Oxide Nanomaterials (ZnO)*

Various metal oxide nanomaterials have been used as an antimicrobial agent. But the main advantage of zinc oxide nanomaterials than the other nanomaterials is their non-toxicity, biocompatibility, and photochemical stability. Additionally, zinc oxide is listed safe material recommended by the US Food and Drug Administration to Food and Administration (Dimapilis et al. 2018). Nanosized zinc oxide nanomaterials (not more than 100 nm) have shown prominent antimicrobial properties than macro-sized nanomaterials. Due to small size (less than 100 nm), they have high surface/volume ratio which allow more interaction between the nanomaterials and bacteria cell. For the last few years, several studies have been shown that zinc oxide nanomaterials selectively destroy the bacteria and also revealed the minimum effects on the human cells (Bhuyan et al. 2015; Dimapilis et al. 2018). Antimicrobial properties of

Table 1.2 Synthesis of AgNPs using different reducing/capping agents and their antimicrobial activity

S. N.	Name of the reducing or capping agent	Shape and size	Antimicrobial activity against bacteria cell	MIC[a]	References
1.	*Sesbania grandiflora* leaf extract	Spherical and 20 nm	Gram-negative (*E. coli, Pseudomonas* spp.) and Gram-positive bacteria (*Bacillus* spp., *Staphylococcus* spp.), and fungi (*Aspergillus niger* subsp., *A. flavus* subsp. and *Penicillium* spp.)	2–6 µL	Ajitha et al. (2016)
2.	Banana peel extract	Spherical and 23.7 nm	Bacteria (*B. subtilis, S. aureus, P. aeruginosa, E. coli*) and yeast (*Candida albicans*)	1.70–6.80 µg/L	Ibrahim (2015)
3.	Seed extract of *Coffea arabica*	Spherical, ellipsoidal and 10–150 nm	*S. aureus* and *E. coli*	0.02–0.1 M	Dhand et al. (2016)
4.	Tanic acid and $C_{76}H_{52}O_{46}$	Spherical and 33.3–69.8 nm	Bacteria (*E. coli* and *Bacillus*) and fungi (*Candida albicans*)	–	Hassanien and Khatoon (2019)
5.	Sesame oil cake	Spherical and 6.6–14.80	*P. aeruginosa, E. coli,* and *K. pneumonia*	0.5 µg/mL	Alfuraydi et al. (2019)
6.	Seed extract of *Trigonella foenum-graecum L*	Spherical and 39.3 nm	*Staphylococcus aureus, E. coli, Klebsiella pneumonia, Aspergillus flavus, Trichophyton rubrum,* and *Trichoderma viridiae*	62.5, 125 and last three each 250 µg/mL	Varghese et al. (2017)
7.	*Eucalyptus camaldulensis*	–	*P. aeruginosa, E. coli, S. aureus,* and *B. subtilis*	–	Sulaiman et al. (2013)
8.	Matricaria chamomile plant extract	–	Bacteria (*S. aureus, E. coli*) and fungus (*Candida albicans*)	1–100 ppm	Padalia et al. (2015)
9.	Flower broth of *Tagetes erecta*	Spherical, hexagonal, and irregular and 10–90 nm	Bacteria (*E. coli, P. aeruginosa, S. aureus, B. cereus*), fungi (*Candida albicans, Candida glabrata, Cryptococcae neoformans*)	–	Palanisamy et al. (2017)

(continued)

Table 1.2 (continued)

S. N.	Name of the reducing or capping agent	Shape and size	Antimicrobial activity against bacteria cell	MIC[a]	References
10.	*Gelidium amansii*	Spherical and 27–54 nm	*Staphylococcus aureus, Pseudomonas aeruginosa, Bacillus pumilus, E. coli, Aeromonas hydrophila,* and *Vibrio parahaemolyticus*	25–100 µg	Rao et al. (2016)
11.	Extract of neem (*Azadirachta indica*) leaves	Spherical	*S. aureus* and *E. coli*	0–120 µg/mL	Verma and Mehata (2016)
12.	*Artocarpus altilis* leaf extract	Different shapes and 34–38 nm	Bacteria (*E. coli, P. aeruginosa, S. aureus*) and fungi (*A. versicolor*)	0–120 µg/mL	Ravichandran et al. (2016)
13.	NaBH₄	Spherical	*E. coli, S. aureus, Klebsiella* spp., *B. subtilis*	–	Nzekwe et al. (2016)
14.	Pod extract of *Cola nitida*	Spherical and 10–80 nm	*P. aeruginosa, K. granulomatis, E. coli, S. aureus, A. flavus, Aspergillus niger,* and *A. fumigatus*	50–150 µg/mL	Jena et al. (2015)

[a]*MIC* Minimum inhibition concentration

zinc oxide nanomaterials have been well explored against both Gram-positive bacteria and Gram-negative bacteria and also for *Escherichia coli, Salmonella typhimurium, B. subtilis, S. enteritidis, Streptococcus pyogenes, E. faecalis, Aeromonas hydrophila, S. aureus, L. monocytogenes, Klebsiella pneumonia,* and *P. aeruginosa* (Kumar et al. 2017). In order to demonstrate the activity of nanomaterials toward microorganism, several step mechanisms are proposed, which involves metal ion discharge/release from nanomaterials, production of ROS, cell membrane dysfunction, disturbance of electron transport, and penetration of nanomaterials into the cell membrane, which are directly involved in the cell death. The procedure of generation of ROS by zinc oxide nanomaterials is shown in Fig. 1.5 (Dimapilis et al. 2018). In addition, the zinc oxide nanomaterials have band gap of 3.28 eV, which reflects their high binding energy of 60 meV, which may also lead to their strong binding with bacterial cell (Kumar et al. 2017).

It has also been found that antibacterial property of ZnO depends on several factors. For example, in the year of 2013 Heniti and Umar investigated that how surafce area of zinc oxide nanomaterials affect the antimicrobial activity and they found propotional relationship between surface area of nanomaterials (ZnO) and antimicrobial activity, which means with increasing surface area of nanomaterials antimicrobial activities also increased and vice versa (Al-Heniti and Umar 2013). Rodrigo et al. have reported that antimicrobial activities of ZnO have been greatly affected by their size, shape, or surface morphology and surface area of the zinc

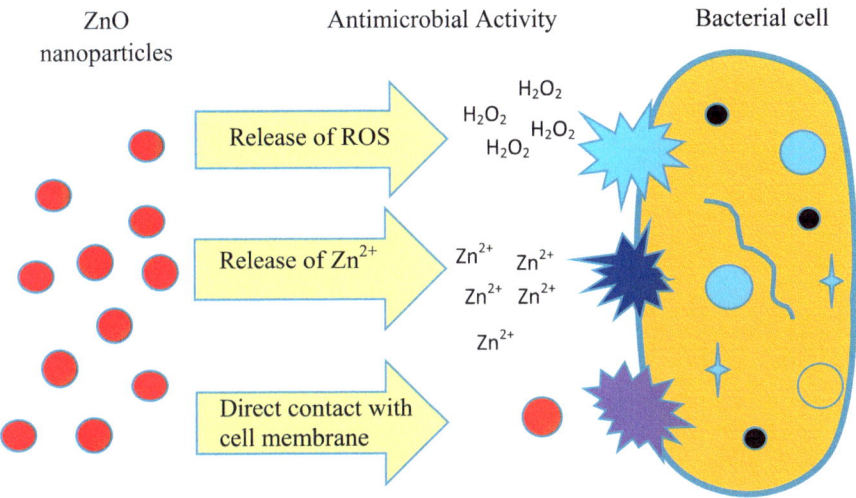

Fig. 1.5 Production of ROS and Zn2+ from ZnO nanomaterials (Dimapilis et al. 2018)

oxide nanomaterials (Rodrigo et al. 2013). Jain et al. have synthesized ZnO NPs via simple one-step hydrothermal method and used as an antimicrobial agent for the Gram-positive bacteria (such as *Staphylococcus aureus*, *Enterobacter aerogenes*, and *Bacillus subtilis*) and Gram-negative bacteria (such as *Escherichia coli* and *Aerobacter aerogenes*) (Jain et al. 2013). Authors have also studied antimicrobial activity of ZnO NPs for both the bacteria (Gram positive and Gram negative) and observed higher antimicrobial activities for Gram-positive (MIC = below 100 µg/mL) than Gram-negative bacteria (MIC = 500 µg/mL). Another research group in year 2013 prepared different morphologies of ZnO nanomaterials (nanorods and nanoflowers) by simple hydrothermal method and used as antimicrobial agent against *Staphylococcus* and *E. coli* (Talebian et al. 2013). The antimicrobial activity has been found higher for nanoflower-shaped ZnO than other shapes of nanomaterials against *E. coli* bacteria. In addition, several other methods have also been reported for the synthesis of various morphologies of ZnO nanomaterials such as nanorods, nanospheres, nanoflowers, thin film, etc. Some of the ZnO nanomaterials along with their antimicrobial activities are summarized in Table 1.3 (Stankovic et al. 2013; Shinde et al. 2014; Aal et al. 2015; Liu et al. 2009; Janaki et al. 2015; Ramesh et al. 2015; Umar et al. 2013; Ambika and Sundrarajan 2015; Huang et al. 2008; Raghupathi et al. 2011; Premanathan et al. 2011; Nair et al. 2009; Padmavathy and Vijayaraghavan 2008; Dwivedi et al. 2014; Kumar et al. 2013).

Table 1.3 Synthesis of ZnO NPs using different methods and their antimicrobial activity

S. N.	Method of preparation	Shape of ZnO NPs	Antimicrobial activity against bacteria cell	Minimum inhibition concentration (MIC)	References
1.	Hydrothermal method	Prismatic, nanorods, nanospheres and nano-ellipsoid	*E. coli* and *S. aureus*	Nanospheres showed highest antimicrobial activity against both the bacteria	Stankovic et al. (2013)
2.	Microwave method	Microspheres	*E. coli* and *S. aureus*	25 µg/mL	Shinde et al. (2014)
3.	Hydrothermal method	Nanotubes	*A. baumannii, E. coli, K. pneumonia, P. mirabilis, P. aeruginosa, S. typhi, B. subtilis, M. luteus, S. aureus,* MRSA, *S. epidermidis,* and *S. pneumonia*	0.55 ± 0.04, 0.45 ± 0.04, 0.55 ± 0.04, 0.45 ± 0.04, 0.55 ± 0.04, 0.65 ± 0.04, 0.65 ± 0.04, 0.75 ± 0.04, 0.75 ± 0.04, 0.75 ± 0.04, 0.60 ± 0.04, and 0.60 ± 0.04 µg/mL, respectively.	Aal et al. (2015)
4.	–	Nanopowders	*E. coli*	3 µg/mL	Liu et al. (2009)
5.	Green synthesis using ginger rhizome extract	Nanoparticles	*K. pneumonia* and *S. aureus*	–	Janaki et al. (2015)
6.	Green synthesis using *Solanum nigrum* leaf extract	Nanoparticles	*S. aureus, S. paratyphi, V. cholera,* and *E. coli*	–	Ramesh et al. (2015)
7.	Solvothermal method	Nanoflowers	*E. coli*	25 µg/mL	Umar et al. (2013)
8.	Green synthesis using *Vitex negundo* extract	Nanoparticles	*E. coli* and *S. aureus*	–	Ambika and Sundrarajan (2015)
9.	Precipitation method	Different shape	*S. agalactiae* and *S. aureus*	–	Huang et al. (2008)

(continued)

Table 1.3 (continued)

S. N.	Method of preparation	Shape of ZnO NPs	Antimicrobial activity against bacteria cell	Minimum inhibition concentration (MIC)	References
10.	Solvothermal method	Nanoparticles	*S. aureus,* *S. epidermidis,* *S. pyogenes,* *E. faecalis,* *B. subtilis, E. coli,* *P. vulgaris*, and *S. typhimurium*	–	Raghupathi et al. (2011)
11.	Wet chemical method	Nanoparticles	*E. coli,* *P. aeruginosa,* and *S. aureus*	500, 500, and 125 µg/mL	Premanathan et al. (2011)
12.	Wet chemical method	Nanoparticles	*E. coli* and *S. aureus*	–	Nair et al. (2009)
13.	Precipitation method	Nanoparticles	*E. coli*	–	Padmavathy and Vijayaraghavan (2008)
14.	Wet chemical method	Nanoparticles	*Pseudomonas aeruginosa*	–	Dwivedi et al. (2014)
15.	Solution method	Nanoflowers	*Pseudomonas aeruginosa*	25 µg/mL	Kumar et al. (2013)

1.4.3 Antimicrobial Activity of Titanium Oxide (TiO_2) Nanomaterials

For the last few years, TiO_2-based nanomaterials or NPs have shown broad range of antimicrobial activity against the various Gram-positive as well as Gram-negative bacteria cell. Presently TiO_2 NPs have become the first choice in various fields and industries like food industry, cosmetics industry, medical field, waste water treatment, drug delivery, etc. TiO_2 NPs have distinctive photocatalytic property and small particles size which create the TiO_2 NPs a model antimicrobial agent against various bacteria cells (Fernando et al. 2018). The action mechanism of TiO_2 NPs is slightly different than the silver or zinc oxide nanomaterials, due to unique photocatalytic properties of TiO_2 NPs. Bacteria cells get damaged by the light, which is coming from TiO_2 NPs (Fernando et al. 2018). After that TiO_2 NPs get oxidized to form reactive oxygen species, which can destroy the cell membrane.

Some of the literatures showing the role of TiO_2 NPs as antimicrobial agent are discussed below. For example, Haghi et al. have synthesized TiO_2 NPs and checked their antimicrobial activity against *E. coli* bacteria using various concentrations of TiO_2 nanomaterials (Haghi et al. 2012). It was found that with increase in concentration of TiO_2 NPs, their antimicrobial activity also increased. Similarly, Saraschandra et al. have prepared TiO_2 NPs coated with high-density polyethylene by sol-gel followed by ultrasonication method (Saraschandraa et al. 2013). The prepared NPs show good antimicrobial activity against *Staphylococcus aureus* and *Escherichia coli*.

1.4.4   Copper Nanomaterials as an Antimicrobial Agent

Copper (Cu) is a key element for metabolism of animals as well as plant cells. More than 30 proteins contain copper element/ion in their composition; however, almost every living organisms also contain trace amount of copper element. For the last few years, copper nanoparticles (CuNPs) have shown ideal biological application in low cost. Recently, copper and copper-based nanomaterials have shown their potential as antimicrobial activity against various bacteria cells (Holubnycha et al. 2017). For example, Rafi et al. have reported copper sulfate, copper hydroxide, and aqua complex of copper as an antimicrobial agent (Raffi et al. 2010). According to the American Environmental Protection agency, copper has been listed at the top for their antimicrobial activity against bacteria cell (Prado et al. 2012). Sometimes, copper nanomaterials have shown higher antimicrobial activity in comparison to costly and popular noble metal-based nanoparticles like AgNPs (Holubnycha et al. 2017).

The action mechanism of copper nanomaterials possesses multiple steps, which includes electrostatic interaction/attraction between the copper nanomaterials and bacterial cell. As CuNPs itself have positive charge on the surface and bacteria cell contains negative charge, results in their strong binding with cell membrane, which later on results in protein denaturation. Sometimes, CuNPs bind with cell DNA also through the binding interaction between phosphorous and copper (Mahmoodi et al. 2018). It is also reported in some of the literature that CuNPs also lead to generation of ROS which lead to cell membrane destruction, protein oxidation, and DNA denaturation, which finally kill the cell (Din et al. 2017; Yadav et al. 2017). The corresponding mechanism is shown in Fig. 1.6 (Din et al. 2017). Bogdanovic et al. have prepared CuNPs having size 5 nm and tested their antimicrobial activity against different bacteria cell such as Gram-negative bacteria (*Escherichia coli*) and Gram-positive bacteria (*Staphylococcus aureus* and *Candida albicans*) (Bogdanovic et al. 2014). Authors calculated the percentage reduction in their growth after treatment of CuNPs and found 99% reduction. The reduction in cell growth has been calculated following the equation given below:

$$R(\%) = \frac{C_o - C}{C_o} * 100 \qquad (1.1)$$

where, R, C_o, and C indicate the percentage reduction of microbial growth, initial number of microbial colonies (CFU), and number of microbial colonies, after treatment with CuNMs, respectively. Holubnycha et al. have synthesized copper

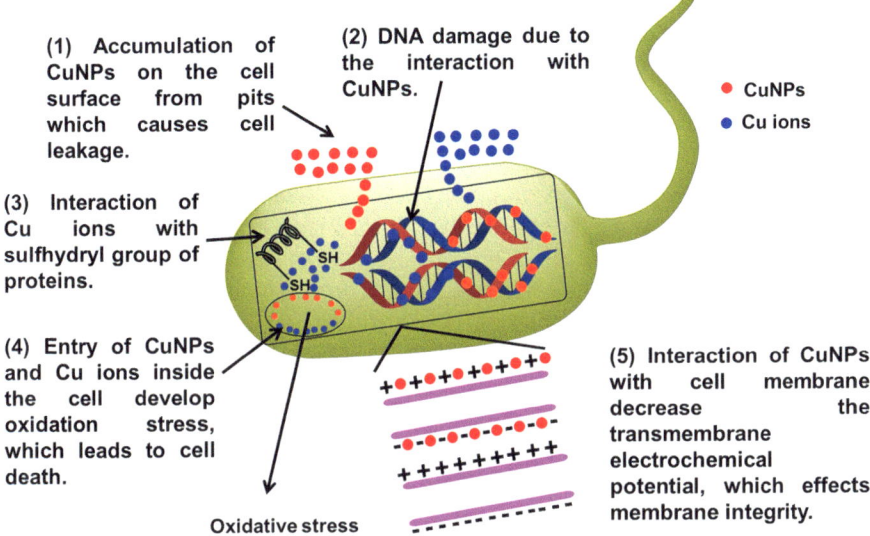

(1) Accumulation of CuNPs on the cell surface from pits which causes cell leakage.

(2) DNA damage due to the interaction with CuNPs.

● CuNPs
● Cu ions

(3) Interaction of Cu ions with sulfhydryl group of proteins.

(4) Entry of CuNPs and Cu ions inside the cell develop oxidation stress, which leads to cell death.

Oxidative stress

(5) Interaction of CuNPs with cell membrane decrease the transmembrane electrochemical potential, which effects membrane integrity.

Fig. 1.6 Mechanisms showing antimicrobial properties of copper nanomaterials (Din et al. 2017)

nanomaterials with particle size in the range of 15–20 nm by chemical reduction method (Holubnycha et al. 2017). The prepared copper nanomaterials along with different concentration of chitosan solution (molecular weight 200 kDa and 500 kDa) were tested against *E. coli* and *S. aureus* (Holubnycha et al. 2017). Interestingly authors observed that copper nanomaterials with 200 kDa chitosan show good antimicrobial activities, but copper nanomaterials with 500 kDa chitosan solution have less or no antimicrobial activity against the bacteria cell (*E. coli* and *S. aureus*). It is assumed that copper nanomaterials with low molecular weight chitosan solution (200 kDa) have shown good antimicrobial activity because of their strong binding with mRNA block present inside the cell membrane. However, high molecular weight chitosan solution (500 kDa) probably made strong binding with copper nanomaterials and therefore not get free to bind with bacterial cell and showed no or decreased antimicrobial activity.

1.4.5 Carbon-Based Nanomaterials as an Antimicrobial Agent

Carbon is the most important element in the world and similarly their nanomaterials are also the most demanded material. Carbon nanostructures exist in mainly three dimensions: (1) zero dimensional, fullerene; (2) one dimensional, carbon nanotube; and (3) two dimensional, graphene. In 1985, Kroto et al. first discovered the zero dimension fullerenes (Kroto 1997). Thereafter in the year of 1991, Sumio Iijima first time discovered the carbon nanotubes (Iijima 1991). Afterward, single layer graphene was first isolated from graphite powder by Andre Geim and Nosovelov in the year of 2014 (Novoselov et al. 2004). Presently, carbon-based nanomaterials such as graphene oxide, fullerene, carbon nanotube (mainly single-walled carbon nanotube), and activated carbon nanomaterials have been greatly used as an antimicrobial agent and show ideal antimicrobial activity against different bacteria cells. It has been well known that size and surface area of the any nanomaterials can greatly affect their antimicrobial properties. Since carbon-based nanomaterials have small size and large surface to volume ratio, they showed higher interaction with bacteria cell. Although, the antimicrobial properties of carbon based nanomaterials also depend on the composition of the nanomaterials, functionalization of carbon based nanomaterials and type of the bacteria (Dizaj et al. 2015).

If we discuss about their mode of action, in the literature, multiple-step mechanisms have been involved for the bacteria cell death. Carbon-based nanomaterials mainly destroy the cell membrane by the oxidative stress. But present studies showed that the physical interaction between the carbon-based nanomaterials and bacteria cell are the primary mechanism rather than oxidative stress (Dizaj et al. 2015). Herein, we have discussed the antimicrobial properties of carbon-based nanomaterials as well as their mode of action.

1.4.5.1 Fullerene

Fullerenes are the allotrope of carbon with football-like shape. Fullerenes have been used as a potent antimicrobial agent against various bacteria cell like Gram-negative bacteria (*E. coli*) and Gram-positive bacteria (*Streptococcus* spp. and *Salmonella*). Several studies have suggested that fullerene and their derivative can destroy the cell membrane by damaging the respiratory system by affecting the oxygen uptake (Deryabin et al. 2014). It has been also observed that with altering fullerene concentration (i.e., from low to high level) leads to change in oxygen uptake also, i.e., from low to high. According to the literature, fullerenes have hydrophobic surface, which can easily bind with the membrane cells and results in cell disruption (Deryabin et al. 2014). Fullerene derivatives can be synthesized with any charge (i.e., positive or negative) or no charge, i.e., neutral charged, but among the three classes of fullerene, only cationic charged showed potent antimicrobial activity against *E. coli* and *Shewanella oneidensis*, while anionic fullerene derivatives do not have any antimicrobial properties. Because, fullerene derivatives bearing cationic charge can interact with negatively charge bacteria cell, while other failed to do so (Tegos et al. 2005). Deryabin and co-workers (2014) have reported the antimicrobial activity of two water-soluble fullerene derivatives (Deryabin et al. 2014). They have prepared deprotonated carboxylic acid and protonated amine functional group containing fullerene derivative using organic linkers and tested their antimicrobial activity against *E. coli* bacteria. It was found that fullerene derivative with protonated amine exhibited potent antimicrobial activity, but deprotonated carboxylic containing fullerene derivative did not show any antimicrobial activity. Finally authors have concluded that the protonated amine containing fullerene derivatives possesses positive charge which can interact with negatively charged *E. coli* bacteria, but another one did not able to develop such kind of interaction. Additionally, it is well known that solubility of fullerene derivatives increases by functionalizing hydrophilic functional group and soluble fullerene can also be used in photodynamic therapy as a photosensitizer's agent, reported by Lu et al. (2010). Moreover water-soluble fullerene and its derivative easily reduced by the biological reducing agent lead to generate superoxide, which can destroy the microbial cells (Lu et al. 2010). In year 2005, Tegos et al. have prepared fulleropyrrolidinium by irradiation and investigated their antimicrobial activity. The fullerene-based nanomaterial showed 99% death of different bacterial and fungal cells (Tegos et al. 2005). Similarly, Cataldo and Da Ros have also prepared fullerene derivative, i.e., by sulfobutyl fullerene photoirradiation, and used them as an antimicrobial agent against bacteria cell (Cataldo and Da Ros 2008).

1.4.5.2 Carbon Nanotubes (CNTs)

Carbon nanotubes are the nanosized hollow cylinder-shaped arrangement of carbon atoms. CNTs are classified in two types, i.e., single-walled carbon nanotubes (SWCNTs) and multiwalled carbon nanotubes (MWCNTs). However, SWCNTs are able to easily penetrate into the bacteria cell, as consequence showed potent antimicrobial applications in comparison to the MWCNTs. In 2007, Kang et al. first time used SWCNTs for antimicrobial applications and found as an ideal antimicrobial agent against *E. coli* bacteria (Kang et al. 2007). The same group in the year of 2008 prepared both SWCNTS and MWCNTs and tested their antimicrobial activity against *E. coli* bacteria (Kang et al. 2007). They found that SWCNTs were more toxic toward bacterial cells than MWCNTs; as a result SWCNTs showed better antimicrobial activity against *E. coli* than MWCNTs. They also reported that it was the direct contact between the CNTs and bacteria cell which affected the cell membrane, metabolism process, and also the surface morphology of *E. coli*. The small size of SWCNTs is able to easily penetrate into the cell membrane and their large surface area helps to provide superior interaction.

In other study, Liju and co-workers have prepared hydroxyl and carboxylic functional group modified SWCNTs and MWCNTs and tested their antimicrobial activity against Gram-positive and Gram-negative bacteria (Yang et al. 2010). They have found that SWCNT-OH and SWCNT-COOH exhibited strong antimicrobial activity toward both Gram-positive and Gram-negative bacterial cell, but MWCNT-OH and MWCNT-COOH did not exhibit any antimicrobial activity. Arias et al. have also tested the antimicrobial activity of different length SWCNTs and used mainly three different lengths, i.e., less than 1 μm, 1–5 μm, and 5 μm (Arias and Yang 2009). It was found that the SWCNTs with longest size showed strong antimicrobial activity than the others. It may be because longer-length SWCNTs can effectively aggregate with bacterial cells, while short-length SWCNTs get aggregated to each other but not to the bacterial cell. Dong and co-workers prepared SWCNTs dispersed in various surfactants like sodium dodecyl sulfate, sodium cholate, and sodium dodecyl benzene sulfonate by simple ultrasonication and tested their antimicrobial activity against *E. coli*, *Salmonella enterica*, and *Enterococcus faecium* (Dong et al. 2012). They found better antimicrobial activity against above bacterial cells using SWCNTs dispersed in sodium cholate and also found that antimicrobial activity increased with increase in the concentration of SWCNTs.

1.4.5.3 Graphene Oxide (GO)

Graphene is an allotrope of carbon with hexagonal structure. Graphene oxide is usually prepared from graphite using strong oxidizing agent. Graphene oxide has been greatly used in various fields such as waste water treatment, medical diagnosis, drug delivery, sensing, etc. From the last few years, graphene oxide has showed their potent antimicrobial activities against bacteria cells. Several literatures studied showed that direct contact between graphene and graphene oxide with bacteria cell resulted in membrane stress and their killing (Dizaj et al. 2015). Akhavan and Elham have prepared graphene oxide and tested their antimicrobial activity against Gram-positive and Gram-negative bacteria cells (Akhavan and Ghaderi 2010). They have found that bacteria cells were damaged by the direct contact between bacteria cells and graphene oxide nanosheets. Gurunathan and co-workers prepared graphene and reduced graphene oxide and tested the antimicrobial activity against *S. aureus* bacteria (Gurunathan et al. 2012). They found better antimicrobial activity, which suggested that the bacterial cell killing mechanism proceeds through direct contact between the bacteria cells and sharp edge of the graphene and reduced graphene oxide.

In the year 2014, Karimnezhad and co-workers have synthesized functionalized graphene oxide such as graphene oxide-chlorophyllin-Zn and graphene oxide-chlorophyllin and tested their antimicrobial properties against *E. coli* (Azimi et al. 2014). They found that graphene oxide-chlorophyllin-Zn showed better antimicrobial activity than graphene oxide-chlorophyllin against *E. coli* bacteria. This is mainly due to the surface chemistry and toxicity of zinc metal, which resulted to damage the integrity of cell, followed by their killing. Yun et al. have synthesized two nanocomposites of CNTs and GO with AgNPs, i.e., CNT-Ag and GO-Ag, and studied their antimicrobial activity against Gram-positive and Gram-negative bacteria (Yun et al. 2013). Authors found that CNT-Ag nanocomposite exhibited better antimicrobial activity than GO-Ag nanocomposite. This is because the silver nanoparticles dispersed more effectively in CNT than the others.

1.4.5.4 Activated Carbon-Based Nanomaterials (ACNMs)

Activated carbon nanomaterials are the nanosized carbonaceous nanomaterials having large surface area to volume ratio and high chemical reactivity (Lakshmi et al. 2018). ACNMs are usually synthesized by hydrothermal, chemical vapor deposition and solution combustion methods (Marsh and Reinoso 2006). Actually, carbonaceous nanomaterials are activated by various physical and chemical activation methods to form activated carbon nanomaterials. Physical activation method consists of two steps, i.e., (1) carbonization of carbon species at high temperature under inert atmosphere and (2) carbonized products are activated using activating agent like air, steam, and carbon dioxide. But in chemical methods, simple chemical reaction between the carbon precursor and chemical agents like salts, acids, and bases takes place to activate the carbonaceous nanomaterials (Nor et al. 2013). The activated carbon nanomaterials prepared through physical activation method have large surface area than the chemical methods. Activated carbon nanomaterials have been widely used as/in adsorbents (Wickramaratne and Jaroniec 2013), supercapacitors (Liu et al. 2016), catalyst (Bian et al. 2012), lithium ion battery (Peng et al. 2014), and sensor fabrication (Claussen et al. 2012). But, from the last few decades, activated carbon nanomaterials have been largely applied as antimicrobial agent. ACNMs interact with bacteria cell which resulted in cell death. The mechanism of action followed few steps: (a) inhibition of the cell wall synthesis, (b) inhibition of protein synthesis, (c) destruction of the nucleic acids replication and transcription, and (d) destruction of the cell membrane.

Zhao et al. have synthesized silver-containing activated carbon nanomaterials (Ag/ACNMs) by physical activated method from coconut shell with particle size 2–4 nm and tested their antimicrobial activity against *E. coli* (Zhao et al. 2013). Soely et al. have synthesized activated carbon from pyrolyzed sugarcane bagasse (ACPB) by physical activation methods, followed by silver loading on the activated carbon nanomaterials (AgNM/ACPB) with pore size in the range of 1.0–3.5 and higher surface area (1200–1400 m^2/g) than commercially available activated carbon nanomaterials (Gonçalves et al. 2016). Authors have tested antimicrobial activity of synthesized materials (ACPB and ACPB/AgNM) toward *E. coli* bacteria cell and found that ACPB/AgNM showed better antimicrobial activity than ACPB. Synthesis of activated carbon nanomaterials from various biowastes and their antimicrobial activity toward particular bacterial cells are summarized in Table 1.4 (Lakshmi et al. 2018; Zhao et al. 2013; Gonçalves et al. 2016; Varghese et al. 2013; Sekaran et al. 2013; Yallappa et al. 2017; Das Purkayastha et al. 2014; Shi et al. 2007; Basker et al. 2016; Saravanan et al. 2016).

1.5 Challenges of Nanomaterials in Antibacterial Treatments

Based on the literatures we have discussed in this chapter, it is very clear that nanomaterials are promising candidate toward antimicrobial treatments. But, as discussed in the chapter, the antimicrobial activity of nanomaterials depends on several parameters like their concentration, temperature, pH, etc. Similarly, having several advantages and superior performances than ancient antimicrobial agents, nanomaterials have to face several challenges to secure their top most places in antibacterial treatments and translate this technology in real clinical use. In the list of challenges, the first and foremost one is study of complete mechanism of interaction between

Table 1.4 Synthesis of activated carbon nanomaterials from various biowastes and their antimicrobial activity

S. N.	Carbon precursor	Microorganism	MIC[a]	References
1.	Coconut	*E. coli*	10^7 CFU/mL	Zhao et al. (2013)
2.	Sugarcane bagasse	*E. coli*	10^7 CFU/mL	Gonçalves et al. (2016)
3.	Kitchen soot	*S. aureus, S. haemolyticus, P. refrigere,* and *P. aeruginosa*	–	Varghese et al. (2013)
4.	Rice husk	*Bacillus* sp.	3.2×10^7 cells/mL	Sekaran et al. (2013)
5.	Groundnut shell	*B. cereus, E. coli,* and *C. violaceum*	–	Yallappa et al. (2017)
6.	Sandalwood bark	*P. notatum, B. cereus, E. coli,* and *C. violaceum*	–	Lakshmi et al. (2018)
7	Rapeseed oil-cake	*E. coli, S. aureus, B. cereus, L. monocytogenes, S. enterica, C. albicans, P. diminuta, K. pneumonia, M. smegmatis*	5–15, 31–62, 31–62, 31–62, 7–15, 7–15, 7–15, 7–15, 31–62 µg/mL, respectively	Das Purkayastha et al. (2014)
8.	–	*E. coli* and *S. aureus*	–	Shi et al. (2007)
9.	*Passiflora foetida*	*B. cereus, S. aureus, Enterococcus* sp., *M. luteus, Corynebacterium* sp., *E. coli, K. pneumonia, P. aeruginosa, S. flexneri, S. typhimurium, P. vulgaris,* and *A. faecalis*	–	Basker et al. (2016)
10.	Fishtail palm seeds	*S. aureus, S. epidermidis, P. aeruginosa, E. coli,* and *C. albicans*	–	Saravanan et al. (2016)

[a]*MIC* Minimum inhibition concentration

nanomaterials and bacterial cells. Similarly, their effect on the living organism, like living cells (i.e., normal and disease), tissues, and organs, should be studied too, to find the optimum dose as well as fastest route of administration of nanomaterials as antibacterial agent. Therefore, to make the nanotechnology applicable in real clinical field, their use in vivo condition is very much required. As we have discussed in the Sect. 1.2, the role of nanomaterials as antibacterial agent can be studied using their particles or solutions in general as in vitro condition. But to study their in vivo role, we are still not in very advanced stage. We can always agree everyone on the fact that nanomaterials and nanotechnology have lots of potential as antimicrobial agents, but it is also very clear that we are still not aware of several facts and needs a lot of things to know, prior to using these nanomaterials in real clinical field with day-to-day use.

Acknowledgement Author Declaration, Mr. Majhi and Ms. Karfa have given the major contribution in writing this book chapter along with drawing the figures and tables, taking the copyright permission, etc.

References

Aal NA, Al-Hazmi F, Al-Ghamdi AA, Al-Ghamdi AA, El-Tantawy F, Yakuphanoglu F (2015) Novel rapid synthesis of zinc oxide nanotubes via hydrothermal technique and antibacterial properties. Spectrochim Acta A Mol Biomol Spectrosc 135:871–877

Ajitha B, Reddy YA, Rajesh KM, Reddy PS (2016) Sesbania grandiflora leaf extract assisted green synthesis of silver nanoparticles: antimicrobial activity. Materials Today: Proceedings 3(6):1977–1984

Akhavan O, Ghaderi E (2010) Toxicity of graphene and graphene oxide nanowalls against bacteria. ACS Nano 4(10):5731–5736

Alfuraydi AA, Devanesan S, Al-Ansari M, AlSalhi MS, Ranjitsingh AJ (2019) Eco-friendly green synthesis of silver nanoparticles from the sesame oil cake and its potential anticancer and antimicrobial activities. J Photochem Photobiol B 192:83–89

Al-Heniti S, Umar A (2013) Structural, optical and field emission properties of urchin-shaped ZnO nanostructures. J Nanosci Nanotechnol 13(1):86–90

Ambika S, Sundrarajan M (2015) Antibacterial behaviour of Vitex negundo extract assisted ZnO nanoparticles against pathogenic bacteria. J Photochem Photobiol B Biol 146:52–57

Arias LR, Yang L (2009) Inactivation of bacterial pathogens by carbon nanotubes in suspensions. Langmuir 25(5):3003–3012

Aruguete DM, Kim B, Hochella MF, Ma Y, Cheng Y, Hoegh A, Liu J, Pruden A (2013) Antimicrobial nanotechnology: its potential for the effective management of microbial drug resistance and implications for research needs in microbial nanotoxicology. Environ Sci: Processes Impacts 15(1):93–102

Azimi S, Behin J, Abiri R, Rajabi L, Derakhshan AA, Karimnezhad H (2014) Synthesis, characterization and antibacterial activity of chlorophyllin functionalized graphene oxide nanostructures. Sci Adv Mater 6(4):771–781

Aziz N, Fatma T, Varma A, Prasad R (2014) Biogenic synthesis of silver nanoparticles using Scenedesmus abundans and evaluation of their antibacterial activity. J Nanopart 2014:689419. http://dx.doi.org/10.1155/2014/689419

Aziz N, Faraz M, Pandey R, Sakir M, Fatma T, Varma A, Barman I, Prasad R (2015) Facile algae-derived route to biogenic silver nanoparticles: Synthesis, antibacterial and photocatalytic properties. Langmuir 31:11605–11612. https://doi.org/10.1021/acs.langmuir.5b03081

Aziz N, Pandey R, Barman I, Prasad R (2016) Leveraging the attributes of Mucor hiemalis-derived silver nanoparticles for a synergistic broad-spectrum antimicrobial platform. Front Microbiol 7:1984. https://doi.org/10.3389/fmicb.2016.01984

Aziz N, Faraz M, Sherwani MA, Fatma T, Prasad R (2019) Illuminating the anticancerous efficacy of a new fungal chassis for silver nanoparticle synthesis. Front Chem 7:65. https://doi.org/10.3389/fchem.2019.00065

Balouiri M, Sadiki M, Ibnsouda SK (2016 Apr 1) Methods for in vitro evaluating antimicrobial activity: a review. J Pharmaceut Anal 6(2):71–79

Basker S, Karthik K, Karthik S (2016) In vitro evaluation of antibacterial efficacy using Passiflora foetida activated carbon. Res Pharm 22:6

Bian X, Zhu J, Liao L, Scanlon MD, Ge P, Ji C, Girault HH, Liu B (2012) Nanocomposite of MoS2 on ordered mesoporous carbon nanospheres: a highly active catalyst for electrochemical hydrogen evolution. Electrochem Commun 22:128–132

Bhuyan T, Mishra K, Khanuja M, Prasad R, Varma A (2015) Biosynthesis of zinc oxide nanoparticles from Azadirachta indica for antibacterial and photocatalytic applications. Mater Sci Semicond Process 32:55–61

Bogdanovic U, Lazic V, Vodnik V, Budimir M, Markovic Z, Dimitrijevic S (2014) Copper nanoparticles with high antimicrobial activity. Mater Lett 128:75–78

Cambau E, Saunderson P, Matsuoka M, Cole ST, Kai M, Suffys P, Rosa PS, Williams D, Gupta UD, Lavania M, Cardona-Castro N (2018) Antimicrobial resistance in leprosy: results of the first prospective open survey conducted by a WHO surveillance network for the period 2009–15. Clin Microbiol Infect 24(12):1305–1310

Cataldo F, Da Ros T (2008) Medicinal chemistry and pharmacological potential of fullerenes and carbon nanotubes. Springer, Trieste

Chernousova S, Epple M (2013) Silver as antibacterial agent: ion, nanoparticle, and metal. Angew Chem Int Ed 52(6):1636–1653

Chiriac V, Stratulat DN, Calin G, Nichitus S, Burlui V, Stadoleanu C, Popa M, Popa IM (2016) Antimicrobial property of zinc based nanoparticles. In: IOP conference series: materials science and engineering, vol 133, no. 1. IOP Publishing, Bristol, p 012055

Claussen JC, Shi J, Rout CS, Artiles MS, Roushar MM, Stensberg MC, Porterfield DM, Fisher TS (2012) Nano-sized biosensors for medical applications. In: Biosensors for medical applications. Woodhead Publishing, Cambridge, pp 65–102

Cui P, Li X, Zhu M, Wang B, Liu J, Chen H (2017) Design, synthesis and antimicrobial activities of thiouracil derivatives containing triazolo-thiadiazole as SecA inhibitors. Eur J Med Chem 127:159–165

Das Purkayastha M, Manhar AK, Mandal M, Mahanta CL (2014) Industrial waste-derived nanoparticles and microspheres can be potent antimicrobial and functional ingredients. J Appl Chem 2014:1

Deryabin DG, Davydova OK, Yankina ZZ, Vasilchenko AS, Miroshnikov SA, Kornev AB et al (2014) The activity of [60] fullerene derivatives bearing amine and carboxylic solubilizing groups against Escherichia coli: a comparative study. J Nanomater 2014:1–9

Deus RC, Cilense M, Foschini CR, Ramirez MA, Longo E, Simões AZ (2013) Influence of mineralizer agents on the growth of crystalline CeO2 nanospheres by the microwave-hydrothermal method. J Alloys Compd 550:245–251

Dhand V, Soumya L, Bharadwaj S, Chakra S, Bhatt D, Sreedhar B (2016) Green synthesis of silver nanoparticles using Coffea arabica seed extract and its antibacterial activity. Mater Sci Eng C 58:36–43

Diez-Pascual AM (2018) Antibacterial activity of nanomaterials. Nanomaterials 8:359

Dimapilis EA, Hsu CS, Mendoza RM, Lu MC (2018) Zinc oxide nanoparticles for water disinfection. Sustain Environ Res 28(2):47–56

Din MI, Arshad F, Hussain Z, Mukhtar M (2017) Green adeptness in the synthesis and stabilization of copper nanoparticles: catalytic, antibacterial, cytotoxicity, and antioxidant activities. Nanoscale Res Lett 12(1):638

Dizaj SM, Mennati A, Jafari S, Khezri K, Adibkia K (2015) Antimicrobial activity of carbon-based nanoparticles. Adv Pharmaceut Bull 5(1):19

Dong L, Henderson A, Field C (2012) Antimicrobial activity of single-walled carbon nanotubes suspended in different surfactants. J Nanotechnol 2012:1

Dwivedi S, Wahab R, Khan F, Mishra YK, Musarrat J, Al-Khedhairy AA (2014) Reactive oxygen species mediated bacterial biofilm inhibition via zinc oxide nanoparticles and their statistical determination. PLoS One 9(11):111289

Fernando SS, Gunasekara TD, Holton J (2018) Antimicrobial nanoparticles: applications and mechanisms of action. Sri Lankan J Infect Dis 8(1):2–11

Gonçalves SP, Strauss M, Delite FS, Clemente Z, Castro VL, Martinez DS (2016) Activated carbon from pyrolysed sugarcane bagasse: silver nanoparticle modification and ecotoxicity assessment. Sci Total Environ 565:833–840

Gurunathan S, Han JW, Dayem AA, Eppakayala V, Kim JH (2012) Oxidative stress-mediated antibacterial activity of graphene oxide and reduced graphene oxide in Pseudomonas aeruginosa. Int J Nanomedicine 7:5901–5914

Haghi M, Hekmatafshar M, Janipour MB, Gholizadeh SS, Faraz MK, Sayyadifar F, Ghaedi M (2012) Antibacterial effect of TiO2 nanoparticles on pathogenic strain of E. coli. Int J Adv Biotechnol Res 3(3):621–624

Hajipour MJ, Fromm KM, Ashkarran AA, de Aberasturi DJ, de Larramendi IR, Rojo T, Serpooshan V, Parak WJ, Mahmoudi M (2012) Antibacterial properties of nanoparticles. Trends Biotechnol 30:499–511

Hassanien AS, Khatoon UT (2019) Synthesis and characterization of stable silver nanoparticles, Ag-NPs: discussion on the applications of Ag-NPs as antimicrobial agents. Phys B Condens Matter 554:21–30

Holubnycha V, Pogorielov M, Korniienko V, Kalinkevych O, Ivashchenko O, Peplinska B, Jarek M (2017) Antibacterial activity of the new copper nanoparticles and Cu NPs/chitosan solution. In: Nanomaterials: application and properties (NAP), 2017 IEEE 7th international conference 2017 Sep 10. IEEE, pp. 04NB10-1

Huang Z, Zheng X, Yan D, Yin G, Liao X, Kang Y, Yao Y, Huang D, Hao B (2008) Toxicological effect of ZnO nanoparticles based on bacteria. Langmuir 24(8):4140–4144

Ibrahim HM (2015) Green synthesis and characterization of silver nanoparticles using banana peel extract and their antimicrobial activity against representative microorganisms. J Radiat Res Appl Sci 8(3):265–275

Iijima S (1991) Helical microtubules of graphitic carbon. Nature 354(6348):56

Jain A, Bhargava R, Poddar P (2013) Probing interaction of Gram-positive and Gram-negative bacterial cells with ZnO nanorods. Mater Sci Eng C 33(3):1247–1253

Janaki AC, Sailatha E, Gunasekaran S (2015 Jun 5) Synthesis, characteristics and antimicrobial activity of ZnO nanoparticles. Spectrochim Acta A Mol Biomol Spectrosc 144:17–22

Janeczko M, Demchuk OM, Strzelecka D, Kubinski K, Masłyk M (2016) New family of antimicrobial agents derived from 1, 4-naphthoquinone. Eur J Med Chem 124:1019–1025

Jena J, Pradhan N, Dash BP, Panda PK, Mishra BK (2015) Pigment mediated biogenic synthesis of silver nanoparticles using diatom Amphora sp. and its antimicrobial activity. J Saudi Chem Soc 19(6):661–666

Kang S, Pinault M, Pfefferle LD, Elimelech M (2007) Single-walled carbon nanotubes exhibit strong antimicrobial activity. Langmuir 23(17):8670–8673

Karaman DS, Manner S, Fallarero A, Rosenholm JM (2017) Current approaches for exploration of nanoparticles as antibacterial agents. In: Antibacterial agents. InTech, London

Kroto H (1997) Symmetry, space, stars and C 60. Rev Mod Phys 69(3):703

Kumar KM, Mandal BK, Naidu EA, Sinha M, Kumar KS, Reddy PS (2013) Synthesis and characterisation of flower shaped zinc oxide nanostructures and its antimicrobial activity. Spectrochim Acta A Mol Biomol Spectrosc 104:171–174

Kumar R, Umar A, Kumar G, Nalwa HS (2017) Antimicrobial properties of ZnO nanomaterials: a review. Ceram Int 43(5):3940–3961

Lakshmi SD, Avti PK, Hegde G (2018) Activated carbon nanoparticles from biowaste as new generation antimicrobial agents: a review. Nano-Struct Nano-Object 16:306–321

Lemire JA, Harrison JJ, Turner RJ (2013) Antimicrobial activity of metals: mechanisms, molecular targets and applications. Nat Rev Microbiol 11(6):371

Li WR, Xie XB, Shi QS, Zeng HY, You-Sheng OY, Chen YB (2010) Antibacterial activity and mechanism of silver nanoparticles on Escherichia coli. Appl Microbiol Biotechnol 85(4):1115–1122

Liu Y, He L, Mustapha A, Li H, Hu ZQ, Lin M (2009) Antibacterial activities of zinc oxide nanoparticles against Escherichia coli O157: H7. J Appl Microbiol 107(4):1193–1201

Liu C, Wang J, Li J, Zeng M, Luo R, Shen J, Sun X, Han W, Wang L (2016) Synthesis of N-doped hollow-structured mesoporous carbon nanospheres for high-performance supercapacitors. ACS Appl Mater Interfaces 8(11):7194–7204

Lu Z, Dai T, Huang L, Kurup DB, Tegos GP, Jahnke A et al (2010) Photodynamic therapy with a cationic functionalized fullerene rescues mice from fatal wound infections. Nanomedicine (Lond) 5(10):1525–1533

Mahmoodi S, Elmi A, Hallaj-nezhadi S (2018) Copper nanoparticles as antibacterial agents. J Mol Pharm Org Process Res 6(1):1–7

Marsh H, Reinoso RF (2006) Activated carbon, first edn. Elsevier, Amsterdam, pp 1–554

Nair S, Sasidharan A, Rani VD, Menon D, Nair S, Manzoor K, Raina S (2009) Role of size scale of ZnO nanoparticles and microparticles on toxicity toward bacteria and osteoblast cancer cells. J Mater Sci Mater Med 20(1):235

Nor NM, Lau LC, Lee KT, Mohamed AR (2013) Synthesis of activated carbon from lignocellulosic biomass and its applications in air pollution control—a review. J Environ Chem Eng 1(4):658–666

Novoselov KS, Geim AK, Morozov SV, Jiang D, Zhang Y, Dubonos SV, Grigorieva IV, Firsov AA (2004) Electric field effect in atomically thin carbon films. Science 306:666–669

Nzekwe IT, Agubata CO, Umeyor CE, Okoye IE, Ogwueleka CB (2016) Synthesis of silver nanoparticles by sodium borohydride reduction method: optimization of conditions for high anti-staphylococcal activity. Br J Pharmaceut Res 14(5):1–9

Padalia H, Moteriya P, Chanda S (2015) Green synthesis of silver nanoparticles from marigold flower and its synergistic antimicrobial potential. Arab J Chem 8(5):732–741

Padmavathy N, Vijayaraghavan R (2008) Enhanced bioactivity of ZnO nanoparticles—an antimicrobial study. Sci Technol Adv Mater 9(3):035004

Palanisamy S, Rajasekar P, Vijayaprasath G, Ravi G, Manikandan R, Prabhu NM (2017) A green route to synthesis silver nanoparticles using Sargassum polycystum and its antioxidant and cytotoxic effects: an in vitro analysis. Mater Lett 189:196–200

Patil MP, Kim GD (2017) Eco-friendly approach for nanoparticles synthesis and mechanism behind antibacterial activity of silver and anticancer activity of gold nanoparticles. Appl Microbiol Biotechnol 101(1):79–92

Peng X, Fu J, Zhang C, Tao J, Sun L, Chu PK (2014) Rice husk-derived activated carbon for Li ion battery anode. Nanosci Nanotechnol Lett 6(1):68–71

Prado JV, Vidal AR, Duran TC (2012) Application of copper bactericidal properties in medical practice. Rev Med Chil 140:1325–1332

Prasad R, Jha A, Prasad K (2018) Exploring the Realms of Nature for Nanosynthesis. Springer International Publishing (ISBN 978-3-319-99570-0) https://doi.org/www.springer.com/978-3-319-99570-0

Prasad R, Swamy VS (2013) Antibacterial activity of silver nanoparticles synthesized by bark extract of Syzygium cumini. J Nanoparticles 2013:431218. https://doi.org/10.1155/2013/431218

Premanathan M, Karthikeyan K, Jeyasubramanian K, Manivannan G (2011) Selective toxicity of ZnO nanoparticles toward Gram-positive bacteria and cancer cells by apoptosis through lipid peroxidation. Nanomed Nanotechnol Biol Med 7(2):184–192

Raffi M, Mehrwan S, Bhatti TM, Akhter JI, Hameed A, Yawar W, ul Hasan MM (2010) Investigations into the antibacterial behavior of copper nanoparticles against Escherichia coli. Ann Microbiol 60(1):75–80

Raghupathi KR, Koodali RT, Manna AC (2011) Size-dependent bacterial growth inhibition and mechanism of antibacterial activity of zinc oxide nanoparticles. Langmuir 27(7):4020–4028

Raimondi MV, Listro R, Cusimano MG, La Franca M, Faddetta T, Gallo G, Schillaci D, Collina S, Leonchiks A, Barone G (2019) Pyrrolomycins as antimicrobial agents. Microwave-assisted organic synthesis and insights into their antimicrobial mechanism of action. Bioorg Med Chem 27:721–728

Ramesh M, Anbuvannan M, Viruthagiri G (2015) Green synthesis of ZnO nanoparticles using Solanum nigrum leaf extract and their antibacterial activity. Spectrochim Acta A Mol Biomol Spectrosc 136:864–870

Rao NH, Lakshmidevi N, Pammi SV, Kollu P, Ganapaty S, Lakshmi P (2016) Green synthesis of silver nanoparticles using methanolic root extracts of Diospyros paniculata and their antimicrobial activities. Mater Sci Eng C 62:553–557

Ravichandran V, Vasanthi S, Shalini S, Shah SA, Harish R (2016) Green synthesis of silver nanoparticles using Artocarpus altilis leaf extract and the study of their antimicrobial and antioxidant activity. Mater Lett 180:264–267

Rodrigo R, Libuy M, Feliu F, Hasson D (2013) Molecular basis of cardioprotective effect of antioxidant vitamins in myocardial infarction. Biomed Res Int 2013:1

Santos CL, Albuquerque AJ, Sampaio FC, Keyson D (2013) Nanomaterials with antimicrobial properties: applications in health sciences. In: Microbial pathogens and strategies for combating them: science, technology and education, Microbiology book series, vol 4. Formatex Research Center, Badajoz, Spain, p 2

Saraschandraa N, Pavithrab M, Sivakumar A (2013) Antimicrobial applications of TiO2 coated modified polyethylene (HDPE) films. Arch Appl Sci Res 5(1):189–194

Saravanan A, Kumar PS, Devi GK, Arumugam T (2016) Synthesis and characterization of metallic nanoparticles impregnated onto activated carbon using leaf extract of Mukia maderasapatna: evaluation of antimicrobial activities. Microb Pathog 97:198–203

Seil JT, Webster TJ (2012) Antimicrobial applications of nanotechnology: methods and literature. Int J Nanomedicine 7:2767

Sekaran G, Karthikeyan S, Gupta VK, Boopathy R, Maharaja P (2013) Immobilization of Bacillus sp. in mesoporous activated carbon for degradation of sulphonated phenolic compound in wastewater. Mater Sci Eng C 33(2):735–745

Sekhar MM, Nagarjuna U, Padmavathi V, Padmaja A, Reddy NV, Vijaya T (2018) Synthesis and antimicrobial activity of pyrimidinyl 1, 3, 4-oxadiazoles, 1, 3, 4-thiadiazoles and 1, 2, 4-triazoles. Eur J Med Chem 145:1–10

Sharma VK, Yngard RA, Lin Y (2009) Silver nanoparticles: green synthesis and their antimicrobial activities. Adv Colloid Interf Sci 145(1–2):83–96

Shi Z, Neoh KG, Kang ET (2007) Antibacterial and adsorption characteristics of activated carbon functionalized with quaternary ammonium moieties. Ind Eng Chem Res 46(2):439–445

Shinde VV, Dalavi DS, Mali SS, Hong CK, Kim JH, Patil PS (2014) Surfactant free microwave assisted synthesis of ZnO microspheres: study of their antibacterial activity. Appl Surf Sci 307:495–502

Stankovic A, Dimitrijevic S, Uskokovic D (2013) Influence of size scale and morphology on antibacterial properties of ZnO powders hydrothermally synthesized using different surface stabilizing agents. Colloids Surf B: Biointerfaces 102:21–28

Sulaiman GM, Mohammed WH, Marzoog TR, Al-Amiery AA, Kadhum AA, Mohamad AB (2013) Green synthesis, antimicrobial and cytotoxic effects of silver nanoparticles using Eucalyptus chapmaniana leaves extract. Asian Pac J Trop Biomed 3(1):58–63

Talebian N, Amininezhad SM, Doudi M (2013) Controllable synthesis of ZnO nanoparticles and their morphology-dependent antibacterial and optical properties. J Photochem Photobiol B Biol 120:66–73

Tegos GP, Demidova TN, Arcila-Lopez D, Lee H, Wharton T, Gali H et al (2005) Cationic fullerenes are effective and selective antimicrobial photosensitizers. Chem Biol 12(10):1127–1135

Umar A, Chauhan MS, Chauhan S, Kumar R, Sharma P, Tomar KJ, Wahab R, Al-Hajry A, Singh D (2013) Applications of ZnO nanoflowers as antimicrobial agents for Escherichia coli and enzyme-free glucose sensor. J Biomed Nanotechnol 9(10):1794–1802

Varghese S, Kuriakose S, Jose S (2013) Antimicrobial activity of carbon nanoparticles isolated from natural sources against pathogenic Gram-negative and Gram-positive bacteria. J Nanosci 6:2013

Varghese R, Almalki MA, Ilavenil S, Rebecca J, Choi KC (2017) Silver nanoparticles synthesized using the seed extract of *Trigonella foenum-graecum* L. and their antimicrobial mechanism and anticancer properties. Saudi J Biol Sci 26:148–154

Verma A, Mehata MS (2016) Controllable synthesis of silver nanoparticles using neem leaves and their antimicrobial activity. J Radiat Res Appl Sci 9(1):109–115

Whitesides GM (2005) Nanoscience, nanotechnology, and chemistry. Small 1(2):172–179

Wickramaratne NP, Jaroniec M (2013 Feb 27) Activated carbon spheres for CO2 adsorption. ACS Appl Mater Interfaces 5(5):1849–1855

Yadav L, Tripathi RM, Prasad R, Pudake RN, Mittal J (2017) Antibacterial activity of Cu nanoparticles against E. coli, Staphylococcus aureus and Pseudomonas aeruginosa. Nano Biomed Eng. 9(1):9–14. https://doi.org/10.5101/nbe.v9i1

Yallappa S, Deepthi DR, Yashaswini S, Hamsanandini R, Chandraprasad M, Kumar SA, Hegde G (2017) Natural biowaste of groundnut shell derived nano carbons: synthesis, characterization and its in vitro antibacterial activity. Nano-Struct Nano-Object 12:84–90

Yang C, Mamouni J, Tang Y, Yang L (2010) Antimicrobial activity of single-walled carbon nanotubes: length effect. Langmuir 26(20):16013–16019

Yun H, Kim JD, Choi HC, Lee CW (2013) Antibacterial activity of CNT-Ag and GO Ag nanocomposites against Gram-negative and Gram-positive bacteria. Bull Kor Chem Soc 34(11):3261–3264

Zhao Y, Wang ZQ, Zhao X, Li W, Liu SX (2013) Antibacterial action of silver-doped activated carbon prepared by vacuum impregnation. Appl Surf Sci 266:67–72

Chapter 2
A Review on Next-Generation Nano-Antimicrobials in Orthopedics: Prospects and Concerns

Archita Gupta, Abhimanyu Dev, Vinod Kumar Nigam,
Padmini Padmanabhan, and Sneha Singh

Contents

Abstract Recently an increase in occurrence of bone implant infections leading to implant failure and need for revision surgery has been observed. The major reason being development of antibiotic resistant microbes (MRSA and MRSE), while the withdrawal of antibiotic and suboptimal dose are further increasing the chances of antibiotic resistance. In recent time, nanotechnology interventions have led to emergence of a new generation of microbicidal agents the "nano-antimicrobials" showing potent antimicrobial efficacy even against MRSA. To circumvent this problem

A. Gupta · V. K. Nigam · P. Padmanabhan · S. Singh (✉)
Department of Bio-Engineering, Birla Institute of Technology, Mesra, Ranchi, India
e-mail: snehasingh@bitmesra.ac.in

A. Dev
Department of Pharmaceutical Sciences and Technology, Birla Institute of Technology, Mesra, Ranchi, India

© Springer Nature Switzerland AG 2020
R. Prasad et al. (eds.), *Nanostructures for Antimicrobial and Antibiofilm Applications*, Nanotechnology in the Life Sciences,
https://doi.org/10.1007/978-3-030-40337-9_2

33

in orthopedics, multiple efforts are being made to make potent antimicrobials that can be nontoxic for cells and able to counter resist microbes simultaneously. Owing to the unique physicochemical properties, the nanomaterials exhibit tremendous capabilities and hence utility in the field of orthopedics. However, the long-term implications of these nanomaterials upon health are feebly understood. This review focuses on the current scenario, mechanism of action, and future perspective and concerns of this emerging field of "nanobiotics" to alleviate the postsurgical infections to pave the way for better clinical orthopedic requirements.

Keywords Nanobiotics · Nano-antimicrobials · Orthopedics · Microbicidal · Osteoinductivity

2.1 Introduction

Orthopedic surgery is perturbed with the musculoskeletal disorders (MSD) which are the main cause of the concern and are majorly associated with half of the chronic conditions faced by aged population (Weinstein 2000; Leveille 2004). The estimated expenditure for MSD includes the cost for preventive care; direct care; care by hospitals, therapists, physicians, and caregivers; and the secondary costs incurring the effects on families and loss of productivity. In India approximately 17% of total population suffers from MSD and 70% of the total domestic demand is being catered through import of orthopedic products at premium prices. With the whooping market size of USD 375 million for Indian orthopedic devices, a 20% increase is anticipated every year, and by 2030 this market share is expected to soar high to USD 2.5 billion (Haralson and Zuckerman 2009). The metallic implants such as steel or titanium alloys are the most commonly used biomaterials in the orthopedic surgery because of their competent mechanical strength. In orthopedic surgery, biomaterials like screws, plates, and nails are used in artificial bones, joint diseases, spine implants, and hip and knee prostheses for the treatment of fractures. As the advancements increase in the field of orthopedic surgery, the number of patients needing the implants and their dependency on numerous medical devices is growing rapidly. The rate of usage of these implants is increasing nowadays and this trend is expected to multiply many folds in the coming decades due to aging population and improvement in medical prudence (Iorio et al. 2008; Holzwarth and Cotogno 2012). During surgeries many adverse causalities result from the endogenous or exogenous infectious agents leading to localized or systemic infections. These agents are present from the very beginning of the patient's hospitalization till his discharge from the medical services. Among the severe causalities, the surgical site infection (SSI) is the most extreme condition which occurs in the tissues handled during the surgery or during surgical incision (NHSN Manual 2009; Mangram et al. 1999).

SSI is one of the most common nosocomial infections apart from urinary tract infection, pneumonia, gastroenteritis, meningitis, and blood-related infections (NNIS Report 2004). The commonly identified etiological agents are *Staphylococcus aureus* (21–43%), followed by coagulase-negative staphylococci (17–39%), streptococci (7–12%), Gram-negative bacilli (2–12%), *Enterococcus* spp. (1–8%), and anaerobic bacteria (2–6%) (Kapadia et al. 2016). SSI is difficult to treat and leads to worse outcomes and thus becomes most vulnerable through orthopedic surgery and emerges with signs of inflammation within 30 days to 12 months depending on the place of implant affecting the superficial or the tissues in the deeper parts of the body (Khosravi et al. 2009; Campoccia et al. 2006; Mardanpour et al. 2017). There are three levels of SSI: first being the *superficial incisional* which affect skin and subcutaneous tissues and majorly identified by the localized inflammational reactions, second the *deep incisional* affecting the fascial and muscle layer, and third represents the *organ or space infections* involving any part of the body other than the site of surgical incision like joints or peritoneum. SSIs majorly become evident between 10th and 20th postoperative days as far as the superficial tissues are involved. However, in case of implants, SSIs affect the deeper tissues occurring several months of post-operation (NICE Clinical Guidelines 2008).

Despite the increasing awareness for aseptic surgical conditions, SSIs account half of the total nosocomial infections, representing a significant burden to the society medically as well as economically (Gomez et al. 2003; Schierholz and Beuth 2001). For controlling SSIs, it mainly requires device removal, multiple debridement surgeries, and long-term systemic antibiotic therapy (Winkler et al. 2006). However, these surgical operations not only increase the healthcare cost but also lead to multiple surgeries mainly due to difficulty in removing infections from devascularized bone and other necrotic tissues. Also, multiple antibiotic dosing develops adaptability in the causing agents, thereby increasing the risk of infections through methicillin-resistant *Staphylococcus aureus* (MRSA) and methicillin-resistant *Staphylococcus epidermidis* (MRSE) (Bhattacharya et al. 2016). Moreover, as soon as the implant is introduced within the body, layer of adhesions (including fibrinogen, fibronectin, collagen, etc.) is imposed on the implant surface promoting the adherence of planktonic bacteria and leading to the development of biofilms (Darouiche 2004). The biofilms impose problems for the treatment of SSIs because they block the accessibility for immune cells and antibiotics, therefore promoting the further survival of the bacteria. As prevention is always better than cure, strategies should be aimed at preventing the formation of SSI and biofilms in the immediate postoperative stage. The advancements should be in developing implant materials or coatings that serve as an effective alternative to resist infection and simultaneously promote bone growth. Thus, in this regard, the review focuses on the various preventive techniques at different levels for eradicating surgical site specific infections and the current approaches of nanobiotics particularly advantageous for orthopedic surgery applications.

2.2   Orthopedic Implants and Infections

Bone, a biocomposite material, is made up of calcium and phosphate minerals in the form of hydroxyapatite, collagen type I fibrils and water. The components of bone tissue are of nanometric scale (Rho et al. 1998). Since the last 50 years, numerous advances have been made in the field of biomaterials wherein these materials in consideration to biomedical applications have been categorized into three different generations (Navarro et al. 2008). First generation represents the bioinert materials, second generation constitutes biodegradable and bioactive materials, and third generation incorporates the materials basically designed to generate specific responses at the molecular level.

The selection of materials is generally on the basis of the final application wherein the material is to be incorporated. For biomaterials the biological properties are enumerated with the physical, chemical, and mechanical properties. In the case of orthopedic implants, the general properties desired in any biomaterial are biocompatibility, bioactivity, osteoconduction, stress shielding, and osteoinduction. The only requirement for the synthetic material used in biomedical applications was to obtain the physical properties of the material mimicking to that of the replaced tissue with minimum toxic effects (Hench and Jones 2005). These materials are named to be first-generation biomaterials due to their inertness and reduced foreign body reactions. Titanium and its alloys represent the first-generation biomaterials because of their biocompatibility, lightness, strength, and resistance toward corrosion (Niinomi 2008; Disegi 2000). The biocompatibility of these materials is mainly due to the titanium dioxide layer coated on the surface of the bulk material, thereby improving its cell adhesion and osseointegration property (Yang et al. 2008; Wang et al. 2013). Bone formation was observed within 12 months of the implantation only on the coated materials as compared to the non-coated smooth surfaces (Fujibayashi et al. 2004). The 3D microporous structures have influenced the osteoinductivity of this material. In 1997 FDA-approved acetabular cups consisting trabecular metal made up of the porous tantalum are highly porous biomaterial with resemblance with the trabecular bone structurally and mechanically (Matassi et al. 2013). The potential of this material has been primarily used in total hip arthroplasty components with excellent clinical results (Elganzoury and Bassiony 2013; Moličnik et al. 2014). The osteoinductivity of tantalum material as bulk has not been reported, but after modifying its surface with tantalum oxide nanotube films, its biocompatibility, anti-corrosion, and osteoinductive potential have been increased (Wang et al. 2012). Magnesium and its alloys had also attained attention as first-generation materials in orthopedic application as magnesium ions have been reported as one of the substituting minerals for the bone and its release from the material can influence the cell migration, attachment, and proliferation (Wei et al. 2010). All these materials are considered biologically active and nonreactive as foreign materials but with some disadvantages of proper tissue adherence on metallic surface leading to the loosening of the implants with a risk of second surgery to remove them and toxicity risk because of the metal ion accumulation during corrosion of the materials (Jacobs et al. 2003; Hallab et al. 2001).

In the years around 1980–2000, second-generation biomaterials were introduced with the ability to interact with the biological systems for enhanced cell growth, enhancing the cell adherence and with potential bioresorbable ability. Mainly bioactive glasses, ceramics, glass ceramics, and their composites showed great bone regeneration capability by the end of the 1980s. However the bioactivity of the materials can also be increased by modifying the material's surface by adsorbing proteins, polymers, or biomolecules onto the material surface for osseointegration. Then natural polymers came into the market in the late 1990s wherein the materials like gelatin, hyaluronic acid, starch, chitosan, collagen, and fibrin offer great biocompatibility and biodegradability due to their native structure likeness to that of extracellular matrix of the tissues. Osteoconductivity of collagen matrices has been reported demonstrating bone formation, but the osteoinductive property is still under investigation (Rocha et al. 2002). The synthetic polymers like poly-lactic acid (PLA), poly-(ε-caprolactone), and poly-glycolic acid (PGA) have different porosity, pore size, mechanical properties, and degradation rates and forms (Nampoothiri et al. 2010; Liu and Ma 2004; Karageorgiou and Kaplan 2005). Degradation products of these polymers like lactic acid and glycolic acid are naturally found in the body and hence can be removed by metabolic processes. Poly-(lactide-co-glycolide) (PLGA) the copolymer of PLA and PGA has been approved by FDA and has better bone regeneration capability as compared to its monomeric units (Lanao et al. 2013). Poly-hydroxyethylmethacrylate (poly-HEMA) is the only synthetic polymer having osteoinductivity and has shown subcutaneous bone formation after implantation (Winter and Simpson 1969). During the use of poly-HEMA, it was observed that calcification preceded the cell differentiation suggesting the significance of calcification and calcium phosphates in osteoinduction (Barradas et al. 2011). Since many years, ceramics of natural or synthetic origin have been in use for biomedical or clinical applications and can be found in various forms, pore sizes, porosities, or topographies. The natural ceramics are coralline hydroxyapatite (HA) and calcium phosphate (CaP), while synthetic ceramics in applications are synthetic HA or β-tricalciumphosphate (β-TCP) (Karageorgiou and Kaplan 2005; Samavedi et al. 2013; Liu and Ma 2004). CaP materials have great biodegradability, biocompatibility, and osteoconductivity as in its presence bioactive apatite layer is formed on its surface, thereby enhancing osseointegration (Blokhuis et al. 2000; LeGeros 2008). Also its surface allows adsorbance of cytokines and fibronectin which are further essential for adherence of cell onto the surface (Hing 2005). Bioglasses and glass ceramics have extraordinary properties, some representing biodegradability and controlled degradation, in spite of them being brittle. They release ions as they degrade which helps in angiogenesis and osteogenesis. Moreover, they convert to hard and soft tissue binding biologically active carbonated HA materials (Kaur et al. 2014; Rahaman et al. 2011). There are different types of bioglasses including silicate glasses like 45S5 or 13-93 or borate/borosilicate glasses, namely, 13-93B2, 13-93B3, or Pyrex®, which support tissue infiltration, cell proliferation, and differentiation (Fu et al. 2008, 2010). Aforementioned polymers apart from great advantages also share disadvantages like appalling mechanical properties, high rate of degradation, locally concentrated degraded product, and inflammatory reactions caused by the degraded small particles (Sachlos and Czernuszka 2003; Kohn et al. 2002).

The third-generation biomaterials are considered bioactive as well as bioresorbable and at the same time stimulate cellular activity and behavioral response at molecular level (Hench and Polak 2002). They include 3D porous structures stimulating cell invasion, penetration, attachment, proliferation, and differentiation through functionalizing surfaces with extracellular matrix mimicking peptide sequences triggering cellular responses (Hutmacher et al. 1996; Temenoff and Mikos 2000; Agrawal and Ray 2001). They help in the delivery of growth factors or drugs controlling the cell behavior through mechano-transduction. This generation materials appeared in the market with the tissue engineering scaffolds. These materials tend to inflence the surrounding cells, signaling molecules and angiogenesis to stimulate the orthopedic tissue regeneration. Bioactive glasses and macroporous foams are commonly used as third-generation materials stimulating regeneration of living tissues. These tailored biomaterials use two alternative routes of tissue repair. *First*, tissue engineering where progenitor cells are seeded along with the modified scaffolds and allowed to differentiate outside the body to mimic natural tissues and finally implanted into the patients' body for replacing damaged tissues. The scaffolds get resorbed and replaced by host tissues with time and adapt to physiological environment providing long-lasting repair platform. *Second*, in situ tissue regeneration involving localized tissue repair by utilizing biomaterials in the form of solutions, powders, or doped particles. Biomaterials incorporated with bone morphogenetic protein (BMP) release ions upon dissolution for activation of cells in contact. The cells further produce their native growth factors to stimulate nearby cell growth and cell assembly. Osteogenic components like bone marrow aspirate (BMA) or osteoinductive signals like rhBMP-2 and rhBMP-7 (e.g., INFUSE® bone graft, OP-1 implant and putty) actively guide stem cell homing and augment cell differentiation (Bongio et al. 2011). Hydrogels made of dextran have represented neovascularization promoting osteogenesis. Akermanite, a silicate bioceramic, provided Si ions for stimulation of aortic endothelial cell proliferation and gene expression during bone regeneration (Zhai et al. 2012). Biomaterials such as heparan sulfate, HA, and fibrin show high affinity toward endothelial growth factor (EGF), vascular endothelial growth factor (VEGF), and basic fibroblast growth factor (bFGF), thereby improving bone formation via regulation of vascularization (Yu et al. 2014). The regenerative effect of these growth factors can be enhanced by synergistic signaling through utilization of fibronectin-based hydrogels and/or heparin-based glycosaminoglycans (Hudalla and Murphy 2011).

The ion channels and pumps present in the cell membrane generate bioelectrical signals for communicating with the neighboring cells. Recently advancements in this field of electrophysiological behavior of cells and tissues have led to the hypothesis for the development of new-generation biomaterials termed fourth-generation biomaterials (Ning et al. 2016). These biomaterials need to fulfill the requirements for manipulation of cellular bioelectric cues for bone regeneration and for monitoring cell stimulus and communications with nearby tissues using these signals. Mainly biomaterials that are electrically active like carbon-based materials, conductive polymers, and piezoelectric materials are considered efficient for delivering electrical responses. Conducting polymers like poly(pyrrole) (PPy) are the commonly used electro-active biomaterial and possess biocompatible, physical, and electrical properties. Graphene sheets, carbon-based material, present an excellent platform for stimulating electrical signals and help in bone regeneration (Kuzum et al. 2014).

Orthopedic implant failure and revision surgeries have decreased their lifetime in the patients who require complicated and expensive surgeries. Due to this by increasing the lifespan of the implants, the financial and physical burden on the patients could be reduced to some extent. It has been reported that majority of implant-related surgeries get complicated due to infections (Goodman et al. 2013). So, to enhance the survival and favorable output of the implants, its osseointegration potential and absence of infects near to it need to be determined. The infections on the one hand led implant failures while on the other hand hinder the progress of bone healing. More than 50% of the infections are caused by *Staphylococcus aureus* and *Staphylococcus epidermidis* (Kose and Kose 2015).

In the case of SSI, there are two main paths for the entry and homing of the infectious agents (McConoughey et al. 2014). First is the perioperative time where the infections are most commonly caused during the surgical incisional procedures through bacteria from the personnel, patient's endogenous flora, or environment of the operating room. Second is the postoperative duration where infections are caused by hematogenous spread (Greene 2012). The presence of foreign material in the body can also cause infections maturing into biofilms. The SSI can be acute or chronically remain for years and their diagnosis and treatment has become a difficult task for clinicians with requirement of high-end equipment and time-intensive procedures (Zimmerli et al. 2004; Parvizi et al. 2013). High-virulence bacteria present symptoms of early infection, while low-virulence bacteria develop infections later in time (Willenegger and Roth 1986). Once bacteria have entered through the surgical sites, it can remain and multiply within the body in three different phenotypic modes.

2.2.1 Planktonic

It represents the single cell form of bacteria which can be identified easily and removed through patient's immune responses or antibiotic therapies. If not cleared and has colonized the patient's surgical site tissue or the implant itself, it can also serve as the source for the biofilm generation. It can also grow in the joint fluids during acute contact of infection.

2.2.2 Biofilm

When the bacteria adhere within the patient's body, it develops into the colonies encased into a structural matrix. The bacteria in the matrix attain a specific characteristic to evade patient's defense mechanism and the antibiotic therapies. The implanted material provides space for the adherence of the bacteria and formation of biofilms. Some clumps of biofilm also reside in the joint fluids while many adhere to the bone substitutes and tissues (Stoodley et al. 2011). They can further change the local environment mainly by shifting the pH of the physiological conditions within few seconds because of their metabolic activity and limited mass transfer (Von Ohle et al. 2010). Over 99% of the bacteria in the biofilm commonly represent the population of staphylococci, while *Pseudomonas aeruginosa* with staphylococci make up to 75% of the biofilm population (Choong et al. 2007; Jämsen et al. 2009).

2.2.3 *Invasive and Intracellular*

Many bacteria invade, survive, and multiply within patient's tissue, mainly epithelial and osteoblast cells. The bacteria adopt this strategy to avoid the effect of antibiotic and immune responses. Mainly the small colonies of *S. aureus* associated with biofilm have been reported to enter the fibroblasts in some cases of SSI, thereby suggesting the association of surface-adhered biofilms and invasive bacteria (Sendi et al. 2006).

The implant surface, patient tissue, and the overlying fluid represent the reservoir for the bacteria of different phenotypic state, from which they can proliferate and cause infections if they are not eradicated completely through any means of aseptic conditions or antibiotic therapy.

2.3 Bacterial Growth and Related Clinical Complications

SSI can occur through different ways such as external contaminants either through operational environment or the implant itself, blood supply carrying pathogens from other body parts, and recurrent infection. First bacteria attached to the surface of the implant then create micro-colonies and finally encase within the matrix of glycocalyx. Through this matrix the bacteria survive longer bypassing the host defenses. The pathogen species that vary from patient to patient, risk factors, comorbidity, infection time, and bacteria characteristics also influence the effects of SSI. The development of biofilm on the implant surface can be outlined through cell adhesion, aggregation, and maturation to detachment (Fig. 2.1).

2.3.1 *Cell Adhesion*

First stage of biofilm formation begins by the adherence of planktonic bacteria onto the surface of the implants. They then produce extracellular polymeric slime matrix comprising of proteins, polysaccharides, and DNA. This process starts within few seconds of pathogen contact and extends for about 2 h (O'Neill et al. 2008). The implant surface properties like its charge, topography or hydrophobicity, as well as

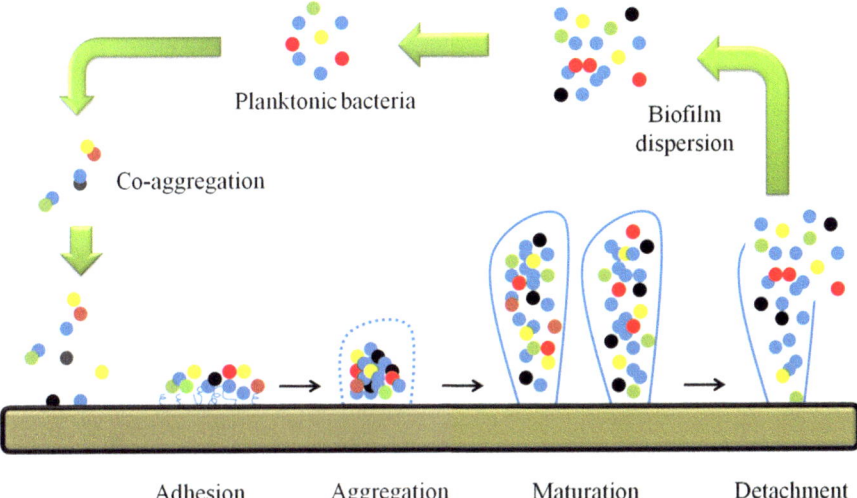

Planktonic bacteria

Biofilm dispersion

Co-aggregation

Adhesion   Aggregation   Maturation   Detachment

Fig. 2.1 Outline depicting stages of implant-associated bacterial infection

the exposure time depict the characteristic of the biofilm formation (Rochford et al. 2012). Bacteria passively adsorb onto the surface nonspecifically and are driven by electrostatic, hydrophobic, and Van der Waals forces. The conditioning film is formed by binding of bacteria with the help of host body proteins such as autolysin, fibronectin, vitronectin, and fibrinogen (Heilmann et al. 1997; Watnick and Kolter 2000; Rochford et al. 2012). Bacterial proteins interact with these proteins and enhance their colonization (Heilmann et al. 1997; Legeay et al. 2006). The matrix formed then recruits other bacteria in the vicinity to grow and self-organize in well-defined and structured manner. The staphylococcal bacteria secrete positively charged homopolymer, intercellular adhesion, for abetting planktonic bacterial aggregation. This positively charged polysaccharide creates a barrier against anti-bacterial peptide and phagocytosis by immune cells. *Escherichia coli*, the most motile strain, also forms biofilms very rapidly exploiting their phenomenon of motility as the flagella used in the motility of the bacteria can also render the initial adhesion to the surface (Wood et al. 2006). Once they are attached, bacterial movement shifts from swimming to gliding or twitching along the surface via type IV pilus motility. Thus motility on the surface can be considered significant for the emergence of the larger biofilm but is not considered essential.

2.3.2   Cellular Aggregation

After the formation of the matrix, the bacteria proliferate to colonize and show cell-to-cell adhesion leading to the formation of micro-colonies of one or more strains. The matrix get more stable by the formation of outer polysaccharide slime layer in which high microbial density is encased (Høiby et al. 2011). Colonization is mediated by the microbial surface components recognizing matrix adhesives and the polysaccharide intercellular adhesins (Patti et al. 1994). The strains which do not form adhesins and their cell-to-cell adhesion are mediated by accumulation-associated protein (AAP), protein A, extracellular matrix protein (EMP), and *Staphylococcus aureus* surface protein G (SASG) (Conrady et al. 2008; Christner et al. 2010; Geoghegan et al. 2010). Even after the formation of slime layer, the biofilm is still unstable and susceptible to eradication.

2.3.3 Biofilm Maturation

After colonization, physiological changes in the bacterial strains take place in the way of regulating flagellae, pilli, and exopolysaccharide for the maturation within biofilm (Lee et al. 2011). The extracellular slime plays a crucial role during the maturation of the biofilm as it provides a physical barrier and limits the transport of ions or chemicals in/out of the matrix. The main constituents of slime are the polysaccharide and extracellular DNA (eDNA), wherein this eDNA mimics the strategy of neutrophils by forming net-like structures for their prevention from eradication (Papayannopoulos and Zychlinsky 2009). The release of eDNA is regulated by *cidA* gene of *S. aureus* which cause subpopulation cell lysis, while in case of *P. aeruginosa*, it helps in developing the biofilm (Rice et al. 2007; Gloag et al. 2013). In case of *Haemophilus influenza*, a network is generated between extracellular proteins and the eDNA for biofilm maturation (Goodman et al. 2011). The cell contact, high cell density, and limited transport create a quorum sensing system (QSS) governing the cell communication through accessory gene regulator (agr) (Periasamy et al. 2012). QSS in microbes allow them to regulate gene transcription on the basis of the cell density through the utilization of auto-inducing peptide (AIP). As bacteria aggregate, the AIP begin to bind to the membrane receptors initiating the signaling cascade finally activating the *agr* gene and consequent expression of secreted virulence factors (Boles and Horswill 2008). Any mutations in *agr* enhance the biofilm formation and reduce their dispersion, thereby increasing the bacterial colonies within the matrix. *P. aeruginosa* follow a different mechanism for QSS through signaling molecule of homoserine lactone family (Galloway et al. 2012). Cell communications are conducive for gene transfer, transmitting virulence gene, as well as the antibiotic resistance gene within the biofilm. During maturation biofilms attain sessile form for resisting high levels of antibiotics and eradication. The bacteria on outside of the biofilm consume oxygen and glucose depriving the biofilm center residing bacteria from the nutrients making them metabolically inactive; thereby it becomes very difficult for their eradication (Høiby et al. 2011).

2.3.4   Cellular Detachment

On maturation, the planktonic bacteria present in the biofilms disperse for localized invasion and infect different sites to begin a new cycle of biofilm formation. QSS also helps in detachment of the bacterial cells through proteases and *agr*-mediated dispersal of the cells (Boles and Horswill 2008). *P. aeruginosa* dispersal is mediated by nitric oxide and *cis*-2-decanoic acid messenger (Amari et al. 2013). As the nutrient and the space get depleted, then bacteria follow a coordinated dispersal appearing as chronic infectious stages. The matrix of biofilm, layer on the cells, metabolic inactivity, and anoxic environment lead to the antibiotic degradation and resistance is generated toward the host's immune defense mechanisms. The extracellular polysaccharides and/or proteins secreted by the bacteria can bind to the antibiotics and degrade them to nonfunctionality. Also due to diffusion-limited transport, a gradient for antibiotic concentration is generated with its very low concentration toward subpopulation, hence generating an antibiotic tolerance and development of MRSA and/or MRSE strains (Fux et al. 2004; Jefferson et al. 2005).

During the early stages of biofilm development, it is less stable and more susceptible to antibiotic and host immune responses; however as biofilm matures, the bacteria become sessile and tolerant. Bacteria release β-lactamases to break down extracellular β-lactams, common antibiotic family for treating orthopedic infections (Høiby et al. 2010a, b; Ciofu et al. 2002). They also overexpress the efflux pumps to remove the antibiotics that have penetrated into the bacterial cells (Høiby et al. 2010a, b). Apart from tolerating and avoiding the antibiotic therapies, the bacteria also inhibit innate and adaptive immune responses of the host. Due to lack of blood supply to the area of infections, immune cells remain unavailable for the bacteria and their mere presence is not sufficient for the activity. Moreover, the layer of scar tissue formed around the implant site is hypovascular and compromised in similar manner. The extracellular matrix of the biofilm restricts the entry of polymorphonuclear leukocytes for their action. Innate immune response further activates adaptive immunity by the generation of antibodies which also is not able to penetrate the biofilm matrix. However, ineffective response by antibodies can lead to chronic inflammation and damage to the host tissues (Høiby et al. 2010a, b). Even during SSI, the elevated inflammation and damaged tissues release the nutrients that become available for the biofilm bacteria, exacerbating the infectious condition (McConoughey et al. 2014).

2.4   Conventional Techniques for Treating Infections

To prevent SSI, the first preventive method is to inhibit the initial adhesion of the bacteria onto the implant surface (Fig. 2.2). This can be achieved by coating the surface with some antimicrobials. Infection-resistant coatings are the recent advances in the science of implant development. Coatings can be passive or active

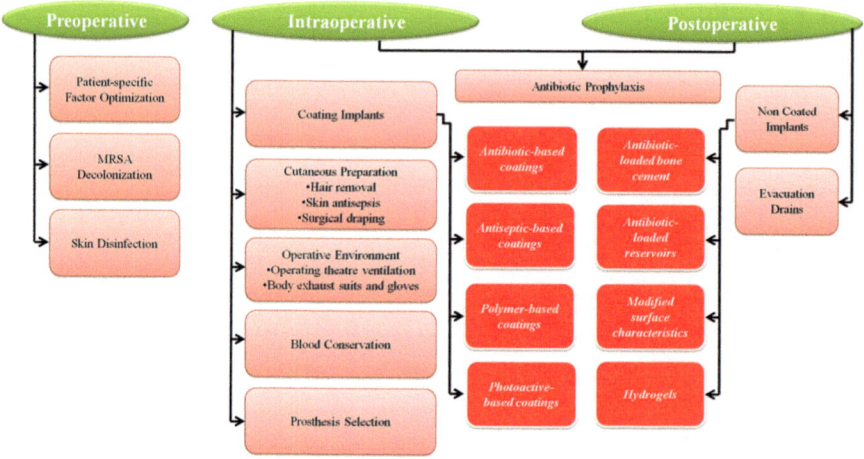

Fig. 2.2 Scheme for conventional methods for prevention of SSI

depending on whether the antimicrobials are distributed locally or not. For instance, in passive coating, bactericidal agents are not released to the surrounding tissues; they prevent the bacterial adhesion and directly kill them on contact. However, in case of active coatings, antimicrobial agents are released from the reservoir for treating infections. The anti-infective biologically active coatings rely on different mechanisms such as generation of anti-adhesive coatings, biomolecules restricting biofilm formation, and bioactive materials with bactericidal property (Campoccia et al. 2013).

Foremost possibility for preventing SSI is the application of local delivery of antimicrobial agents. Antibiotic coatings provide a sustained release of the antibiotic according to the pharmacokinetic drug delivery mechanism (Wu and Grainger 2006). A number of antibiotics have been studied like vancomycin, amoxicillin, tobramycin, and cephalothin, but the most commonly used one is gentamicin (Veerachamy et al. 2014). The antibiotic reservoirs commonly used and considered biocompatible are PLGA, polymethylmethacrylate (PMMA), polyethylene glycol, poly(lactic acid), and poly(D,L)lactide (Luo et al. 2006). HA has been recently reported to be an effective carrier for gentamicin (Avés et al. 2009). Polymers such as chitosan and collagen are dipped into the solution of antibiotic for simple and rapid loading of them. The delivery of antibiotics from these is very quick and burst release is obtained within 4 days of application (Wachol-Drewek et al. 1996; Noel et al. 2008). This burst release is not beneficial for implant as well as the host tissues. For local delivery, antibiotic loaded PMMA beads are used. They can treat acute infections, but since beads are non-biodegradable and there is no complete release of the antibiotics (75% remain unreleased), a second surgery is required to remove the beads (Wenke et al. 2006; Kose and Kose 2015). Through these types of coatings, fixed predetermined doses of antibiotics are used with limited release time and the risk of antibiotic resistance generation (Arciola et al. 2006). Due to improper release of antibiotics, there is requirement of adequate doses to be delivered at the infection site in a therapeutic range without development of antibiotic resistance. To achieve this antibiotic-encapsulated bioresorbable microspheres have been developed for their sustained release in specific time duration (Ambrose et al. 2014). Through this method, the side effects were reduced as high concentration of antibiotics cannot be obtained; also the polymers used are biodegradable and biocompatible.

Since antibiotics are targeted toward the specific strains, antiseptic-based coatings can be used for the wide range of bacteria and target them through general chemical agents, and hence the bacteria would not be able to induce resistance against them (Reading et al. 2000). The commonly used antiseptics include chlorhexidine and chloroxylenol which work through the mechanism of the interaction between their cationic nature and the anionic nature of the bacterial cell membrane lipids. This led to the destruction of the cell membrane, thus limiting the adherence of the bacterial cells. These antiseptics possess general toxicity and therefore they are only utilized as topical supplement.

Chitosan, a polymer made of chitin, is known to have antimicrobial properties. It has been evidenced that composites of chitosan have an effective antimicrobial action for titanium orthopedic implants. A vancomycin-chitosan composite coating on implants has shown lesser biofilm formation and increased the adhesion and proliferation of osteoblast cells (Yang et al. 2013). Also the mixture of chitosan and cinnamon oil has also been proven to possess antimicrobial properties against *Staphylococcus epidermidis* on titanium implants (Magetsari et al. 2014). In the similar manner, calcium phosphate bioceramics has also improved to biological properties of the metallic implants (Kose and Kose 2015).

Photocatalyst coatings on the titanium implants displayed bactericidal effect as they degrade the membranes of the bacteria on exposure to ultraviolet irradiation (Jamal et al. 2014). The commonly used photocatalytic agent includes titanium oxide (TiO_2) as it is non-toxic, is chemically stable, and has strong oxidizing strength. Along with the acceptable physical and chemical properties, they have represented an effective antibacterial effect against *S. aureus* and *S. epidermis* providing added benefit of easy scalability and cost effectiveness (Villatte et al. 2015).

Apart from the coatings, other antimicrobial strategies adopted during orthopedic surgeries include utilization of antibiotic-loaded bone cements. PMMA is widely used cement by orthopedic clinicians to fill bone voids and treat fractures (Samuel 2012). Antibiotic-loaded cements were first established in 1970s for in situ release of antibiotics (Buchholz and Engelbrecht 1970). Despite its widespread application, their efficacy is limited as the antibiotic release through these cements is very slow and not controlled. Due to this, cements are loaded with high concentration of antibiotics thereby posing the risk for generating resistance in the pathogens (Belt et al. 2000; Passuti and Gouin 2003). The advancements were made to obtain the sustained release of the antibiotics from the cements. The researches proposed fixation pins with tubular reservoirs for antibiotic loading wherein the antibiotics will release in a controlled manner depending on the number and size of the orifices (Gimeno et al. 2015).

To broaden the activity range of bactericidal treatment, the capability of hydrogels has been explored for delivering multiple antibiotics to surgical sites. The implant is coated with the resorbable hydrogels containing antibiotic cocktail with maintained doses and impediment to osseointegration (Drago et al. 2014). Incorporation of antibiotics is effective to some extent, but their overuse can exacerbate the problem of SSI as observed in the case of *Carbapenem* (Peleg and Hooper 2010; Nordmann et al. 2012; Citorik et al. 2014). Further, SSI caused by microbes of different strains cannot be effectively treated using this strategy due to drug resistance conferred from one strain to the other via horizontal gene transfer (Peleg and Hooper 2010).

Rather than coating or loading with antibiotics, a different way of preventing bacterial adhesion was to modify the implant surface characteristics itself. The mixture of polyethylene oxide and polyethylene glycol when applied onto the surface of the implants has repelled the binding of the adhesin proteins, thereby inhibiting the adhesion and formation bacterial biofilms (Kaper et al. 2003; Zhang et al. 2008). There are other studies also where increasing the roughness of the implant has reduced the biofilm-forming capability of the bacteria (Singh et al. 2011). However, some studies have reported that modifying surface also interferes with the osseointegration of implants making the technology a challenging one; also there is some threshold up to which the modified topography reduces biofilm (Braem et al. 2014).

2.5 Nanomaterials in Eradicating Infections

Despite technological advancements, SSI still remains a clinical issue; thereby to reduce the risk, some additional improvements are necessitated. As already seen, the conventional approach uses non-biodegradable antibiotic reservoirs for bone regeneration and alleviation of the infections; they are proved to be not as effective and the issue still persists with an increased antibiotic resistance in the strains. Therefore, there is an urgent need for discovering new technologies for developing an antimicrobial agent which efficiently irradiates the bacterial infections without hindering the osseointegration capability of the implants and reducing the requirement of second surgeries mainly used to remove nonresorbable implants. Recently nanotechnology has already attained its significance in developing antibiotic delivery for extended release and increased efficacy. This promising technology on horizon has answered the questions suggesting a synergistic application of these technologies for orthopedic surgeries.

Nanotechnology is an emerging field of interest wherein tailored materials at atomic level as that of natural systems attain unique properties desirable for multiple applications (Roco et al. 2011; Prasad et al. 2016, 2018). This field provides a unique and sophisticated approach in medical implant technology. Nanomaterial-based bone substitute is majorly favored nowadays because of their mimicking characteristic of natural tissues, osteoconductive and/or osteoinductive capability, mechanical support, and 3D scaffold surface for the growth of osteocytes and helps in the differentiation of bone marrow stem cells. Nanoformulations have been developed in the form of scaffolds or injections for sustained and extended release of antibiotics while promoting bone formation.

Research has reported the development of antibiotic material into the nanoparticles with increased surface area, porosity, and biocompatibility with extended release for the antibiotics (Shi et al. 2006). One of the widely used materials in orthopedics is PMMA cement beads for delivering antibiotics. PMMA cement loaded with aminoglycoside is the most commonly used nano-bead-based antibiotic therapy. Due to this, sustained, extended, and effective action of antibiotics was achieved. But the non-biodegradability, toxic byproducts, creation of inflammation, and elicit immune response damage host tissues and hinder bone formation (Yang et al. 2009). Hence there was the requirement for biodegradable nanomaterials that can possibly promote bone regeneration without needing any second surgery (Wilberforce et al. 2011; Uskoković and Desai 2014).

Researchers developed micro and nanoparticles of calcium phosphate and PLGA to determine their bone regeneration capability (Bastari et al. 2014). To this formulation autologous plasma was also added as the source of growth factors and cytokines. The antibiotic, tigecycline, was also added through chemical bottom-up dissolution. The nano-sized formulation produced greater bone regeneration after infection than micro-formulation. It was observed that 0.6% tigecycline nanoparticle was able to release satisfactory amount of antibiotic along with simultaneous bone regeneration; also the systemic administration of antibiotics was not required. The hydrophobic nature of PLGA makes it less biocompatible and difficult in antibiotic loading. This property can be overcome by incorporating a hydrophilic polymer like mono-methoxypoly(ethylene glycol) (mPEG). The research on the effectiveness and release of antibiotics using these compositions revealed that effectiveness of teicoplanin and its release kinetic were better observed with the amphipathic mixture.

HA nanoparticles produced through wet synthesis, proceeding at low temperature, can be used for the better delivery of drugs as compared to other materials used in orthopedics. The composite of HA and chitosan helped in alleviating crystallization problem, burst release of antibiotics, and benefitted in infection treatment by increasing bone biocompatibility (Yeh et al. 2011). Later new technique like electrospinning was used to generate the nano-fibers that provide the backbone for the release of antibiotics. The loading and release kinetics of amoxicillin were improved by using nano-fibers of PLGA with HA (Zheng et al. 2013).

Recently, gelatin and hydrogels are used for incorporating nanoparticles due to their biodegradability and permeability for nutrients and can be easily manipulated. The use of hydrogels has enhanced the osteoinductive property and strength. These properties can further be improved by adding HA and chitosan to the scaffolds (Im et al. 2012). Even the incorporation of nanotubes helped in osteoinduction by promoting extracellular matrix through their elongated structure (Harrison and Atala 2007). It was also discovered that osteoblast proliferation increased by incorporating magnetized nanotubes. The charged polarity added due to nanotubes also increased the antibacterial property of the scaffolds.

Some researchers are trying to implant these polymers through injections to eliminate the need for the surgery. Incorporating hydrogels and gelatins are the first choice of polymers due to their good thermoresponsive properties and research is moving in this direction to increase their water solubility and manipulation of their physical state temperatures (Chen and Cheng 2006; Peng et al. 2010). This technique was implemented through the composite of PLGA-PEG and later incorporating HA to it. It was observed by the researchers that the incorporation of HA increased strength, storage capability, preserved thermoresponsive property, increased biocompatibility, and sustained release kinetics of antibiotics (Lin et al. 2012). Also, during in vivo studies it was investigated that PLGA-PEG matrix when incorporated with gentamycin and clindamycin eradicated infection caused by *S. aureus* through a combination of gentamycin and clindamycin antibiotics. The matrix showed initial fast release followed by slower stable release of antibiotics for over a week inhibiting bacterial growth and promoting bone regeneration (McLaren et al. 2014). Similarly, chitosan was adopted to be used with carbon nanotubes as injectible materials wherein it was believed to increase the cross linking with maintained thermoresponsive properties. Chitosan in combination with bioglass has proven merit for thermoregulation, structural support, biocompatibility, osseointegration, particle size regulation, and antibiotic release through network structure (Yasmeen et al. 2014; Sohrabi et al. 2014). But a problem exists about the gelation points being sensitive to increasing concentrations of these compounds; therefore new advancements are needed for the same (Couto et al. 2009). Later, injectibles made of calcium phosphate and poly(DL-lactide-coglycolide) in the form of microparticles and nanoparticles were analyzed and provided better surface properties, bone void filling capability, release kinetics, absorption, adhesion, and distribution profile in solution due to their very small size (Ignjatović et al. 2007). Implanting linezolid loaded silica microparticles into porous stainless steel pins provided two sets of variables for investigation. These loaded pins were co-cultured with *S. aureus* and after 48 h drug loaded pins showed reduced bacterial growth three times better than the nonloaded ones (Perez et al. 2011). The photocatalytic property of TiO$_2$ particles helped in inhibiting bacterial growth and can be used in treating SSI. Also, TiO$_2$ nanotubes produced by anodization of titanium can be used for bone regeneration with reduced cell adhesion or drug loading capability (Wang et al. 2011; Crear et al. 2013). The growth of *Streptococcus mutans* and *Porphyromonas gingivalis* was lower on the surface of the TiO$_2$ nanotubes and nanoparticle composite as compared to the individual materials with enhanced mesenchymal stem cell differentiation for osteogenesis (Liu et al. 2015a, b). Moreover, coating TiO$_2$ nanotubes with ZnO has considerably prevented SSI (Liu et al. 2015a, b). Silver nanoparticle coating onto titanium implants controlled release rate and duration. The electrical impulses generated by using silver nanoparticles as cathode and titanium matrix as anode in presence of electrolytes caused bacterial membrane morphology changes and DNA damage leading to cell death (Veerachamy et al. 2014; Zheng et al. 2012). Silver doping provided antibacterial efficacy against

E. coli, *S. aureus*, and *S. epidermidis* (Feng et al. 2000; Tran et al. 2015). In the case of osteomyelitis, the coating led to better clinical and histology outcomes with reduced SSI as compared to the uncoated pins (Tran et al. 2013). In another study the infection decrease to about 11% was observed by using silver nanoparticles coated materials (Wafa et al. 2015; Eltorai et al. 2016). The silver nanocoatings could be a promising tool against MRSA and MRSE due to their efficacious release kinetics ion formulation of silver nanoparticles. However, cytotoxicity on mammalian bone cells has been observed at long-term implications but is much lower than the antimicrobial threshold suggesting a mild toxicity on long-term usage (Park et al. 2011). Silver nanoparticles have a problem of proper distribution which was improved by attaching these nanoparticles onto carbon nanotubes (CNTs) due to latter's better lattice structure, stable structure, easy loading, and biocompatibility. This also improves the antibacterial properties of nanoparticles. The lattice allowed researchers to load antibiotics in similar pattern as adopted that for osteons and osteocytes (Afzal et al. 2013). The single-walled CNTs are more effective in bactericidal activity allowing them to be used in treating SSI (Liu et al. 2009). Also in a study with silver hydroxyapatite, coating the rate of colonization was very low implementing lower rate of infections associated with implants (Kose and Kose 2015). Diamond nanoparticles with strong and stable structures of small size were observed to release antibacterial for more than a month (Simchi et al. 2011). These nanoparticles were able to increase the gene expressions promoting bone regeneration while inhibiting, at the same time, the production of interleukin and cytokines involved in inflammation and cell damages (Huang et al. 2008). The production of diamond nanofilms is very tedious and expensive that required more efficient method. The integration of growth factors and macrophages into the magnetically charged materials was established. They promoted osteogenic differentiation and antibacterial property by providing magnetic polarity (Durmus and Webster 2013). When the composite was incubated with the infection complex, a dramatic effect was observed with increased bone regeneration as they enhanced body's natural immunity and growth mechanism (Goodman et al. 2013; Ignjatovic et al. 2010). Another approach adopted recently is to modify the implant surfaces at nanoscale level providing antibacterial properties by inhibition of pathogen adherence onto the surface. But this approach also interfered with the osseointegration ability of implants suggesting the threshold for surface micro-topography for reducing SSI. Electrospinning nano-fibers of PLGA with nanometric diameter and high surface area to volume ratio were loaded with antibiotics. This matrix can easily adhere to the implants with increased flexibility, porosity, and antibiotic release (Gilchrist et al. 2013). A polyelectrolyte multilayer film composed of polyarginine, hyaluronic acid, and nano-silver coating showed combined antimicrobial and immunomodulatory strategy. The film inhibited production of inflammatory cytokines and combated bacterial growth opening an opportunity for bacteria-specific customization using antimicrobials (Özçelik et al. 2015).

2.6   Mechanism of Action for Nanobiotics

The nanomaterials incorporated into the implant material by creating nano-porous structure or complete dissolution in the material. They have proven surface antimicrobial activity and antibiofilm formation even against MRSA or MRSE (van Hengel et al. 2017). The nanomaterials follow different mechanisms as shown in Fig. 2.3 for eradicating bacterial infections during SSI providing increased bone formation capability.

The surface modification using nanomaterials can inhibit bacterial adhesion and transcription of *ica*AD, thereby reducing biofilm formation (Qin et al. 2014). Silver nanoparticles show multiple ways for their antimicrobial effect like binding to cell membrane having sulfur-containing proteins and phosphorus-containing structures like DNA. Due to this, NPs led conformational changes in the bacterial cell leading to cell death (Rai et al. 2009). Moreover, silver nanoparticles ionize and activate DNA defense mechanism of bacteria compromising the replication process leading to reduced growth of the bacteria (Seil and Webster 2012). They also bind to thiol groups of enzymes, nucleic acids, and cell and/or organelle membrane leading to change in morphology, cell envelope damage, and impeding cell division (Jung et al. 2008). The galvanic coupling between silver nanoparticles and titanium matrix in presence of electrolytes can also lead to cell death (Cao et al. 2011). The silver ions form stable bonds with proteins of cell membrane involved in transmembrane ATP generation and cell membrane transport system, resulting in protein deactivation (Klueh et al. 2000). They also obstruct the membrane penetrability of protons and phosphates. These ions can intercalate with purines and pyrimidines disrupting the hydrogen bonding between them which prevent cell reproduction and division (Monteiro et al. 2012). Bacteria use disulfide reductase system against oxidative

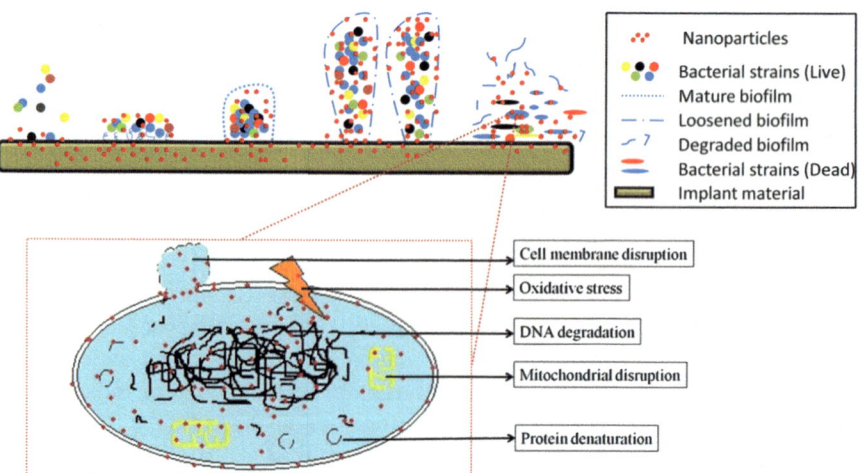

Fig. 2.3 Proposed mechanism of action of nanobiotics

stress involving thioredoxin, thioredoxin reductase, and nicotinamide adenine dinu-cleotide phosphate. Silver ions bind to active sites of thioredoxin and thioredoxin reductase leading to the oligomerization of the two leading to functional disruption. Bacterial thiol-redox homeostasis is also disrupted by silver ions of the nanoparti-cles (Liao et al. 2017). Gold nanoparticles show antibacterial effect by diminishing bacterial cell metabolism, altering bacterial membrane potential, inhibiting ribo-somal tRNA binding, inhibiting ATP synthesis, and promoting cell lysis (Li et al. 2014; Rai et al. 2010). In addition, irradiating gold nanoparticles with laser radia-tions can decrease bacterial cell viability and mechanical disruption through photo-thermal cell lysis and acoustic waves, respectively (Millenbaugh et al. 2015; Tikariha et al. 2017). Implant surface modification using thiol-modified gold nanoparticles revealed reduced expression of tRNA in bacteria and increased expression of chemotaxis genes in *E. coli* and *S. aureus* present at surgical sites (Cui et al. 2012). The bacterial dispersion in biofilm formation can be dissipated by action of nanomaterials onto the EPS components like eDNA (McConoughey et al. 2014). Nanoparticles can infiltrate into bacterial cells causing physical changes in the membrane leading to cell material leakage and finally cell death (Seong and Lee 2017; Khalandi et al. 2017). Their antibacterial effect on Gram-negative bacteria is more prominent than that on Gram-positive bacteria because of easy penetration into the former because of only peptidoglycan layer of cell wall (Chatterjee et al. 2015). The positively charged nanomaterials can easily attach the negatively charged bacterial cell membrane and change the membrane potential making the main cause for cell disruption. Small-sized nanoparticles can easily penetrate the cell mem-brane as compared to the larger ones and interact with cellular structures and bio-molecules leading to cellular dysfunction and finally cell death (Khalandi et al. 2017). Nanoparticles interact with ribosomes causing translation and protein syn-thesis hindrance. The interaction with carboxyl and thiol groups of β-galactosidase does not allow bacterial cell to function properly and become cause for the cell death (You et al. 2012; Singh et al. 2015). Nanomaterials generate oxidative stress, thereby inactivating respiratory chain dehydrogenases, inhibiting respiration, and leading to cell death (Quinteros et al. 2016; Su et al. 2009). In addition, they can downregulate the expression of enzymes involved in reducing reactive oxygen spe-cies concentration like superoxide dismutase, glutathione, and catalase. This results in accumulated ROS levels leading to apoptosis, lipid peroxidation, and DNA dam-age (Korshed et al. 2016; Yuan et al. 2017). Nitric oxide (NO) plays critical role in natural immune response against SSI. It increases the activity of immune cells stim-ulating the immune response. NO destabilizes DNA, inhibits DNA repair, and gen-erates hydrogen peroxide and genotoxic alkylating agents (Wink et al. 1991). NO as a donor molecule can be encapsulated in PLGA nanoparticles. They showed signifi-cant eventual effect against both Gram-positive and Gram-negative bacteria. However, chitosan encapsulation of NO demonstrated penetration of *P. aeruginosa* biofilms (Lu et al. 2015; Reighard et al. 2015). NO releasing sol gel coatings on implants significantly reduced the levels of *S. aureus* and *P. aeruginosa* at surgical sites (Nablo et al. 2005). NO releasing silica nanoparticles have reduced the bacte-rial burden while on the other hand represented enhanced angiogenesis and wound healing. Additionally, efficacy of NO silica nanomaterials can be determined through prevention and treatment of MRSA.

2.7 Future Perspectives and Concerns

SSI is a disease state in orthopedics that has no limit and can affect any person going surgical interventions with costly and devastating implications. Not only the patient but the socioeconomic impact is so high leading to enervation and morbidity. Even after advancement in the technologies and scientific development, SSI continues to be a hindrance in orthopedic surgery and sufferings for the patient. Currently, treatment through prophylactic antibiotic therapy has succored the incidence of SSI, but the risk is not eliminated completely. In most of the SSI-related problems, removal of implant becomes a necessity followed by many debridements. Moreover, the antibiotic access to the site is not adequate and bacteria generate antibiotic resistance, therefore increasing the need for new biomaterials to overcome this problem. The research was focused toward the use of nanomaterials as promising field in the field of medicine. Nanomaterial-based antibacterial therapies have shown efficacy toward a wide variety of pathogenic bacteria locally, but their long-term effects need to be elaborated. In addition, effect of nanomaterial-based antibiotic needs to reach an optimal levels and translation to human trials is limited. These insufficient ways require strong potential research for the costly and morbid surgical complications. For antibacterial agents the therapy should sufficiently start from inhibiting bacterial growth to minimizing antibacterial resistance development. The advancements in nanotechnology can promise a significant antibacterial agent, but incomplete knowledge of nano-toxicology could hinder the progress.

In the area of prevention and treatment of SSI, it has been proposed that future work related to nanomaterials must be properly characterized in terms of chemistry, toxicity, and impurities as they represent the collaborative work of engineers, clinical professionals, and environmental scientists. Early research efforts have shown incredible potential of nanomedicine, but some barriers remain during orthopedic clinical implementations. Long-term implications of nanomedicine on human health are not clearly understood, but some early research has demonstrated cytotoxicity, oxidative stress, and systemic inflammation associated with nanomaterial (Polyzois et al. 2012). However there are also reports suggesting beneficial effects of nanomaterial metabolism on bone tissues (Sato and Webster 2004). Therefore, due to uncertainties, intense regulatory processes put forward by US FDA come with every nanomaterial commercial use. In addition to this, the costs incurred in clinical trials can en masse billions of dollars. Hence, the companies do not make efforts to invest such a high cost on these new nanomaterial-based implant devices when adequate implants already exist in the market. Since the past decade, only about 3% nanotechnology-based research has gone toward the clinical trials (Sullivan et al. 2014). Considering these issues, substantial research for investigating nano-toxicity is needed before their clinical uses. Another challenge remains with the scalability of nanomaterial production due to non-reproducibility of very small-sized nanomaterials (Smith et al. 2018). The structural properties vary by decreasing the size, hence nanomaterials with low cost, non-toxicity, and scalability with high reproducibility. All the parameters cannot be achieved without sacrificing the other. Therefore, with the safe usage guidelines, nanomaterials can be safely used for SSI. The future for nanotechnology involves excitement in the collaborative anti-infection therapy.

References

Afzal MAF, Kalmodia S, Kesarwani P, Basu B, Balani K (2013) Bactericidal effect of silver-reinforced carbon nanotube and hydroxyapatite composites. J Biomater Appl 27(8):967–978

Agrawal CM, Ray RB (2001) Biodegradable polymeric scaffolds for musculoskeletal tissue engineering. J Biomed Mater Res 55(2):141–150

Amari DT, Marques CN, Davies DG (2013) The putative enoyl-CoA hydratase DspI is required for production of the *Pseudomonas aeruginosa* biofilm dispersion autoinducer, cis-2-decenoic acid. J Bacteriol 195(20):4600–4610

Ambrose CG, Clyburn TA, Mika J, Gogola GR, Kaplan HB, Wanger A, Mikos AG (2014) Evaluation of antibiotic-impregnated microspheres for the prevention of implant-associated orthopaedic infections. JBJS 96(2):128–134

Arciola CR, Campoccia D, An YH, Baldassarri L, Pirini V, Donati ME, Montanaro L (2006) Prevalence and antibiotic resistance of 15 minor staphylococcal species colonizing orthopedic implants. Int J Artif Organs 29(4):395–401

Avés EP, Estévez GF, Sader MS, Sierra JCG, Yurell JCL, Bastos IN, Soares GDA (2009) Hydroxyapatite coating by sol–gel on Ti–6Al–4V alloy as drug carrier. J Mater Sci Mater Med 20(2):543–547

Barradas, A. M., Yuan, H., van Blitterswijk, C. A., & Habibovic, P. (2011). Osteoinductive biomaterials: current knowledge of properties, experimental models and biological mechanisms. Eur Cell Mater, 21 407 429.

Bastari K, Arshath M, Ng ZHM, Chia JH, Yow ZXD, Sana B, Loo SCJ (2014) A controlled release of antibiotics from calcium phosphate-coated poly (lactic-co-glycolic acid) particles and their in vitro efficacy against Staphylococcus aureus biofilm. J Mater Sci Mater Med 25(3):747–757

Belt HVD, Neut D, Schenk W, Horn JRV, Mei HCVD, Busscher HJ (2000) Gentamicin release from polymethylmethacrylate bone cements and *Staphylococcus aureus* biofilm formation. Acta Orthop Scand 71(6):625–629

Bhattacharya S, Pal K, Jain S, Chatterjee SS, Konar J (2016) Surgical site infection by methicillin resistant *Staphylococcus aureus*–on decline? J Clin Diagn Res 10(9):DC32

Blokhuis TJ, Termaat MF, den Boer FC, Patka P, Bakker FC, Henk JTM (2000) Properties of calcium phosphate ceramics in relation to their in vivo behavior. J Trauma Acute Care Surg 48(1):179

Boles BR, Horswill AR (2008) *Agr*-mediated dispersal of *Staphylococcus aureus* biofilms. PLoS Pathog 4(4):e1000052

Bongio M, Van den Beucken JJJP, Nejadnik MR, Leeuwenburgh SCG, Kinard LA, Kasper FK, Jansen JA (2011) Biomimetic modification of synthetic hydrogels by incorporation of adhesive peptides and calcium phosphate nanoparticles: in vitro evaluation of cell behavior. Eur Cell Mater 22(3):59–76

Braem A, Van Mellaert L, Mattheys T, Hofmans D, De Waelheyns E, Geris L, Vleugels J (2014) Staphylococcal biofilm growth on smooth and porous titanium coatings for biomedical applications. J Biomed Mater Res A 102(1):215–224

Buchholz HW, Engelbrecht H (1970) Depot effects of various antibiotics mixed with Palacos resins. Chirurg 41(11):511–515

Campoccia D, Montanaro L, Arciola CR (2006) The significance of infection related to orthopedic devices and issues of antibiotic resistance. Biomaterials 27(11):2331–2339

Campoccia D, Montanaro L, Arciola CR (2013) A review of the biomaterials technologies for infection-resistant surfaces. Biomaterials 34(34):8533–8554

Cao H, Liu X, Meng F, Chu PK (2011) Biological actions of silver nanoparticles embedded in titanium controlled by micro-galvanic effects. Biomaterials 32(3):693–705

Centers for Disease Control and Prevention (2009) The National Healthcare Safety Network (NHSN) Manual. Patient safety component protocol [Internet]. Atlanta (GA): CDC [cited 2019 Jan 2]. Available from: http://www.cdc.gov/nhsn/PDFs/pscManual/pscManual_current.pdf

Chatterjee T, Chatterjee BK, Majumdar D, Chakrabarti P (2015) Antibacterial effect of silver nanoparticles and the modeling of bacterial growth kinetics using a modified Gompertz model. Biochim Biophys Acta 1850(2):299–306

Chen JP, Cheng TH (2006) Thermo-responsive chitosan-graft-poly (N-isopropylacrylamide) injectable hydrogel for cultivation of chondrocytes and meniscus cells. Macromol Biosci 6(12):1026–1039

Choong PF, Dowsey MM, Carr D, Daffy J, Stanley P (2007) Risk factors associated with acute hip prosthetic joint infections and outcome of treatment with a rifampin based regimen. Acta Orthop 78(6):755–765

Christner M, Rohde H, Wolters M, Sobottka I, Wegscheider K, Aepfelbacher M (2010) Rapid identification of bacteria from positive blood culture bottles by use of matrix-assisted laser desorption-ionization time of flight mass spectrometry fingerprinting. J Clin Microbiol 48(5):1584–1591

Ciofu O, Bagge N, HØiby N (2002) Antibodies against β-lactamase can improve ceftazidime treatment of lung infection with β-lactam-resistant *Pseudomonas aeruginosa* in a rat model of chronic lung infection. APMIS 110(12):881–891

Citorik RJ, Mimee M, Lu TK (2014) Sequence-specific antimicrobials using efficiently delivered RNA-guided nucleases. Nat Biotechnol 32(11):1141

Conrady DG, Brescia CC, Horii K, Weiss AA, Hassett DJ, Herr AB (2008) A zinc-dependent adhesion module is responsible for intercellular adhesion in staphylococcal biofilms. Proc Natl Acad Sci 105(49):19456–19461

Couto DS, Hong Z, Mano JF (2009) Development of bioactive and biodegradable chitosan-based injectable systems containing bioactive glass nanoparticles. Acta Biomater 5(1):115–123

Crear J, Kummer KM, Webster TJ (2013) Decreased cervical cancer cell adhesion on nanotubular titanium for the treatment of cervical cancer. Int J Nanomedicine 8:995

Cui Y, Zhao Y, Tian Y, Zhang W, Lü X, Jiang X (2012) The molecular mechanism of action of bactericidal gold nanoparticles on Escherichia coli. Biomaterials 33(7):2327–2333

Darouiche RO (2004) Treatment of infections associated with surgical implants. N Engl J Med 350(14):1422–1429

Disegi JA (2000) Titanium alloys for fracture fixation implants. Injury 31:D14–D17

Drago L, Boot W, Dimas K, Malizos K, Hänsch GM, Stuyck J, Romanò CL (2014) Does implant coating with antibacterial-loaded hydrogel reduce bacterial colonization and biofilm formation in vitro? Clin Orthop Relat Res 472(11):3311–3323

Durmus NG, Webster TJ (2013) Eradicating antibiotic-resistant biofilms with silver-conjugated superparamagnetic iron oxide nanoparticles. Adv Healthc Mater 2(1):165–171

Elganzoury I, Bassiony AA (2013) Early results of trabecular metal augment for acetabular reconstruction in revision hip arthroplasty. Acta Orthop Belg 79(5):530–535

Eltorai AE, Haglin J, Perera S, Brea BA, Ruttiman R, Garcia DR, Daniels AH (2016) Antimicrobial technology in orthopedic and spinal implants. World J Orthop 7(6):361

Feng QL, Wu J, Chen GQ, Cui FZ, Kim TN, Kim JO (2000) A mechanistic study of the antibacterial effect of silver ions on *Escherichia coli* and *Staphylococcus aureus*. J Biomed Mater Res 52(4):662–668

Fu Q, Rahaman MN, Bal BS, Brown RF, Day DE (2008) Mechanical and in vitro performance of 13–93 bioactive glass scaffolds prepared by a polymer foam replication technique. Acta Biomater 4(6):1854–1864

Fu Q, Rahaman MN, Bal BS, Bonewald LF, Kuroki K, Brown RF (2010) Bioactive glass scaffolds with controllable degradation rates for bone tissue engineering applications, II: In vitro and in vivo biological evaluation. J Biomed Mater Res A 95:172–179

Fujibayashi S, Neo M, Kim HM, Kokubo T, Nakamura T (2004) Osteoinduction of porous bioactive titanium metal. Biomaterials 25(3):443–450

Fux CA, Wilson S, Stoodley P (2004) Detachment characteristics and oxacillin resistance of Staphyloccocus aureus biofilm emboli in an in vitro catheter infection model. J Bacteriol 186(14):4486–4491

Galloway WR, Hodgkinson JT, Bowden S, Welch M, Spring DR (2012) Applications of small molecule activators and inhibitors of quorum sensing in Gram-negative bacteria. Trends Microbiol 20(9):449–458

Geoghegan JA, Corrigan RM, Gruszka DT, Speziale P, O'Gara JP, Potts JR, Foster TJ (2010) Role of surface protein SasG in biofilm formation by Staphylococcus aureus. J Bacteriol 192(21):5663–5673

Gilchrist SE, Lange D, Letchford K, Bach H, Fazli L, Burt HM (2013) Fusidic acid and rifampicin co-loaded PLGA nanofibers for the prevention of orthopedic implant associated infections. J Control Release 170(1):64–73

Gimeno M, Pinczowski P, Pérez M, Giorello A, Martínez MÁ, Santamaría J, Luján L (2015) A controlled antibiotic release system to prevent orthopedic-implant associated infections: An in vitro study. Eur J Pharm Biopharm 96:264–271

Gloag ES, Turnbull L, Huang A, Vallotton P, Wang H, Nolan LM, Monahan LG (2013) Self-organization of bacterial biofilms is facilitated by extracellular DNA. Proc Natl Acad Sci 110(28):11541–11546

Gomez J, Rodriguez M, Banos V, Martinez L, Claver MA, Ruiz J, Clavel M (2003) Orthopedic implant infection: prognostic factors and influence of long-term antibiotic treatment on evolution. Prospective study, 1992–1999. Enferm Infecc Microbiol Clin 21(5):232–236

Goodman SD, Obergfell KP, Jurcisek JA, Novotny LA, Downey JS, Ayala EA, Tjokro N, Li B, Justice SS, Bakaletz LO (2011) Biofilms can be dispersed by focusing the immune system on a common family of bacterial nucleoid-associated proteins. Mucosal Immunol 4(6):625

Goodman SB, Yao Z, Keeney M, Yang F (2013) The future of biologic coatings for orthopaedic implants. Biomaterials 34(13):3174–3183

Greene LR (2012) Guide to the elimination of orthopedic surgery surgical site infections: an executive summary of the Association for Professionals in Infection Control and Epidemiology elimination guide. Am J Infect Control 40(4):384–386

Hallab N, Merritt K, Jacobs JJ (2001) Metal sensitivity in patients with orthopaedic implants. J Bone Joint Surg 83(3):428–436

Haralson RH, Zuckerman JD (2009) Prevalence, health care expenditures, and orthopedic surgery workforce for musculoskeletal conditions. JAMA 302(14):1586–1587

Harrison BS, Atala A (2007) Carbon nanotube applications for tissue engineering. Biomaterials 28(2):344–353

Heilmann C, Hussain M, Peters G, Götz F (1997) Evidence for autolysin-mediated primary attachment of Staphylococcus epidermidis to a polystyrene surface. Mol Microbiol 24(5):1013–1024

Hench L, Jones J (eds) (2005) Biomaterials, artificial organs and tissue engineering. Elsevier, Cambridge

Hench LL, Polak JM (2002) Third-generation biomedical materials. Science 295(5557):1014–1017

Hing KA (2005) Bioceramic bone graft substitutes: influence of porosity and chemistry. Int J Appl Ceram Technol 2(3):184–199

Høiby N, Bjarnsholt T, Givskov M, Molin S, Ciofu O (2010a) Antibiotic resistance of bacterial biofilms. Int J Antimicrob Agents 35(4):322–332

Høiby N, Ciofu O, Bjarnsholt T (2010b) *Pseudomonas aeruginosa* biofilms in cystic fibrosis. Future Microbiol 5(11):1663–1674

Høiby N, Ciofu O, Johansen HK, Song ZJ, Moser C, Jensen PØ, Bjarnsholt T (2011) The clinical impact of bacterial biofilms. Int J Oral Sci 3(2):55

Holzwarth U, Cotogno G (2012) Total hip arthroplasty. European Commission, Brussels

Huang H, Pierstorff E, Osawa E, Ho D (2008) Protein-mediated assembly of nanodiamond hydrogels into a biocompatible and biofunctional multilayer nanofilm. ACS Nano 2(2):203–212

Hudalla GA, Murphy WL (2011) Biomaterials that regulate growth factor activity via bioinspired interactions. Adv Funct Mater 21(10):1754–1768

Hutmacher D, Hürzeler MB, Schliephake H (1996) A review of material properties of biodegradable and bioresorbable polymers and devices for GTR and GBR applications. Int J Oral Maxillofac Implant 11(5):667–678

Ignjatović NL, Liu CZ, Czernuszka JT, Uskoković DP (2007) Micro-and nano-injectable composite biomaterials containing calcium phosphate coated with poly (dl-lactide-co-glycolide). Acta Biomater 3(6):927–935

Ignjatovic NL, Ajdukovic ZR, Savic VP, Uskokovic DP (2010) Size effect of calcium phosphate coated with poly-DL-lactide-co-glycolide on healing processes in bone reconstruction. J Biomed Mater Res B Appl Biomater 94(1):108–117

Im O, Li J, Wang M, Zhang LG, Keidar M (2012) Biomimetic three-dimensional nanocrystalline hydroxyapatite and magnetically synthesized single-walled carbon nanotube chitosan nanocomposite for bone regeneration. Int J Nanomedicine 7:2087

Iorio R, Robb WJ, Healy WL, Berry DJ, Hozack WJ, Kyle RF, Parsley BS (2008) Orthopaedic surgeon workforce and volume assessment for total hip and knee replacement in the United States: preparing for an epidemic. JBJS 90(7):1598–1605

Jacobs JJ, Hallab NJ, Skipor AK, Urban RM (2003) Metal degradation products: a cause for concern in metal-metal bearings? Clin Orthop Relat Res 417:139–147

Jamal R, Osman Y, Rahman A, Ali A, Zhang Y, Abdiryim T (2014) Solid-state synthesis and photocatalytic activity of polyterthiophene derivatives/TiO$_2$ nanocomposites. Materials 7(5):3786–3801

Jämsen E, Huhtala H, Puolakka T, Moilanen T (2009) Risk factors for infection after knee arthroplasty: a register-based analysis of 43,149 cases. JBJS 91(1):38–47

Jefferson KK, Goldmann DA, Pier GB (2005) Use of confocal microscopy to analyze the rate of vancomycin penetration through *Staphylococcus aureus* biofilms. Antimicrob Agents Chemother 49(6):2467–2473

Jung WK, Koo HC, Kim KW, Shin S, Kim SH, Park YH (2008) Antibacterial activity and mechanism of action of the silver ion in *Staphylococcus aureus* and *Escherichia coli*. Appl Environ Microbiol 74(7):2171–2178

Kapadia BH, Berg RA, Daley JA, Fritz J, Bhave A, Mont MA (2016) Periprosthetic joint infection. Lancet 387(10016):386–394

Kaper HJ, Busscher HJ, Norde W (2003) Characterization of poly (ethylene oxide) brushes on glass surfaces and adhesion of *Staphylococcus epidermidis*. J Biomater Sci Polym Ed 14(4):313–324

Karageorgiou V, Kaplan D (2005) Porosity of 3D biomaterial scaffolds and osteogenesis. Biomaterials 26(27):5474–5491

Kaur G, Pandey OP, Singh K, Homa D, Scott B, Pickrell G (2014) A review of bioactive glasses: their structure, properties, fabrication and apatite formation. J Biomed Mater Res A 102(1):254–274

Khalandi B, Asadi N, Milani M, Davaran S, Abadi AJN, Abasi E, Akbarzadeh A (2017) A review on potential role of silver nanoparticles and possible mechanisms of their actions on bacteria. Drug Res 11(02):70–76

Khosravi AD, Ahmadi F, Salmanzadeh S, Dashtbozorg A, Montazeri EA (2009) Study of bacteria isolated from orthopedic implant infections and their antimicrobial susceptibility pattern. Res J Microbiol 4:158–163

Klueh U, Wagner V, Kelly S, Johnson A, Bryers JD (2000) Efficacy of silver-coated fabric to prevent bacterial colonization and subsequent device-based biofilm formation. J Biomed Mater Res 53(6):621–631

Kohn DH, Sarmadi M, Helman JI, Krebsbach PH (2002) Effects of pH on human bone marrow stromal cells in vitro: implications for tissue engineering of bone. J Biomed Mater Res 60(2):292–299

Korshed P, Li L, Liu Z, Wang T (2016) The molecular mechanisms of the antibacterial effect of picosecond laser generated silver nanoparticles and their toxicity to human cells. PLoS One 11(8):e0160078

Kose N, Kose AA (2015) Application of nanomaterials in prevention of bone and joint infections. In: Nanotechnology in diagnosis, treatment and prophylaxis of infectious diseases, pp 107–117

Kuzum D, Takano H, Shim E, Reed JC, Juul H, Richardson AG, Coulter DA (2014) Transparent and flexible low noise graphene electrodes for simultaneous electrophysiology and neuroimaging. Nat Commun 5:5259

Lanao RPF, Jonker AM, Wolke JG, Jansen JA, van Hest JC, Leeuwenburgh SC (2013) Physicochemical properties and applications of poly (lactic-co-glycolic acid) for use in bone regeneration. Tissue Eng Part B Rev 19(4):380–390

Lee JH, Park JH, Kim JA, Neupane GP, Cho MH, Lee CS, Lee J (2011) Low concentrations of honey reduce biofilm formation, quorum sensing, and virulence in Escherichia coli O157: H7. Biofouling 27(10):1095–1104

Legeay G, Poncin-Epaillard F, Arciola CR (2006) New surfaces with hydrophilic/hydrophobic characteristics in relation to (no) bioadhesion. Int J Artif Organs 29(4):453–461

LeGeros RZ (2008) Calcium phosphate-based osteoinductive materials. Chem Rev 108(11):4742–4753

Leveille SG (2004) Musculoskeletal aging. Curr Opin Rheumatol 16(2):114–118

Li X, Robinson SM, Gupta A, Saha K, Jiang Z, Moyano DF, Rotello VM (2014) Functional gold nanoparticles as potent antimicrobial agents against multi-drug-resistant bacteria. ACS Nano 8(10):10682–10686

Liao X, Yang F, Li H, So PK, Yao Z, Xia W, Sun H (2017) Targeting the thioredoxin reductase–thioredoxin system from Staphylococcus aureus by silver ions. Inorg Chem 56(24):14823–14830

Lin G, Cosimbescu L, Karin NJ, Tarasevich BJ (2012) Injectable and thermosensitive PLGA-g-PEG hydrogels containing hydroxyapatite: preparation, characterization and in vitro release behavior. Biomed Mater 7(2):024107

Liu X, Ma PX (2004) Polymeric scaffolds for bone tissue engineering. Ann Biomed Eng 32(3):477–486

Liu S, Wei L, Hao L, Fang N, Chang MW, Xu R, Chen Y (2009) Sharper and faster "nano darts" kill more bacteria: a study of antibacterial activity of individually dispersed pristine single-walled carbon nanotube. ACS Nano 3(12):3891–3902

Liu W, Su P, Chen S, Wang N, Wang J, Liu Y, Webster TJ (2015a) Antibacterial and osteogenic stem cell differentiation properties of photoinduced TiO2 nanoparticle-decorated TiO2 nanotubes. Nanomedicine 10(5):713–723

Liu W, Su P, Arthur Gonzales SC III, Wang N, Wang J, Li H, Webster TJ (2015b) Optimizing stem cell functions and antibacterial properties of TiO2 nanotubes incorporated with ZnO nanoparticles: experiments and modeling. Int J Nanomedicine 10:1997

Lu Y, Shah A, Hunter RA, Soto RJ, Schoenfisch MH (2015) S-Nitrosothiol-modified nitric oxide-releasing chitosan oligosaccharides as antibacterial agents. Acta Biomater 12:62–69

Luo J, Chen Z, Sun Y (2006) Controlling biofilm formation with an N-halamine-based polymeric additive. J Biomed Mater Res A 77(4):823–831

Magetsari R, Dewo P, Saputro BK, Lanodiyu Z (2014) Cinnamon oil and chitosan coating on orthopaedic implant surface for prevention of staphylococcus epidermidis biofilm formation. Malays Orthop J 8(3):11

Mangram AJ, Horan TC, Pearson ML, Silver LC, Jarvis WR, Hospital Infection Control Practices Advisory Committee (1999) Guideline for prevention of surgical site infection, 1999. Infect Control Hosp Epidemiol 20(4):247–280

Mardanpour K, Rahbar M, Mardanpour S, Mardanpour N (2017) Surgical site infections in orthopedic surgery: incidence and risk factors at an Iranian teaching hospital. Clin Trials Orthop Disord 2(4):132

Matassi F, Botti A, Sirleo L, Carulli C, Innocenti M (2013) Porous metal for orthopedics implants. Clinical cases in mineral and bone. Metabolism 10(2):111

McConoughey SJ, Howlin R, Granger JF, Manring MM, Calhoun JH, Shirtliff M, Stoodley P (2014) Biofilms in periprosthetic orthopedic infections. Future Microbiol 9(8):987–1007

McLaren JS, White LJ, Cox HC, Ashraf W, Rahman CV, Blunn GW, Scammell BE (2014) A biodegradable antibiotic-impregnated scaffold to prevent osteomyelitis in a contaminated in vivo bone defect model. Eur Cell Mater 27:332–349

Millenbaugh NJ, Baskin JB, DeSilva MN, Elliott WR, Glickman RD (2015) Photothermal killing of *Staphylococcus aureus* using antibody-targeted gold nanoparticles. Int J Nanomedicine 10:1953

Moličnik A, Hanc M, Rečnik G, Krajnc Z, Rupreht M, Fokter SK (2014) Porous tantalum shells and augments for acetabular cup revisions. Eur J Orthop Surg Traumatol 24(6):911–917

Monteiro DR, Gorup LF, Takamiya AS, de Camargo ER, Filho ACR, Barbosa DB (2012) Silver distribution and release from an antimicrobial denture base resin containing silver colloidal nanoparticles. J Prosthodont 21(1):7–15

Nablo BJ, Rothrock AR, Schoenfisch MH (2005) Nitric oxide-releasing sol–gels as antibacterial coatings for orthopedic implants. Biomaterials 26(8):917–924

Nampoothiri KM, Nair NR, John RP (2010) An overview of the recent developments in polylactide (PLA) research. Bioresour Technol 101(22):8493–8501

National Collaborating Centre for Women's and Children's Health (UK) (2008) Surgical site infection: prevention and treatment of surgical site infection. RCOG Press, London. (NICE Clinical Guidelines, No. 74.) 3, Definitions, surveillance and risk factors. Available from: https://www.ncbi.nlm.nih.gov/books/NBK53724/

National Nosocomial Infections Surveillance System (2004) National Nosocomial Infections Surveillance (NNIS) system report, data summary from January 1992 through June 2004, issued October 2004. Am J Infect Control 32:470–485

Navarro M, Michiardi A, Castano O, Planell JA (2008) Biomaterials in orthopaedics. J R Soc Interface 5(27):1137–1158

Niinomi M (2008) Mechanical biocompatibilities of titanium alloys for biomedical applications. J Mech Behav Biomed Mater 1(1):30–42

Ning CY, Zhou L, Tan GX (2016) Fourth-generation biomedical materials. Mater Today 19(1):2–3

Noel SP, Courtney H, Bumgardner JD, Haggard WO (2008) Chitosan films: a potential local drug delivery system for antibiotics. Clin Orthop Relat Res 466(6):1377–1382

Nordmann P, Dortet L, Poirel L (2012) Carbapenem resistance in Enterobacteriaceae: here is the storm! Trends Mol Med 18(5):263–272

O'neill JS, Maywood ES, Chesham JE, Takahashi JS, Hastings MH (2008) cAMP-dependent signaling as a core component of the mammalian circadian pacemaker. Science 320(5878):949–953

Özçelik H, Vrana NE, Gudima A, Riabov V, Gratchev A, Haikel Y, Klüter H (2015) Harnessing the multifunctionality in nature: a bioactive agent release system with self-antimicrobial and immunomodulatory properties. Adv Healthc Mater 4(13):2026–2036

Papayannopoulos V, Zychlinsky A (2009) NETs: a new strategy for using old weapons. Trends Immunol 30(11):513–521

Park MV, Neigh AM, Vermeulen JP, de la Fonteyne LJ, Verharen HW, Briedé JJ, de Jong WH (2011) The effect of particle size on the cytotoxicity, inflammation, developmental toxicity and genotoxicity of silver nanoparticles. Biomaterials 32(36):9810–9817

Parvizi J, Gehrke T, Chen AF (2013) Proceedings of the international consensus on periprosthetic joint infection. Bone Joint J 95(11):1450–1452

Passuti N, Gouin F (2003) Antibiotic-loaded bone cement in orthopedic surgery. Joint Bone Spine 70(3):169–174

Patti JM, Allen BL, McGavin MJ, Höök M (1994) MSCRAMM-mediated adherence of microorganisms to host tissues. Annu Rev Microbiol 48(1):585–617

Peleg AY, Hooper DC (2010) Hospital-acquired infections due to Gram-negative bacteria. N Engl J Med 362(19):1804–1813

Peng KT, Chen CF, Chu IM, Li YM, Hsu WH, Hsu RWW, Chang PJ (2010) Treatment of osteomyelitis with teicoplanin-encapsulated biodegradable thermosensitive hydrogel nanoparticles. Biomaterials 31(19):5227–5236

Perez LM, Lalueza P, Monzon M, Puertolas JA, Arruebo M, Santamaría J (2011) Hollow porous implants filled with mesoporous silica particles as a two-stage antibiotic-eluting device. Int J Pharm 409(1-2):1–8

Periasamy S, Joo HS, Duong AC, Bach THL, Tan VY, Chatterjee SS, Otto M (2012) How Staphylococcus aureus biofilms develop their characteristic structure. Proc Natl Acad Sci 109(4):1281–1286

Polyzois I, Nikolopoulos D, Michos I, Patsouris E, Theocharis S (2012) Local and systemic toxicity of nanoscale debris particles in total hip arthroplasty. J Appl Toxicol 32(4):255–269

Prasad R, Pandey R, Barman I (2016) Engineering tailored nanoparticles with microbes: quo vadis. WIREs Nanomed Nanobiotechnol 8:316–330. https://doi.org/10.1002/wnan.1363

Prasad R, Jha A, Prasad K (2018) Exploring the Realms of Nature for Nanosynthesis. Springer International Publishing (ISBN 978-3-319-99570-0) https://www.springer.com/978-3-319-99570-0

Qin H, Cao H, Zhao Y, Zhu C, Cheng T, Wang Q, Jiang Y (2014) In vitro and in vivo anti-biofilm effects of silver nanoparticles immobilized on titanium. Biomaterials 35(33):9114–9125

Quinteros MA, Aristizábal VC, Dalmasso PR, Paraje MG, Páez PL (2016) Oxidative stress generation of silver nanoparticles in three bacterial genera and its relationship with the antimicrobial activity. Toxicol In Vitro 36:216–223

Rahaman MN, Day DE, Bal BS, Fu Q, Jung SB, Bonewald LF, Tomsia AP (2011) Bioactive glass in tissue engineering. Acta Biomater 7(6):2355–2373

Rai M, Yadav A, Gade A (2009) Silver nanoparticles as a new generation of antimicrobials. Biotechnol Adv 27(1):76–83

Rai A, Prabhune A, Perry CC (2010) Antibiotic mediated synthesis of gold nanoparticles with potent antimicrobial activity and their application in antimicrobial coatings. J Mater Chem 20(32):6789–6798

Reading AD, Rooney P, Taylor GJS (2000) Quantitative assessment of the effect of 0.05% chlorhexidine on rat articular cartilage metabolism in vitro and in vivo. J Orthop Res 18(5):762–767

Reighard KP, Hill DB, Dixon GA, Worley BV, Schoenfisch MH (2015) Disruption and eradication of *P. aeruginosa* biofilms using nitric oxide-releasing chitosan oligosaccharides. Biofouling 31(9-10):775–787

Rho JY, Kuhn-Spearing L, Zioupos P (1998) Mechanical properties and the hierarchical structure of bone. Med Eng Phys 20(2):92–102

Rice KC, Mann EE, Endres JL, Weiss EC, Cassat JE, Smeltzer MS, Bayles KW (2007) The *cidA* murein hydrolase regulator contributes to DNA release and biofilm development in *Staphylococcus aureus*. Proc Natl Acad Sci 104(19):8113–8118

Rocha LB, Goissis G, Rossi MA (2002) Biocompatibility of anionic collagen matrix as scaffold for bone healing. Biomaterials 23(2):449–456

Rochford ETJ, Richards RG, Moriarty TF (2012) Influence of material on the development of device-associated infections. Clin Microbiol Infect 18(12):1162–1167

Roco MC, Mirkin CA, Hersam MC (2011) Nanotechnology research directions for societal needs in 2020: retrospective and outlook, vol 1. Springer Science & Business Media, New York

Sachlos E, Czernuszka JT (2003) Making tissue engineering scaffolds work. Review: the application of solid freeform fabrication technology to the production of tissue engineering scaffolds. Eur Cells Mater 5:29–40

Samavedi S, Whittington AR, Goldstein AS (2013) Calcium phosphate ceramics in bone tissue engineering: a review of properties and their influence on cell behavior. Acta Biomater 9(9):8037–8045

Samuel S (2012) Antibiotic loaded acrylic bone cement in orthopaedic trauma. In: Osteomyelitis. InTech, London

Sato M, Webster TJ (2004) Nanobiotechnology: implications for the future of nanotechnology in orthopedic applications. Expert Rev Med Devices 1(1):105–114

Schierholz JM, Beuth J (2001) Implant infections: a haven for opportunistic bacteria. J Hosp Infect 49(2):87–93

Seil JT, Webster TJ (2012) Antimicrobial applications of nanotechnology: methods and literature. Int J Nanomedicine 7:2767

Sendi P, Rohrbach M, Graber P, Frei R, Ochsner PE, Zimmerli W (2006) Staphylococcus aureus small colony variants in prosthetic joint infection. Clin Infect Dis 43(8):961–967

Seong M, Lee DG (2017) Silver nanoparticles against *Salmonella enterica* serotype typhimurium: role of inner membrane dysfunction. Curr Microbiol 74(6):661–670

Shi Z, Neoh KG, Kang ET, Wang W (2006) Antibacterial and mechanical properties of bone cement impregnated with chitosan nanoparticles. Biomaterials 27(11):2440–2449

Simchi A, Tamjid E, Pishbin F, Boccaccini AR (2011) Recent progress in inorganic and composite coatings with bactericidal capability for orthopaedic applications. Nanomedicine 7(1):22–39

Singh AV, Vyas V, Patil R, Sharma V, Scopelliti PE, Bongiorno G, Milani P (2011) Quantitative characterization of the influence of the nanoscale morphology of nanostructured surfaces on bacterial adhesion and biofilm formation. PLoS One 6(9):e25029

Singh S, Vidyarthi AS, Dev A (2015) Microbial synthesis of nanoparticles: An overview. In: Singh OV (ed) Bio-nanoparticles: biosynthesis and sustainable biotechnological implications. Wiley, New York, pp 155–186

Smith WR, Hudson PW, Ponce BA, Manoharan SRR (2018) Nanotechnology in orthopedics: a clinically oriented review. BMC Musculoskelet Disord 19(1):67

Sohrabi M, Hesaraki S, Kazemzadeh A (2014) The influence of polymeric component of bioactive glass-based nanocomposite paste on its rheological behaviors and in vitro responses: hyaluronic acid versus sodium alginate. J Biomed Mater Res B Appl Biomater 102(3):561–573

Stoodley P, Conti SF, DeMeo PJ, Nistico L, Melton-Kreft R, Johnson S, Kathju S (2011) Characterization of a mixed MRSA/MRSE biofilm in an explanted total ankle arthroplasty. FEMS Immunol Med Microbiol 62(1):66–74

Su HL, Chou CC, Hung DJ, Lin SH, Pao IC, Lin JH, Lin JJ (2009) The disruption of bacterial membrane integrity through ROS generation induced by nanohybrids of silver and clay. Biomaterials 30(30):5979–5987

Sullivan MP, McHale KJ, Parvizi J, Mehta S (2014) Nanotechnology: current concepts in orthopaedic surgery and future directions. Bone Joint J 96(5):569–573

Temenoff JS, Mikos AG (2000) Tissue engineering for regeneration of articular cartilage. Biomaterials 21(5):431–440

Tikariha S, Banerjee S, Dev A, Singh S (2017) Growth phase-dependent synthesis of gold nanoparticles using *Bacillus Licheniformis*. In: Mukhopahyay K, Kumar M, Sachan A (eds) Applications of biotechnology for sustainable development. Springer Nature, Singapore, pp 1–8

Tran N, Tran PA, Jarrell JD, Engiles JB, Thomas NP, Young MD, Born CT (2013) In vivo caprine model for osteomyelitis and evaluation of biofilm-resistant intramedullary nails. Biomed Res Int 2013:674378

Tran N, Kelley MN, Tran PA, Garcia DR, Jarrell JD, Hayda RA, Born CT (2015) Silver doped titanium oxide–PDMS hybrid coating inhibits *Staphylococcus aureus* and *Staphylococcus epidermidis* growth on PEEK. Mater Sci Eng C 49:201–209

Uskoković V, Desai TA (2014) In vitro analysis of nanoparticulate hydroxyapatite/chitosan composites as potential drug delivery platforms for the sustained release of antibiotics in the treatment of osteomyelitis. J Pharm Sci 103(2):567–579 van Hengel IA, Riool M, Fratila-Apachitei LE, Witte-Bouma J, Farrell E, Zadpoor AA, Apachitei I (2017) Selective laser melting porous metallic implants with immobilized silver nanoparticles kill and prevent biofilm formation by methicillin-resistant *Staphylococcus aureus*. Biomaterials 140:1–15

Veerachamy S, Yarlagadda T, Manivasagam G, Yarlagadda PK (2014) Bacterial adherence and biofilm formation on medical implants: a review. Proc Inst Mech Eng H J Eng Med 228(10):1083–1099

Villatte G, Massard C, Descamps S, Sibaud Y, Forestier C, Awitor KO (2015) Photoactive TiO_2 antibacterial coating on surgical external fixation pins for clinical application. Int J Nanomedicine 10:3367

Von Ohle C, Gieseke A, Nistico L, Decker EM, Stoodley P (2010) Real-time microsensor measurement of local metabolic activities in ex vivo dental biofilms exposed to sucrose and treated with chlorhexidine. Appl Environ Microbiol 76(7):2326–2334

Wachol-Drewek Z, Pfeiffer M, Scholl E (1996) Comparative investigation of drug delivery of collagen implants saturated in antibiotic solutions and a sponge containing gentamicin. Biomaterials 17(17):1733–1738

Wafa H, Grimer RJ, Reddy K, Jeys L, Abudu A, Carter SR, Tillman RM (2015) Retrospective evaluation of the incidence of early periprosthetic infection with silver-treated endoprostheses in high-risk patients: case-control study. Bone Joint J 97(2):252–257

Wang N, Li H, Lü W, Li J, Wang J, Zhang Z, Liu Y (2011) Effects of TiO_2 nanotubes with different diameters on gene expression and osseointegration of implants in minipigs. Biomaterials 32(29):6900–6911

Wang N, Li H, Wang J, Chen S, Ma Y, Zhang Z (2012) Study on the anticorrosion, biocompatibility, and osteoinductivity of tantalum decorated with tantalum oxide nanotube array films. ACS Appl Mater Interfaces 4(9):4516–4523

Wang LN, Lin LX, Lin CJ, Shen C, Shinbine A, Luo JL (2013) Anodic TiO_2 nanotubular arrays with pre-synthesized hydroxyapatite—an effective approach to enhance the biocompatibility of titanium. J Nanosci Nanotechnol 13(8):5316–5326

Watnick P, Kolter R (2000) Biofilm, city of microbes. J Bacteriol 182(10):2675–2679

Wei J, Jia J, Wu F, Wei S, Zhou H, Zhang H, Liu C (2010) Hierarchically microporous/macroporous scaffold of magnesium–calcium phosphate for bone tissue regeneration. Biomaterials 31(6):1260–1269

Weinstein SL (2000) 2000–2010: the bone and joint decade, 1–3. Ann Rheum Dis 59(2):81–82

Wenke JC, Owens BD, Svoboda SJ, Brooks DE (2006) Effectiveness of commercially-available antibiotic-impregnated implants. J Bone Joint Surg 88(8):1102–1104

Wilberforce SI, Finlayson CE, Best SM, Cameron RE (2011) A comparative study of the thermal and dynamic mechanical behaviour of quenched and annealed bioresorbable poly-L-lactide/α--tricalcium phosphate nanocomposites. Acta Biomater 7(5):2176–2184

Willenegger H, Roth B (1986) Treatment tactics and late results in early infection following osteosynthesis. Unfallchirurgie 12(5):241–246

Wink DA, Kasprzak KS, Maragos CM, Elespuru RK, Misra M, Dunams TM, Allen JS (1991) DNA deaminating ability and genotoxicity of nitric oxide and its progenitors. Science 254(5034):1001–1003

Winkler H, Kaudela K, Stoiber A, Menschik F (2006) Bone grafts impregnated with antibiotics as a tool for treating infected implants in orthopedic surgery–one stage revision results. Cell Tissue Bank 7(4):319–323

Winter GD, Simpson BJ (1969) Heterotopic bone formed in a synthetic sponge in the skin of young pigs. Nature 223(5201):88

Wood TK, Barrios AFG, Herzberg M, Lee J (2006) Motility influences biofilm architecture in Escherichia coli. Appl Microbiol Biotechnol 72(2):361–367

Wu P, Grainger DW (2006) Drug/device combinations for local drug therapies and infection prophylaxis. Biomaterials 27(11):2450–2467

Yang Y, Park S, Liu Y, Lee K, Kim HS, Koh JT, Ong JL (2008) Development of sputtered nanoscale titanium oxide coating on osseointegrated implant devices and their biological evaluation. Vacuum 83(3):569–574

Yang C, Plackett D, Needham D, Burt HM (2009) PLGA and PHBV microsphere formulations and solid-state characterization: possible implications for local delivery of fusidic acid for the treatment and prevention of orthopaedic infections. Pharm Res 26(7):1644–1656

Yang CC, Lin CC, Liao JW, Yen SK (2013) Vancomycin–chitosan composite deposited on post porous hydroxyapatite coated Ti6Al4V implant for drug controlled release. Mater Sci Eng C 33(4):2203–2212

Yasmeen S, Lo MK, Bajracharya S, Roldo M (2014) Injectable scaffolds for bone regeneration. Langmuir 30(43):12977–12985

Yeh TH, Hsu LW, Tseng MT, Lee PL, Sonjae K, Ho YC, Sung HW (2011) Mechanism and consequence of chitosan-mediated reversible epithelial tight junction opening. Biomaterials 32(26):6164–6173

You C, Han C, Wang X, Zheng Y, Li Q, Hu X, Sun H (2012) The progress of silver nanoparticles in the antibacterial mechanism, clinical application and cytotoxicity. Mol Biol Rep 39(9):9193–9201

Yu X, Khalil A, Dang PN, Alsberg E, Murphy WL (2014) Multilayered inorganic microparticles for tunable dual growth factor delivery. Adv Funct Mater 24(20):3082–3093

Yuan YG, Peng QL, Gurunathan S (2017) Effects of silver nanoparticles on multiple drug-resistant strains of *Staphylococcus aureus* and *Pseudomonas aeruginosa* from mastitis-infected goats: an alternative approach for antimicrobial therapy. Int J Mol Sci 18(3):569

Zhai W, Lu H, Chen L, Lin X, Huang Y, Dai K, Chang J (2012) Silicate bioceramics induce angiogenesis during bone regeneration. Acta Biomater 8(1):341–349

Zhang F, Zhang Z, Zhu X, Kang ET, Neoh KG (2008) Silk-functionalized titanium surfaces for enhancing osteoblast functions and reducing bacterial adhesion. Biomaterials 29(36):4751–4759

Zheng Y, Li J, Liu X, Sun J (2012) Antimicrobial and osteogenic effect of Ag-implanted titanium with a nanostructured surface. Int J Nanomedicine 7:875

Zheng F, Wang S, Wen S, Shen M, Zhu M, Shi X (2013) Characterization and antibacterial activity of amoxicillin-loaded electrospun nano-hydroxyapatite/poly (lactic-co-glycolic acid) composite nanofibers. Biomaterials 34(4):1402–1412

Zimmerli W, Trampuz A, Ochsner PE (2004) Prosthetic-joint infections. N Engl J Med 351(16):1645–1654

Chapter 3
Antibacterial Activity by Functionalized Carbon Nanotubes

Devanabanda Mallaiah

Contents

Abstract The nanoscaled unique physicochemical properties of carbon nanotubes are offered many types of applications in biomedical field. Due to high hydrophobicity and less dispersibility in the biological medium, carbon nanotubes are functionalized with many bioactive compounds. The functionalized carbon nanotubes had less toxicity problem and became efficient, nontoxic drug delivery vehicles. The antibacterial activities of carbon nanotubes are also studied well along with other biomedical applications. Many antibiotics and drugs are using for the control of bacterial infections. However, bacterial infections have developed resistance to these traditional antibiotics, drugs, and host immunity through various mechanisms. The continuous increase of bacterial drug resistance has been challenged to develop novel antibacterial agents. The functionalized carbon nanotubes will become promising antibacterial agents due to their nanosized properties. The functionalized carbon nanotubes are exploited in various medical and non-medical fields. Many studies have been reported on functionalized carbon nanotubes role in drug delivery, diagnosis, and therapy including antibacterial activity by various mechanisms. This book chapter focuses on different carbon nanotubes and their functionalization, antibacterial activity, and mechanisms of functionalized carbon nanotubes.

D. Mallaiah (✉)
Department of Biotechnology and Bioinformatics, Yogi Vemana University,
Kadapa, Andhra Pradesh, India

© Springer Nature Switzerland AG 2020
R. Prasad et al. (eds.), *Nanostructures for Antimicrobial and Antibiofilm Applications*, Nanotechnology in the Life Sciences,
https://doi.org/10.1007/978-3-030-40337-9_3

Keywords Functionalized carbon nanotubes · Anti-bacterial · Drug-delivery ·
Single walled carbon nanotubes · Multi-walled carbon nanotubes

3.1   Introduction

Among carbon-based allotropes, the carbon nanotubes (CNT) were best studied
nanomaterials. CNT were described as "graphite sheets rolled on above the other"
by Sumio Iijima after their discovery in 1991 (Iijima 1991). The attractive physicochemical properties of CNT have been exploited in many applications of biomedical field. The CNT physicochemical properties include high ratio of surface area/
volume, thermal stability, and structural and mechanical properties. CNT possess
other beneficial characteristics such as ultra-light weight, elasticity, chemical stability, more tensile strength, and electrical conductivity (Jia and Wei 2017).

CNT are considered as 1D nanomaterials or 1D nanocylinders, in which the
coaxial tubes are formed from rolled up 2D graphene sheets. The large portions of
CNT are made up from sp^2-hybridized carbon atoms, but at defects very less amount
of sp^3-hybridized carbon atoms are present. The single-walled carbon nanotubes
(SWCNT) and multi-walled carbon nanotubes (MWCNT) are the two types of CNT
that have been classified. SWCNT is a rolled cylindrical tube made up by single
graphene sheet, whereas MWCNT contains a number of concentric rolled graphite
sheets that vary from 2 to 50. The diameter and lengths were 0.4–3 nm and
20–1000 nm for SWCNT and 2–100 nm and 1–50 μm for MWCNT, respectively.
The CNT have empty space inside and walls are made up by carbon atoms with
hexagonal lattice, which is analogous to graphite atomic planes (Fig. 3.1). In
MWCNT, the distance between rolled graphene sheets is 0.340 nm and sheets held
together strongly by non-covalent interactions. These non-covalent interactions
allow the formation of large aggregates, which in turn leads to insolubility of CNT
in both aqueous and organic solvents. The limited solubility and toxicity of CNT
impact their applications in many bio-medical fields. However, the functionalization of CNT improves the solubility and reduces their toxic effects. The CNT were
functionalized with various organic molecules such as folic acid, PEG, antibodies,
chitosan, aptamers, and peptides for specific targeting (Karimi et al. 2015).

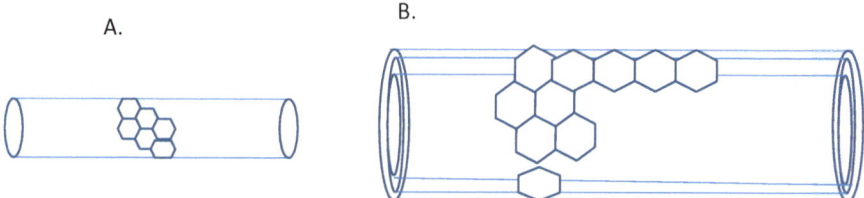

Fig. 3.1 (**a**) Single walled carbon nanotubes (**b**) Multi walled carbon nanotubes

Nowadays treating bacterial infections becomes a challenge due to their drug resistance. Drug resistance has developed to almost all traditional antibiotics by innate (mutations) or acquired (resistance genes) mechanisms. The development of multidrug resistance infections causes serious threat to human health and economy. Currently, the combination therapy is used for treating drug resistance conditions, which are again facing new problems such as potential antagonism with antibiotics and side effects. New modified antibiotics are also developing from the known antibiotics and proved to be not much effective against multidrug resistance (MDR) microorganisms (Xu et al. 2014).

Therefore, synthesis of new antibacterial agents, which works differently from traditional antibiotics to overcome drug resistance problems, is urgently needed. Recently, nanomaterials have become alternative promising agents for drug resistance bacterial infection treatment. Due to their simultaneous multiple antibacterial mechanisms, nanomaterials do not induce drug resistance in bacteria as compared with antibiotics. The in vitro and in vivo studies have been shown antibacterial activity of many nanomaterials including carbon-based nanomaterials. However, many disadvantages are associated with nanomaterials, which are to be addressed carefully. For example, many studies show silver nanoparticles (SNP) as excellent antibacterial agents, but it causes harmful effects on some key components of human cells by releasing Ag^+ ions. Other nanomaterials like semiconductor NP and polymeric nanostructures are also shown antibacterial activity, but their complete long-term biocompatibility and toxic effects have not been studied (Xin et al. 2018).

The insolubility and agglomeration properties of bare or unfunctionalized CNT are also causes the contact mediated cytotoxicity. But, the functionalized carbon nanotubes (f-CNT) had shown more biocompatibility and less toxicity. The f-CNT are proved to be excellent drug delivery vehicles and efficient antibacterial agents (Dumortier et al. 2006). This chapter deals with f-CNT synthesis and characterization, functionalization, and their various studies on antimicrobial activities and mechanisms.

3.2 Synthesis of Functionalized Carbon Nanotubes

The laser ablation, electric arc discharge, and chemical vapor deposition are some methods used to prepare CNT. Apart from synthesis, CNT has to purify from carbon-based materials and metals, which are the by-products using oxidation and acid treatment, respectively. The structural imperfections and impurities were also implicated partly in the toxicity of CNT. Electric discharge was used by Ijima in 1991 to prepare MWCNT. In the same technique, metal catalyst was used by Ijima and Ichihashi and Bethune to prepare SWCNT (Karimi et al. 2015).

The CNT are functionalized by two methods: (1) covalent bonding and (2) non-covalent interactions (Fig. 3.2). In covalent bonding, polar functional groups (–COOH, $-NH_2$, –OH) are attached to the surface of CNT by standard synthetic chemistry methods, which allows further reactions with CNT, including

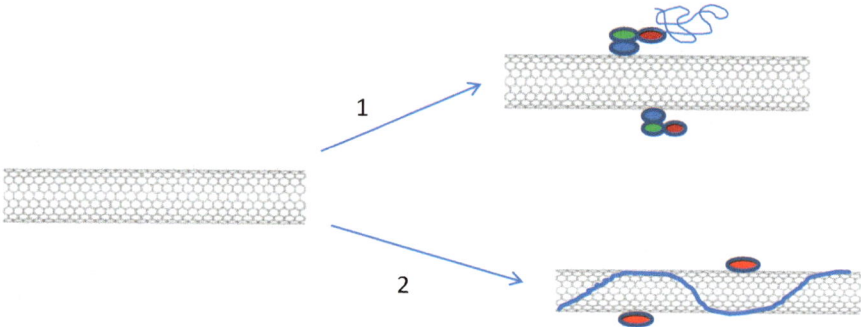

Fig. 3.2 Functionalization of carbon nanotubes (1) Covalent (2) Non covalent bonding. ⬤⬤= Functional groups, ∿=polymers, ⬤=surfactants, ꙮ=Biologically active molecule

esterification, alkylation, and thiolation. Functionalization by covalent bonding enables preparation of stable conjugates with exceptional mechanical properties. The disadvantage of covalent functionalization is hybridization changes from SP^2 to SP^3, which results π-conjugation loss and consequently decreases in the electrical conductivity of CNT (Harun-Or Rashid and Ralph 2017; and Karimi et al. 2015).

The alternative functionalization by non-covalent interactions has also both advantage and disadvantages. The advantage of non-covalent functionalization is it preserves the physicochemical properties of CNT. However, poor solubility and biocompatibility are the disadvantages. In non-covalent interaction, the functionalized molecules attach to surface of CNT through adsorption. The functionalized molecules wrap around CNT through interactions called π–π stacking and van der Waals. The van der Waals interactions forms between CNT and aromatic rings of molecules decreases the van der Waals interactions of CNT and thereby increases their solubility in aqueous solutions (Harun-Or Rashid and Ralph 2017; Mallakpour and Soltanian 2016).

3.3 Characterization of Functionalized Carbon Nanotubes

Different types of techniques are needed to characterize functionalized carbon nanotubes. The common techniques used are listed in Table 3.1.

Table 3.1 Techniques for characterization of functionalized carbon nanotubes

S.No	Technique	Characteristics
1	UV-visible spectroscopy	Functionalization
2	Transmission electron microscopy	Morphology, size, and aggregation %
3	Scanning electron microscopy	Morphology, size, and aggregation %
4	Atomic force microscopy	Surface structure
5	Fourier-transform infrared spectroscopy	Chemical structure
6	Zetasizer	Zeta potential
7	Raman spectroscopy	Chemical structure
8	X-ray diffraction	Crystalline structure, elemental composition
9	Thermogravimetric analysis	% of functionalization
10	Nanoparticle tracking analysis	Hydrodynamic size, nanoparticles number

Compiled from Mehra and Palakuthi (2016) and Mourdikoudis et al. (2018)

3.4 Antimicrobial Activity of Functionalized Carbon Nanotubes

Functionalization has many advantages than bare CNT. Functionalizations of CNT prevent nonspecific or unwanted adsorption of molecules in biological medium. Functionalization also reduces hydrophobicity and improves adhesion and stability of CNT. Functionalized SWCNT and MWCNT offered excellent antibacterial activity on Gram-positive and Gram-negative bacteria.

3.4.1 Single-Walled Carbon Nanotubes

Both pegylated and non-pegylated SNP-coated SWCNT damaged the outer cell membrane and downregulated genes involved in metabolic transporter system of membrane, biosynthesis of amino acids, and outer membrane integrity significantly in *Salmonella*. As compared with plain SNP-coated SWCNT, pegylated silver nanoparticles-coated SWCNT were safe to human cells at bacterial concentration ~63 µg/mL. The pegylated SNP-coated SWCNT down regulated genes expression exclusively associated with normal physiological processes, virulence, outer membrane structure, invasion, and quorum sensing in *Salmonella typhimurium* (Chaudhari et al. 2015). Antimicrobial peptides fabricated silver-coated carbon nanotubes that were used against *S. aureus* and proved their high antibacterial effect using three-dimensional human skin model (Chaudhari et al. 2019). Some studies also found antibacterial effect of pegylated silver-coated SWCNT was due to differential expression of proteins in *Salmonella enterica* serovar *typhimurium*. The proteins involved in biofilm formation, energy metabolism, and quorum sensing were decreased, but the proteins associated with DNA protection, oxygen stress, and starvation were induced significantly (Park et al. 2018). Silver-coated SWCNT are functionalized with antimicrobial peptide, TP359 by covalently and non-covalently, and investigated their effect on bacteria. The covalently functionalized conjugate showed higher antibacterial activity than non-covalent conjugate and plain conjugate with less toxicity to eukaryotic cells (Chaudhari et al. 2016).

Functionalization of SWCNT (diameter ~1.5 nm and length ~10 μm) with –OH, –COOH, and –NH$_2$ showed strong antimicrobial activity toward *B. subtilis*, rod-shaped bacteria; *S. aureus*, sphere-shaped bacteria; and *S. typhimurium*, rod-shaped bacteria. The antibacterial activity is buffer, concentration, and time dependent, but not surface functional groups are critical factors. In water or 0.9% Nacl, the functionalized SWCNT (f-SWCNT) exhibited high antibacterial activity but is not shown in PBS buffer or brain heart infusion (BHI) broth. Functionalization with the anionic –OH and –COOH groups exhibited more antibacterial activity than cationic –NH$_2$ groups. Neither the difference in cell wall structure nor the bacterial cell shape influenced antimicrobial efficiency of f-SWCNT (Arias and Yang 2009). Functionalization of SWCNT with monosaccharide species (galactose) attributed with strong cell adhesion property and agglutinated pathogenic *E.coli* efficiently (Gu et al. 2005). Other studies reported that monosaccharide-f-SWCNT bound strongly to *B.anthracis* spores and effectively mediated their aggregation by divalent ions (wang et al. 2006). A systematic study performed on the interaction between sugar (galactose and mannose) f-SWCNT and *Bacillus* spores. The differences and selectivity were observed between sugars, in their configurations, and between *Bacillus* spores significantly (Luo et al. 2009).

The SWCNT were dispersed in Tween-20 and bile salt and visualized in solution as moving nanodarts. The antibacterial activity of dispersed SWCNT was more than SWCNT alone (Liu et al. 2009). SWCNT were dispersed individually in Tween-20-saline solution and observed their antibacterial effect using atomic force microscopy (Liu et al. 2010). Some studies proved that antibacterial activity of SWCNT mainly depends on the dispersed medium; for example, the SWCNT dispersed in Triton X-100 was toxic to *E. coli* but SWCNT dispersed in natural dispersing agents and synthetic dispersing agent was not affected by the cell viability of *E. coli* (Alpatova et al. 2010). Some other studies proved that functionalization affects the antibacterial activity of SWCNT; for example, f-SWCNT prepared with nine groups, (1) n-propylamine, (2) phenylhydrazine, (3) hydroxyl, (4) phenyldicarboxy, (5) phenyl, (6) salfonic acid, (7) n-butyl, (8) diphenylcyclopropyl, and hydrazine, are different in their physicochemical properties. The impact of these nine f-SWCNT on bacterial cell viability was correlated with aggregate size distribution, dispersity, and morphology. The cell viability was decreased significantly by n-butyl f-SWCNT, diphenylcyclopropyl f-SWCNT, and hydrazine f-SWCNT. Highly compact and narrowly aggregate size distribution of f-SWCNT result in high cell viability (Pasquini et al. 2012). Antibacterial activity of SWCNT dispersed in PLGA (poly(lactic-co-glycolic acid) was compared with pure PLGA. Approximately 98% of bacteria were killed in 1 h by SWCNT-PLGA than only 15–20% by PLGA alone (Aslan et al. 2010).

Some studies reported that SWCNT were more toxic to *E. coli* and *B. atrophaeus* due to presence of some toxic nickel catalysts in the as-received CNT (Cortes et al. 2014). The covalent binding of SWCNT on surfaces of thin-film composite polyamide membranes (SWCNT-TFC) provides novel properties in addition to exhibiting continued high performance in the processes of water separation. The functionalization also enhances their bactericidal activity up to 60%, which was demonstrated with *E. coli* (Tiraferri et al. 2011). The magnetic nanoparticles (Fe$_3$O$_4$) were prepared and conjugated with monoclonal antibody (mAb@MNP), which specifically binds to *E. coli*. The capturation and pre-concentration of *E. coli* by mAb@MNP follow binding with Raman-tagged concanavalin-A-AuNP@SWCNT for detection. In addition to detection, drug-resistant pathogens photothermal ablation was enhanced by MNP/AuNP@SWCNT nanoconjugate (Ondera and Hamme 2015). The SWCNT were functionalized (aptamer–SWCNT and aptamer–ciprofloxacin–SWCNT) and targeted against *Pseudomonas aeruginosa* biofilms. Both conjugates were showed higher antibiofilm activity than SWCNT alone (Wang et al. 2018a, b). The antibiotic ciprofloxacin was covalently functionalized to SWCNT and studied their antibacterial activity. The study showed that the synthesized nanoantibiotic showed enhanced antibacterial activity than free antibiotic (Assali et al. 2017).

Porphyrin-SWCNT nanoconjugate was prepared, which exhibits more light absorption in the visible region. The conjugate was utilized in antimicrobial photodynamic therapy against *S. aureus* (Sah et al. 2018). The acid SWCNT prepared and assessed their antibacterial activity on *S. aureus* and *E. coli*. The study found that f-SWCNT exhibited better antibacterial activity on Gram-positive bacteria than Gram-negative bacteria (Deokar et al. 2013). The SWCNT were also functionalized by herbal extract of Hempedu bumi and studied antibacterial activity toward *Bacillus* sp., *E. coli*, and *Aspergillus niger*. The f-SWCNT exhibited the inhibition of bacterial species (Foo et al. 2018). The SWCNT were dispersed in different surfactant solutions and evaluated their effect on bacteria *S.enterica*, *E.coli*, and *E. faecium*. The study demonstrated that among the three surfactants, sodium cholate showed less antibacterial activity (Dong et al. 2012). SNP were attached on the surface of SWCNT through functional linkages and investigated their bactericidal effect on *E. coli* and *S. aureus*. The SNP-modified SWCNT showed strong antibacterial activity (Kosaka et al. 2009).

3.4.2 Multiwalled Carbon Nanotubes

The MWCNT were functionalized non-covalently with ionic (sodium dodecylbenzenesulfonate, sodium cholate, dodecyltrimethylammonium bromide, cetyltrimethylammonium) and non-ionic (polyvinylpyrrolidone) surfactants and used to test their antibacterial activity against *E. coli*. The MWCNT dispersed in cationic surfactants were efficiently killed the bacteria than MWCNT dispersed in anionic and non-ionic surfactants, which shown desirable biocompatibility. In between biocompatible surfactants, the MWCNT dispersed in anionic surfactants exhibited more antibacterial effect than MWCNT dispersed in non-ionic surfactants (Khazaee et al. 2016). The broad-spectrum antibacterial activity of hydroxyl and carboxyl functionalized MWCNT is shown against *E. coli, S. aureus, L. acidophilus, E. faecalis*, and *B. adolescentis*, which are mostly present in human digestive system (Chen et al. 2013).

The Acacia extract (AE) was used to disperse the MWCNT due to its surfactant properties and observed AE-MWCNT effect on *E. coli*. The AE-MWCNT nanocomposite reduced cell viability of bacteria and offered an alternative to chemical surfactant to disperse MWCNT (Yadav et al. 2015). The nisin (antibacterial peptide from *Lactococcus lactis*) was immobilized covalently onto MWCNT by polyethylene glycol (PEG) as cross-linking agent to form nisin-MWCNT nanocomposite, which improved the leaching of PEG and nisin. When compared with pristine MWCNT, the nisin-MWCNT nanocomposite showed several fold bactericidal activity on Gram-positive and Gram-negative bacteria. The nanocomposite also inhibited the biofilm formation ~100 fold when compared with pristine MWCNT (Qi et al. 2011). The vertically aligned MWCNT were fabricated on Ni/Si substrates by plasma-enhanced chemical vapor deposition. A wide hollow core was formed by removal of Ni seeds at the tip of MWCNT in a purification process. SNP were deposited on both Ni-removed and unpurified MWCNT and studied their effect on *E. coli*. The purified Ag-MWCNT showed more antibacterial activity against *E. coli* when compared with Ni-MWCNT and Ag-Ni/MWCNT (Akhavan et al. 2011a). The chitosan-MWCNT hydrogels were prepared and examined their bactericidal effect. The chitosan-MWCNT composite showed efficient antibacterial activity in dose-dependent manner (Venkatesan et al. 2014).

Both functionalized and unfunctionalized MWCNT were used to prepare MWCNT-ZnO nanocomposites and investigated their effect on photoinactivation of *E. coli*. The functionalized MWCNT-ZnO nanocomposite that contains 10 wt% MWCNT killed 100% of bacteria, while unfunctionalized MWCNT inactivated only 63% of bacteria by UV-visible irradiation treatment for 10 min. The Zn-O-C carbonaceous bonds formed between Zn and carboxylic oxygen are involved in charge transfer to kill bacteria (Akhavan et al. 2011b). Two amino acids lysine and arginine were used separately to functionalize MWCNT and tested antibacterial efficiency against *E. coli, S. aureus*, and *S. typhymurium*. The lysine functionalized MWCNT recorded 3.02, 2.28, and 1.83, and arginine-functionalized MWCNT recorded 3.04, 2.33, and 1.84 times more bactericidal activity against *E. coli, S. aureus*, and *S. typhymurium*, respectively (Zardini et al. 2012). The MWCNT were functionalized with lysozyme by immobilization and observed higher antibacterial effect toward Gram-positive *S. aureus* (Merli et al. 2011). The MWCNT were functionalized with four ethanalomine groups and studied their antibacterial effect on four Gram-negative and four Gram-positive bacteria. The results showed the antibacterial activity of MWCNT functionalized by ethanolamine groups in the order of TEA >, DEA > MEA > pristine (Zardini et al. 2014). SNP and CuNP were fabricated on MWCNT by simple chemical reduction method and studied their effect on Gram-negative *E. coli*. The SNP-MWCNT and CuNP-MWCNT killed 97 and 75%, respectively, whereas MWCNT-COOH killed only 20% of the bacteria

(Mohan et al. 2011). SNP were coated on MWCNT and prepared Ag/MWCNT hybrid nanoparticles and studied their antibacterial activity. The hybrid nanoparticles showed higher antibacterial activity when compared with nanoparticles alone (Jung et al. 2011; Rangari et al. 2010; Li et al. 2011). The antibiotic (vancomycin hydrochloride) modified MWCNT were fabricated on thermoplastic polyurethane and showed their excellent antibacterial activity (Liu et al. 2017).

The MWCNT were dispersed in three different surfactants and examined their antibacterial effect against *Streptococcus mutants*. The CTAB and TX-100 provided the maximum and minimum dispersion of MWCNT, respectively. Among three surfactants, the CTAB dispersed MWCNT showed more antibacterial activity than TX-100 and SDD-dispersed MWCNT in time- and concentration-dependent manner (Bai et al. 2011). The MWCNT also is modified by different types of surfactants such as dioctyl sodium sulfosuccinate and 9:1 ratio of sodium dodecyl benzenesulfate and hexadecyltrimethylammonium bromide. The study observed improved antibacterial activity than surfactants alone (Bai et al. 2012, 2016).

Anti-group A streptococcus antibodies were functionalized on MWCNT and used in the specific targeting of planktonic bacteria and biofilms of group A streptococcus along with laser irradiation. The combination therapy, which includes antibody-functionalized MWCNT and photothermal ablation, killed both types of bacteria more significantly than MWCNT and photothermal ablation therapy alone (Levi-Polyachenko et al. 2014). A conductive nanocomposite membrane was fabricated by mixing both carboxy-functionalized MWCNT and polysulfone polymer. The conductive nanocomposite membrane inactivated (99.99%) the model Gram-positive pathogenic bacteria by providing electric current for 40 min at 5 mA (Zhu et al. 2015). The carboxy-MWCNT were functionalized with dapsone derivative (anti-leprosy drug) and examined for their inherent antibacterial activity on *E. coli* and *S. aureus*. The study reported that dapsone-fabricated MWCNT showed strong antibacterial activity than carboxy-functionalized MWCNT alone (Azizian et al. 2014). The thermally active nanomaterials such as MWCNT were functionalized covalently with specific IgG antibodies against methicillin-resistant *Staphylococcus aureus*. Following treatment with MWCNT-IgG, the cultures were irradiated and observed highly efficacious, targeted killing of methicillin-resistant bacteria (Mocan et al. 2016). Functionalization with lysine increased the water dispersion of MWCNT and also improved bactericidal activity on Gram-positive and Gram-negative bacteria (Amiri et al. 2012). The nanocomposites were prepared by using hydrogel-MWCNT and chitosan and examined their effect on bacterial growth and adhesion. The study found that the nanocomposite was not affected bacterial cell viability, but it decreased bacterial cell adhesion or colonization (Bellingeri et al. 2018).

The MWCNT were functionalized on ceramic-based membrane and used in the purification of drinking water. The composite mullite-CNT membrane removed the bacteria *E. coli* and *S. aureus* from drinking water with high flux (Zhu et al. 2019). The methylene blue conjugated MWCNT prepared and studied their photodynamic antibacterial, antibiofilm activities on *E. coli* and *S. aureus*, which are commonly infecting medical devices. The conjugate showed higher photoinactivation in

Gram-positive bacteria *S. aureus* than *E. coli* (Parasuraman et al. 2019). The phyto-chemically conjugated MWCNT with SNP and CuNP showed highest inhibitory activity against bacteria and cancer cells (Yallappa et al. 2016). The MWCNT were incorporated with other carbon quantum dots (Cdot) and used in the water filter membranes. The MWCNT-Cdot efficiently removed the bacteria and inactivated captured bacteria (Dong et al. 2018).) SNP and copper nanoparticles (CuNP) were functionalized on the surface of MWCNT and used against *Methylobacterium* spp. The conjugates inhibited the growth of *Methylobacterium* spp. by different mecha-nisms (Seo et al. 2018, 2014). The MWCNT were functionalized with aminohydra-zide cross-linked chitosan and studied their effect on two Gram-positive and Gram-negative bacteria. The composite exhibited more potent bactericidal activity on Gram-positive than Gram-negative bacteria without being toxic to human cells (Mohamed and Abd El-Ghany 2018). In other study, novel trimellitic anhydride isothiocyanate-chitosan hydrogels fabricated with MWCNT and studied their effect on bacteria. The conjugate exhibited better bactericidal activity against Gram-positive than Gram-negative bacteria (Mohamed et al. 2019). Quercetin was immo-bilized on MWCNT/TiO$_2$ nanocomposite and studied their effect on *Bacillus subtilis* biofilm development. The quercetin conjugated nanocomposite decreased the surface-attached bacteria (Raie et al. 2018).

The nanocomposite of Fe-doped TiO2-MWCNT prepared and observed their higher photoinactivation of bacteria (Koli et al. 2016). The fluorine-co-doped TiO2-MWCNT nanocomposite prepared and observed higher antibacterial effect (Sangari et al. 2015). The oxygenated carbon nanotubes (o-CNT) were synthesized and uti-lized their peroxidase enzyme activity. The admirable enzyme activity of o-CNT exploited for bacterial infections treatment (Wang et al. 2018a, b). The MWCNT/epilson-polylysine synthesized and showed excellent antibacterial activity against *P. aeruginosa* and *S. aureus* (Zhou and Qi 2011). The MWCNT/porphyrin conju-gate prepared and observed that the conjugate deactivated *S. aureus* efficiently upon exposure with visible light (Banerjee et al. 2010).

3.5   Mechanisms of Action by Functionalized Carbon Nanotubes

The antibacterial activity of carbon-based nanomaterials is related with their various physicochemical properties such as average size, morphology, surface charge, and number of layers. Along with these physicochemical properties, antibacterial activ-ity also depends on surface-functionalized molecules such as surfactants, polymers, bioactive agents, and efficient bacterial targeting agents. Further, the efficiency of antibacterial activity can be increased by doping and decorating surface with vari-ous nanomaterials such as metal or semiconductor NP (Mocan et al. 2016). Many studies have reported different antibacterial mechanisms of action by functionalized carbon nanomaterials (Fig. 3.3).

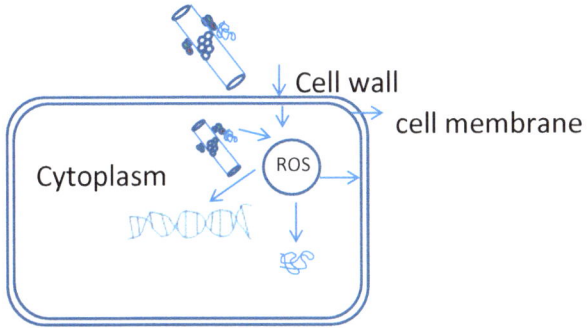

Fig. 3.3 Anti-bacterial mechanisms of functionalized carbon nanotubes.  = f-CNT, = Protiens, = Nucleoid

The antibacterial activity of f-SWCNT results from its physical mode of action, where f-SWCNT directly interacts and damages the outer membrane. Due to loss of membrane integrity, intracellular essential components were released and membrane potential disrupted. Some studies showed that piercing cell membrane and wrapping around microbes' cell walls or membranes by f-SWCNT depends on diameter and length, respectively, which induces membrane damage, loss of membrane potential, and leakage of cytoplasmic as well as nuclear components (Chen et al. 2013). Some experiments under controlled conditions suggested that physical puncture and release of intracellular components were the major reasons for antibacterial activity of f-SWCNT (Liu et al. 2009, 2010). Some studies also showed molecular-scale interaction plays important role in antibacterial activity of f-SWCNT other than physical contact and piercing membrane (Deokar et al. 2013).

The electrostatic interactions between bacterial membrane and surfactants followed by surfactants dispersed MWCNT penetration into membranes and surfactants cooperated toxicity are the main mechanisms in antibacterial activity of f-MWCNT (Khazaee et al. 2016). Some studies showed that f-MWCNT induces free radical (ROS) generation inside bacterial cells, which in turn disrupts the bacterial cell membrane (Parasuraman et al. 2019). Some other studies proved that f-MWCNT (Cu/MWCNT conjugate) inhibited the growth of bacterial cells both by physical damage of cell wall and production of reactive oxygen species. The copper ions released from conjugate inhibit biofilm formation or quorum sensing in *Methylobacterium* spp. (Seo et al. 2018).

3.6 Conclusion

The SWCNTand MWCNT were functionalized by many molecules or biologically active compounds through covalent or non-covalent bonding to improve their solubility and biocompatibility. The functionalized carbon nanotubes showed efficient bactericidal activity by mainly two types of antibacterial mechanisms such as directly by cell membrane physical damage and indirectly through reactive oxygen species (ROS) induction, which in turn affect cell viability. The functionalization proved higher antibacterial activity than their bare carbon nanotubes without being toxic to human cells. More studies still needed to examine antibacterial activity and biocompatibility of functionalized carbon nanotubes in in vivo models also. The detailed in vitro and in vivo studies on f-CNT help in the development of very efficient and safe alternative antibacterial agents.

References

Akhavan O, Abdolahad M, Abdi Y, Mohajerzadeh S (2011a) SNP within vertically aligned multi-wall carbon nanotubes with open tips for antibacterial purposes. J Mater Chem 21:387–393

Akhavan O, Azimirad R, Safa S (2011b) Functionalized carbon nanotubes in ZnO thin films for photoinactivation of bacteria. Mater Chem Phys 130(1–2):598–602

Alpatova AL, Shan W, Babica P, Upham BL, Rogensues AR, Masten SJ, Drown E, Mohanty AK, Alocilja EC, Tarabara VV (2010) Single-walled carbon nanotubes dispersed in aqueous media via non-covalent functionalization: effect of dispersant on the stability, cytotoxicity, and epigenetic toxicity of nanotube suspensions. Water Res 44(2):505–520

Amiri A, Zardini HZ, Shanbedi M, Maghrebi M, Baniadam MBT (2012) Efficient method for functionalization of carbon nanotubes by lysine and improved antimicrobial activity and water-dispersion. Mater Lett 72:153–156

Arias LR, Yang L (2009) Inactivation of bacterial pathogens by carbon nanotubes in suspensions. Langmuir 25(5):3003–3012

Aslan S, Loebick CZ, Kang S, Elimelech M, Pfefferle LD, Van Tassel PR (2010) Antimicrobial biomaterials based on carbon nanotubes dispersed in poly (lactic-co-glycolic acid). Nanoscale 2(9):1789–1794

Assali M, Zaid AN, Abdallah F, Almasri M, Khayyat R (2017) Single-walled carbon nanotubes-ciprofloxacin nanoantibiotic: strategy to improve ciprofloxacin antibacterial activity. Int J Nanomedicine 12:6647–6659

Azizian J, Hekmati M, Dadras OG (2014) Functionalization of carboxylated multiwall nanotubes with dapsone derivatives and study of their antibacterial activities against E. coli and S. aureus. Orient J Chem 30:667–673

Bai YPI, Lee SJ, Bae TS, Watari F, Uo M, Lee MH (2011) Aqueous dispersion of surfactant-modified multiwalled carbon nanotubes and their application as an antibacterial agent. Carbon 49(11):3663–3671

Bai Y, Park IS, Lee SJ, Wen PS, Bae TS, Lee MH (2012) Effect of AOT-assisted multi-walled carbon nanotubes on antibacterial activity. Colloids Surf B Biointerfaces 89:101–107

Bai Y, Gao J, Wang C, Zhang R, Ma W (2016) Mixed surfactant solutions for the dispersion of multiwalled carbon nanotubes and the study of their antibacterial activity. J Nanosci Nanotechnol 16(3):2239–2245

Banerjee I, Mondal D, Martin J, Kane RS (2010) Photoactivated antimicrobial activity of carbon nanotube-porphyrin conjugates. Langmuir 26(22):17369–17374

Bellingeri R, Mulko L, Molina M, Picco N, Alustiza F, Grosso C, Vivas A, Acevedo DF, Cesar A, Barbero CA (2018) Nanocomposites based on pH-sensitive hydrogels and chitosan decorated carbon nanotubes with antibacterial properties. Mater Sci Eng C 90:461–467

Chaudhari AA, Jasper SL, Dosunmu E, Miller ME, Arnold RD, Singh SR, Pillai S (2015) Novel pegylated silver coated carbon nanotubes kill Salmonella but they are non-toxic to eukaryotic cells. J Nanobiotechnol 13:23

Chaudhari AA, Ashmore D, Nath SD, Kate K, Dennis V, Singh SR, Owen DR, Palazzo C, Arnold RD, Miller ME, Pillai SR (2016) A novel covalent approach to bio-conjugate silver coated single walled carbon nanotubes with antimicrobial peptide. J Nanobiotechnol 14(1):58

Chaudhari AA, Joshi S, Vig K, Sahu R, Dixit S, Baganizi R, Dennis VA, Singh SR, Pillai S (2019) A three-dimensional human skin model to evaluate the inhibition of Staphylococcus aureus by antimicrobial peptide-functionalized silver carbon nanotubes. J Biomater Appl 33(7):924–934

Chen H, Wang B, Gao D, Guan M, Zheng L, Ouyang H, Chai Z, Zhao Y, Feng W (2013) Broad-spectrum antibacterial activity of carbon nanotubes to human gut bacteria. Small 9(16):2735–2746

Cortes P, Deng S, Smith GB (2014) The toxic effects of single wall carbon nanotubes on E. coli and a spore-forming Bacillus species. Nanosci Nanotechnol Lett 6(1):26–30

Deokar AR, Lin LY, Chang CC, Ling YC (2013) Single-walled carbon nanotube coated antibacterial paper: preparation and mechanistic study. J Mater Chem B 1:2639–2646

Dong L, Henderson A, Field C (2012) Antimicrobial activity of single-walled carbon nanotubes suspended in different surfactants. J Nanotechnol 2012:1–7

Dong X, Awak AM, Wang P, Sun YP, Yang L (2018) Carbon dot incorporated multi-walled carbon nanotube coated filters for bacterial removal and inactivation. RSC Adv 8:8292–8301

Dumortier H, Lacotte S, Pastorin G, Marega R, Wu W, Bonifazi D, Briand JP, Prato M, Muller S, Bianco A (2006) Functionalized carbon nanotubes are non-cytotoxic and preserve the functionality of primary immune cells. Nano Lett 6(7):1522–1528

Foo ME, Anbu P, Gopinath SCB, Lakshmipriya T, Lee CG, Yun HS, MNA U, ARW Y (2018) Antimicrobial activity of functionalized single-walled carbon nanotube with herbal extract of Hempedubumi. Wiley Online Library 50(3):354–361

Gu L, Elkin T, Jiang X, Li H, Lin Y, Qu L, Tzeng TR, Joseph R, Sun YP (2005) Single-walled carbon nanotubes displaying multivalent ligands for capturing pathogens. Chem Commun (Camb) 7:874–876

Iijima S (1991) Helical microtubules of graphitic carbon. Nature 354:56–58

Jia X, Wei F (2017) Advances in production and applications of carbon nanotubes. Top Curr Chem (Cham) 375(1):18

Jung JH, Hwang GB, Lee JE, Bae GN (2011) Preparation of airborne Ag/CNT hybrid nanoparticles using an aerosol process and their application to antimicrobial air filtration. Langmuir 27(16):10256–10264

Karimi M, Solati N, Amiri M, Mirshekari H, Mohamed E, Taheri M, Hashemkhani M, Saeidi A, Estiar MA, Kiani P, Ghasemi A, Basri SM, Aref AR, Hamblin MR (2015) Carbon nanotubes part I: preparation of a novel and versatile drug-delivery vehicle. Expert Opin Drug Deliv 12(7):1071–1087

Khazaee M, Ye D, Majumder A, Baraban L, Opitz J, Cuniberti G (2016) Non-covalent modified multi-walled carbon nanotubes: dispersion capabilities and interactions with bacteria. Biomed Phys Eng Express 2:005008

Koli VB, Delekar SD, Pawar SH (2016) Photoinactivation of bacteria by using Fe-doped TiO_2-MWCNTs nanocomposites. J Mater Sci Mater Med 27(12):177

Kosaka T, Ohgami A, Nakamura T, Ohana T, Ishihara M (2009 Sep) The preparation of Ag nanoparticle-modified single-walled carbon nanotubes and their antibacterial activity. Biocontrol Sci 14(3):133–138

Levi-Polyachenko N, Young C, MacNeill C, Braden A, Argenta L, Reid S (2014) Eradicating group A streptococcus bacteria and biofilms using functionalised multi-wall carbon nanotubes. Int J Hyperth 30(7):490–501

Li Z, Fan L, Zhang T, Li K (2011) Facile synthesis of Ag nanoparticles supported on MWCNTs with favorable stability and their bactericidal properties. J Hazard Mater 187(1–3):466–472

Liu S, Wei L, Hao L, Fang N, Chang MW, Xu R, Yang Y, Chen Y (2009) Sharper and faster "nano darts" kill more bacteria: a study of antibacterial activity of individually dispersed pristine single-walled carbon nanotube. ACS Nano 3(12):3891–3902

Liu S, Ng AK, Xu R, Wei J, Tan CM, Yang Y, Chen Y (2010) Antibacterial action of dispersed single-walled carbon nanotubes on Escherichia coli and Bacillus subtilis investigated by atomic force microscopy. Nanoscale 2(12):2744–2750

Liu C, Shi H, Yang H, Yan S, Luan S, Li Y, Teng M, Khan AF, Yin J (2017) Fabrication of antibacterial electrospun nanofibers with vancomycin-carbon nanotube via ultrasonication assistance. Mater Des 120:128–1354

Luo PG, Wang H, Gu L, Lu F, Lin Y, Christensen KA, Yang ST, Sun YP (2009) Selective interactions of sugar-functionalized single-walled carbon nanotubes with Bacillus spores. ACS Nano 3(12):3909–3916

Mallakpour S, Soltanian S (2016) Surface functionalization of carbon nanotubes: fabrication and applications. RSC Adv 6:109916–109935

Mehra NK, Palakurthi S (2016) Interaction between carbon nanotubes and bioactives: a drug delivery perspective. Drug Discov Today 21(4):585–597

Merli D, Ugonino M, Profumo A, Fagnoni M, Quartarone E, Mustarelli P, Visai L, Grandi MS, Galinetto P, Canton P (2011) Increasing the antibacterial effect of lysozyme by immobilization on multi-walled carbon nanotubes. J Nanosci Nanotechnol 11(4):3100–3106

Mocan L, Ilie I, Tabaran FA, Iancu C, Mosteanu O, Pop T, Zdrehus C, Bartos D, Mocan T, Matea C (2016) Selective laser ablation of methicillin-resistant Staphylococcus aureus with IgG functionalized multi-walled carbon nanotubes. J Biomed Nanotechnol 12(4):781–788

Mohamed NA, Abd El-Ghany NA (2018) Novel aminohydrazide cross-linked chitosan filled with multi-walled carbon nanotubes as antimicrobial agents. Int J Biol Macromol 115:651–662

Mohamed NA, Al-Harby NF, Almarshed MS (2019) Synthesis and characterization of novel trimellitic anhydride isothiocyanate-cross linked chitosan hydrogels modified with multi-walled carbon nanotubes for enhancement of antimicrobial activity. Int J Biol Macromol 132:416–428

Mohan R, Shanmugharaj AM, Sung Hun R (2011) An efficient growth of silver and copper nanoparticles on multiwalled carbon nanotube with enhanced antimicrobial activity. J Biomed Mater Res B Appl Biomater 96(1):119–126

Mourdikoudis S, Pallares RM, Thanh NTK (2018) Characterization techniques for nanoparticles: comparison and complementarity upon studying nanoparticle properties. Nanoscale 10(27):12871–12934

Ondera TJ, Hamme AT II (2015) Magnetic-optical nanohybrids for targeted detection, separation, and photothermal ablation of drug-resistant pathogens. Analyst 140(23):7902–7911

Parasuraman P, Anju VT, Sruthil Lal SB, Sharan A, Busi S, Kaviyarasu K, Arshad M, Dawoud TMS, Syed A (2019) Synthesis and antimicrobial photodynamic effect of methylene blue conjugated carbon nanotubes on E. coli and S. aureus. Photochem Photobiol Sci 18(2):563–576

Park SB, Steadman CS, Chaudhari AA, Pillai SR, Singh SR, Ryan PL, Willard ST, Feugang JM (2018) Proteomic analysis of antimicrobial effects of pegylated silver coated carbon nanotubes in Salmonella enterica serovar Typhimurium. J Nanobiotechnol 16(1):31

Pasquini LM, Hashmi SM, Sommer TJ, Elimelech M, Zimmerman JB (2012) Impact of surface functionalization on bacterial cytotoxicity of single-walled carbon nanotubes. Environ Sci Technol 46:6297–6305

Qi X, Poernomo G, Wang K, Chen Y, Chan-Park MB, Xu R, Chang MW (2011) Covalent immobilization of nisin on multi-walled carbon nanotubes: superior antimicrobial and anti-biofilm properties. Nanoscale 3(4):1874–1880

Raie DS, Mhatre E, El-Desouki DS, Labena A, El-Ghannam G, Farahat LA, Youssef T, Fritzsche W, Kovács ÁT (2018) Effect of novel quercetin titanium dioxide-decorated multi-walled carbon nanotubes nanocomposite on Bacillus subtilis biofilm development. Materials (Basel) 11(1):E157

Rangari VK, Mohammad GM, Jeelani S, Hundley A, Vig K, Singh SR, Pillai S (2010) Synthesis of Ag/CNT hybrid nanoparticles and fabrication of their nylon-6 polymer nanocomposite fibers for antimicrobial applications. Nanotechnology 21(9):095102

Rashid MH-O, Ralph SF (2017) Carbon nanotube membranes: synthesis, properties, and future filtration applications. Nanomaterials (Basel) 7(5):99

Sah U, Sharma K, Chaudhri N, Sankar M, Gopinath P (2018) Antimicrobial photodynamic therapy: single-walled carbon nanotube (SWCNT)-porphyrin conjugate for visible light mediated inactivation of Staphylococcus aureus. Colloids Surf B Biointerfaces 162:108–117

Sangari M, Umadevi M, Mayandi J, Pinheiro JP (2015) Photocatalytic degradation and antimicrobial applications of F-doped MWCNTs/TiO2 composites. Spectrochim Acta A Mol Biomol Spectrosc 139:290–295

Seo Y, Hwang J, Kim J, Jeong Y, Hwang MP, Choi J (2014) Antibacterial activity and cytotoxicity of multi-walled carbon nanotubes decorated with silver nanoparticles. Int J Nanomedicine 9:4621–4629

Seo Y, Hwang J, Lee E, Kim YJ, Lee K, Park C, Choi Y, Jeon H, Choi J (2018) Engineering copper nanoparticles synthesized on the surface of carbon nanotubes for anti-microbial and anti-biofilm applications. Nanoscale 10(33):15529–15544

Tiraferri A, Vecitis CD, Elimelech M (2011) Covalent binding of single-walled carbon nanotubes to polyamide membranes for antimicrobial surface properties. ACS Appl Mater Interfaces 3(8):2869–2877

Venkatesan J, Jayakumar R, Mohandas A, Bhatnagar I, Kim S-K (2014) Antimicrobial activity of chitosan-carbon nanotube hydrogels. Materials (Basel) 7(5):3946–3955

Wang H, Gu L, Lin Y, Lu F, Meziani MJ, Luo PG, Wang W, Cao L, Sun YP (2006) Unique aggregation of anthrax (Bacillus anthracis) spores by sugar-coated single-walled carbon nanotubes. J Am Chem Soc 128(41):13364–13365

Wang S, Mao B, Wu M, Liang J, Deng L (2018a) Influence of aptamer-targeted antibiofilm agents for treatment of Pseudomonas aeruginosa biofilms. Antonie Van Leeuwenhoek 111(2):199–208

Wang H, Li P, Yu D, Zhang Y, Wang Z, Liu C, Qiu H, Liu Z, Ren J, Qu X (2018b) Unraveling the enzymatic activity of oxygenated carbon nanotubes and their application in the treatment of bacterial infections. Nano Lett 18(6):3344–3351

Xin Q, Shah H, Nawaz A, Xie W, Akram MZ, Batool A, Tian L, Jan SU, Boddula R, Guo B, Liu Q, Gong JR (2018) Antibacterial carbon-based nanomaterials. Adv Mater 31(45):e1804838

Xu ZQ, Flavin MT, Flavin J (2014 Feb) Combating multidrug-resistant gram-negative bacterial infections. Expert Opin Investig Drugs 23(2):163–182

Yadav T, Mungray AA, Mungray AK (2015) Dispersion of multiwalled carbon nanotubes in Acacia extract and its utility as an anti-microbial agent. RSC Adv 5:103956–103963

Yallappa S, Manjanna J, Dhananjaya BL, Vishwanatha U, Ravishankar B, Gururaj H, Niranjana P, Hungund BS (2016) Phytochemically functionalized Cu and Ag nanoparticles embedded in MWCNTs for enhanced antimicrobial and anticancer properties. Nano-Micro Lett 8(2):120–130

Zardini HZ, Amiri A, Shanbedi M, Maghrebi M, Baniadam M (2012) Enhanced antibacterial activity of amino acids-functionalized multi walled carbon nanotubes by a simple method. Colloids Surf B Biointerfaces 92:196–202

Zardini HZ, Davarpanah M, Shanbedi M, Amiri A, Maghrebi M, Ebrahimi L (2014 Jun) Microbial toxicity of ethanolamines--multiwalled carbon nanotubes. J Biomed Mater Res A 102(6):1774–1781

Zhou J, Qi X (2011) Multi-walled carbon nanotubes/epilson-polylysine nanocomposite with enhanced antibacterial activity. Lett Appl Microbiol 52(1):76–83

Zhu A, Liu HK, Long F, Su E, Klibanov AM (2015) Inactivation of bacteria by electric current in the presence of carbon nanotubes embedded within a polymeric membrane. Appl Biochem Biotechnol 175(2):666–676

Zhu L, Rakesh KP, Xu M, Dong Y (2019) Ceramic-based composite membrane with a porous network surface featuring a highly stable flux for drinking water purification. Membranes (Basel) 9(1):5

Chapter 4
Nanoparticle-Based Antimicrobial Coating on Medical Implants

Birru Bhaskar, Jintu Dutta, Shalini, Ponnala Vimal Mosahari, Jonjyoti Kalita, Papori Buragohain, and Utpal Bora

Contents

B. Bhaskar (✉)
Brien Holden Eye Research Centre, L V Prasad Eye Institute, Banjara Hills, Hyderabad, India
e-mail: bbhaskar@lvpei.org

J. Kalita · P. Buragohain · U. Bora
Department of Biosciences and Bioengineering, Indian Institute of Technology Guwahati, Guwahati, Assam, India

J. Dutta · P. V. Mosahari
Center for the Environment, Indian Institute of Technology Guwahati, Guwahati, Assam, India

Shalini
Department of Chemical Engineering, National Institute of Technology Warangal, Warangal, Telangana, India

© Springer Nature Switzerland AG 2020   79
R. Prasad et al. (eds.), *Nanostructures for Antimicrobial and Antibiofilm Applications*, Nanotechnology in the Life Sciences,
https://doi.org/10.1007/978-3-030-40337-9_4

Abstract Infections after implantation are the major concern in the field of clinical surgery. The microbial adhesion on the surface of implant favors for colonization and biofilm formation. The host immune system and systemic antibiotics could have not inhibited the microbial adhesion and growth, which resulted in sepsis. Thus, surface modification of implants has gained attention to prevent the bacterial adhesion on the surface of the implant. Various coating methods have been developed, namely, passive surface coating, active surface coating, and local carriers or coatings, with the aim of preventing microbial adhesion and biofilm formation. Nanoparticle coating is one potential alternative for successful clinical outcomes. This book chapter aims to give a brief report on the need of coating on the implants against the bacterial colonization on the surface of the implant and especially the role of nanoparticle-based antimicrobial coated medical devices in the prevention of sepsis.

Keywords Antimicrobial · Coating · Implant · Microbial contamination · Microbial adhesion · Osteointegration

4.1 Introduction

The first tooth implant was successfully implanted and proved that it has osteointegrationability in the year 1998 (Crubzy et al. 1998). The advances in engineering technology have produced versatility of medical implants, which drew the attention of clinical biology to set up aseptic surgical procedures and anesthetics for pain relief during the surgery. Various medical implants include orthopedic, dental, and cardiac stents, and valves have been widely used for raising the quality life standards. Initially, the larger interest was given to develop material to fit the mechanical strength of human tissue and microenvironment of tissue-specific region, and the successful dental and orthopedic implants by considering the aforesaid factors were shown the best results in retaining the function of native tissue after implantation (Hickok et al. 2018). The aseptic surgical procedures prior to the implantation have reduced the surgical site infections (SSI), and it is accounted for the reduction of health-care cost. However, prosthetic infections after surgery were found to be considered most significant to avoid the multiple surgery revisions and prolonged antibiotic treatment, which contributes for reducing the cost on health-care system (Hexter et al. 2018). The increase in medical implants usage drastically improved the quality of health care and life time. Prior to the prosthesis, prevention procedures like antibiotic prophylaxis have reduced the SSI drastically and make easier to proceed for surgery in aseptic conditions (Romanò et al. 2015). Besides, implant-associated infections also reported mostly in revision surgery. The integration of material science and biology, improved diagnostics, preventive measures for SSI, and innovative strategies for inhibition of bacterial adhesion on the surface of implants are the essential factors to be considered to combat with SSI.

The implants are designed and developed to integrate with human tissue by adsorption of proteins and biomolecules from the body fluids. The bacterial fouling on the implant surface also occurs after prosthesis; it causes the infections by biofilm formation. The immune system and antibiotics also can't give the protection against the prosthesis infection because of bacterial plaque resistance (Zeng et al. 2018). The prevalence of prosthetic joint infection (PJI) has enormously increased and raised more complications. The implant removal would be imperative, which leads to high mortality and morbidity with accelerated health care and social cost. Implant infections depend on various factors including the type of the implant material, surgical preventions and technique, specific to microorganism and tissue, and antibacterial prophylaxis (Romanò et al. 2015). It is quite challenging to execute the surgery completely in a disinfected environment and mostly operation theaters usually contaminated during surgery. The improved host immune response, sequnetial antibacterial prophylaxis and preventive strategy for inhibition of bacterial adhesion on the implants are the possible ways to get protection from the infections. Though well-developed strategies in terms of surgical techniques and procedure to control the SSIs, septic cases were reported in high-risk patient after prosthesis (Illingworth et al. 2013). In orthopedic surgery, several implants loaded with antibiotics can be used as a bone graft. Implant material, size, shape, and host tissue type play a vital role in successful implantation.

In the case of dental and orthopedic implants, surface modification is the key approach to develop a protein adsorption strategy on the implant, which accounts for implant integration with bone. During this process of integration, bacterial colonization could be easier due to the adsorbed proteins on the implant surface. The immune response will act immediately against the bacterial colonized implant. However, there are reports on the biofilm formation on implants. Thus, the antimicrobial coating on the implants is the potential approach to inhibit adhesion of bacteria and its colonization and proliferation around the tissue.

Given this outline, this chapter aims to discuss on the antimicrobial coatings, importance of nanoparticle coating over other coating approaches, implant-related infections, geno and cytotoxic effects of nanoparticles, nanoparticle-coated implants used for grafting, and global market trend on nanoparticle-coated implants for preventing PJIs.

4.2 Implant-Related Infections

Microbial contamination during operation leads to prosthetic infection. The time period for the occurrence of infection after implantation is considered for classification of orthopedic implant infections. Postoperative infection means infection occurs after completion of surgery which is less than 3 months period. Infection occurs between 3 and 24 months after operation called as delayed infection. After 24 months of surgery, also infections may occur by hematogenous spread, which is called as late infection (Arciola et al. 2018). It was reported that early infections usually apparent in 1 month post-surgery, and chronic infection could transform into acute haematogeneous infection within 3 weeks after surgery. The infection depends on the type of microorganism, implant site, and time period (Panteli and Giannoudis 2016). Virulent microorganisms cause early infection, whereas low virulence microorganisms cause for delayed infections. Bacteria cause the late hematogeneous infections, which accounts for urinary tract, skin, dental, and respiratory infections (Honkanen et al. 2019).

Mostly the scientific reports were confirmed that staphylococci cause orthopedic implant infections. However, the reported literature identified that different species of the same genus family, i.e., staphylococci caused orthopedic implantation in the USA and Europe. The prevalence of *S. epidermis* and *S. aureus* was reported in Europe and the USA, respectively. The pathogen type and its prevalence, antibiotic resistance, aseptic environment, and prophylactic measurements are the significant factors in causing the implant-related infections. The host immune system usually responds to the causative agents; however, microorganisms have developed the antibiotic resistance which led to infections. It was reported that higher antibiotic resistance has developed in *S. aureus* and caused early orthopedic implant infections by methicillin-resistant*S. aureus*. The delayed and late infections are caused by methicillin-sensitive *S. aureus*.

4.3 Clinical Aspect of Antimicrobial Coating

Microorganisms were adhered to the implant surface and form a protective layer against the host immune system and antibiotics called as a biofilm. The antibiotic treatment at the right time of prior and post surgery and the sustained, localized delivery of antibiotics at the surgery site are pivotal factors in preventing the biofilm formation. (Romanò et al. 2015). After implantation, the colonization of the host cell on the surface only can avoid the bacterial adhesion on the surface. This is the best way to overcome a chance of implant infection in early stages. However, it is essential to design and develop the implant in such a way to improve the host cell adhesion and inhibit the bacterial attachment.

Bacteria have the ability to adopt and colonize in adverse conditions, and adhesion and survival of bacteria on implant surface either natural or synthetic are quite easy. The implant material surface chemistry plays an important role in the bacteria adhesion, especially electrostatic charge, roughness, and hydrophobicity. Generally, the implant surface is possessing protein layer which has potential ligand binding receptors for bacterial adhesion, which quite commonly occurs immediately after implantation in the host body (Guo et al. 2017; Busscher and Mei 2012). The adhesion of bacteria on the implant surface can be done either reversible phase or irreversible phase. In the case of irreversible phase, adhesion is usually assisted by cellular or molecular interactions, whereas reversible phase refers to nonspecific reactions between bacterial cells and implant surface.

Mostly, the physical contact of medical professionals and/or patient during the surgery or migration of bacteria to the surgery site causes surgical site infections, wherein the development of biofilm mediates the multidrug resistance to the given antibiotics. The formation of biofilm after bacterial adhesion takes place in two steps. Initially, the bacteria multiply after irreversible adhesion on the implant surface, and gelatinous insoluble exopolymer matrix formation occurs which aids for microcolony formation. In a later stage, the formed microcolonies matured further, and intensification of the microbial population drives for mature biofilm formation, which is highly resistant to systematic antibiotics (Percival et al. 2015). Through this, it can be confirmed that initial general bacterial adhesion and colonization occur which does not cause infection, and the formation of insoluble matrix only leads to colonization of pathogenic bacteria and eventually forms a biofilm. The process of bacterial adhesion and biofilm formation on the implant surface will occur within 24 h. Thus, it is essential to have antibacterial nature on the implant surface to resist bacterial adhesion and colonization, and also the host immune system will take advantage of bacterial cells for adhesion and colonization. This is the most important strategy in implantation irrespective of immunity of the patient and also in the revision of surgery. The research inclined toward the development of antibacterial coating on implants to resist the bacterial adhesion. At the same time, this coating should facilitate for host tissue integration; cost of implant should be affordable and also have the potential of antibacterial functionality in the long run. Interestingly, there is a great need of resisting the bacterial adhesion on the implant surface within 3 h of surgery; mostly there is a greater possibility of occurring bacterial infection after immediate surgery within 3–5 h (Holá et al. 2006). The protection from bacteria at the time and after surgery is very crucial which decides the success of implantation.

4.4 Antimicrobial Coating: Classification

The antimicrobial coating on the implant can be done in different ways. However, so far a universal coating method has not been standardized. Currently, significant research has been taken place on antimicrobial coating technologies, and regulatory principles for clinical use are the prime concern of research in this field. So far, the antimicrobial coatings were classified into three groups:

1. Passive surface coating
2. Active surface coating
3. Active-passive combined surface coating

4.4.1 Passive Surface Coating/Modification (PSM)

The implant surface modification through physical or chemical means does not permit the bacterial adhesion and colonization. Surface chemistry, roughness, hydrophilic nature, surface potential or energy, and material conductivity are the important factors which play a vital role in adhesion of bacteria; later on, biofilm formation takes place on the implant surface. This approach is cost-effective, very easy to modify the surface through physical or chemical means, and easy for large-scale production. Surface modification aimed to provide good interaction of implant with host tissue and also to avoid the pathogen adhesion. Titanium has been widely accepted implant for orthopedic and dental applications, and surface modification of titanium and its alloys mostly opted to increase roughness. This provides a large surface area compared over the smooth surface. Cell adhesion is usually higher due to the increased surface area and colonization of host cell has blocked the microbial adhesion on the surface (Bosco et al. 2012). Polyethylene glycol, polyethylene oxide, and polyurethane were successfully used for passive surface modification. However, this passive surface coating is not effective and limited to certain bacteria, and it was noticed that adhesion on the implant surface is also varied on the bacterial species (Rodrigues 2011). It was observed that super hydrophobic and hydrophobic treatment for surface modification did not allow bacteria to adhere to the surface of the implant. The target of bacteria and protein interaction by studying the protein adsorption is a highly potential approach to protect from implant-related infections. Different techniques have been developed for passive surface modification including acid etching and grit blasting. Implant surface treated with acids like sulfuric, nitric, and hydrofluoric acids is called acid-etching treatment. Usually, a variety of particles fixed on the implant surface through the bombardment of particles include silica, titanium, alumina, and hydroxyapatite (Bosco et al. 2012). Nanotechnology has been applied for topographical surface modification at the nanoscale level, which has significantly affected biomolecule, cell, and ion interaction. A variety of nanoparticles have been synthesized and characterized for various application; the range of particles is 1–100 nm. Surface modification using nanotechnology approach is the emerging concept in the field of antimicrobial coating and proved that these nanoparticle-based antimicrobial medical implants are highly protected from pathogenic bacteria and also enhanced the living cells/tissue interaction with the implant surface. Mostly, nanocoatings were restricted to a thickness below 100 nm. The surface modification using nanoparticle coating is advantageous over other conventional approach because it will not alter the mechanical strength of the material and also improved the osteointegration. Thus, these implants were highly recommended for bone and dental applications (Mendonça et al. 2008).

4.4.2 Active Surface Coating/Modification

The active surface coating can be done using organic, inorganic, antiseptic, and antibiotic compounds. This coated compound actively protects from pathogens and drives the material from passive to an active surface coated implant. This approach is highly effective in the short term, and effect in the long term has not been fully elucidated. The active compounds may elute the drugs or kill bacteria through direct contact. This coating was classified into two categories based on the durability, namely, non-degradable and degradable coatings. The antibacterial surface can be achieved by coating with organic (proteins, peptides, enzymes), inorganic (calcium phosphate, hydroxyapatite, silica), and metals (copper, zinc, silver). The active surface with these compounds promotes cell adhesion on the surface and helps for host tissue integration. Several signaling molecules also have been tested to improve the cell attachment on the implant surface, thereby affecting the bacterial adhesion. Silver is a widely used metal for medical implant coating. The metabolism of bacteria is downregulated and alters the permeability of the cell wall due to the silver coating on the surface. However, silver used as a coating material also has limitations due to its toxic nature. The silver ions can be optimally used for the surface coating to make them nontoxic without altering its potential of antimicrobial activity. In this way, functional silver-coated implants can be developed for real-time applications.

4.4.3 Active-Passive Combined Surface Coating

Bone is a composite of organic and inorganic substances. There is a great need to develop a combined surface coated implant for orthopedic application. Biomolecules can be delivered systematically using nanotechnology approach to provide good interaction between the implant surface and living cells/tissue. The research was focused to develop variety for combined coated implants for different tissue engineering applications. The collagen and calcium phosphate nanocomposites have been evaluated for the bone regeneration and osteointegration. These composites were able to mimic native composition and structure of bone, which facilitates for higher cell adhesion and proliferation, subseqently helps in the integration with the host bone tissue (Stock 2015). Growth factors have been loaded in a variety of implant material including collagen, hydroxyapatite, polymers, and calcium phosphate. The combined coating approach for surface modification is highly supportive of cell adhesion. Titanium-collagen composite treated with growth factors showed higher osteointegration compared over untreated implant surface (Bosco et al. 2012).

It was reported that many implants caused infection after surgery; antibiotics can be loaded on implant prior to implantation. Hydrogels preloaded in such a way to deliver the antibiotics to protect from infections, and recently various hydrogels have been introduced into the global market (Pitarresi et al. 2013). Hyaluronan and polylactide composites were coated with antibacterial, and composition was optimized to degrade completely after implantation within 72 h. During this degradation period, antibacterial agents were systematically released including amino glycosides, glycol peptides, and fluoroquinolones (Romanò et al. 2015).

4.5   Mechanism of Antimicrobial Action of Nanoparticles

The increase of antibiotic resistance of pathogenic bacteria becomes a threat to the human population which leads to the investigation for novel antimicrobial agents. In recent times, nanoscale materials have accelerated the research for finding novel antimicrobial agents due to their exclusive chemical and physical properties (Morones et al. 2005; Kim et al. 2008; Aziz et al. 2015, 2016). The branch of nanotechnology is dealing with the materials at nanoscale level and application of these materials in different scientific fields (Albrecht et al. 2006). Nanotechnology term was defined by Professor Norio Taniguchi of Tokyo Science University, 1974, to demonstrate the accurate manufacturing of nanoscale level materials (Taniguchi 1983). In nanotechnology, the synthesis of nanoparticles (NPs) is an important part and it has various ways to synthesize. In present time, interdisciplinary subjects are getting more attention as it has the ability to solve the real-time problems. The bio-nanotechnology is one of those subjects where biotechnology is integrated with nanotechnology for developing eco-friendly technology and biosynthesis of nanomaterials. The use of nanoparticles is gaining momentum in the present century as they possess defined chemical, optical, and mechanical properties. Nanoparticles and nanoparticles-derived materials have obtained more attention for their potential antimicrobial effects. Metallic nanoparticles are showing potential antibacterial properties because of their large surface area to volume ratio (Gong et al. 2007; Prasad et al. 2016). Different metal nanoparticles are documented as good antimicrobial agents such as silver (Ag), silver oxide (Ag_2O), titanium dioxide (TiO_2), silicon (Si), copper oxide (CuO), zinc oxide (ZnO), Au, calcium oxide (CaO), and magnesium oxide (MgO) (Dizaj et al. 2014). The antimicrobial mechanism of nanoparticles is illustrated in Fig. 4.1. Nanoparticles generate reactive oxygen species (ROS) which enter into the bacterial cell to cause disruption of cell. Nanoparticles would enter into cell in different ways and then damage DNA and denaturation of proteins and enzymes lead to cell death (Prasad 2019).

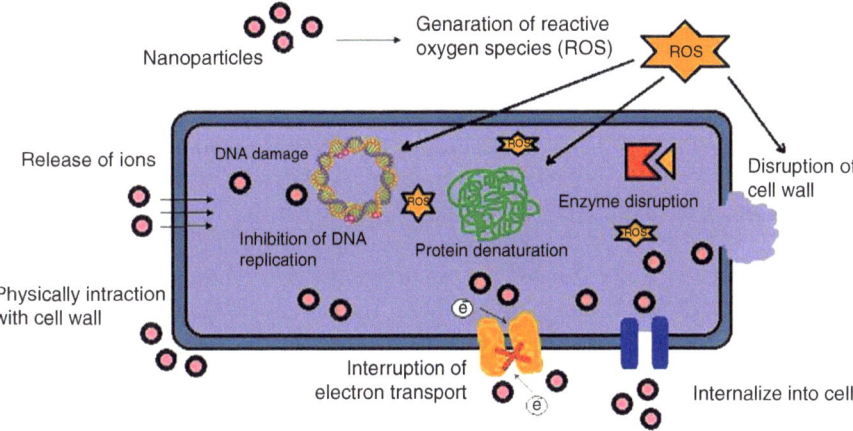

Fig. 4.1 Antimicrobial mechanism of nanoparticles. Reproduced with permission from Dizaj et al. (2014). Elsevier © 2014

4.5.1   Silver Nanoparticles

Silver nanoparticles (Ag-NPs) are considered as very efficient antimicrobial agent due to their large surface area, which offers better attachment with microorganisms. These nanoparticles can get attached to the cell membrane and also penetrate inside the bacteria. The Ag-NPs interact with the sulfur-containing protein in the membrane of bacteria also with the phosphorus-containing compounds like DNA. The nanoparticles usually attack the respiratory chain and cell division of bacteria which leads to cell death. Moreover, bactericidal activity was increased due to the release of silver ions inside the bacterial cells by Ag-NPs (Feng et al. 2000; Sondi and Salopek-sondi 2004; Morones et al. 2005; Aziz et al. 2014, 2015, 2016).

Ag-NPs are also used for coating medical devices and surgical masks (Li et al. 2006). Coating Ag-NPs on medical devices can be used for promising antimicrobial activity. The advantage of impregnation of medical devices with Ag-NPs is that it protects both outer and inner surfaces of devices and there is continuous release of silver ions providing antimicrobial activity (Darouiche and Raad 1999; Wilcox et al. 1998). Though Ag-NPs are used for coating different medical devices, sometimes it is proved to be unsatisfactory in clinical tests. This might be caused due to the inactivation of metallic silver when it comes in contact with blood plasma or the lack of robustness of nanoparticles coating. However, metallic silver has failed to induce the antimicrobial activity in some cases (Riley et al. 1995; Furno et al. 2004). Qasim et al. (2018) reported the antimicrobial effect of fabricated Ag-NPs, which is encapsulated in poly-N-isopropyl acrylamide. However, they found that the antibacterial efficiency of Ag-NPs reduces after washing (Qasim et al. 2018).

4.5.2 Zinc Oxide Nanoparticles

The antimicrobial activities of zinc oxide nanoparticles (ZnO-NPs) are generating a great scientific and technological interest worldwide. It has received a significant attention globally. The exact mechanism of ZnO-NPs for antimicrobial activity is not well known, but these nanoparticles are observed to release some antimicrobial ions (Kasemets et al. 2009) or formation of ROS which attributed to its antimicrobial action (Jalal et al. 2010). Moreover, it was reported that ZnO-NPs can increase specific surface area to enhance particle surface reactivity for antimicrobial action (Sirelkhatim et al. 2015; Bhuyan et al. 2015). The nanoparticles interact with microorganisms and damage the bacterial cell integrity due to the formation of ROS (Yang Zhang et al. 2009; Bhuyan et al. 2015). The release of Zn^{2+} ions in the medium from ZnO-NPs lead to the inclusion of antimicrobial activity (Kasemets et al. 2009). He reported the antimicrobial effect of ZnO-NPs against *Saccharomyces cerevisiae*. However, solubility of the metal oxides such as ZnO and Al_2O_3 depends upon the concentration of that particular metal oxide in the solution (Wang et al. 2009). Therefore, it always depends upon the Zn^{2+} ion concentration in the media for antimicrobial action. Reddy et al. (2007) studied the antimicrobial activity of ZnO-NPs for *Escherichia coli* and *Staphylococcus aureus*. They reported the complete growth inhibition of *E. coli* and *S. aureus* at concentrations of ≥ 3.4 mM and ≥ 1 mM, respectively. This observation clearly demonstrated the differences in toxicity thresholds of ZnO-NPs. Interestingly, they also observed that colony-forming units (CFU) count of *E. coli* cells increased when ZnO concentration was low as 1 mM but *S. aureus* showed sensitivity when it exposes to the same concentration. A similar result of ZnO-NPs was also observed by Padmavathy and Vijayaraghavan (2008) and showed that the antimicrobial effect of ZnO-NPs against *E. coli* is significantly less when there is low concentration (0.01–1 mM) of nanoparticles. While metals and metallic oxides are known to be toxic at relatively high concentrations, ZnO has shown no toxicity at low concentrations. Moreover, Zhang et al. (2008) stated that the antimicrobial activity of ZnO-NPs occurred due to the direct contact of nanoparticles with bacterial membrane and production of ROS close to the bacterial membrane. They observed the interaction between ZnO-NPs and *E. coli* cells is caused by electrostatic forces. ZnO-NPs have a positive charge, while due to the excess carboxylic groups, the charge of bacterial cell surface is negative at biological pH (Stoimenov et al. 2002). The electrostatic forces are generated due to the presence of opposite charges between the bacteria and nanoparticles. Thus, the bacterial cell membrane is disrupted due this strong interaction. The internalization of nanoparticles due to the disruption of the *E. coli* cell wall was also observed with the interaction of ZnO-NPs (Brayner et al. 2006). Similar results were also found on *E. coli* by Zhang et al. (2007) where they observed the presence of ZnO-NPs caused by the collapse of *E. coli* bacterial cell membrane.

4.5.3   Titanium Dioxide Nanoparticles

The crystal structure, shape, and size of titanium dioxide (TiO_2) are highly attributed for its antimicrobial property (Haghighi et al. 2013). The crucial antimicrobial mechanism of TiO_2-NPs is the formation of ROS due to the oxidative stress. The formation of ROS is responsible for the site-specific DNA damage. It was reported that TiO_2-NPs enhanced the antimicrobial activity of diverse classes of antibiotics such as beta lactums, cephalosporins, aminoglycosides, glycopeptides, macrolides, lincosamides, and tetracycline against methicillin-resistant *S. aureus*(MRSA). It was also showed that the presence of TiO_2-NPs relatively decreased the antimicrobial resistance of MRSA (Roy et al. 2010). Antifungal effect of TiO_2-NPs was also observed on the fungal biofilms of fluconazole-resistant standard strains of *Candida albicans* (Haghighi et al. 2013). They synthesized TiO_2-NPs which showed the antifungal activity against fluconazole-resistant strain *C. albicans* biofilms. Moreover, the results also suggested that the TiO_2-NPs can effectively control the fungal biofilms formed especially on the surface of medical devices.

TiO_2-NPs also have the photocatalytic properties which help them to efficiently eradicate the bacteria. Besides, TiO_2-NPs produce ROS under UV light. The antibacterial activity of TiO_2-NPs due to its photocatalytic effect has occurred by lipid peroxidation which finally leads to enhance membrane fluidity and disrupt the cell integrity (Carré et al. 2014). Moreover, it was observed that the effect of antimicrobial activity and photocatalysis of TiO_2-NPs are significantly enhanced by doping with metal ions. The doping of TiO_2-NPs with metal ions is very essential as it shifts TiO_2-NPs light absorption range to visible light and there is no need to irradiate them with UV light (Allahverdiyev et al. 2011). Moreover, it is important to not use UV light for TiO_2-NPs because it may cause genetic damage in human cells and tissues (Allahverdiyev et al. 2011).

4.5.4   Gold Nanoparticles

Gold nanoparticles (AuNPs) are considered to be biologically inert, but it can be a promising candidate for diagnosis and treatment of bacterial infection. A spectrum of research is going on to engineer the chemical and photothermal activity of AuNPs. The attributes which makes the AuNPs a potential carrier in drug delivery system are its ability to bind and/or penetrate the cell wall followed by delivery of larger number of antibiotic into a highly localized volume (Nirmala Grace and Pandian 2007; Pissuwan et al. 2010; Saha et al. 2007). Most bactericidal antibiotics and nanomaterials induce cell death by ROS-related mechanism, whereas AuNPs function is independent of any ROS-related process. The antibacterial activity of AuNPs is carried out by two ways: by declining the general metabolism process which includes an alteration in the membrane potential and an inhibition of ATP synthase activity and by downregulating the tRNA binding function of ribosomal protein S10 (Cui et al. 2012).

The preferential target molecules of AuNPs are the sulfur-containing proteins harbored by the membrane or cellular cytoplasm and the phosphorus-containing DNA molecules. A contrasting view on ROS-related process responsible for antifungal activity of AuNPs is supported by the binding of thiol group present in the enzymes (Ahmad et al. 2013). A study on the antifungal effect of AuNPs on clinically important *Candida* sp. reveals the inhibition of proton pumping activity mediated by H⁺ATPase. A comparison of the fluconazole and vanadate effect suggests the possibility that AuNPs make direct interaction with the enzyme (Wani and Ahmad 2013). Zhou et al. (2012) demonstrated significant antibacterial potential of AuNPs and AgNPs against Gram-negative (*E. coli*) and the Gram-positive bacteria (BCG). The antibacterial activity of NPs depends on several factors like composition, types, and surface modification. The capping agents were used to modify the surface and aimed to provide different inhibitory concentrations of nanoparticles. For example, weakly bound (citrate) causes more aggregation thereby reducing surface area for interaction on the contrary strongly bound capping agent PAH (poly-allylamine hydrochloride) AuNPs aggregate into 4–5 μm long chains thereby a strong interaction is established. They claimed that Au and Ag-NPs exhibited significant antibacterial activity against both. The negative charge of the cell wall of bacteria plays an important role for strong attraction to PAH which in turn facilitates delivery of large number of AuNPs (Goodman et al. 2004).

A number of researches devoted to modify the surface layer AuNPs for enhancing its aqueous solubility, biocompatibility, and biorecognition. A conjugate consisting of toluidine blue and tiopronin AuNPs was used against *S. aureus* and an enhanced antimicrobial efficacy was reported (Gil-Tomás et al. 2007). Similar studies were carried out by Selvaraj and Alagar (2007) to evaluate the antibacterial and antifungal potential of 5Fluororuracil colloidal gold complex (5FU-Au). This 5FU-Au complex exhibit appreciable antifungal activity against *Aspergillus niger* and *A. fumigatus*. Further on comparing the efficacies of 5FU-Au complex against different Gram-positive and Gram-negative bacteria like *Micrococcus luteus*, *S. aureus*, *Pseudomonas aeruginosa*, and *E. coli*, it can be inferred that it exhibits higher activity against Gram-negative bacteria due to easier internalization of the presence of thin cell wall.

4.5.5   Silicon and Silicon Dioxide Nanoparticles

The antimicrobial activity of silicon dioxide (SiO_2) is very significant at nanoscale level due to its larger surface area (Dhapte et al. 2014). The silicon dioxide nanoparticles are also popularly known as silica nanoparticles or nanosilica. These nanoparticles are widely used in the biomedical field especially in drug delivery and gene therapy. It was reported that silicon nanoparticles (Si-NPs) significantly reduced the bacterial adherence to oral biofilms (Cousins et al. 2007). The combination of different biocidal metal ions such as Ag with Si-NPs is widely studied. For example, the synthesis of novel Ag-Si nanocomposite for antimicrobial activity was reported by Egger et al. (2009). They observed the significant increase in antimicrobial activity of Ag-Si nanocomposite against a wide range of microorganisms compared to conventional materials like silver nitrate and silver zeolite. Similarly, the antimicrobial activity of Ag-SiO$_2$ and Au-SiO$_2$ nanocomposite was also investigated. In the study, it was found that Ag-SiO$_2$ nanocomposites enhanced the antimicrobial activity against *E. coli*, *S. aureus*, and *C. albicans* as compared to the Au-SiO$_2$ nanocomposites (Mukha et al. 2010).

4.5.6 Copper and Copper Oxide Nanoparticles

Copper (Cu) and copper oxide nanoparticles (CuO-NPs) were mostly evaluated for their antimicrobial activity against various microorganisms living in different habitats. Scientists are giving great attention to CuO-NPs due to their exclusive physical, chemical, and biological properties for antimicrobial activities and low cost of preparation (Dizaj et al. 2014). The antimicrobial effect of CuO-NPs was reported against human and fish pathogenic Gram-positive bacteria *Staphylococcus aureus* (MTCC 3160) and Gram-negative bacteria *Proteus mirabilis* (MTCC 425), *Bacillus subtilis* (MTCC 1305), *Aeromonas caviae* (MTCC 7725), *Aeromonas hydrophila* (MTCC 1739), *Edwardsiella tarda* (MTCC 2400), and *Vibrio anguillarum* (Nabila and Kannabiran 2018). However, the rapid oxidation of the Cu-NPs on exposure to the air limits their application (Nabila and Kannabiran 2018; Usman et al. 2013). The antimicrobial effect of CuO-NPs was also reported against *Klebsiella pneumoniae*, *Pseudomonas aeruginosa*, *Salmonella paratyphi*, and *Shigella* strains (Mahapatra et al. 2008). Herein, the CuO- NPs penetrates through the bacterial cell membrane and inhibiting the vital enzymes activity of bacteria.

4.6 Medical Implants Coated with Nanoparticles Used in Grafting

Medical devices are implanted in patients' body for a short period of time (up to 3 months) or for a very longer or indefinite time. Contact lenses, surgical suture, wound healing skin patch, central venous catheters, etc. stay in the body for a short period of time, while artificial heart valves, prosthetic hip and joints, pacemaker, etc. stay for a long period. Generally, the complexity in patient arises due to nosocomial infection associated with the implants (N. Tran and Tran 2012). Systemic application of antibiotics is conventionally used in medical practices to counter these infections. However antibiotic resistance of bacteria and associated side effects of antibiotics like cytotoxicity has rendered the effectiveness of applying antibiotics in implant-related infections.

Metallic implants of metals like titanium and its alloys are used for orthopedic and dental application which provides excellent mechanical properties like biocompatibility, relatively low modulus, toughness, ductility, formability, low density, etc. However, due to poor biorecognition of the implant surface, it can cause micromotion at the bone-implant interface leading to loosening. This may also lead to inflammation in adjacent tissue due to wear of implant and bio-corrosion causing leaching of metal particles or ions (Javadi et al. 2019; Bartkowiak et al. 2018). Medical implants coated with nanoparticle and applications were given in Table 4.1.

Nanoparticles can provide advantages in many ways for enhancing tissue acceptance and especially antimicrobial property. Many studies have established that the presence of nanoscale roughness on the surface of the implant can significantly improve the growth of tissue cell in tissue engineering applications. Nanoparticle-coated implant surface in case of bone tissue engineering applications imitates the same natural biological environment in which the osteoblast bone-forming cells grow (Javadi et al. 2019). The presence of nanostructure on the surface of implants increases the surface area for host proteins and cells to interact with the implant. Cell differentiation and mineralization are achieved advantageously in vitro over the rough surface of implants when osteoblast cell interacts (Agarwal and García 2015).

Nanoparticulate systems can be attached on the implant surface for multifunctional coating approach which can hold drug molecules and release them with varied releasing kinetics. Nanoparticles used in coating modify the surface morphology of implant creating an unfavorable ambience for bacterial adhesion. This reduces the infection around implants and increases the healing rate (Baghdan et al. 2018). Nano Spray Dryer B-90 has been recently developed and being used for the production of nanoparticles of desired biomaterials and coating the surface of medical implants directly (Baghdan et al. 2018).

One of the most important nanoparticles that have found extensive use in coatings over implants is silver nanoparticle. Silver is known for its antimicrobial property and it has been used in the coating over many medical implants in earlier days. But silver metal coating has found to be cytotoxic to adjacent tissues in many studies due to concentrated silver concentration over a small region of implant and tissue interaction. The silver nanoparticles we found to be convenient in coatings over implants as the nanosilver concentration was potent against microorganisms without any significant cytotoxicity and also found to be anti-angiogenic (Gonz et al. 2015; Taheri et al. 2014; Aziz et al. 2019). In most of the studies concerning implants for orthopedics and dental applications, silver nanoparticles have been found to be used extensively over implant materials like titanium and its alloys. Ag-nanoparticles were also used in diverse medical implants coatings, for example, artificial heart valve made of silicone, medical stents, poly(methyl methacrylate) ((PMMA) bone cement based hip and knee joint implants, (poly-(−3-hydroxybutyrate-co-3-hydroxyvalerate) PHBV nanofibrous scaffolds for bone and skin tissue engineering applications, etc. (Haider and Kang 2015).

Table 4.1 Few examples of medical implants coated with nanoparticles

Medical implant	Nanoparticles used for coating	Applications	References
Polyethylene terephthalate (PET) cardiovascular grafts	PLGA nanoparticles	Treatment of the early thrombosis, inflammation, or bacterial infection via local delivery of therapeutic agents	Al et al. (2018)
Surface-modified titanium implant	Ag nanoparticles	Control of peri-implant infection in dental or endo-osseous application	Dong et al. (2017)
Surface-mocified titanium implant	Ag nanoparticles loaded TiO2 nanotubes array	Enhancement of antibacterial property for orthopedic titanium implants	Dong et al. (2017)
Chitosan-based skin graft	K-doped zinc oxide (ZnO) nanoparticles	Enhancement of angiogenesis in skin tissue regeneration applications	Shahzadi et al. (2018)
Medical imp ants made of polycaprolactone (PCL) polymer	Ag nanoparticles	Enhance antimicrobial efficacy by modifying hydrophobic polycaprolactone surface with hydrophilic Ag-nanoparticles	Tran et al. (2015)
Silver nanoparticle/poly(dl-lactic-co-glycolic acid) (PLGA)-coated stainless steel alloy(SNPSA)-based orthopedic implant	Ag nanoparticle	Enhancement antimicrobial and osteoinductive properties for possible application in orthopedics.	Liu et al. (2012)
Hydroxyapatite coated metallic implant	Ag nanoparticles	Enhancement of osteoinductivity and antibacterial properties	Xie et al. (2014)
Plastic catheters	Ag nanoparticles	Targeted and sustained release of bactericidal silver at the implantation site	Roe et al. (2008)
Titanium dental implant	Chlorhexidine (CHX) hexametaphosphate (HMP) nanoparticles (NPs)	Prevention and treatment of peri-implant infection	Wood et al. (2015)
Biostable fibre-reinforced composites (FRC)-based bone implants	Ag nanoparticles	Prevention of bacterial adhesion and biofilm formation on bone implant	Nganga et al. (2013)
Stainless steel stent	TiO2 nanoparticles	Enhancement of hemocompatibility of the stent	Elsawy et al. (2018)
Methacrylate hydrogels	Ag nanoparticles	Bone graft applications	Gonz et al. (2015)

4.7   Global Market: Nanoparticle-Coated Medical Implants

Engineered nanomaterials already have a huge economic impact globally owing to its application in various industries, products, and processes. The biomedical industries have been most affected by the advent of nanomaterials, both economically and in novelty and innovation. Nanoparticles and nanomaterials have been used to develop new drugs, implants, drug delivery systems, and medical devices. Nanoparticles are continuously being explored to design novel implants and implantable devices, which form one of the important revenue generators in the health-care sector.

The modern medicine focuses on treating disease; hence patients don't visit doctors until physiological symptoms show, and subsequently, diagnostics happens when the disease is at the advanced stage. The scenario is changing with the advent of novel technologies including nanotechnology, which is now moving toward a more preventive approach. Implantable devices can be used not only for reinstating the functions of organs but also for monitoring the various vitals of the subject. Implantable devices will play a huge role in "point of care" technology, for preventive diagnostics in the near future. "Point of care" technology is evolving to make even smaller devices for diagnostics and monitoring. "Point of care" technology market will be worth USD 38.13 billion by 2022 (www.marketsandmarkets.com).

Engineered nanomaterials have helped overcome technological barriers in biomedical engineering. Nanocoating is an evolving area; further research can exploit potential novel applications in implants and implantable devices. Coating of nanoparticles or nanocoating of medical implants adds value to the functionality and reliability of the implants. This value addition is the driving force of nanocoated implants in the global market. The predicted market growth of antibacterial coating is about USD 343.16 million by 2023 (www.medgadet.com). The coatings can be segregated into metallic and non-metallic, due to high applicability metallic coatings accounted for 75% of the global market in 2016 (www.medgadet.com). This trend is expected to continue for the foreseeable near future.

It has been predicted that medical implants market will worth about USD 171.5 billion by 2023 (Market Research Future). The predicted growth is based on the growing number of population suffering from congenital diseases, heart ailments, and orthopedic complexities. The other factors that will drive this growth are continuously increasing the number of the aged population, innovations in the health sector, and health awareness among the public in general.

The global orthopedic biomaterials implant market, which includes implants slated to grow by 7% by 2022 (Technavio). Growing application of a wide range of nanomaterials in orthopedic implants is also predicted to increase. Increasing muscoskeletal disorders are driving this growth, along with a growing number of war injuries and increasing traffic and occupational accidents.

Major established brands like Johnson & Johnson and Boston Scientific Corporation are highly competitive in their product development as well as product placements. Heavy investments are being done in the research and development of the products. Small players are continuously penetrating the market by offering cheaper variants. Apart from the novelty and innovativeness, marketing of the product is a critical component of growth in this sector. North America at present holds 40% of the market share of the medical device coating. Improved health sector and increase in income could further boost the market in this region. Europe at present is the second largest market followed by Asia. India and China will continue to be the fastest-growing market in this sector driven by population and cheap labor.

A short review shows that a wide range of nano-engineered materials is being used for nanocoating. Gold and silver nanoparticles are being used in implantable scaffolds, functional implants, and also implantable devices. The market is still growing and looking at the present trends; it will continue to grow encouraging more innovations and applications in the future.

References

Agarwal R, García AJ (2015) Biomaterial strategies for engineering implants for enhanced osseo-integration and bone repair. Adv Drug Deliv Rev 94:53–62

Ahmad T, Wani IA, Manzoor N, Ahmed J, Asiri AM (2013) Biosynthesis, structural characterization and antimicrobial activity of gold and silver nanoparticles. Colloids Surf B: Biointerfaces 107:227–234

Al BM, Mahmoud GF, Bakowsky U (2018) Immobilization and characterization of PLGA nanoparticles on polyethylene terephthalate cardiovascular grafts for local drug therapy of associated graft complications. J Drug Delivery Sci Technol 47:144–150

Albrecht MA, Evans CW, Raston CL, Evans C (2006) Green chemistry and the health implications of nanoparticles. Green Chem 8:417–432

Allahverdiyev AM, Abamor ES, Bagirova M, Rafailovich M (2011) Antimicrobial effects of TiO(2) and Ag(2)O nanoparticles against drug-resistant Bacteria and leishmania parasites. Future Microbiol 6:933–940

Arciola CR, Campoccia D, Montanaro L (2018) Implant infections: adhesion, biofilm formation and immune evasion. Nat Rev Microbiol 16(7):397–409

Aziz N, Fatma T, Varma A, Prasad R (2014) Biogenic synthesis of silver nanoparticles using Scenedesmus abundans and evaluation of their antibacterial activity. J Nanopart 2014:689419. https://doi.org/10.1155/2014/689419

Aziz N, Faraz M, Pandey R, Sakir M, Fatma T, Varma A, Barman I, Prasad R (2015) Facile algae-derived route to biogenic silver nanoparticles: synthesis, antibacterial and photocatalytic properties. Langmuir 31:11605–11612. https://doi.org/10.1021/acs.langmuir.5b03081

Aziz N, Pandey R, Barman I, Prasad R (2016) Leveraging the attributes of *Mucor hiemalis*-derived silver nanoparticles for a synergistic broad-spectrum antimicrobial platform. Front Microbiol 7:1984. https://doi.org/10.3389/fmicb.2016.01984

Aziz N, Faraz M, Sherwani MA, Fatma T, Prasad R (2019) Illuminating the anticancerous efficacy of a new fungal chassis for silver nanoparticle synthesis. Front Chem 7:65. https://doi.org/10.3389/fchem.2019.00065

Baghdan E, Pinnapireddy SR, Vögeling H, Schäfer J, Eckert AW, Bakowsky U, (2018) Nano spray drying: A novel technique to prepare welldefined surface coatings for medical implants. J Drug Deliv Sci Technol 48:145–151

Bartkowiak A, Suchanek K, Szaraniec B, Lekki J, Perzanowski M, Marsza M (2018) Biological effect of hydrothermally synthesized silica nanoparticles within crystalline hydroxyapatite coatings for titanium implants. Mater Sci Eng C 92:88–95

Bhuyan T, Mishra K, Khanuja M, Prasad R, Varma A (2015) Biosynthesis of zinc oxide nanoparticles from Azadirachta indica for antibacterial and photocatalytic applications. Mater Sci Semicond Process 32:55–61

Bosco R, Van Den Beucken J, Leeuwenburgh S, Jansen J (2012) Surface engineering for bone implants: a trend from passive to active surfaces. Coatings 2:95–119

Brayner R, Ferrari-iliou R, Brivois N, Djediat S, Benedetti MF, Fie F, Cedex P, De Ge L (2006) Toxicological impact studies based on Escherichia coli bacteria in ultrafine ZnO nanoparticles colloidal medium. Nano 6:866–870

Busscher HJ, Van Der Mei HC (2012) How do Bacteria know they are on a surface and regulate their response to an adhering state ? PLoS Pathog 8(1):e100240

Carré G, Hamon E, Ennahar S, Estner M, Lett M-c, Horvatovich P, Gies J-p (2014) TiO2 photocatalysis damages lipids and proteins in. Appl Environ Microbiol 80(8):2573–2581

Cousins BG, Allison HE, Doherty PJ, Edwards C, Garvey MJ, Martin DS (2007) Effects of a nanoparticulate silica substrate on cell attachment of Candida albicans. J Appl Microbiol I 102:757–765

Crubzy E, Murail P, Girard L, Bernadou J-P (1998) False teeth of the Roman World. Nature 391:29–29

Cui Y, Zhao Y, Tian Y, Zhang W, Lü X, Jiang X (2012) The molecular mechanism of action of bactericidal gold nanoparticles on Escherichia coli Q. Biomaterials 33:2327–2333

Darouiche RO, Raad I (1999) Antimicrobial impregnated catheters and other medical implants. US Patent 5,902,283

Dhapte V, Kadam S, Moghe A, Pokharkar V, Bhavan V, Marg BS (2014) Probing the wound healing potential of biogenic silver nanoparticles. J Wound Care 23:431–441

Dizaj SM, Lotfipour F, Barzegar-jalali M, Zarrintan MH, Adibkia K (2014) Antimicrobial activity of the metals and metal oxides nanoparticles. Mater Sci Eng C 44:278–284

Dong Y, Ye H, Liu Y, Xu L, Wu Z, Hu X, Ma J, Pathak JL, Liu J, Wu G (2017) PH dependent silver nanoparticles releasing titanium implant: a novel therapeutic approach to control Peri-implant infection. Colloids Surf B: Biointerfaces 158:127–136

Egger S, Lehmann RP, Height MJ, Loessner MJ, Schuppler M (2009) Antimicrobial properties of a novel silver-silica nanocomposite material. Appl Environ Microbiol 75(9):2973–6

Elsawy AM, Attia NF, Mohamed HI, Mohsen M, Talaat MH (2018) Innovative coating based on graphene and their decorated nanoparticles for medical stent applications. Mater Sci Eng C 96:708–715

Feng QL, Wu J, Chen GQ, Cui FZ, Kim TN, Kim JO (2000) A mechanistic study of the antibacterial effect of silver ions on Escherichia coli and Staphylococcus aureus. J Biomed Mater Res 52:662–668

Furno F, Morley KS, Wong B, Sharp BL, Arnold PL, Steven M, Bayston R, Brown PD, Winship PD, Reid HJ (2004) Silver nanoparticles and polymeric medical devices: a new approach to prevention of infection? J Antimicrob Chemother 54(6):1019–1024

Gil-Tomás J, Tubby S, Parkin IP, Narband N, Dekker L, Nair SP, Wilson M, Street C (2007) Lethal photosensitisation of Staphylococcus aureus using a toluidine blue O–tiopronin–gold nanoparticle conjugate. J Mater Chem 17:3739–3746

Gong P, Li H, He X, Wang K, Hu J, Tan W, Zhang S, Yang X (2007) Preparation and antibacterial activity of Fe_3O_4@Ag nanoparticles. Nanotechnology 18:285604

Gonz MI, Perni S, Tommasi G, Morris G, Karl Hawkins LE, Perni S et al (2015) Silver nanoparticles based antibacterial methacrylate hydrogels potentially for bone graft applications. Mater Sci Eng C 50:332–340

Goodman CM, Mccusker CD, Yilmaz T, Rotello VM (2004) Toxicity of gold nanoparticles functionalized with cationic and anionic side chains. Bioconjug Chem 15:897–900

Guo G, Wang J, You Y, Tan J, Shen HAO (2017) Distribution characteristics of Staphylococcus Spp. in different phases of periprosthetic joint infection: a review. Exp Ther Med 13:2599–2608

Haghighi F, Roudbar Mohammadi S, Mohammadi P, Hosseinkhani S (2013) Antifungal activity of TiO(2) nanoparticles and EDTA on Candida albicans biofilms. Infect Epidemiol Microbiol 1:33–38

Haider A, Kang I-k (2015) Preparation of silver nanoparticles and their industrial and biomedical applications : a comprehensive review. Adv Mater Sci Eng 2015:1–5

Hexter AT, Hislop SM, Blunn GW, Liddle AD (2018) The effect of bearing surface on risk of periprosthetic joint infection in total hip arthroplasty: a systematic review and meta-analysis. Bone Joint J 100:134–142

Hickok NJ, Shapiro IM, Chen AF (2018) The impact of incorporating antimicrobials into implant surfaces. J Dent Res 97(1):14–22

Holá V, Růžička F, Votava M (2006) The dynamics of Staphylococcus epidermis biofilm formation in relation to nutrition, temperature, and time. Scr Med 79(July):169–174

Honkanen M, Jämsen E, Karppelin M, Huttunen R, Eskelinen A, Syrjänen J (2019) Periprosthetic Joint Infections as a Consequence of Bacteremia. Open Forum Infect. Dis 6(6):1–6

Illingworth KD, Mihalko WM, Parvizi J, Sculco T, McArthur B, el Bitar Y, Saleh KJ (2013) How to minimize infection and thereby maximize patient outcomes in total joint arthroplasty. J Bone Joint Surg 50:1–13

Jalal R, Goharshadi EK, Abareshi M, Moosavi M, Yousefi A, Nancarrow P (2010) ZnO nanofluids: green synthesis, characterization, and antibacterial activity. Mater Chem Phys 121(1–2):198–201

Javadi A, Solouk A, Nazarpak MH, Bagheri F (2019) Surface engineering of titanium-based implants using electrospraying and dip coating methods. Mater Sci Eng C 99:620

Kasemets K, Ivask A, Dubourguier H-c, Kahru A (2009) Toxicology in vitro toxicity of nanoparticles of ZnO, CuO and TiO(2) to yeast Saccharomyces cerevisiae. Toxicol In Vitro 23(6):1116–1122

Kim GC, Kim GJ, Park SR, Jeon SM, Seo HJ, Iza F, Lee JK (2008) Air plasma coupled with antibody-conjugated nanoparticles. J Phys D Appl Phys 42:032005

Li Y, Leung P, Yao L, Song QW, Newton E (2006) Antimicrobial effect of surgical masks coated with nanoparticles. J Hosp Infect 62:58–63

Liu Y, Zheng Z, Zara JN, Hsu C, Soofer DE, Lee KS, Siu RK et al (2012) The antimicrobial and osteoinductive properties of silver nanoparticle/poly(D, L-lactic-co-glycolic acid) -coated stainless steel. Biomaterials 33(34):8745–8756

Mahapatra O, Bhagat M, Gopalakrishnan C, Arunachalam KD (2008) Ultrafine dispersed CuO nanoparticles and their antibacterial activity. J Exp Nanosci 3(3):185–93

Mendonça G, Mendonça DBS, Araga FJL (2008) Biomaterials advancing dental implant surface technology – from micron- to nanotopography. Biomaterials 29:3822–3835

Morones JR, Elechiguerra JL, Camacho A, Holt K, Kouri JB, Ram JT, Yacaman MJ (2005) The bactericidal effect of silver nanoparticles. Nanotechnology 16:2346

Mukha I, Eremenko A, Korchak G, Michienkova A (2010) Antibacterial action and physicochemical properties of stabilized silver and gold nanostructures on the surface of disperse silica. J Water Res Protect 2:131–136

Nabila MI, Kannabiran K (2018) Biosynthesis, characterization and antibacterial activity of copper oxide nanoparticles (CuO NPs) from Actinomycetes. Biocatal Agric Biotechnol 15:56–62

Nganga S, Travan A, Marsich E (2013) In vitro antimicrobial properties of silver – polysaccharide coatings on porous fiber-reinforced composites for bone implants. J Mater Sci Mater Med 24:2775–2785

Nirmala Grace A, Pandian K (2007) Antibacterial efficacy of aminoglycosidic antibiotics protected gold nanoparticles-a brief study. Colloids Surf A Physicochem Eng Asp 297(1–3):63–70

Padmavathy N, Vijayaraghavan R (2008) Enhanced bioactivity of ZnO nanoparticles an antimicrobial study. Sci Technol Adv Mater 9:035004

Panteli M, Giannoudis PV (2016) Chronic osteomyelitis: what the surgeon needs to know. EFORT Open Reviews 1(5):128–135

Percival SL, Suleman L, Vuotto C, Donelli G (2015) Healthcare-associated infections, medical devices and biofilms: risk, tolerance and control. J Med Microbiol 64(4):323–334

Pissuwan D, Cortie CH, Valenzuela SM, Cortie MB (2010) Functionalised gold nanoparticles for controlling pathogenic bacteria. Trends Biotechnol 28(4):207–213

Pitarresi G, Salvatore F, Calascibetta F, Fiorica C, Di M, Giammona G (2013) Medicated hydrogels of hyaluronic acid derivatives for use in orthopedic field. Int J Pharmaceut 449:84–94

Prasad R, Pandey R, Barman I (2016) Engineering tailored nanoparticles with microbes: quo vadis. Wiley Interdiscip Rev Nanomed Nanobiotechnol 8:316–330. https://doi.org/10.1002/wnan.1363

Prasad R (2019) Plant Nanobionics: Approaches in Nanoparticles Biosynthesis and Toxicity. Springer International Publishing (ISBN 978-3-030-16379-2). https://www.springer.com/gp/book/9783030163785

Qasim M, Udomluck N, Chang J, Park H, Kim K (2018) Antimicrobial activity of silver nanoparticles encapsulated in poly- N -isopropylacrylamide-based polymeric nanoparticles. Int J Nanomedicine 54:235–249

Reddy KM, Feris K, Bell J, Wingett DG, Hanley C, Feris K, Bell J, Wingett DG, Hanley C (2012) Selective toxicity of zinc oxide nanoparticles to prokaryotic and eukaryotic systems. Appl Phys Lett 213902(2007):10–13

Riley DK, Classen DC, Stevens LE, Burke JP (1995) A large randomized clinical trial of a silver-impregnated urinary catheter: lack of efficacy and staphylococcal superinfection. Am J Med 98:349–356

Rodrigues LR (2011) Inhibition of bacterial adhesion on medical devices. In: Linke D, Goldman A (eds) Bacterial adhesion, Advances in experimental medicine and biology. Springer, Dordrecht, pp 351–367

Roe D, Karandikar B, Bonn-savage N, Gibbins B, Roullet J-b (2008) Antimicrobial surface functionalization of plastic catheters by silver nanoparticles. J Antimicrob Chemother 61:869–876

Romanò CL, Scarponi S, Gallazzi E, Romanò D, Drago L (2015) Antibacterial coating of implants in orthopaedics and trauma: a classification proposal in an evolving panorama. J Orthop Surg Res 10:57

Roy AS, Parveen A, Koppalkar AR, Ambika Prasad MVN (2010) Effect of Nano – titanium dioxide with different antibiotics against methicillin-resistant Staphylococcus aureus. J Biomater Nanobiotechnol 1:37–41

Saha B, Bhattacharya J, Mukherjee A, Ghosh AK, Santra CR, Dasgupta AK, Karmakar P (2007) In vitro structural and functional evaluation of gold nanoparticles conjugated antibiotics. Nanoscale Res Lett 2(12):614–622

Selvaraj V, Alagar M (2007) Analytical detection and biological assay of antileukemic drug 5-fluorouracil using gold nanoparticles as probe. Int J Pharm 337(1–2):275–281

Shahzadi L, Chaudhry AA, Aleem AR (2018) Development of K-doped ZnO nanoparticles encapsulated crosslinked chitosan based new membranes to stimulate angiogenesis in tissue engineered skin grafts. Int J Biol Macromol 120:721–728

Sirelkhatim A, Mahmud S, Seeni A (2015) Review on zinc oxide nanoparticles: antibacterial activity and toxicity mechanism. Nano-Micro Lett 7:219–242

Sondi I, Salopek-sondi B (2004) Silver nanoparticles as antimicrobial agent: a case study on E. coli as a model for gram-negative bacteria. J Colloid Interface 275:177–182

Stock SR (2015) The mineral – collagen interface in bone. Calcif Tissue Int 97:262–280

Stoimenov PK, Klinger RL, Marchin GL, Klabunde KJ (2002) Metal oxide nanoparticles as bactericidal agents. Langmuir 18:6679–6686

Taheri S, Cavallaro A, Christo SN, Smith LE, Majewski P, Barton M, Hayball JD, Vasilev K (2014) Biomaterials substrate independent silver nanoparticle based antibacterial coatings. Biomaterials 35(16):4601–4609

Taniguchi N (1983) Current status in, and future trends of, ultraprecision machining and ultrafine materials processing. CIRP Ann 32:573–582

Tran N, Tran PA (2012) Nanomaterial-based treatments for medical device-associated infections. Chem Phys Chem 13:2481–2494

Tran PA, Hocking DM, Connor AJO (2015) In situ formation of antimicrobial silver nanoparticles and the impregnation of hydrophobic polycaprolactone matrix for antimicrobial medical device applications. Mater Sci Eng C 47:63–69

Usman MS, El Zowalaty ME, Shameli K, Zainuddin N, Salama M, Ibrahim NA (2013) Synthesis, characterization, and antimicrobial properties of copper nanoparticles. Int J Nanomedicine 2013(8):4467–4479

Wang H, Wick RL, Xing B (2009) Toxicity of nanoparticulate and bulk ZnO, Al_2O_3 and TiO_2 to the nematode Caenorhabditis elegans. Environ Pollut 157(4):1171–1177

Wani IA, Ahmad T (2013) Size and shape dependant antifungal activity of gold nanoparticles: a case study of Candida. Colloids Surf B: Biointerfaces 101:162–170

Wilcox MW, Nash E, Martin TW, Rodriguez A, Rogers RF (1998) Cartridge dispensing system for dental material. United States Patent US 5,722,829

Wood NJ, Jenkinson HF, Davis SA, Mann S, Sullivan DJO, Barbour ME (2015) Chlorhexidine hexametaphosphate nanoparticles as a novel antimicrobial coating for dental implants. J Mater Sci Mater Med 26:201

Xie C-m, Lu X, Wang K-f, Meng F-z, Jiang O, Zhang H-p, Zhi W (2014) Silver nanoparticles and growth factors incorporated hydroxyapatite coatings on metallic implant surfaces for enhancement of osteoinductivity and antibacterial properties. ACS Appl Mater Interfaces 6:8580–8589

Zeng Q, Zhu Y, Yu B, Sun Y (2018) Antimicrobial and antifouling polymeric agents for surface functionalization of medical implants antimicrobial and antifouling polymeric agents for surface functionalization of medical implants. Biomacromolecules 19:2805–2811

Zhang L, Jiang Y, Ding Y, Povey M, York D, (2007) Investigation into the antibacterial behaviour of suspensions of ZnO nanoparticles (ZnO nanofluids). J Nanoparticle Res 9(3):479–489

Zhang Y, Mo Y, Portney NG (2008) Zeta potential: a surface electrical characteristic to probe the interaction of nanoparticles with normal and cancer human breast epithelial cells. Biomed Microdevices 10:321–328

Zhang Y, Chen Y, Westerhoff P, Crittenden J (2009) Impact of natural organic matter and divalent cations on the stability of aqueous nanoparticles. Water Res 43(17):4249–4257

Zhou Y, Kong Y, Kundu S, Cirillo JD, Liang H (2012) Antibacterial activities of gold and silver nanoparticles against Escherichia coli and Bacillus. J Nanotechnol 10:19

Chapter 5
Novel Antimicrobial Compounds from Indigenous Plants and Microbes: An Imminent Resource

Deepika Jothinathan, Lavanyasri Rathinavel, Prabhakaran Mylsamy, and Kiyoshi Omine

Contents

Abstract Resistance of microbes towards the antibiotics has agitated the scientific community to look for authentic compounds from plants and microbes. They also make sure that these metabolites are strong enough to fight back the diseases in the long run. Traditional medicines that were used during the ancient times remain as unsung heroes until today. Only few medicines are of plant origin, and their usage is limited due to efficacy, treatment time and toxicity factors. But still, it can be overcome by the modern methods. Due to the increasing health problems all over

D. Jothinathan (✉)
Department of Life Sciences, Central University of Tamil Nadu,
Thiruvarur, Tamil Nadu, India

L. Rathinavel · P. Mylsamy
Post Graduate and Research Department of Botany, Pachaiyappa's College,
Chennai, Tamil Nadu, India

K. Omine
Geo-Tech Laboratory, Department of Civil Engineering, Graduate School of Engineering,
Nagasaki University, Nagasaki, Japan

© Springer Nature Switzerland AG 2020
R. Prasad et al. (eds.), *Nanostructures for Antimicrobial and Antibiofilm Applications*, Nanotechnology in the Life Sciences,
https://doi.org/10.1007/978-3-030-40337-9_5

the world, people are now switching over to the pharmaceutical products from plants and microbes. Considering these facts, this chapter has been framed to give a deep understanding of the current and forthcoming antimicrobial compounds from indigenous plants and microbes. It also discusses about the modern methods that are employed in the processing of antimicrobial compounds and their effects against different groups of pathogens.

Keywords Antimicrobial compounds · Phytochemicals · Techniques · Herbal medicine

5.1   Introduction

From the past decade, we have been receiving many alerts on emerging and re-emerging diseases. Though the traditional medicine obtained from plants and microbes was neglected for various reasons, now people have started to pay attention to the conventional herbal medicine. There are very many incidents where plant-derived metabolites have served as a preventive and curative drug. For instance, Nilavembu concoction was advised to consume during dengue outbreak in India. These are instant remedies to patients and have been found to be effective in preventing the disease. The side effects are not much harmful and provide remedy in the long run.

The secondary metabolites obtained from plants do not play a vital part in the plant's reproduction, growth, etc. (Fraenkel 1959). However, there are evidences that these products serve in the defence mechanism of plants (Stamp 2003). With this cue, researchers are inspecting the derivatives of herbal plants. Plants have a specialized pathway to synthesize the compounds. The study of these compounds has been elucidated with the help of various techniques, namely solvent extraction; microwave-assisted extraction; supercritical fluid extraction (SFE); solid phase micro extraction (SPME); chromatography methods such as HP-TLC, TLC, UPLC, GC-MS, LC-MS; phytochemical screening assays, nuclear magnetic resonance (NMR), Fourier-transform infrared spectroscopy (FTIR), etc.

Plant-derived compounds are comparatively safer than their chemically synthesized counterparts (Upadhyay et al. 2014; Rajeh et al. 2010). These compounds are primarily responsible for their antimicrobial activity (Savoia 2012). Phenolics and polyphenols such as flavonoids, quinones, coumarins and other types like alkaloids and terpenoids do possess the antimicrobial properties from the secondary metabolite group (Cowan 1999; Savoia 2012). The mechanisms by which plants control the microbial population are: (a) inhibition of biofilm formation, (b) inhibition of capsule production of bacteria, (c) microbial membrane disruption, (d) weakening of cell wall and metabolism and (e) by toxins production (Upadhyay et al. 2014).

Microbial-derived metabolites on the other side perpetuate their properties in the form of antibacterial, antifungal, antitumour and anticancer compounds. The major group of microbes that had a remarkable place in the pharmaceutical industry is Actinomycetes (Bérdy 1989). It was followed by fungi, bacteria and cyanobacteria, which had a notable role in metabolite production (Prasad et al. 2016, 2018).

This chapter enlightens the readers about the novel plant metabolites and microbial metabolites in the recent years. The merits and demerits of the recent extraction and analytical techniques with respect to the metabolite production are discussed in detail in Sect. 5.2.

5.2 Traditional Medicine

Traditional medicine has been a magical potion since prehistoric times for human population. It has been followed in many countries all over the world such as traditional medicine in China (Fabricant and Farnsworth 2001; Qi et al. 2013), Ayurveda and Unani in India (Parasuraman et al. 2014), Kampo in Japan (Yakubo et al. 2014), Sasang constitutional medicine in Korea (Kim and Noble 2014), traditional Aboriginal medicine in Australia (Oliver 2013), traditional medicine in Africa (Boakye et al. 2015) and Russian herbal medicine (Shikov et al. 2014). Traditional medicine has been a platform for formulating the modern pharmaceutical products. Plants which are used in traditional medicine can be derived from a single or mixed plant material and might contain a range of phytoconstituents. These components might act solely or in integration with other drugs to treat a particular disease (Parasuraman et al. 2014). The World Health Organization (WHO) recognized that traditional medicine plays a vital part in meeting the essential health maintenance requirement of the global population and emphasize definite types of the medicine (WHO 2014).

The medicinal components from both flora and fauna, despite being used in traditional medicine, have also been given much value as a primary source during the development of herbal preparations and modern medicines (Kang and Phipps 2003). Animal parts like hooves, skins, bones, feathers, tusks and their by-products such as honey, venom, wax and ambergris serve as the key constituents in the preparation of curative, protective and preventive medicine (Adeola 1992). Further, a major portion of bioactive compounds are derived from plant sources like herbs, shrubs, trees and higher plants followed by microorganisms and others (Farnsworth and Morris 1976). Figure 5.1 represents the group of antimicrobial compounds that are produced from plants.

India has been known to possess the rich source of medicinal plants. Around 8000 medical preparations from plants have been classified in Ayurveda. The famous Vedas have acknowledged more medicinal plants, accounting for 67 medicinal plants in Rig Veda around 5000 BC, 290 species in Atharva Veda during the 4500–2500 BC, 81 species in Yajur Veda, 1270 species in Sushrut Samhita and 1100 plant varieties in Charak Samhita for the drug preparation and its use. To our sur-

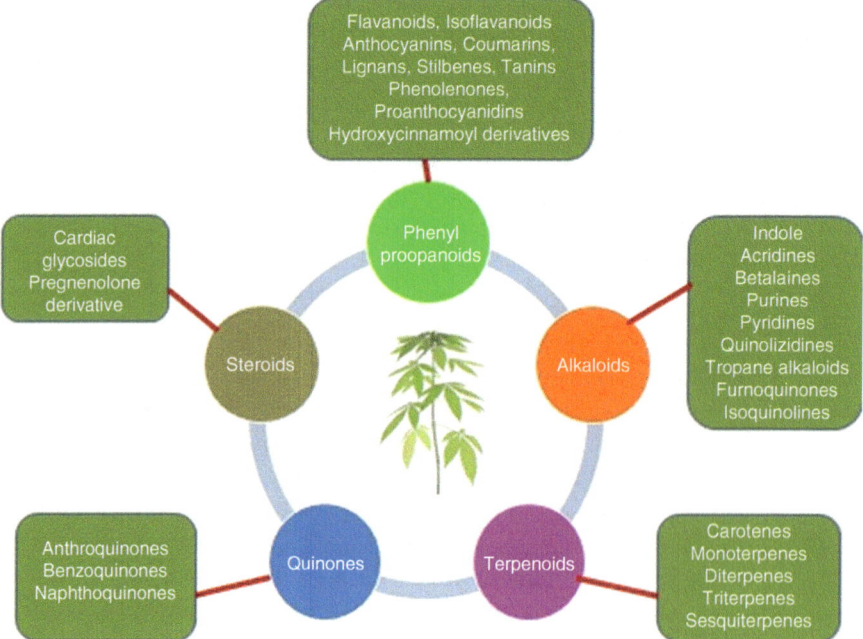

Fig. 5.1 Plant metabolite classification

prise, these plant preparations are still in use as a standard material in Ayurvedic medicine (Joy et al. 2001).

5.2.1 Technologies/Methodologies for the Compounds from Plants and Microbes

The assays and assessment techniques used to validate the antimicrobial potential of the secondary metabolites against pathogens need to be addressed in an elaborate manner. In this way, the choice of method for plants/microbes can be made, and accordingly, the precise results are obtained (Rios et al. 1988). Though the plant/microbial derived drugs are in a successful rate, their screening procedures in the past were strenuous, expensive and sluggish. This has eventually led to the increase of high-end technologies to screen and identify the metabolites. Table 5.1 discusses the recent methods in the identification of bioactive components and phytochemical analysis.

The standard antimicrobial screening methods availed currently comprise diffusion, dilution and bioautographic methods. Both diffusion and bioautographic method accomplish a qualitative analysis of the samples by revealing the presence or absence of inhibitory compounds. In contrast, dilution method defines the

Table 5.1 Merits and demerits of phytochemical analysis

Technique	Advantage	Disadvantage	References
Ultra-high-performance liquid chromatography (UHPLC)	Less time taken for analysis, minimum amount of samples and solvents used and amplified peak capacity	–	Gaudêncio and Pereira (2015)
High-performance liquid chromatography-nuclear magnetic resonance (HPLC-NMR)	Versatile separation technique providing structural information on separated compounds	(a) Needs ample concentration of the sample based on the availability of the metabolites in eluates (b) NMR spectrum's achievement time	Wolfender (2010)
Ultrafiltration high-performance liquid chromatography (HPLC)	Screens intricate combination of bioactive components	(a) Since the target protein should be analogous to the affinity (K_d) of the weakest ligand, a high sample concentration is required (b) False-positive results due to non-specific enzyme–ligand interactions (c) Not suitable for transmembrane proteins	van Breemen et al. (1997), Johnson et al. (2002)
Ultrafiltration liquid chromatography-mass spectrometry (LC-MS) with enzyme channel blocking assay (ECB)	Inhibits the possible ligands to arrive at the primary binding region	–	Song et al. (2014)
Bioaffinity chromatography	The synchronized examination of the fingerprint and biological activity allows the spotting of peaks with the preferred activity	(a) The buffer and solvent used in the analysis should be suitable for the extraction of the desired protein (b) The instrumentation is too complicated, and the reaction time between the components and the protein is minimal	Potterat and Hamburger (2013)
Ligand fishing	Chances of identification of low-affinity ligands	(a) The necessary quantity of protein for effective fishing is 50 µg (b) Possibility of retaining false positives	Vanzolini et al. (2013)
Cell membrane affinity chromatography	(a) The active compounds can be identified directly (b) The desired transmembrane protein can be subjected to immobilization	(a) One major demerit is that there is a chance of co-immobilization of numerous receptor types (b) The mobile phase is mostly aqueous, and there is a need to run the parallel control column, to check the nonexistence of non-specific interaction	Moaddel and Wainer (2009)

quantity of the compound (Valgas et al. 2007). Table 5.2 reviews the common and recent technologies in antimicrobial activity screening and evaluation methods.

5.2.2 Novel Metabolites from Plants

Gymnema sylvestre from family Apocynaceae has been reported to have larvicidal, anticancer, hypolipidemic, antimicrobial, antiviral and antioxidant activities. Ten percent aqueous extract of the leaves inhibited 9 out of 13 microbial strains, with a wide-spectrum activity showing inhibition zone of 14–23 mm. Among the gram-negative bacteria, *Klebsiella pneumoniae* was the most sensitive with 22.1 mm of inhibition zone, followed by *Salmonella typhimurium* with 21.2 mm and *Escherichia coli* with 19.6 mm. *Pseudomonas aeruginosa* was the most vulnerable bacterium for the plant metabolites with 23.3 mm of inhibition zone followed by *Candida albicans* with 22.6 mm of inhibition zone. The zones of inhibition of *Staphylococcus aureus* and methicillin-resistant *Staphylococcus aureus* (MRSA) were 20 mm and 19 mm, respectively. *Staphylococcus epidermidis* and *Enterococcus faecalis* were unresponsive to the plant metabolites (Arora and Sood 2017).

Ficus species from family Moraceae have been shown to have antifungal (Hassan et al. 2006), anti-diarrhoeal (Ahmadu et al. 2007), anti-inflammatory and anti-nociceptive activities (Amos et al. 2002). In addition to possessing the common metabolites, the ethanolic extracts of leaf and stem bark of *Ficus sycomorus* and *Ficus platyphylla* showed prominent antimicrobial activity against *Trichophyton mentagrophytes*, *Staphylococcus aureus* and ciprofloxacin-resistant *Salmonella typhi* (Adeshina et al. 2009). Table 5.3 lists the antibacterial activity of the medicinal plants.

Recently, an interesting study on green tea and Korean mixed tea subjected to antibacterial and antifungal activities was done, and the results indicated that green tea possessed substantial antibacterial and antifungal activities, while the mixed tea presented less activity. It was evident that epigallocatechin gallate (EGCG) was the responsible compound for the property (Muthu et al. 2016).

Moringa oleifera is known to possess immense pharmacological properties such as antioxidant, antidiabetic, antihypertensive, antitumor, antiulcer, cholesterol-lowering agent and is used in the traditional medicine. Though most of the plant parts are potent in producing metabolites, seed coat has been analysed for antimicrobial activity. *Candida albicans* was the most sensitive microbe for the ethanolic extract. *Staphylococcus epidermidis* and *Staphylococcus aureus* were the most sensitive Gram-positive bacteria. Among Gram-negative bacteria, *Klebsiella pneumoniae* was the most sensitive (Arora and Onsare 2014). A potent bioactive compound from turmeric was found to be effective against *Helicobacter pylori*. The metabolite curcumin (diferuloylmethane) was anticipated as a promising antimicrobial compound to solve gastroduodenal diseases like peptic ulcer, gastritis, etc. The minimum inhibitory concentration (MIC) of curcumin against different strains of

H. pylori was in the range of 5–50 μg/mL (De et al. 2009). Tables 5.4 and 5.5 list the antifungal and antiviral activities of the medicinal plants, respectively.

5.3 Novel Metabolites from Microbes

5.3.1 Antibacterial Activity

Penicillium sp. is a well-renowned fungus that produces an array of secondary metabolites like antibacterial substances (Lucas Esther et al. 2007), immunosuppressants, cholesterol-lowering agents (Kwon et al. 2002), antiviral substances (Nishihara et al. 2000), antifungal substances (Nicoletti et al. 2007) and also strong

Table 5.2 Common and recent methods to evaluate the antimicrobial activity

Technique/method	Advantages	Disadvantages	References
Agar disk diffusion	Minimal cost, simple to perform, tests a wide variety of microbes and antimicrobial agents, result interpretation is very simple	No differentiation of bacteriostatic and bactericidal effects	Reller et al. (2009), Kreger et al. (1980)
Antimicrobial gradient method (Etest)	(a) Minimum inhibitory concentration (MIC) can be done (b) The antimicrobial interface among two drugs can be studied	Expensive	White et al. (1996)
Thin layer chromatography (TLC)—bioautography	(a) Simplest method to identify the antifungal compounds, especially spore forming fungi (b) Can effectively separate a complicated mixture of compounds		Dewanjee et al. (2015)
Dilution test (a) Broth macro dilution (b) Broth micro dilution	MIC is accurately measured	Laborious, manual errors, inaccuracy of the antimicrobial solutions during preparation, lot of space requirement Results are not effectively reproduced	CLSI (2012)
ATP Bioluminescence assay	(a) Less time (3–4 days) (b) Antimicrobial analysis in situ or in vivo	Rapidity	Beckers et al. (1985), Vojtek et al. (2014)
Flow cytofluorometric method	Reproduce results rapidly in short time (2–6 h)	Inaccessibility of the flow cytometry equipment	Ramani and Chaturvedi (2000)

Table 5.3 Antibacterial secondary metabolites from medicinal plants

Medicinal plants	Endophyte/ actinomycetes	Organism/ species	Secondary metabolite	References
Acrostichum aureum	Endophyte	Penicillium sp.	Cyclo(Pro-Thr)	Cui et al. (2008)
Acrostichum aureum	Endophyte	Penicillium sp.	Cyclo (Pro-Tyr)	Cui et al. (2008)
Alpinia oxyphylla	Actinomycetes	Streptomyces sp.	2,6-dimethoxy terephthalic acid	Zhou et al. (2014)
Bruguiera gymnorrhiza	Endophyte	Streptomyces sp. HKI0576	Divergolides	Ding et al. (2011a, b)
Cynodon dactylon	Endophyte	Aspergillus fumigatus CY018	Asperfumoid	Wang et al. (2008)
Garcinia dulcis (Roxb.) Kurz.	Endophyte	Phomopsis sp. PSU-D15	Phomoenamide	Rukachaisirikul et al. (2008)
Garcinia dulcis (Roxb.) Kurz.	Endophyte	Phomopsis sp. PSU-D15	Phomonitroester	Rukachaisirikul et al. (2008)
Ginkgo biloba L	Endophyte	Xylaria	7-amino-4-methylcoumarin	Liu et al. (2008)
Kandelia candel	Endophyte	Streptomyces sp. HKI0595	Xiamycin B	Ding et al. (2011a, b)
Kandelia candel	Endophyte	Streptomyces sp. HKI0595	Indosespene	Ding et al. (2011a, b)
Kandelia candel	Endophyte	Streptomyces sp. HKI0595	Sespenine	Ding et al. (2011a, b)
Porteresia coarctata (Roxb.)	Endophyte	Penicillium chrysogenum, (MTCC 5108)	3,1″-didehydro-3[2″(3″,3″-dimethylprop-2-enyl)-3″-indolylmethylene]-6-methyl pipera-zine-2,5-dione	Devi et al. (2012)
Zea mays	Endophyte	Acremonium zeae	Pyrrocidines A–B	Wicklow and Poling (2009)

mycotoxins (Frisvad Jens and Samson Robert 2004). A group of scientists from Brazil has reported three metabolites from *Penicillium* sp. Among the three compounds, methyl 6-acetyl-4-methoxy-5,7,8-trihydroxynaphthalene-2-carboxylate worked as a wonder substance having a broad spectrum of antimicrobial activity against *Candida albicans* with a minimum inhibitory concentration of 32 µg/mL and *Bacillus cereus* and *Listeria monocytogenes* with MIC of 64 µg/mL (Petit et al. 2009). Another novel compound extracted from *Penicillium* sp. is 7-methoxy2,2-dimethyl-4-octa-4′,6′-dienyl-2*H*-napthalene-1-one which exhibited antimicrobial activity against *K. pneumoniae,* gram-negative bacteria and MRSA, *S. aureus, S. epidermidis,* Gram-positive bacteria. This study has also proven results in cytotoxicity and mutagenicity assay, revealing the suitability of the compound for further use (Kaur et al. 2015).

Table 5.4 Antifungal secondary metabolite from medicinal plants

Medicinal plants	Endophyte/ actinomycetes	Organism/species	Secondary metabolite	References
Alpinia galanga a (L.)	Endophyte	Streptomyces sp. LJK109	3-Methylcarbazole	Taechowisan et al. (2012)
Alpinia galanga a (L.)	Endophyte	Streptomyces sp. LJK109	1-Methoxy-3-methylcarbazole	Taechowisan et al. (2012)
Alpinia oxyphylla	Actinomycetes	Streptomyces sp.	2,6-Dimethoxy terephthalic acid	Zhou et al. (2014)
Catunaregam tomentosa	Endophyte	Curvularia geniculata	Curvularides A–E	Chomcheon et al. (2010)
Emblica officinalis Gaertn	Actinomycetes	Streptomyces griseofuscus	Indole acetic acid (IAA), siderophores	Gangwar et al. (2014)
Emblica officinalis Gaertn	Actinomycetes	Streptomyces cinereus	(IAA), siderophores	Gangwar et al. (2014)
Emblica officinalis Gaertn	Actinomycetes	Streptomyces flavus	Indole acetic acid (IAA)	Gangwar et al. (2014)
Ginkgo biloba	Endophyte	Chaetomium globosum No. 04	Chaetoglobosin A, G, D, R	Zhang et al. (2013)
Gloriosa superba	Endophyte	Aspergillus sp.	KL-4	Budhiraja et al. (2013)
Melia azedarach	Endophyte	Aspergillus fumigatus LN-4	12β-Hydroxy-13α-methoxyverruculogen TR-2	Li et al. (2012)
Melia azedarach	Endophyte	Aspergillus fumigatus LN-4	3-Hydroxyfumiquinazoline A	Li et al. (2012)
Phyllanthus niruri	Actinomycetes	Actinomadura sp.	Volatile organic compounds (VOCs)	Priya (2012)
Xinglong Chinese tree	Endophyte	Pestalotiopsis adusta	Pestalachloride A	Li et al. (2008)

Table 5.5 Antiviral secondary metabolites from medicinal plants

Medicinal plants	Endophyte/ actinomycetes	Organism/species	Secondary metabolite	References
Acer truncatum Bunge	Endophyte	*Paraconiothyrium brasiliense*	Brasilamides A–D	Liu et al. (2010)
Aegiceras corniculatum	Endophyte	*Emericella* sp.	Merimidine A	Zhang et al. (2011)
Aegiceras corniculatum	Endophyte	*Emericella* sp.	Emerimidine B	Zhang et al. (2011)
Aegiceras corniculatum	Endophyte	*Emericella* sp.	Emeriphenolicin A–F	Zhang et al. (2011)
Bruguiera gymnorrhiza	Endophyte	*Streptomyces* sp.	Xiamycin A	Ding et al. (2011a, b)
Xylocarpus granatum	Endophyte	*Jishengella endophytica*	Perlolyrine, 1-hydroxy-β-carboline	Wang et al. (2014)

Two metabolites elucidated as 4-Chromanone, 6-hydroxy-2-methyl-(5CI) and Modiolide A were isolated from a unique endophytic fungus *Periconia siamensis* extracted from leaves of *Thysanoleana latifolia*. These compounds were found to be effective against *Listeria monocytogenes*, MRSA, *Bacillus cereus* and *Pseudomonas aeruginosa* (Bhilabutra et al. 2007). Table 5.6 provides the list of bacterial metabolites from various microbes.

The lactic acid bacterial cultures were treated against *Mycobacterium avium* subsp. *Paratuberculosis* to test for the production of bacteriocin (Kralik et al. 2018). Recently, volatile organic compounds with antimicrobial activity were extracted from the bacterium *Bacillus amyloliquefaciens* strain S13. These compounds were found to be effective against *Saccharomyces cerevisiae, Agrobacterium tumefaciens* and *Bacillus subtilis* (Sonia Hamiche et al. 2019).

Bacillus subtilis isolated from soil sample served as a potent biological agent by controlling the fatal phytopathogens, namely *Botrytis cinerea, Monilia linhartiana, Alternaria solani, Rhizoctonia* sp. and *Phytophthora cryptogea*. This paves the way to control several plant diseases. The same bacterium also showed antibacterial activity against two pathogens *Pseudomonas syringae* and *Xanthomonas campestris* (Todorova and Kozhuharova 2010). There is more impending work from the researchers in this field, and the microbial-derived drugs should be used wisely by the human population.

5.3.2 Antifungal Activity

Very few antifungal drugs are successful in the market, owing to their poor efficacy and other reasons like azole resistance (Carledge et al. 1997). However, there are recent enhancements involving polyene lipid formulation with less toxicity and

Table 5.6 Bacterial secondary metabolites with antimicrobial activity

Isolated organism	Source of organism	Identified secondary metabolite	Antimicrobial activity against	Inhibition concentrationIC 50 valuesMg mL⁻¹	References
Streptomyces sp.	Marine water	Fijimycins A–C, etamycin A	Methicillin-resistant *Staphylococcus aureus*	4–16	Peng et al. (2011)
Streptomyces sp.	Marine water	Heronamycin A	*Bacillus subtilis*	14 and 18	Raju et al. (2012)
Halobacillus litoralis YS3106	Marine water	Halolitoralin A, B and C	*Candida albicans*	20, 30 and 30	Yang et al. (2002)
			Trichophyton rubrum	25, 35 and 40	
Pseudomonas sp.	Marine water	1-acetyl-beta-carboline	Methicillin-resistant *Staphylococcus aureus*	32–128	Lee et al. (2013)
Streptomyces sp. Merv8102	Marine water	Essramycin	*Staphylococcus aureus*, *Escherichia coli*, *Micrococcus luteus* and *Pseudomonas aeruginosa*	1.0–8.0	El-Gendy et al. (2008)
Bacillus sp. 05ID194	Marine water	Macrolactin W	*Bacillus subtilis*, *Staphylococcus aureus*, *Escherichia coli* and *Pseudomonas aeruginosa*	64	Mondol et al. (2011)
Pseudomonas sp.	Algae: *Diginea* sp.	Cyclic tetrapeptides	*Bacillus subtilis*, *Vibrio anguillarum*	No activity	Rungprom et al. (2008)
Rhodococcuserythropolis and*Pseudomonas* sp.	Petrosia ficiformis	Nil	*Staphylococcus aureus*	Minimal inhibition zone was observed	Chelossi et al. (2004)
Enterobacter sp.	*Dysidea granulose*	Crude extracts	*Salmonella typhi*, *Escherichia coli* and *Klebsiella pneumonia*	Minimal inhibition about 5	Gopi et al. (2012)

(continued)

Table 5.6 (continued)

Isolated organism	Source of organism	Identified secondary metabolite	Antimicrobial activity against	Inhibition concentrationIC 50 valuesMg mL^{-1}	References
Pseudoalteromonasflavipulchra	*Montipora aequituberculata*	Nil	15 strainsof methicillin-resistant *Staphylococcus aureus*	Minimal inhibition zone was observed	Chen et al. (2010)
Flavobacteria, α- and *γ-Proteobacteria* and*actinobacteria*	Unidentified	Nil	*Escherichia coli, Bacillus subtilis, Staphylococcus lentus, Candida glabrata*	Minimal inhibition zone was observed for 3 days	Heindl et al. (2010)
*Pseudoalteromonas*Phenolica O-BC30	Marine water	2, 2′,3-Tribromophenyl-4,4′-dicarboxylic acid	Methicillin-resistant *Staphylococcus aureus*	1–4	Isnansetyo and Kamei (2009)
*Brevibacillus*laterosporus PNG276	Marine water	Tauramamide as its methyland ethyl esters	*Enterococcus* sp.	0.1	Desjardine et al. (2007)
Streptomyces strain	Marine water	Marinopyrroles A and B	Methicillin-resistant *Staphylococcus aureus*	0.61 and 1.1	Hughes et al. (2008)
Streptomyces sp. BCC45,596	Marine water	Urdamycinone E,urdamycinone G anddehydroxyaquayamycin	*Mycobacterium tuberculosis*	3.13–12.50	Supong et al. (2012)
Rapidithrix sp.	Marine water	Ariakemicins A and B	*Brevibacterium* sp.	83	Oku et al. (2008)
			Staphylococcus aureus	0.46	
			Bacillus subtilis	83	
			Cytophaga marinoflava	>700	
			Pseudovibrio sp.	>700	
			Enterococcus sp.	>700	
			Pseudomonas aeruginosa	>700	
			Candida albicans	>700	

Isolated organism	Source of organism	Identified secondary metabolite	Antimicrobial activity against	Inhibition concentrationIC 50 valuesMg mL^{-1}	References
Fischerella sp.	Marine water	Ambiguine H isonitrile and Ambiguine I isonitrile	Candida albicans	6.25 and 0.39	Smitka et al. (1992)
			Staphylococcus albus	0.625 and 0.078	
			Saccharomyces cerevisiae	1.25 and 0.312	
			Bacillus subtilis	5 and 0.312	
			Escherichia coli	10 and 2.5	
Fischerella ambigua	Marine water	Ambiguine K and M isonitriles	Mycobacterium tuberculosis	6.6 and 7.5	Mo et al. (2009)
Micrococcus sp.,Vibrio sp., and Pseudoalteromonas and Bacillus sp.	Sponge: Aplysina aerophobaand Aplysina cavernicola	Nil	Gram-positive bacteria, Methicillin-resistant Staphylococcus aureus, gram-negative bacteria, Staphylococcus epidermidis	Minimal inhibition zone was determined	Hentschel et al. (2001)

novel triazoles possessing a wide range of action against many fungal isolates (Granier 2000). The drug target is presumably on the fungal cell wall, which comprises three major constituents: chitin, glucan and mannoproteins. Hence, the drug's action is determined by the nature of the fungal cell wall composition. There are also natural fungal inhibitors isolated from microbes which are covered in detail in the subsequent sections.

A potent strain named *Lactococcus lactis* LI4 strain isolated from dairy foods was found to prevent the growth of *Candida albicans* DMST 5239. Surprisingly, the activity of the culture was persistent in pH 2.0–4.0 (Lertcanawanichakul 2011). It was reported that by combining both natural and engineered lactococci strains, a group of compounds termed alkyl ketones were produced, and they were responsible for the antifungal activity against yeast genera *Candida* and *Rhodotorula* and *Aspergillus* and *Fusarium* (Stoyanova et al. 2006, 2010). Few reports have suggested that these strains can be efficiently used in the preservation of vegetables and fruits from fungal contamination (Trias et al. 2008).

In 2007, Gai et al. studied antifungal compounds in the ethanolic extracts of *Fusarium* sp. Based on the identification studies of the bioactive compound, it was interpreted as Fusarielin E (Gai et al. 2007). In a recent study, two glycolipids isolated from *Bacillus licheniformis* exhibited antifungal activities against *Rhizoctonia solani*, *Colletotrichum acutatum*, *Aspergillus niger*, *Candida albicans* and *Botrytis cinerea*. They were identified as leodoglucomide and leodoglycolipid (Tareq et al. 2015).

Haliangicin, a beta-methoxyacrylate antibiotic, exhibited effective inhibition of growth against *Pythium ultimum*, *B. cinerea* and *Saprolegnia parasitica* (Fudou et al. 2001a, b). *Bacillus laterosporus* isolated from Papua produced Basiliskamides A and B and showed potent antifungal action against *Aspergillus fumigatus* and *Candida albicans* (Barsby et al. 2002).

A novel compound named Saadamycin isolated from *Streptomyces* sp. exhibited antifungal action against *A. fumigatus*, *T. mentagrophytes*, *A. niger*, *F. oxysporum*, *Epidermophyton floccosum*, *Microsporum gypseum* and *C. albicans* (El-Gendy and El-Bondkly 2010). A novel hexapeptide, Sclerotide B, derived from marine-based *Aspergillus sclerotiorum* PT06-1 possessed antifungal action against *Candida albicans* (Zheng et al. 2009). The same organism from an anonymous source exhibited antimycotic activity against *A. flavus* (Bao et al. 2010). Two sesquiterpene type compounds isolated from *Penicillium bilaiae* MA-267 showed antifungal activity against phytopathogenic fungi such as *Colletotrichum gloeosporioides*, *Gaeumannomyces graminis*, *F. graminearum* and *Alternaria brassicae* (Meng et al. 2014). Peniciadametizine A isolated from *Penicillium adametzioides* AS-53, presented selective antimycotic activity against the phytopathogenic fungus *A. brassicae* (Liu et al. 2015). Table 5.7 shows the list of fungal secondary metabolites.

5.3.3 Antiviral Activity

At present, there are many well reputable microbial-driven secondary metabolites used for antiviral activity. Among the Actinomycetes, *Streptomyces roseus* played a promising role in producing leupeptin, a serine and cysteine protease (Aoyagi et al. 1969). This enzyme has proven to impede the glycoprotein-mediated admittance of *Marburg virus* (Gnirss et al. 2012). Similarly, protease inhibitors like Antipain and Elastatinal were utilized to hinder poliovirus protease (George A. Belov 2004). An interesting analogue of sialic acid, siastatin B, extracted from *Streptomyces verticillus* showed anti-influenza activity and in vitro propagation (Yoshio Nishimura et al. 1993).

There is an increasing need to focus on drugs for Hepatitis B virus (HBV) since the number of individuals who are infected with HBV is on steep escalation. The existing approved antiviral drugs comprise immunomodulators and nucleotide analogues that help to interfere in the disease advancement. Nevertheless, these drugs have their limitations such as side effects, limited efficacy and development of drug-resistant varieties. In contrast to the phytochemicals, there are very few microbial metabolites to act against HBV (Zhang and Wang 2014; Wu 2016).

To treat influenza virus, the following synthetic antiviral drugs have been approved: (a) amantadine (b) ribavirin (c) favipiravir (d) rimantadine (e) peramivir (f) and laninamivir octanoate (g) oseltamivir and (h) zanamivir. The drugs from natural products are currently under research, and the targets of these drugs are neuraminidase and haemagglutinin. Stachyflin, a novel compound isolated from *Stachybotrys sp.* RF-7260 provides conformational changes in haemagglutinin, thus preventing the fusion of host cells with virus (Nakatani et al. 2002; Minagawa et al. 2002).

The antiviral drugs that are derived from natural products exhibited a mild anti-HIV activity: disorazol, tubulysin and stigmatellin. Two myxobacterial strains named *Polyangium* sp. and *Myxococcus stipitatus* produced compounds such as thiangazole, phenalamide A1 and phenoxan with anti-HIV activity (Martinez et al. 2013). Myriocin and NA255, potent serine palmitoyl transferase inhibitors, were isolated from *Myriococcum albomyces* and *Fusarium incarnatum* and known to prevent the propagation of HBV, HCV and influenza virus (Kluepfel et al. 1972; Sakamoto et al. 2005; Tafesse et al. 2013).

5.4 Challenge to Antimicrobial Resistance

Antimicrobial resistance has turned up as a growing threat to the human health. It is affecting both the developing and developed nations, and its needs to be addressed immediately to save several lives. In the past 10 years, increasing awareness over this issue has been observed, and this is the right time that we use this occasion to obviate the disaster projected on health.

Table 5.7 Fungal secondary metabolites with antimicrobial activity

Isolated organism	Source of organism	Identified secondary metabolite	Antimicrobial activity test against	Inhibition concentrationIC 50 valuesMg mL^{-1}	References
Aspergillus protuberus SP1	Marine water	n-Butanol fraction	Pneumococcus mirabilis, Escherichia coli, Klebsiella pneumoniae and Bacillus subtilis	Minimal inhibition zone was observed	Mathan et al. (2011)
Zopfiella latipes	Marine water	Zopfiellamides A and B	Arthrobacter citreus, Arthrobacter citreus, Bacillus brevis, Bacillus subtilis, Corynebacterium insidiosum, Micrococcus luteus, Bacillus licheniformis, Mycobacterium phlei, Streptomyces sp.	2 and 10 for Zopfiellamides A	Daferner et al. (2002)
Daldinia eschscholzii	Gracilaria sp.	Helicascolisides A, B and C	Escherichia coli, Pseudomonas aeruginosa, Bacillus subtilis, Staphylococcus aureus, Cytophaga marinoflava	No antibacterial activity	Tarman et al. (2012)
Eurotium herbariorum	Enteromorpha prolifera	Cristatumin E	Enterobacter aerogenes, Enterococcus sp.	44	Li et al. (2013)
ZZF36	Algae: Sargassum sp.	5-Hydroxy-de-O-methyllasiodiplodin, 6-Oxo-de-methyllasiodiplodin,Lasiodiplodin,de-O-Methyllasiodipodin, (E)-9-Etheno-lasiodiplodin	Bacillus subtilis, Staphylococcus aureus, Escherichia coli, Salmonella enteritidis, Candida albicans, Fusarium oxysporum	6.25–100	Yang et al. (2006)
Penicillium chrysogenum QEN-24S,	Algae: Laurencia sp.	Penicisteroid A	Aspergillus niger, Alternaria brassicae	Minimal inhibition zone was observed	Gao et al. (2011)
Cryptosporiopsis spp.	Sponge (Clidemia hirta)	2-Butyryl-3-hydroxyphenoxy)-6-hydroxyphenyl)-3-hydroxybut-2-en-1-one, 1-(2,6-Dihydroxyphenyl) pentan-1-one and (Z)-1-(2)	Pseudomonas fluoroscens	6, 8–30	Zilla et al. (2013)
Eurotium cristatum KUFC7356	Sponge: Suberites domuncula	Eurocristatine	Escherichia coli, Pseudomonas aeruginosa, Staphylococcus aureus, Candida albicans, Trichophyton rubrum	Nil effects to fungi and bacteria	Gomes et al. (2012)

Isolated organism	Source of organism	Identified secondary metabolite	Antimicrobial activity test against	Inhibition concentration IC 50 values Mg mL^{-1}	References
Aspergillus ochraceus MP2	Unidentified marine sponge	α-Campholene aldehyde and Lucenin-2	*Klebsiella pneumonia, Staphylococcus aureus, Pseudomonas aeruginosa*	Minimal inhibition zone was observed	Meenupriya and Thangaraj (2011)
Aspergillus versicolor LCJ-5-4	Coral: *Cladiella* sp.	Versicolorides and tetraorcinol	*Staphylococcus aureus, Enterobacter aerogenes, Escherichia coli, Bacillus subtilis, Candida albicans, Pseudomonas aeruginosa*	Minimal inhibition zone was observed >150	Zhuang et al. (2011)
Zygosporium sp. *KNC52*	Hard coral	Sulfoalkylresorcinol	*Mycobacterium tuberculosis, Mycobacterium avium, Mycobacterium bovis, Pseudomonas aeruginosa,* methicillin-resistant *Staphylococcus aureus*	166, 50 and 12.5	Kanoh et al. (2008)
Aspergillus carneus KMM 4638	Marine water	Carneamides A–C, carnequinazolines A–C and carnemycin A, B	*Staphylococcus aureus, Escherichia coli, Pseudomonas aeruginosa, Bacillus cereus, Candida albicans*	No effects	Zhuravleva et al. (2012)
Aspergillus carbonarius WZ-4-11	Marine water	Naphtho-γ-pyrones	*Mycobacterium tuberculosis*	43, 21.5	Zhang et al. (2008)

As per the meeting held on 21 September 2016 to combat the antimicrobial resistance, the UN General Assembly has decided to organize a global effort with certain plans. Peter Thomson, president of the UN General Assembly, informed the representatives that the entire world should join hands to fight against this deadly factor with the integrative approaches. Besides endorsing WHO's current antimicrobial resistance plan and 2030 Agenda for Sustainable Development, the chapter represents a pledge by each member country (Muoio and Adalja 2016):

(a) To organize and manage investment into novel therapeutic technologies, research and investigation.
(b) To develop multisectoral programmes and strategies concentrated on antimicrobial resistance.
(c) To raise alertness of antimicrobial resistance to inculcate positive activities from the general public.
(d) To demand the launch of an ad hoc interagency coordination group in consultation with WHO, the Food and Agricultural Organization.

Though antibiotics are treated as a lifesaver globally, the CDC has reported that around 50% of the antibiotics that are prescribed for the patients are actually not desirable (CDC report 2013). The overuse of antibiotics commonly comes from recommending these medications to cure nonbacterial ailments for which they are inadequate. For instance, in USA, more than 60 million cases/year of viral flu are suggested antibiotics (CDC-About Antimicrobial resistance n.d.). In USA, every year, the number of people acquiring antibiotic-resistant bacterial infection is around two million, and 23,000 people die because of the chronic infection. On the other hand, over usage of these drugs might pose yeast infection in intestine, certain allergic reactions and diarrhoea (Bartlett 2002). On the contrary, no or late antibiotic treatment in case of bacterial infections is quite normal (Houck et al. 2004). This may decrease the risk of antibiotic-resistant bacteria but might pose a serious threat to life, leading to mortality (Little 2005; Spiro et al. 2006).

5.5 Future of Natural Medicine

Both microbes and medicinal herbs will endure to play a better part in pharmaceutical industry since only 0.001% of microbes and half a million plants are explored in world till date. Scientists will come up with astonishing metabolites to treat the life-threatening diseases in the near future (Singh 2015). Herbal medicine is extensively used across the globe in a constant way. This has led to the trend in replacing the synthetic drugs with the conventional medicine. However, plants/microbes are prone to contamination or infection, making their recovery and usage a critical issue. Hence, they should be subjected to a set of tests before the extraction process to ensure the safety and efficacy of the drug (Clark 1996; Firenzuoli and Gori, 2007).

One of the challenges that is imposed on herbal therapy is the extensive use of the whole plants leading to the threat of extinction. Habitat destruction, increasing population and wildlife resource exploitation are the serious problems that are associated with this therapy (Bentley 2010). However, they can be solved by wise usage, proper plantation and harvest of the medicinal plants.

5.6 Conclusion

There is no doubt that both plants and microbes had made the entire globe to accept the fact that secondary metabolites derived from them turn to be a lifesaver. This is well correlated with the recent improvements in the research field. The development of new medicines and advanced approaches towards the disease diagnosis will surely combat the threatening antibiotic resistance. On the other hand, personalized medicine is also gaining more attention. Henceforth, both microbial and plant products will serve the pharmaceutical needs of the human population in a safe way.

References

Adeola MO (1992) Importance of wild animals and their parts in the culture, religious festivals, and traditional medicine, of Nigeria. Environ Conserv 19(2):125–134

Adeshina G, Okeke CL, Onwuegbuchulam N, Ehinmidu J (2009) Preliminary studies on antimicrobial activities of ethanolic extracts of *Ficus sycomorus* Linn. and *Ficus platyphylla* Del. (Moraceae). Int J Biol Chem Sci 3(5)

Ahmadu AA, Zezi AU, Yaro AH (2007) Anti-diarrheal activity of the leaf extracts of *Daniellia oliveri* Hutch and Dalz (Fabaceae) and *Ficus sycomorus* Miq (Moraceae). Afr J Tradit Complement Altern Med 4(4):524–528

Amos S, Chindo B, Edmond I, Akah P, Wambebe C, Gamaniel K (2002) Antiinflammatory and antinociceptive effects of *Ficus platyphylla* extracts in mice and rat. J Evol Biol 18(15):1234–1244

Aoyagi T, Takeuchi T, Matsuzaki A, Kawamura K, Kondo S, Hamada M, Maeda K, Umezawa H (1969) Leupeptins, new protease inhibitors from Actinomycetes. J Antibiot (Tokyo) 22:283–286

Arora DS, Onsare JG (2014) Antimicrobial potential of *Moringa oleifera* seed coat and its bioactive phytoconstituents. Korean J Microbiol Biotechnol 42:152–161

Arora DS, Sood H (2017) In vitro antimicrobial potential of extracts and phytoconstituents from *Gymnema sylvestre* R. Br. Leaves and their biosafety evaluation. AMB Express 7(1):115

Bao L, Xu Z, Niu SB, Namikoshi M, Kobayashi H, Liu HW (2010) (−)-Sclerotiorin from an unidentified marine fungus as an anti-meiotic and anti-fungal agent. Nat Prod Commun 5(11):1789–1792

Barsby T, Kelly MT, Andersen RJ (2002) Tupuseleiamides and basiliskamides, new acyldipeptides and antifungal polyketides produced in culture by a *Bacillus laterosporus* isolate obtained from a tropical marine habitat. J Nat Prod 65(10):1447–1451

Bartlett JG (2002) Clinical practice. Antibiotic-associated diarrhea. N Engl J Med 346(5):334–339

Beckers B, Lang HRM, Schimke D, Lammers A (1985) Evaluation of a bioluminescence assay for rapid antimicrobial susceptibility testing of mycobacteria. Eur J Clin Microbiol 4(6):556–561

Belov GA, Lidsky PV, Mikitas OV, Egger D, Lukyanov KA, Bienz K, Agol VI (2004) Bidirectional increase in permeability of nuclear envelope upon poliovirus infection and accompanying alterations of nuclear pores. J Virol 78(18):10166–10177

Bentley RE (2010) Medicinal plants. Domville-Fife Press, London, pp 23–46

Bérdy J (1989) The discovery of new bioactive microbial metabolites: screening and identification. In: Bushell ME, Grafe U (eds) Bioactive metabolites from microorganisms. Elsevier, Amsterdam, pp 3–33

Bhilabutra W, Techowisan T, Peberdy JF, Lumyong S (2007) Antimicrobial activity of bioactive compounds from *Periconia siamensis* CMUGE015. Res J Microbiol 2(10):749–755

Boakye MK, Pietersen DW, Kotzé A, Dalton DL, Jansen R (2015) Knowledge and uses of African pangolins as a source of traditional medicine in Ghana. PLoS One 10:e0117199

Budhiraja A, Nepali K, Sapra S, Gupta S, Kumar S, Dhar KL (2013) Bioactive metabolites from an endophytic fungus of *Aspergillus* species isolated from seeds of Gloriosa superba Linn. Med Chem Res 22(1):323–329

Carledge JD, Midgley J, Gazzard BG (1997) Clinically significant azole cross-resistance in Candida isolates from HIV-positive patients with oral candidiasis. AIDS 11(15):1839–1844

CDC (n.d.) About Antimicrobial Resistance. www.cdc.gov/drugresistance/about.html

CDC Threat Report 2013. Antimicrobial Resistance. www.cdc.gov

Chelossi E, Milanese M, Milano A, Pronzato R, Riccardi G (2004) Characterization and antimicrobial activity of epibiotic bacteria from *Petrosia ficiformis* (Porifera, Demospongiae). J Exp Mar Biol Ecol 309(1):21–23

Chen WM, Lin CY, Chen CA, Wang JT, Sheu SY (2010) Involvement of an L-amino acid oxidase in the activity of the marine bacterium *Pseudoalteromonas flavipulchra* against methicillin resistant *Staphylococcus aureus*. Enzyme Microbiol Tech 47(1–2):52–58

Chomcheon P, Wiyakrutta S, Aree T, Sriubolmas N, Ngamrojanavanich N, Mahidol CS, Ruchirawat S, Kittakoop P (2010) Curvularides A–E: antifungal hybrid peptide–polyketides from the endophytic fungus *Curvularia geniculata*. Chem Eur J 16(36):11178–11185

Clark AM (1996) Natural products as a resource for new drugs. Pharm Res 13(8):1133–1141

CLSI (2012) Methods for dilution antimicrobial susceptibility tests for bacteria that grow aerobically, approved standard, 9thed., CLSI document M07-A9. Clinical and Laboratory Standards Institute, 950 West Valley Road, Suite 2500, Wayne, Pennsylvania 19087, USA

Cowan MM (1999) Plant products as antimicrobial agents. Clin Microbiol Rev 12:564–582

Cui HB, Mei WL, Miao CD, Lin HP, Kui H, Dai HF (2008) Antibacterial constituents from the endophytic fungus *Penicillium* sp. 0935030 of mangrove plant *Acrostichum aureurm*. Chin J Antibiot 7:4

Daferner M, Anke T, Sterner O (2002) Zopfiellamides A and B, antimicrobial pyrrolidinone derivatives from the marine fungus *Zopfiella latipes*. Tetrahedron 58:7781–7784

De R, Kundu P, Swarnakar S, Ramamurthy T, Chowdhury A, Balakrish Nair G, Mukhopadhyay AK (2009) Antimicrobial activity of curcumin against *Helicobacter pylori* isolates from India and during infections in mice. Antimicrob Agents Chemother 53(4):1592–1597

Desjardine K, Pereira A, Wright H, Matainaho T, Kelly M, Andersen RJ (2007) Tauramamide, a lipopeptide antibiotic produced in culture by *Brevibacillus laterosporus* isolated from a marine habitat: structure elucidation and synthesis. J Nat Prod 70(12):1850–1853

Devi P, Rodrigues C, Naik CG, D'souza L (2012) Isolation and characterization of antibacterial compound from a mangrove-endophytic fungus, *Penicillium chrysogenum* MTCC 5108. Indian J Microbiol 52(4):617–623

Dewanjee S, Gangopadhyay M, Bhattacharya N, Khanra R, Dua TK (2015) Bioautography and its scope in the field of natural product chemistry. J Pharm Anal 5(2):75–84

Ding L, Maier A, Fiebig HH, Lin WH, Hertweck C (2011a) A family of multicyclic indolosesquiterpenes from a bacterial endophyte. Org Biomol Chem 9(11):4029–4031

Ding L, Maier HH, Fiebig H, Gorls WH, Lin G, Peschel C, Hertweck C (2011b) Divergolides A–D from a mangrove endophyte reveals an unparalleled plasticity in ansa-macrolide biosynthesis. Chem Int 50(7):1630–1634

El-Gendy MM, El-Bondkly AM (2010) Production and genetic improvement of a novel antimycotic agent, saadamycin, against dermatophytes and other clinical fungi from endophytic *Streptomyces* sp. Hedaya48. J Ind Microbiol Biotechnol 37(8):831–841

El-Gendy MM, Shaaban M, Shaaban KA, El-Bondkly AM, Laatsch H (2008) Essramycin: a first triazolopyrimidine antibiotic isolated from nature. Antibiot 61(3):149–157

Fabricant DS, Farnsworth NR (2001) The value of plants used in traditional medicine for drug discovery. Environ Health Perspect 109:69–75

Farnsworth NR, Morris RW (1976) Higher plants: the sleeping giant for drug development. Am J Pharm 148:46–52

Firenzuoli F, Gori L (2007) Herbal medicine today: clinical and research issues. Evid Based Complement Alternat Med 4(Suppl 1):37–40

Fraenkel GS (1959) The raison d'Être of secondary plant substances these odd chemicals arose as a means of protecting plants from insects and now guide insects to food. Science 129(3361):1466–1470

Frisvad Jens C, Samson Robert A (2004) Polyphasic taxonomy of *Penicillium* subgenus *Penicillium*: a guide to identification of food and air-borne terverticillate *Penicillia* and their mycotoxins. Stud Mycol 49:1–174

Fudou R, Iizuka T, Sato S, Ando T, Shimba N, Yamanaka S (2001a) Haliangicin, a novel antifungal metabolite produced by a marine myxobacterium. J Antibiot (Tokyo) 54(2):153–156

Fudou R, Iizuka T, Yamanaka S (2001b) Haliangicin, a novel antifungal metabolite produced by a marine myxobacterium. J Antibiot (Tokyo) 54(2):149–152

Gai Y, Zhao LL, Hu CQ, Zhang HP (2007) Fusarielin E, a new antifungal antibiotic from *Fusarium* sp. Chin Chem Lett 18(8):954–956

Gangwar M, Khushboo SP, Saini P (2014) Diversity of endophytic actinomycetes in *Musa acuminata* and their plant growth promoting activity. J Biol Chem Sci 1:13–23

Gao SS, Li XM, Li CS, Proksch P, Wang BG (2011) Penicisteroids a and B, antifungal and cytotoxic polyoxygenated steroids from the marine alga-derived endophytic fungus *Penicillium chrysogenum* QEN-24S. Bioorg Med Chem Lett 21:2894–2897

Gaudêncio SP, Pereira F (2015) Dereplication: racing to speed up the natural products discovery process. Nat Prod Rep 32(6):779–810

Gnirss K, Kühl A, Karsten C, Glowacka I, Bertram S, Kaup F, Hofmann H, Pöhlmann S (2012) Cathepsins B and L activate Ebola but not Marburg virus glycoproteins for efficient entry into cell lines and macrophages independent of TMPRSS2 expression. Virology 424(1):3–10

Gomes NM, Dethoup T, Singburaudom N, Gales L, Silva AM, Kijjoa A (2012) Eurocristatine, a new diketopiperazine dimer from the marine sponge-associated fungus *Eurotium cristatum*. Phytochem Lett 5(4):717–720

Gopi M, Kumaran S, Kumar TTA, Deivasigamani B, Alagappan K, Prasad SG (2012) Antibacterial potential of sponge endosymbiont marine *Enterobacter* sp at Kavaratti Island, Lakshadweep archipelago. Asian Pac J Trop Med 5(2):142–146

Granier F (2000) Invasive fungal infections. Epidemiol New Therap Presse Med 29:2051–2056

Hamiche S, Badis A, Jouadi B, Bouzidi N, Daghbouche Y, Utczás M, Mondello L, El Hattab M (2019) Identification of antimicrobial volatile compounds produced by the marine bacterium *Bacillus amyloliquefaciens* strain S13 newly isolated from brown alga *Zonaria tournefortii*. J Essent Oil Res 31:203. https://doi.org/10.1080/10412905.2018.1564380

Hassan SW, Umar RA, Lawal M, Bilbis LS, Muhammad BY (2006) Evaluation of antifungal activity of *Ficus sycomorus* L. (Moraceae). Biol Environ Sci J Tropics 3:18–25

Heindl H, Wiese J, Thiel V, Imhoff JF (2010) Phylogenetic diversity and antimicrobial activities of bryozoans associated bacteria isolated from Mediterranean and Baltic Sea habitats. Syst Appl Microbiol 33(2):94–104

Hentschel U, Schmid M, Wagner M, Fieseler L, Gernert C, Hacker J (2001) Isolation and phylogenetic analysis of bacteria with antimicrobial activities from the Mediterranean sponges *Aplysina aerophoba* and *Aplysina cavernicola*. FEMS Micro Ecol 35(3):305–312

Houck PM, Bratzler DW, Nsa W, Ma A, Bartlett JG (2004) Timing of antibiotic administration and outcomes for medicare patients hospitalized with community acquired pneumonia. Arch Intern Med 164(6):637–644

Hughes CC, Prieto-Davo A, Jensen PR, Fenical W (2008) The marinopyrroles, antibiotics of an unprecedented structure class from a marine *Streptomyces* sp. Org Lett 10(4):629–631

Isnansetyo A, Kamei Y (2009) Anti-methicillin-resistant *Staphylococcus aureus* (MRSA) activity of MC21-B, an antibacterial compound produced by the marine bacterium *Pseudoalteromonas phenolica* O-BC30T. Int J Antimicrob Agents 34(2):131–135

Johnson BM, Nikolic D, van Breemen RB (2002) Applications of pulsed ultrafiltration-mass spectrometry. Mass Spectrom Rev 21(2):76–86

Joy PP, Thomas J, Mathew S, Skaria BP (2001) Medicinal plants. In: Bose TK, Kabir J, Joy PP (eds) Tropical horticulture, vol 2. Naya Prakash, Kolkata, India

Kang S, Phipps MJ (2003) A question of attitude: South Korea's traditional medicine practitioners and wildlife conservation. TRAFFIC East Asia

Kanoh K, Adachi K, Matsuda S, Shizuri Y, Yasumoto K, Kusumi T, Okumura K, Kirikae T (2008) New sulfoalkylresorcinol from marine derived fungus, *Zygosporium* sp. KNC52. J Antibiot 61(3):192–194

Kaur H, Onsare JG, Sharma V, Arora DS (2015) Isolation, purification and characterization of novel antimicrobial compound 7-methoxy-2, 2-dimethyl-4-octa-4′, 6′-dienyl-2 H-napthalene-1-one from *Penicillium* sp. and its cytotoxicity studies. AMB Express 5(1):40–120

Kim JY, Noble D (2014) Recent progress and prospects in Sasang constitutional medicine: a traditional type of physiome-based treatment. Prog Biophys Mol Biol 116:76–80

Kluepfel D, Bagli J, Baker H, Charest MP, Kudelski A, Sehgal SN, Vézina C (1972) Myoriocin, a new antifungal antibiotic from *Myriococcum arbomyces*. J Antibiot (Tokyo) 22:109–115

Kralik P, Babak V, Dziedzinska R (2018) The impact of the antimicrobial compounds produced by lactic acid Bacteria on the growth performance of *Mycobacterium avium* subsp. *paratuberculosis*. Front Microbiol 9:638

Kreger BE, Craven DE, McCabe WR (1980) Gram-negative bacteremia: IV. Re-evaluation of clinical features and treatment in 612 patients. Am J Med 68(3):344–355

Kwon OE, Rho M-C, Song HY, Lee SW, Chung MY, Lee JH, Kim YH, Lee HS, Kim Y-K (2002) Phenylpyropene A and B, new inhibitors of acyl-CoA: cholesterol acyltransferase produced by *Penicillium griseofulvum* F1959. J Antibiot 55(11):1004–1008

Lee DS, Eom SH, Jeong SY, Shin HJ, Je JY, Lee EW, Chung YH, Kim YM, Kang CK, Lee MS (2013) Anti-methicillin-resistant *Staphylococcus aureus* (MRSA) substance from the marine bacterium *Pseudomonas* sp. UJ-6. Environ Toxicol Pharmacol 35(2):171–177

Lertcanawanichakul M (2011) Isolation and selection of anti-*Candida albicans* producing lactic acid Bacteria. Walailak J Sci Tech 2(2):179–187

Li E, Jiang L, Guo L, Zhang H, Che Y (2008) Pestalachlorides A-C, antifungal metabolites from the plant endophytic fungus *Pestalotiopsis adusta*. Bioorg Med Chem 16(17):7894–7899

Li XJ, Zhang Q, Zhang AL, Gao JM (2012) Metabolites from Aspergillus fumigatus, an endophytic fungus associated with Melia azedarach, and their antifungal, antifeedant, and toxic activities. J Agric Food Chem 60(13):3424–3431

Li Y, Sun KL, Wang Y, Fu P, Liu PP, Wang C, Zhu WM (2013) A cytotoxic pyrrolidinoindoline diketopiperazine dimer from the algal fungus *Eurotium herbariorum* HT-2. Chin Chem Lett 24(12):1049–1052

Little P (2005) Delayed prescribing of antibiotics for upper respiratory tract infection. BMJ 331(7512):301–302

Liu X, Dong M, Chen X, Jiang M, Lv X, Zhou J (2008) Antimicrobial activity of an endophytic *Xylaria* sp. YX-28 and identification of its antimicrobial compound 7-amino-4-methylcoumarin. Appl Microbiol Biotechnol 78(2):241–247

Liu L, Gao H, Chen X, Cai X, Yang L, Guo L, Yao X, Che Y (2010) Brasilamides A–D: sesquiterpenoids from the plant endophytic fungus *Paraconiothyrium brasiliense*. Eur J Org Chem 2010(17):3302–3306

Liu Y, Mandi A, Li XM, Meng LH, Kurtan T, Wang BG (2015) Peniciadametizine a, a dithiodiketopiperazine with a unique spiro[furan-2,7′-pyrazino[1,2-b] [1,2]oxazine] skeleton, and a related analogue, peniciadametizine B, from the marine sponge-derived fungus *Penicillium adametzioides*. Mar Drugs 13(6):3640–3652

Lucas Esther MF, Castro Mateus CM, Takahashi Jacqueline A (2007) Antimicrobial properties of sclerotiorin, isochromophilone VI and pencolide, metabolites from a Brazilian cerrado isolate of *Penicillium sclerotiorum* Van Beyma. Braz J Microbiol 38(4):785–789

Martinez JP, Hinkelmann B, Fleta-Soriano E, Steinmetz H, Jansen R, Diez J, Frank R, Sasse F, Meyerhans A (2013) Identification of myxobacteria-derived HIV inhibitors by a high-throughput two-step infectivity assay. Microb Cell Factories 12(1):85

Mathan S, Smith AA, Kumaran J, Prakash S (2011) Anticancer and antimicrobial activity of *Aspergillus protuberus* SP1 isolated from marine sediments of south Indian coast. Chin J Nat Med 9(4):286–292

Meenupriya J, Thangaraj M (2011) Analytical characterization and structure elucidation of metabolites from *Aspergillus ochraceus* MP2 fungi. Asian Pac J Trop Biomed 1(5):376–380

Meng LH, Li XM, Liu X, Wang BG (2014) Penicibilaenes a and B, sesquiterpenes with a tricyclo[6.3.1.0(1,5)]dodecane skeleton from the marine isolate of *Penicillium bilaiae* MA-267. Org Lett 16(23):6052–6055

Minagawa K, Kouzoki S, Yoshimoto J, Kawamura Y, Tani H, Iwata T (2002) Stachyflin and acetylstachyflin, novel anti-influenza a virus substances, produced by *Stachybotrys* sp. RF-7260. J Antibiot (Tokyo) 55:155–164

Mo S, Krunic A, Chlipala G, Orjala J (2009) Antimicrobial ambiguine isonitriles from the cyanobacterium *Fischerella ambigua*. J Nat Prod 72(5):894–899

Moaddel R, Wainer IW (2009) The preparation and development of cellular membrane affinity chromatography columns. Nat Protoc 4(2):197

Mondol MM, Kim JH, Lee HS, Lee YJ, Shin HJ (2011) Macrolactin W, a new antibacterial macrolide from a marine *Bacillus* sp. Bioorg Med Chem Lett 21(12):3832–3835

Muoio D, Adalja AA (2016) UN addresses' great challenges' of antimicrobial resistance. Infect Dis Child 29(10):3

Muthu M, Gopal J, Min SX, Chun S (2016) Green tea versus traditional Korean teas: antibacterial/antifungal or both? Appl Biochem Biotechnol 180(4):780–790

Nakatani M, Nakamura M, Suzuki A, Inoue M, Katoh T (2002) A new strategy toward the total synthesis of stachyflin, a potent anti-influenza a virus agent: concise route to the tetracyclic core structure. Org Lett 4:4483–4486

Nicoletti R, Lopez-Gresa MP, Manzo E, Carella A, Ciavatta ML (2007) Production and fungitoxic activity of Sch 642305, a secondary metabolite of *Penicillium canescens*. Mycopathologia 163(5):295–301

Nishihara YE, Tsujii S, Takase Y, Tsurumi T, Kino M, Hino M et al (2000) FR191512, a novel anti-influenza agent isolated from fungus strain no. 17415. J Antibiot 53:1333–1340

Nishimura Y, Kudo T, Kondo S, Takeuchi T (1993) Synthesis of 3-episiastatin B analogues having anti-influenza virus activity. J Antibiot (Tokyo) 46:1883–1889

Oku N, Adachi K, Matsuda S, Kasai H, Takatsuki A, Shizuri Y (2008) Ariakemicins A and B, novel polyketide-peptide antibiotics from a marine gliding bacterium of the genus Rapidithrix. Org Lett 10(12):2481–2484

Oliver SJ (2013) The role of traditional medicine practice in primary health care within aboriginal Australia: a review of the literature. J Ethnobiol Ethnomed 9:46

Parasuraman S, Thing GS, Dhanaraj SA (2014) Polyherbal formation: concept of ayurveda. Pharmacogn Rev 8:73–80

Peng S, Katherine N, Maloneya SN (2011) Fijimycins A-C, three antibacterial etamycin-class depsipeptides from a marine derived *Streptomyces* sp. Bioorg Med Chem Lett 19(22):6557–6562

Petit P, Lucas E, Abreu L, Pfenning L, Takahashi J (2009) Novel antimicrobial secondary metabolites from a *Penicillium* sp. isolated from Brazilian cerrado soil. Electron J Biotechnol 12(4):8–9

Potterat O, Hamburger M (2013) Concepts and technologies for tracking bioactive compounds in natural product extracts: generation of libraries, and hyphenation of analytical processes with bioassays. Nat Prod Rep 30(4):546–564

Prasad R, Pandey R, Barman I (2016) Engineering tailored nanoparticles with microbes: quo vadis. WIREs Nanomed Nanobiotechnol 8:316–330. https://doi.org/10.1002/wnan.1363

Prasad R, Jha A, Prasad K (2018) Exploring the realms of nature for Nanosynthesis. Springer International Publishing, New York. ISBN 978-3-319-99570-0. https://www.springer.com/978-3-319-99570-0

Priya RM (2012) Open Access. Sci Rep 4:259

Qi FH, Wang ZX, Cai PP, Zhao L, Gao JJ, Kokudo N, Li AY, Han JQ, Tang W (2013) Traditional Chinese medicine and related active compounds: a review of their role on hepatitis B virus infection. Drug Discov Ther 7:212–224

Rajeh MAB, Zuraini Z, Sasidharan S, Latha LY, Amutha S (2010) Assessment of *Euphorbia hirta* L. leaf, flower, stem and root extracts for their antibacterial and antifungal activity and brine shrimp lethality. Molecules 15:6008–6018

Raju R, Piggott AM, Khalil Z, Bernhardt PV, Capon RJ (2012) Heronamycin a: a new benzo-thiazine ansamycin from an Australian marine-derived *Streptomyces* sp. Tetrahedron Lett 53(9):1063–1065

Ramani R, Chaturvedi V (2000) Flow cytometry antifungal susceptibility testing of pathogenic yeasts other than *Candida albicans* and comparison with the NCCLS broth microdilution test. Antimicrob Agents Chemother 44(10):2752–2758

Reller LB, Weinstein M, Jorgensen JH, Ferraro MJ (2009) Antimicrobial susceptibility testing: a review of general principles and contemporary practices. Clin Infect Dis 49(11):1749–1755

Rios J, Recio M, Villar A (1988) Screening methods for natural products with antimicrobial activity: a review of the literature. J Ethnopharmacol 23:127

Rukachaisirikul V, Sommart U, Phongpaichit S, Sakayaroj J, Kirtikara K (2008) Metabolites from the endophytic fungus *Phomopsis* sp.PSU-D15. Phytochem Lett 69(3):783–787

Rungprom W, Siwu ER, Lambert LK, Dechsakulwatana C, Barden MC, Kokpol U, Blanchfield JT, Kita M, Garson MJ (2008) Cyclic tetrapeptides from marine bacteria associated with the seaweed *Diginea* sp. and the sponge *Halisarca ectofibrosa*. Tetrahedron 64(14):3147–3152

Sakamoto H, Okamoto K, Aoki M, Kato H, Katsume A, Ohta A, Kohara M (2005) Host sphingo-lipid biosynthesis as a target for hepatitis C virus therapy. Nat Chem Biol 1:333–337

Savoia D (2012) Plant-derived antimicrobial compounds: alternatives to antibiotics. Future Microbiol 7(8):979–990

Shikov AN, Pozharitskaya ON, Makarov VG, Wagner H, Verpoorte R, Heinrich M (2014) Medicinal plants of the Russian pharmacopoeia; their history and applications. J Ethnopharmacol 154:481–536

Singh R (2015) Medicinal plants: a review. J Plant Sci 3(1–1):50–55

Smitka TA, Bonjouklian R, Doolin L, Jones ND, Deeter JB, Yoshida WY, Prinsep MR, Moore RE, Patterson GM (1992) Ambiguine isonitriles, fungicidal hapalindole-type alkaloids from three genera of blue green algae belonging to the stigonemataceae. J Organomet Chem 57:857–861

Song HP, Zhang H, Fu Y, Mo HY, Zhang M, Chen J, Li P (2014) Screening for selective inhibitors of xanthine oxidase from Flos Chrysanthemum using ultrafiltration LC–MS combined with enzyme channel blocking. J Chromatogr B: anal Technol biomed. Life Sci 961:56–61

Spiro DM, Tay K-Y, Arnold DH, Dziura JD, Baker MD, Shapiro ED (2006) Wait-and-see prescription for the treatment of acute otitis media: a randomized controlled trial. J Am Med Assoc 296(10):1235–1241

Stamp N (2003) Out of the quagmire of plant defense hypotheses. Q Rev Biol 78(1):23–55

Stoyanova LG, Sul'timova TD, Botina SG, Netrusov AI (2006) Isolation and identification of new nisin-producing *Lactococcus lactis* subsp. lactis from milk. Appl Biochem Microbiol 42(5):492–499

Stoyanova LG, Ustyugova EA, Sultimova TD, Bilanenko EN, Fedorova GB, Khatrukha GS, Netrusov AI (2010) New antifungal bacteriocin-synthesizing strains of Lactococcus lactis ssp. lactis as the perspective biopreservatives for protection of raw smoked sausages. Am J Agric Biol Sci 5(4):477–485

Supong K, Thawai C, Suwanborirux K, Choowong W, Supothina S, Pittayakhajonwut P (2012) Antimalarial and antitubercular C-glycosylated benz[α]anthraquinones from the marine-derived *Streptomyces* sp. BCC45596. Phytochem Lett 5(3):651–656

Taechowisan T, Chanaphat S, Ruensamran W, Phutdhawong WS (2012) Antifungal activity of 3-methylcarbazoles from *Streptomyces* sp. LJK109; an endophyte in *Alpinia galanga*. J Appl Pharm Sci 2(3):124–128

Tafesse FG, Sanyal S, Ashour J, Guimaraes CP, Hermansson M, Somerharju P, Ploegh HL (2013) Intact sphingomyelin biosynthetic pathway is essential for intracellular transport of influenza virus glycoproteins. Proc Natl Acad Sci U S A 110:6406–6411

Tareq FS, Lee HS, Lee YJ, Lee JS, Shin HJ (2015) Ieodoglucomide C and ieodoglycolipid, new glycolipids from a marine-derived bacterium *Bacillus licheniformis* 09IDYM23. Lipids 50(5):513–519

Tarman K, Palm GJ, Porzel A, Merzweiler K, Arnold N, Wessjohann LA, Unterseher M, Lindequist U (2012) Helicascolide C, a new lactone from an Indonesian marine algicolous strain of *Daldinia eschscholzii* (Xylariaceae, Ascomycota). Phytochem Lett 5:83–86

Todorova S, Kozhuharova L (2010) Characteristics and antimicrobial activity of *Bacillus subtilis* strains isolated from soil. World J Microbiol Biotechnol 26(7):1207–1216

Trias R, Baneras L, Montesino E, Badosa E (2008) Lactic acid bacteria from fresh fruits and vegetables as biocontrol against phytopathogenic bacteria and fungi. Int Microbiol 11(4):231–236

Upadhyay A, Upadhyaya I, Kollanoor-Johny A, Venkitanarayanan K (2014) Combating pathogenic microorganisms using plant-derived antimicrobials: a Minireview of the mechanistic basis. Biomed Res Int 18

Valgas C, Souza SM, Smania EF, Smania A Jr (2007) Screening methods to determine antibacterial activity of natural products. Braz J Microbiol 38(2):369–380

van Breemen RB, Huang CR, Nikolic D, Woodbury CP, Zhao YZ, Venton DL (1997) Pulsed ultrafiltration mass spectrometry: a new method for screening combinatorial libraries. Anal Chem 69(11):2159–2164

Vanzolini KL, Jiang Z, Zhang X, Vieira LCC, Corrêa AG, Cardoso CL, Cass QB, Moaddel R (2013) Acetylcholinesterase immobilized capillary reactors coupled to protein coated magnetic beads: a new tool for plant extract ligand screening. Talanta 116:647–652

Vojtek L, Dobes P, Büyükgüzel E, Atosuo J, Hyrsl P (2014) Bioluminescent assay for evaluating antimicrobial activity in insect haemolymph. Eur J Entomol 111(3):335–340

Wang FW, Hou ZM, Wang CR, Li P, Shi DH (2008) Bioactive metabolites from *Penicillium* sp., an endophytic fungus residing in *Hopea hainanensis*. World J Microbiol Biotechnol 24(10):2143–2147

Wang P, Kong F, Wei J, Wang Y, Wang W, Hong K, Zhu W (2014) Alkaloids from the mangrove-derived actinomycete *Jishengella endophytica* 161111. Mar Drugs 12(1):477–490

White RL, Burgess DS, Manduru M, Bosso JA (1996) Comparison of three different in vitro methods of detecting synergy: time-kill, checkerboard, and E test. Antimicrob Agents Chemother 40(8):1914–1918

Wicklow DT, Poling SM (2009) Antimicrobial activity of pyrrocidines from *Acremonium zeae* against endophytes and pathogens of maize. Phytopathology 99(1):109–115

Wolfender JC (2010) In: Hajnos MW, Sherma J (eds) High performance liquid chromatography in phytochemical analysis. CRC Press, Boca Raton, pp 287–330

World Health Organization (WHO) (2014) Traditional Medicine Strategy 2014-2023. 2014. http://apps.who.int7irislbitstream7106651924551119789241506090eng.pdf?ua=1.

Wu Y-H (2016) Naturally derived anti-hepatitis B virus agents and their mechanism of action. World J Gastroenterol 22(1):188–204

Yakubo S, Ito M, Ueda Y, Okamoto H, Kimura Y, Amano Y, Togo T, Adachi H, Mitsuma T, Watanabe K (2014) Pattern classification in kampo medicine. Evid Based Complement Alternat Med 2014:535146

Yang L, Tan RX, Wang Q, Huang WY, Yin YX (2002) Antifungal cyclopeptides from *Halobacillus litoralis* YS3106 of marine origin. Tetrahedron Lett 43:6545–6548

Yang RY, Li CY, Lin YC, Peng GT, She ZG, Zhou SN (2006) Lactones from a brown alga endophytic fungus (No. ZZF36) from the South China Sea and their antimicrobial activities. Bioorg Med Chem Lett 16(16):4205–4208

Zhang F, Wang G (2014) A review of non-nucleoside anti-hepatitis B virus agents. Eur J Med Chem 75:267–281

Zhang Y, Ling S, Fang Y, Zhu T, Gu Q, Zhu WM (2008) Isolation, structure elucidation, and antimycobacterial properties of dimeric naphtho-γ-pyrones from the marine-derived fungus *Aspergillus carbonarius*. Chem Biodivers 5(1):93–100

Zhang G, Sun S, Zhu T, Lin Z, Gu J, Li D, Gu Q (2011) Antiviral isoindolone derivatives from an endophytic fungus *Emericella* sp. associated with *Aegiceras corniculatum*. Phytochemistry 72(11–12):1436–1442

Zhang G, Zhang Y, Qin J, Qu X, Liu J, Li X, Pan H (2013) Antifungal metabolites produced by *Chaetomium globosum* no. 04, an endophytic fungus isolated from *Ginkgo biloba*. Indian J Microbiol 53(2):175–180

Zheng J, Zhu H, Hong K, Wang Y, Liu P, Wang X, Peng X, Zhu W (2009) Novel cyclic hexapeptides from marine-derived fungus, *Aspergillus sclerotiorum* PT06-1. Org Lett 11(22):5262–5265

Zhou H, Yang Y, Peng T, Li W, Zhao L, Xu L, Ding Z (2014) Metabolites of *Streptomyces* sp., an endophytic actinomycete from *Alpinia oxyphylla*. Nat Prod Res 28(4):265–267

Zhuang Y, Teng X, Wang Y, Liu P, Wang H, Li J, Li G, Zhu W (2011) Cyclopeptides and polyketides from coral-associated fungus, *Aspergillus versicolor* LCJ-5-4. Tetrahedron 67(37):7085–7089

Zhuravleva OI, Afiyatullov SS, Denisenko VA, Ermakova SP, Slinkina NN, Dmitrenok PS, Kim NY (2012) A secondary metabolites from a marine-derived fungus *Aspergillus carneus* Blochwitz. Phytochemistry 80:123–131

Zilla MK, Qadri M, Pathania AS, Strobel GA, Nalli Y, Kumar S, Guru SK, Bhushan S, Singh SK, Vishwakarma RA, Riyaz-Ul-Hassan S (2013) Bioactive metabolites from an endophytic *Cryptosporiopsis* sp. inhabiting Clidemia hirta. Phytochemistry 95:291–297

Chapter 6
Algal Nanoparticles: Boon for Antimicrobial Therapeutic Applications

Lavanyasri Rathinavel, Deepika Jothinathan, V. Sivasankar, Prabhakaran Mylsamy, Kiyoshi Omine, and Ramganesh Selvarajan

Contents

L. Rathinavel · P. Mylsamy (✉)
Post Graduate and Research Department of Botany, Pachaiyappa's College
(Affiliated to University of Madras), Chennai, Tamil Nadu, India
e-mail: mprabhakaran@pachaiyappascollege.edu.in

D. Jothinathan
Department of Life Sciences, Central University of Tamil Nadu,
Thiruvarur, Tamil Nadu, India

V. Sivasankar
Post Graduate and Research Department of Chemistry, Pachaiyappa's College
(Affiliated to University of Madras), Chennai, Tamil Nadu, India

K. Omine
Geo-Tech Laboratory, Department of Civil Engineering, Graduate School of Engineering,
Nagasaki University, Nagasaki, Japan

R. Selvarajan
Department of Environmental Sciences, College of Agriculture and Environmental Sciences,
University of South Africa-Science Campus, Florida, South Africa

© Springer Nature Switzerland AG 2020
R. Prasad et al. (eds.), *Nanostructures for Antimicrobial and Antibiofilm Applications*, Nanotechnology in the Life Sciences,
https://doi.org/10.1007/978-3-030-40337-9_6

Abstract Nanotechnology is the field that deals with nano-sized structures of different shapes. Nanoparticles are generally synthesized using inorganic materials with restricted usages. However, there has been a great demand for nanoparticles in various medical applications. In the recent times, the usage of antibiotics has increased, which also has adverse effects to the immune system. Nanotechnology paves way to find the alternate resources for antibiotics. Eco-friendly nanoparticles are synthesized using biological organisms such as plants, bacteria, fungi, and algae. Algae are said to be the reservoir of nanoparticles, and hence, they are also called as nanofactories. Algae are used to produce metallic nanoparticles that can be used as antimicrobial agents. This chapter discusses the types of nanoparticles, synthesis of nanoparticles from algae, different kinds of algae that produce antimicrobial nanoparticles, and mechanism of nanoparticle as an antimicrobial agent.

Keywords Algae · Nanoparticles · Antimicrobial agent · Antibiotics · Mechanism

6.1 Introduction to Nanoparticles

In the present arena, technology based on modern material science is found to be the primitive research choice in the field of nanotechnology. Nanotechnology deals with objects with their size ranging in nanometers (Mishra et al. 1993). Nanoparticles and nanostructured materials are considered to be the foundation of nanotechnology (Mukherjee et al. 2002; Gurunathan et al. 2009; Philip 2010; Castro-Longoria et al. 2011; Indira et al. 2012). Nanoparticles have relatively a high surface area when compared to fine particles with less surface area, as the material might tend to change its own physical and chemical parameters. Sometimes, the magnetic and electrical properties also get changed (Moghimi and Kissel 2006). There are different types of nanoparticles such as gold, silver, zirconium, strontium, and titanium, which have a varied range of applications in the field of biomedical sciences, optics, physical sciences, engineering sectors, and chemical industries (Kalabegishvili et al. 2012; Ganesh Kumar et al. 2011; Logeswari et al. 2015; Ahmed et al. 2015, 2016; Raliya et al. 2015; Indira et al. 2014).

Among the above-mentioned nanoparticles, gold, and silver are considered to be most commonly used nanoparticles that can be extensively used in the field of medicine. Since ancient times, silver has been considered to be the choice of metal in the field of medicine for its own extensive property to fight against infections (Parashar et al. 2009). Applications of silver nanoparticles have been used for catalytic activities, electronics, diagnosis, detective analysis, antimicrobial activities, and therapeutic applications. Specifically, silver nanoparticles with own toxicity and high thermal stability have been the desired choice of application in the field of pharmaceutical industries (Silver 2003). In ancient times, Macedonians used silver plates for wound healing property, and further, they were used in surgeries to prevent infections. Later, Hippocrates applied silver for treating ulcers and started promoting its applications in the field of medicine in 69 BCE, which was highlighted in Rome publication *Pharmacopeia* (Hill and Pillsbury 1939).

Different kinds of compounds were treated with silver, and the derivatives obtained were used as antimicrobial agents to treat infections, wounds, and burns (Alexander 2009). Similar to silver, in ancient times of China, Egypt, and India, gold was also used in the field of medicine to treat skin diseases, small pox, measles, and syphilis, which was called as chrysotherapy (Huaizhi and Yuantao 2001; Richards et al. 2002). In 1857, Faraday was the first man to describe the chemical reactions between the metal salts that generated particles with zero valence. Later, researchers in the year 1990 determined the diameter of nanoparticles of about 70 nm, made up of gold and silver at British museum. The uses of both silver and gold nanoparticles were represented in the Lycurgus Cup (Evanoff and Chumanov 2005).

Designing, constructing, and fabricating the morphological features of nanoparticles are setting a lineup for the upcoming researchers. Distinct biological, physical, and chemical methods are optimized to obtain different shapes and sizes of nanoparticles based on their applications. Capping agents during the process of synthesizing nanoparticles include sodium citrate, alcohols, and sodium borohydride in different concentrations (Balagurunathan et al. 2011). Interestingly, the researchers focused on the best suitable bioresources such as plants, microbes, and algae for generating nanoparticles in the recent past (Ravinder Singh et al. 2015). In the yester years, inorganic nanoparticles from different sources had been the source of research interest for their special physical, chemical, and biological features. Hence, recent researches have been highly focusing on the field of biomedical engineering and other industrial sectors. With these advancements aside, very few reports are available with regard to the effects and toxicity of the synthesized nanoparticles. Selecting biological organisms such as bacteria, algae, fungi, and plants for nanoparticle production and implementing suitable biomethodologies will be cost efficient, eco-friendly, and nontoxic (Barwal et al. 2011; Prasad et al. 2016, 2018a, b; Prasad 2014).

6.2   Types and Antimicrobial Properties of Nanoparticles

Nanoparticles are classified into two different types, such as organic and inorganic nanoparticles. Inorganic nanoparticles are those made from metals and metal oxides exhibiting high prominence as antibacterial agents (Loomba and Scarabelli 2013). Inorganic polymers are considered to be the best choice of polymers as antimicrobial agents, thanks to the high thermal stability against organic polymers with less thermal stability (Jain et al. 2014). Different kinds of inorganic and organic nanoparticles are described below.

6.2.1   Types of Inorganic Nanoparticles

6.2.1.1   Silver

Ashok Kumar et al. (2014) reported that silver nanoparticles are photocatalytic by which they generate the reactive oxygenic species abbreviated as ROS (Ninganagouda et al. 2014; Khurana et al. 2014). Silver nanoparticles are predominantly used for their antimicrobial property against bacteria, virus, and fungi (Rai et al. 2009). Alberti et al. (2004), Panacek et al. (2006), and Poulose et al. (2014) reported that the activity of a silver nanoparticle depends on its size—the smaller the size, the higher the efficiency—which has been proved by analyzing the silver nanoparticles in both in vitro and in vivo conditions. Leid et al. (2012) and Chernousova and Epple (2013) have reported that silver nanoparticles are more efficient than the antibiotics. Sondi and Salopek-Sondi (2004) and Beyth et al. (2010) explained that silver nanoparticles invade the cell of *Escherichia coli* by creating a hole and permeate the membrane to inactivate the activity of the cell. Majdalawieh et al. (2014) opined that the research on silver nanoparticles has been increasing day by day, but the mode of action is still unclear and very less has been reported. It was revealed that silver nanoparticles bind to the amino acid groups and destruct the structure of protein.

6.2.1.2   Titanium Oxide

Zan et al. (2007) and Blecher et al. (2011) reported that titanium oxide nanoparticles are considered to exhibit the antimicrobial property against bacteria, parasites, and virus. Hamal et al. (2010) reported that titanium oxide nanoparticles are photocatalytic and generate ROS when they are exposed to UV and visible light. Palanikumar et al. (2014) reported that titanium oxide nanoparticles have sporicidal activity against the bacteria.

6.2.1.3 Zinc Oxide

Zinc oxide inorganic nanoparticle is also extensively used as an antimicrobial agent at different concentrations and various sizes (Jin et al. 2009). The production cost is considered to be less and, they are reported to have inhibition activities against different kinds of bacterial cells of *Klebsiella pneumonia, Salmonella enteritidis* (Liu et al. 2009); besides, that they are harmless to human cells was reported by Espitia et al. (2012). Hamal et al. (2010) reported that titanium oxide nanoparticles also have an extensive property of anti-biofilm activity by inhibiting the UV radiation and is potentially used as a coating material in the medical equipment manufacturing industries. This technique of utilizing titanium oxide nanoparticle as coating material is approved by the Food and Drug Administration (FDA) for the purpose of detecting adulterations in food.

6.2.1.4 Iron Oxide

Brown et al. (2012) reported that iron oxide nanoparticles exhibit a special property such as preventing the colonization of bacteria, by which the infection rate can be decreased. Besides the Iron oxide nanoparticle is a highly active antimicrobial agent when the size ranges in nanometers as compared to the larger sized inactive nanoparticles.

6.2.1.5 Gold

Gold nanoparticles are reported to have very less antimicrobial activity compared to silver nanoparticles. Hence, gold nanoparticles are mixed along with the antibiotics (Varisco et al. 2014). Raji et al. (2011) used gold nanoparticles by combining with pyrimidines and citrate-substituted amino acids for generating ROS. Further, they are used for cancer treatment (Pey et al. 2014).

6.2.1.6 Copper Oxide

Copper oxide nanoparticles are used in high concentration for an efficient antimicrobial activity. Copper oxide is found to have less antibacterial activity when compared to silver and zinc oxide nanoparticles. Ruparelia et al. (2008) reported that bacteria such as *Bacillus subtilis* and *Bacillus anthracis* are very sensitive to copper oxide nanoparticles. This is because the cell walls of both the bacteria are highly enriched with carboxyl and amine groups, which make them more susceptible to the copper oxide nanoparticles. Huh and Kwon (2011) reported that the antibacterial activity takes place when the copper oxide nanoparticles disrupt the membrane and produce ROS.

6.2.1.7   Magnesium Oxide

Hamal et al. (2010) and Slomberg et al. (2013) reported that similar to other inorganic nanoparticles, the antimicrobial behavior of magnesium oxide nanoparticles is potentially high. These nanoparticles possess a special feature for inhibiting the essential enzymes produced by the bacteria and generate ROS by disrupting the cell membrane of the bacteria.

6.2.1.8   Nitric Oxide

Kutner and Friedman (2013) and Han et al. (2009) reported that nitric oxide nanoparticles are potential antimicrobial agents. The efficacy in potentiality increases with respect to the decreasing size of nanoparticles. Most of the nanoparticles produce ROS during the inhibition process, whereas the nitric oxide nanoparticles produce reactive Nitrogen species (RNS) (Hetrick et al. 2009).

6.2.1.9   Aluminum Oxide

As a very mild antibacterial agent, the potentiality of aluminum oxide increases with respect to the increase in concentration (Huh and Kwon 2011). The peculiarity of this nanoparticle's interaction with bacteria lies in its ability of perforation into the membrane to form a pit, leading to a completely disrupted membrane and hence the bacterial cell destruction (Imada et al. 2015).

6.2.2   Types of Organic Nanoparticles

6.2.2.1   Poly-ε-Lysine

Beyth et al. (2015) reported that organic nanoparticle poly-ε-lysine is a homo peptide and of cationic nature (of L-lysine) with a high Gram negative bactericidal property. It is a prominent candidate in destructing the spores of bacteria such as *Bacillus subtilis, B. coagulans,* and *B. stearothermophilus.*

6.2.2.2   Quaternary Ammonium Compounds

Munoz-Bonilla and Fernandez-Garcia (2012) reported that the quaternary ammonium compounds are one among the organic nanoparticles with antimicrobial property, and their potentiality depends on the corresponding chain length. The mechanism of electrostatic interaction is feasible between the charged ends of quaternary ammonium compounds (positive) and bacterial membranes (negative). As a consequence of the insertion of the nanoparticle's tail into the bacterial membrane, the protein-made bacterial structure gets denatured and leads to the cell death.

6.2.2.3 Cationic Quaternary Polyelectrolytes

Methacrylic compounds, an acrylic derivate from the class of major cationic quaternary polyelectrolytes, are reported to have a flexible structure and suitable surface charge to cater to wide and feasible applications (Denyer and Stewart 1998).

6.2.2.4 *N*-Halamine Compounds

Chung et al. (2003) reported that *N*-halamine compounds are generated by the process of halogenation of amine group by interacting with covalent bonds of nitrogen and halogen. This nanoparticle is very stable because of its controlled release of halogens to the environment and thus inhibits the microbes.

6.2.2.5 Chitosan

Chitosan nanoparticles are considered to be nontoxic and biocompatible. The antimicrobial ability of these particles is governed by factors such as pH and the solvent nature (Tavaria et al. 2013). The combination of chitosan nanoparticle with antibiotic molecules displayed an improved efficiency but found to decrease with the addition of metals (Ibrahim et al. 2014). The interaction of chitosan with bacteria mimics that of organic nanoparticles, which further leads to membrane destruction and cell death (Reddy and Urban 2008).

6.3 Algal Nanoparticles for Antimicrobial Applications

Microbial infections due to microbial contamination cause many serious health issues. As a solution to this problem, different methods have been developed to prevent and control the microbial infection (Dutkiewicz and Malinowski 2012) using antibiotics. Nevertheless, the spread of resistant organisms due to the frequent use of antibiotics has become a threat for people. Burmølle et al. (2014) reported that the microbial biofilms formed from abiotic and biotic factors remain highly infectious and are difficult to eliminate. Metabolic derivatives extracted from plants serve as substitutes for reducing the effect of antibiotics and have been evaluated with appreciable efficacy and applied in medical fields. Applications of different kinds of metabolites are best realized in pharmaceutical and nutraceutical industries. Conversely, the organic and inorganic elements in these extracts were found to exhibit less potentiality towards the antimicrobial activity during the chemical interactions (Barra Caracciolo et al. 2015). At this stage, an even better technology and replacement were in great need for the community, and here comes nanotechnology with a wide range of applications by developing materials that exhibit antimicrobial properties. But nanomaterials have their own limitations due to the toxicity in nature and limited applications in medical fields. So, better nontoxic and eco-friendly materials are required for the development of nanomaterials (Narayanan and Sakthivel 2011), which could best be synthesized from biological organisms. Compared to other biological organisms such as bacteria, fungi, and yeast, algae are considered to be the best and important organism for the synthesis of nanoparticles (Prasad et al. 2016, 2018a, b). The production of algae-mediated nanoparticle biosynthesis is termed as "phyconanotechnology" (Thakkar et al. 2010).

Algae are unicellular and significant photosynthetic organisms that grow in fresh water and marine water. There are specific algal species that grow on the surfaces of rocks (Thajuddin and Subramanian 1992). Algae, the most powerful reservoir of nanomaterials, are known for the fixation of carbondioxide and biofuel production (Jacob-Lopes and Franco 2013). Algae are considered as bionanofactories because the live and dried biomass can be utilized for the synthesis of the metallic nanoparticles (Davis et al. 1998). Credibility on algae for the production of nanoparticles has become an emerging trend in the recent years. Algae have the ability to reduce metal ions such as gold and silver by accumulating metals in the form of nanoparticles measuring between 1 and 10 nm (Eom et al. 2012). Algae have many advantages such as easy handling and convenience in the synthesis of less toxic nanoparticles at low temperatures with higher energy efficiency. Algae are considered as the source of commercial products that include biofuel, cosmetics, food products, and natural dyes (LewisOscar et al. (2014). Chelation and trapping of heavy metal species by algae have also been reported (Hameed and Hasnain 2005). Algal species can be used as an effective source for the production of antimicrobial agents by integrating nanotechnology to overcome the infections that are resistant to antibiotics. The enhancement of antimicrobial activity could be made feasible by integrating algae-delivered nanoparticles with nanotubes and nanofibers (De Philippis et al. 2001). These nanofibers are applied for healing wounds with the aid of bandages and medical equipment. In this way, the prevailing medical issues could be solved by coupling nanoparticles with biotechnology.

6.4   Synthesis of Nanoparticles Using Algae

Algae are used as the best source for the production of metallic nanoparticles. Synthesis of algal nanoparticle is explained as follows: Algae extract is prepared using organic solvents or water, and the solvent is subjected to heat for a particular time period. Meanwhile, molar solutions are prepared using the respective metallic compounds. The molar solutions with ionic compound are mixed with the algal extract, and they are subjected to stirring for a certain period and left undisturbed for some time without agitation (Thakkar et al. 2010; Rauwel et al. 2015). Dahoumane et al. (2012) stated that based on the type of algae, the nanoparticle can be synthesized under controlled dose by intracellular or extracellular formation. Singaravelu et al. (2007) reported that the extracellular formation using biomolecules (polysaccharides, peptide chains, and pigments) facilitates the reduction of metals. Capping and stabilizing the metallic nanoparticles are achieved using aqueous extracts of amino acid residues such as cysteine and sulfated polysaccharides. The metallic nanoparticle intracellular formation is based on the metabolism of algae, and the metallic ions get reduced by the respiration and photosynthesis of algae (Jeffryes et al. 2015). Shabnam and Pardha-Saradhi (2013) reported that the photochemical reduction of Au (III) takes place in thylakoids of chloroplast by gaining electrons donated by water molecules. Dahoumane et al. (2014) reported the synthesis of nanoparticles using metallic ions at lower doses where the nanoparticles are stabilized by biomolecules. Oza et al. (2012) evaluated the shape and size of the nanoparticles by altering the physical parameters like pH, temperature, type of the metal used, and the initial concentration of the aqueous extracts (reducing agents). In addition, these physical parameters prevent the aggregation of nanoparticles. The influence of pH on the synthesis of nanoparticles in extracellular mode was studied by Mata et al. (2009) and Parial et al. (2012). In aqueous state, the rate of functional group reduction decreased at high concentrations, preferably at low pH values. On the other hand, at high pH, well-stabilized nanoparticles resulted because of high reduction rate of the functional groups. During the synthesis of gold and silver nanoparticles, majority of the changes with respect to the reaction mixture are said to be a visual marker. Govindaraju et al. (2008) and Jena et al. (2013) reported that the color of the reaction mixture turns brown in the case of silver nanoparticle generation and ruby red, pink, or purple when gold nanoparticle is synthesized. Mulvaney (1996) observed orange and blue colors, respectively, for the particle size of ~20 and ~100 nm. The surface plasmon resonance (SPR) corroborates the optical properties of metallic nanoparticles, and the SPR bands arise with combined simulations of the conduction electrons of metallic nanoparticles under the light, depending on the size and shape of nanoparticles (Mulvaney 1996). Based on the range of the SPR band width, the size of nanoparticles could be determined (Govindaraju et al. 2008), and the spectroscopic range lies in the range of 329–580 nm and 510–560 nm for silver and gold nanoparticles, respectively. In aqueous solution, the increase in the particle size was indicated with the decreased band width along with increased band intensity. The characterization of nanoparticles using Fourier transform infrared (FTIR), scanning electron microscopy (SEM) with energy dispersive analysis (EDS), transmission electron microscopy (TEM), and X-ray diffraction (XRD) studies are also carried out to ensure various functional groups, surface morphology, shape/size determination, and crystalline nature, respectively (Schrofel et al. 2011). Nanoparticle synthesis using algae takes place in a short time as compared to the other methodologies involving other biological organisms (Rauwel et al. 2015).

6.5 Algal Nanoparticles with Antimicrobial Activity

Among various biological organisms, algae are considered to be potential candidates for synthesizing nanoparticles with inevitable applications. Nanoparticles such as gold, silver, zirconia, chloride, copper, and fluorescent carbons were synthesized using different algal species. The nanoparticles synthesized from algae were evaluated for the antimicrobial property against different bacterial species and reported by several researchers. Among all other nanoparticles, gold and silver nanoparticles produced from different algal species were found to be the best candidates with excellent antimicrobial property against different kinds of bacterial species (Aziz et al. 2014, 2015). Table 6.1 represents the different types of algae used for synthesizing different kinds of nanoparticles along with their antimicrobial properties.

6.6 Future Perspectives and Applications of Algal Nanoparticles

Nanotechnology has developed into a premium zone for researchers to develop different kinds of nanomaterials using the biological organisms like algae which is an abundant source from the nature itself. The nanoparticles generated using algae are extensively used in the medical field for various kinds of biomedical applications. The use of nanoparticles in medical field has created expectation and demand among the society.

Nanoparticles synthesized using algae have a wide range of applications in the field of medicine. Nanoparticles are used for labeling biological organisms and to detect the proteins and pathogens as well. Nanoparticles are used for immune assay, which is a rapid technology diagnosing diseases. Nanoparticles are highly used in cancer treatment as drug delivering systems. Biotechnology, in combination with nanoparticles, is used for DNA analysis in humans for gene therapy and to treat genetic disorders. In the image analysis, the iron oxide nanoparticles are used for effective image analysis such as magnetic resonance imaging (MRI). In tissue repairing and treatment of tumors, nanoparticle-coated medicines are of high interest. Especially as an anti-coagulant agent, silver nanoparticles play an eminent role in hemorrhagic treatment by preventing the coagulation of blood. Nanoparticle synthesis from algae has gained attention among researches for its eco-friendly conditions and uses. Meanwhile, alternate sources for antibiotics with similar effects are also developing rapidly using hybrid nanobiotechnology. It is presumed in few years that standard protocols to generate antimicrobial agents using algae nanoparticles will be of utmost interest in pharmaceutical industries.

Table 6.1 Antimicrobial nanoparticles from different types of algal species

Species	Nano particle	Nanoparticle Size (nm)	Antimicrobial activity	References
Turbinaria conoides	Gold	60	*Streptococcus* sp.	Rajeshkumar et al. (2013)
Scenedesmus abundans	Silver	59–66	*Escherichia coli*	Aziz et al. (2014)
Chlorella pyrenoidosa	Silver	5–15	*Escherichia coli*	Aziz et al. (2015)
Sargassum wightii	Zirconia	4.8	*Bacillus subtilis; Escherichia coli; Salmonella typhi*	Kumaresan et al. (2018)
Sargassum plagiophyllum; C. agarth	Silver chloride	NR	*Escherichia coli*	Stalin Dhas et al. (2014)
Turbinaria conoides	Silver	96	*Bacillus subtilis; Klebsiella planticola*	Rajeshkumar et al. (2012)
Sargassum polycystum	Copper	NR	*Pseudomonas aeruginosa*	Sri Vishnu Priya et al. (2016)
Caulerpa serrulata	Silver		*Escherichia coli*	Eman Aboelfetoh et al. (2017)
Botryococcus braunii	Copper and Silver	10–70/40–100	*Staphylococcus aureus; Klebsiella pneumoniae; Escherichia coli; Pseudomonas aeruginosa*	Anju Arya et al. (2018)
Ulva compressa (L.) Kütz	Silver	66.3	*Klebsiella pneumonia; Pseudomonas aeruginosa;*	Fozia T. Minhas et al. (2018)
Cladophora glomerata (L.) Kütz	Silver	81.8	*Escherichia coli; Enterococcus faecium; Staphylococcus aureus*	
Botryococcus braunii	Silver	168	*Pseudomonas aeruginosa; Staphylococcus aureus*	Alejandra Arévalo-Gallegos et al. (2018)
Spirulina platensis	Gold	5	*Bacillus subtilis; staphylococcus aureus*	Uma Suganya et al. (2015)
Chlorella vulgaris	Silver	34.4	*Staphylococcus aureus; Klebsiella pneumoniae*	Veronica da Silva Ferreira et al. (2017)
Cladophora vagabunda	Fluroscent carbon	42.78	*Staphylococcus aureus; Escherichia coli*	Maria Theresa et al. (2018)
Spirogyra varians	Silver	17.6	*Staphylococcus aureus; Listeria monocytogenes Escherichia coli*	Zeinab Salari et al. (2016)
Marine seaweeds	Silver	25–40	*Bacillus cereus, Escherichia coli*	Shiny et al. (2013)
Chaetoceros calcitrans	Silver	53.1	*Klebsiella sps., Pseudomonas aeruginosa; Proteus vulgaris*	Devina Merin et al. (2010)
Enteromorpha flexuosa	Silver	13.8–59.2	*Pseudomonas aeruginosa*	Morteza Yousefzadi et al. (2014)

(continued)

Table 6.1 (continued)

Species	Nano particle	Nanoparticle Size (nm)	Antimicrobial activity	References
Gelidium amansii	Gold	5–25	*Escherichia coli; Staphylococcus aureus*	Paskalis Sahaya Murphin Kumar et al. (2017)
Gracilaria corticata	Gold	NR	*Staphylococcus aureus; Enterococcus faecalis*	Edhaya Naveena et al. (2013)
Amphiroa fragilissima	Silver	104.6	Gram-positive bacteria	Ramalingam et al. (2018)
Gracilaria crassa	Silver	122.7	*Escherichia coli; Proteus mirabilis*	Lavakumar (2015)
Gracilaria birdiae	Silver	20.2–94.9	*Escherichia coli; Staphylococcus aureus*	De Aragao et al. (2019)
Porphyra vietnamensis	Silver	13	*Escherichia coli*	Vinod Venkatpurwar et al. (2011)

6.7   Conclusion

Synthesis of nanoparticles using algae has become a prime area of research in recent days. Nanoparticles from biological organisms are always a boon to the society. In view of expanded research in the field of medicine, the application of nanoparticles became very vast and wide with appealing novelty. Usage of antibiotics is considered as a threat to our own immune system, and hence, a research for the alternate sources has given rise to a field where algae are considered as the nanofactories or the reservoir of nanoparticles. The potential of algal nanoparticles as an antimicrobial agent is a growing area of research in the field of nanobiotechnology. Yet, researches have to find the alternate cost-efficient and eco-friendly methodologies to synthesize nanoparticles from algae that can be used widely in the medical field.

References

Aboelfetoh EF, El-Shenody RA, Ghobara MM (2017) Eco-friendly synthesis of silver nanoparticles using green algae (*Caulerpa serrulata*): reaction optimization, catalytic and antibacterial activities. Environ Monit Assess 189:349

Ahmed R, Mohsin M, Ahmad T, Sardar M (2015) Alpha amylase assisted synthesis of TiO$_2$ nano particles: structural characterization and application as antibacterial agents. J Hazard Mater 283:171–177

Ahmed S, Ahmad M, Swami BL, Ikram S (2016) A review on plants extract mediated synthesis of silver nanoparticles for antimicrobial applications: a green expertise. J Adv Res 7:17–28

Alberti S, Böhse K, Arndt V, Schmitz A, Höhfeld J (2004) The cochaperone HspBP1 inhibits the CHIP ubiquitin ligase and stimulates the maturation of the cystic fibrosis transmembrane conductance regulator. Mol Biol Cell 15(9):4003–4010

Alexander JW (2009) History of the medical use of silver. Surg Infect 10:3

Arévalo-Gallegos A et al (2018) Botryococcus braunii as a bioreactor for the production of nanoparticles with antimicrobial potentialities. Int J Nanomed 13:5591–5604

Arya A, Gupta K, Chundawat TS, Vaya D (2018) Biogenic synthesis of copper and Silver nanoparticles using green alga Botryococcus braunii and its antimicrobial activity. Bioinorg Chem Appl 2018:7879403, 9p

Ashok Kumar D, Palanichamy V, Roopan M (2014) Photocatalytic action of AgCl nanoparticles and its antibacterial activity. J Photochem Photobiol B 138:302–306

Aziz N, Fatma T, Varma A, Prasad R (2014) Biogenic synthesis of silver nanoparticles using *Scenedesmus abundans* and evaluation of their antibacterial activity. J Nanopart 2014:689419. https://doi.org/10.1155/2014/689419

Aziz N, Faraz M, Pandey R, Sakir M, Fatma T, Varma A, Barman I, Prasad R (2015) Facile algae-derived route to biogenic silver nanoparticles: synthesis, antibacterial and photocatalytic properties. Langmuir 31:11605–11612. https://doi.org/10.1021/acs.langmuir.5b03081

Balagurunathan R, Radhakrishnan M, BabuRajendran R, Velmurugan D (2011) Biosynthesis of gold nanoparticles by actinomycete *Streptomyces viridogens* strain HM10. Indian J Biochem Biophys 48:331–335

Barra Caracciolo A, Topp E, Grenni P (2015) Pharmaceuticals in the environment: biodegradation and effects on natural microbial communities- a review. J Pharm Biomed Anal 106:25–36

Barwal I, Ranjan V, Kateriya S, Yadav C (2011) Cellular oxido-reductive proteins of *Chlamydomonas reinhardtii* control the biosynthesis of silver nanoparticles. J Nanobiotechnol 9:56

Beyth N, Yudovin-Farber I, Perez-Davidi M, Domb AJ, Weiss EI (2010) Polyethyleneimine nanoparticles incorporated into resin composite cause cell death and trigger biofilm stress in vivo. Proc Natl Acad Sci U S A 107(51):22038–22043

Beyth N, Houri-Haddad Y, Domb A, Khan W, Hazan R (2015) Alternative antimicrobial approach: nano-antimicrobial materials. Evid Based Complement Alternat Med 2015:246012

Blecher K, Nasir A, Friedman A (2011) The growing role of nanotechnology in combating infectious disease. Virulence 2(5):395–401

Brown AN, Smith K, Samuels TA, Lu J, Obare SO, Scott ME (2012) Nanoparticles functionalized with ampicillin destroy multiple-antibiotic-resistant isolates of *Pseudomonas aeruginosa* and *Enterobacter aerogenes* and methicillin-resistant *Staphylococcus aureus*. Appl Environ Microbiol 78(8):2768–2774

Burmølle M, Ren D, Bjarnsholt T, Sørensen SJ (2014) Interactions in multispecies biofilms: do they actually matter? Trends Microbiol 22:84–91

Castro-Longoria E, Vilchis-Nestor AR, Avalos-Borja M (2011) Biosynthesis of silver, gold and bimetallic nanoparticles using the filamentous fungus Neurospora crassa. Coll Surf B 83:42–48

Chernousova S, Epple M (2013) Silver as antibacterial agent: ion, nanoparticle, and metal. Angew Chem Int Ed 52(6):1636–1653

Chung YC, Wang HL, Chen YM, Li SL (2003) Effect of abiotic factors on the antibacterial activity of chitosan against waterborne pathogens. Bioresour Technol 88(3):179–184

da Silva Ferreira V et al (2017) Green production of microalgae-based silver chloride nanoparticles with antimicrobial activity against pathogenic bacteria. Enzym Microb Technol 97:114–121

Dahoumane SA, Djediat C, Yepremian C, Coute A, Fievet F, Coradin T, Brayner R (2012) Recycling and adaptation of *Klebsormidium flaccidum* microalgae for the sustained production of gold nanoparticles. Biotechnol Bioeng 109:284–288

Dahoumane SA, Wijesekera K, Filipe CDM, Brennan JD (2014) Stoichiometrically controlled production of bimetallic gold-silver alloy colloids using micro-alga cultures. J Colloid Interface Sci 416:67–72

Davis SA, Patel HM, Mayes EL, Mendelson NH, Franco G, Mann S (1998) Brittle bacteria: a biomimetic approach to the formation of fibrous composite materials. Chem Mater 10:2516–2524

De Aragao AP et al (2019) Green synthesis of silver nanoparticles using the seaweed *Gracilaria birdiae* and their antibacterial activity. Arab J Chem 12(8):4182–4188. https://doi.org/10.1016/j.arabjc.2016.04.014

De Philippis R, Sili C, Paperi R, Vincenzini M (2001) Exo polysaccharide producing cyanobacteria and their possible exploitation: a review. J Appl Phycol 13:293–299

Denyer SP, Stewart GSAB (1998) Mechanisms of action of disinfectants. Int Biodeterior Biodegrad 41(3–4):261–268

Devina Merin D et al (2010) Antibacterial screening of silver nanoparticles synthesized by marine micro algae. Asian Pac J Trop Med 1:797–799

Dutkiewicz A, Malinowski P (2012) Aortobifemoral prothesis infection. Pol Ann Med 19:129–133

Edhaya Naveena B et al (2013) Biological synthesis of gold nanoparticles using marine algae *racilaria corticata* and its application as a potent antimicrobial and antioxidant agent. Asian J Pharm Clin Res 6(2):179–182

Eom SH, Kim YM, Kim SK (2012) Antimicrobial effect of phlorotannins from marine brown algae. Food Chem Toxicol 50:3251–3255

Espitia PJP, Soares NFF, Coimbra JSR, de Andrade NJ, Cruz RS, Medeiros EAA (2012) Zinc oxide nanoparticles: synthesis, antimicrobial activity and food packaging applications. Food Bioprocess Technol 5(5):1447–1464

Evanoff D, Chumanov GJR (2005) Synthesis and optical properties of Silver nanoparticles and arrays. ChemPhysChem 6:1221–1231

Ganesh Kumar C, Mamidyala SK, Sreedhar B, Belum Reddy VS (2011) Synthesis and characterization of gold glycol nano particles functionalized with sugars of sweet sorghum syrup. Biotechnol Prog 27:1455–1463

Govindaraju K, Basha SK, Kumar VG, Singaravelu G (2008) Silver, gold and bimetallic nanoparticles production using single-cell protein (Spirulina platensis) Geitler. J Mater Sci 43:5115–5122

Gurunathan S, Kalishwaralal K, Vaidyanathan R, Deepak V, Pandian SRK, Muniyandi J, Hariharan N, Eom SH (2009) Biosynthesis, purification and characterization of silver nanoparticles using *Escherichia coli*. Coll Surf B 74:328–335

Hamal B, Haggstrom JA, Marchin GL, Ikenberry MA, Hohn K, Klabunde KJ (2010) A multifunctional biocide/sporocide and photocatalyst based on titanium dioxide (TiO₂) codoped with silver, carbon, and sulfur. Langmuir 26(4):2805–2810

Hameed A, Hasnain S (2005) Cultural characteristics of chromium resistant unicellular cyanobacteria isolated from local environment in Pakistan. Chin J Oceanol Limnol 23:433–441

Han G, Martinez LR, Mihu MR, Friedman AJ, Friedman JM, Nosanchuk JD (2009) Nitric oxide releasing nanoparticles are therapeutic for Staphylococcus aureus abscesses in a murine model of infection. PLoS One 4(11):e7804

Hetrick EM, Shin JH, Paul HS, Schoenfisch MH (2009) Anti-biofilm efficacy of nitric oxide-releasing silica nanoparticles. Biomaterials 30(14):2782–2789

Hill WR, Pillsbury DM (1939) Argyria–the pharmacology of silver. Williams & Wilkins, Baltimore

Huaizhi Z, Yuantao N (2001) China's ancient gold drugs. Gold Bull 34:24–29

Huh AJ, Kwon YJ (2011) Nanoantibiotics: a new paradigm for treating infectious diseases using nanomaterials in the antibiotics resistant era. J Control Release 156(2):128–145

Ibrahim M, Tao Z, Hussain A et al (2014) Deciphering the role of Burkholderia cenocepacia membrane proteins in antimicrobial properties of chitosan. Arch Microbiol 196(1):9–16

Imada K, Sakai S, Kajihara H, Tanaka S, Ito S (2015) Magnesium oxide nanoparticles induce systemic resistance in tomato against bacterial wilt disease. Plant Pathol 65(4):551–560. https://doi.org/10.1111/ppa.12443

Indira K, Ningshen S, Kamachi Mudali U, Rajendran N (2012) Effect of anodization parameters on the structural morphology of titanium in fluoride containing electrolytes. Mater Charact 71:58–65

Indira K, Kamachi Mudali U, Rajendran N (2014) Invitro bioactivity and corrosion resistance of Zrin corporated TiO₂ nanotube arrays for orthopaedic applications. Appl Surf Sci 316:264–275

Jacob-Lopes E, Franco TT (2013) From oil refinery to microalgal bioreinery. J CO2 Utilizat 2:1–7

Jain A, Duvvuri LS, Farah S, Beyth N, Domb AJ, Khan W (2014) Antimicrobial polymers. Adv Healthc Mater 3:1969–1985

Jeffryes C, Agathos SN, Rorrer G (2015) Biogenic nanomaterials from photosynthetic microorganisms. Curr Opin Biotechnol 33:23–31

Jena J, Pradhan N, Dash BP, Sukla LB, Panda PK (2013) Biosynthesis and characterization of silver nanoparticles using microalga Chlorococcum humicola and its antibacterial activity. Int J Nanomater Bios 3:1–8

Jin T, Sun D, Su JY, Zhang H, Sue HJ (2009) Antimicrobial efficacy of zinc oxide quantum dots against *Listeria monocytogenes, Salmonella enteritidis*, and *Escherichia coli* O157:H7. J Food Sci 74(1):46–52

Kalabegishvili T et al (2012) Synthesis of gold nanoparticles by blue-green algae *Spirulina platensis*. Adv Sci Eng Med 4:1–7

Khurana C, Vala AK, Andhariya N, Pandey OP, Chudasama B (2014) Antibacterial activities of silver nanoparticles and antibiotic-adsorbed silver nanoparticles against biorecycling microbes. Environ Sci Process Impact 16(9):2191–2198

Kumaresan M et al (2018) Seaweed *Sargassum wightii* mediated preparation of zirconia (ZrO_2) nanoparticles and their antibacterial activity against Gram positive and Gram negative bacteria. Microb Pathog 124:311–315

Kutner AJ, Friedman AJ (2013) Use of nitric oxide nanoparticulate platform for the treatment of skin and soft tissue infections. Wiley Interdiscip Rev Nanomed Nanobiotechnol 5(5): 502–514

Lavakumar et al (2015) Promising upshot of silver nanoparticles primed from *Gracilaria crassa* against bacterial pathogens. Chem Cent J 9:42. https://doi.org/10.1186/s13065-015-0120-5

Leid JG et al (2012) In vitro antimicrobial studies of silver carbene complexes: activity of free and nanoparticle carbene formulations against clinical isolates of pathogenic bacteria. J Antimicrob Chemother 67(1):138–148

LewisOscar F, Bakkiyaraj D, Nithya C, Thajuddin N (2014) Deciphering the diversity of micro-algal bloom in wastewater-an attempt to construct potential consortia for bioremediation. JCPAM 3(2):92–96

Liu Y, He L, Mustapha A, Li H, Hu ZQ, Lin M (2009) Antibacterial activities of zinc oxide nanoparticles against *Escherichia coli* O157:H7. J Appl Microbiol 107(4):1193–1201

Logeswari P, Silambarasan S, Abraham J (2015) Synthesis of silver nanoparticles using plants extract and analysis of their antimicrobial property. J Saudi Chem Soc 19:311–317

Loomba L, Scarabelli T (2013) Metallic nanoparticles and their medicinal potential. Part II: aluminosilicates, nanobiomagnets, quantum dots and cochleates. Ther Deliv 4(9):1179–1196

Majdalawieh A, Kanan MC, El-Kadri O, Kanan SM (2014) Recent advances in gold and silver nanoparticles: synthesis and applications. J Nanosci Nanotechnol 14(7):4757–4780

Maria Theresa F et al (2018) Facile synthesis of biologically derived fluorescent carbon nanoparticles (FCNPs) from an abundant marine alga and its biological activities. Orient J Chem 34(2):791–799

Mata YN, Torres E, Blazquez ML, Ballester A, González F, Munoz JA (2009) Gold(III) biosorption and bioreduction with the brown alga *Fucus vesiculosus*. J Hazard Mater 166:612–618

Minhas FT et al (2018) Evaluation of antibacterial properties on polysulfone composite membranes using synthesized biogenic silver nanoparticles with *Ulva compressa* (L.) Kütz. and *Cladophora glomerata* (L.) Kütz extracts. Int J Biol Macromol 107:157–165

Mishra VK, Temelli F, Ooraikul Shacklock PF, Craigie JS (1993) Lipids of the red alga. Palmaria Palmate Bot Mar 36:169–174

Moghimi SM, Kissel T (2006) Particulate nanomedicines. Adv Drug Deliv Rev 58:1451–1455

Mukherjee P, Senapati S, Mandal D, Ahmad A, Khan MI, Kumar R, Sastry M (2002) Extra cellular synthesis of gold nanoparticles by the fungus *Fusarium oxysporum*. Chem Bio Chem 3:461–463

Mulvaney P (1996) Surface plasmon spectroscopy of nanosized metal particles. Langmuir 12:788–800

Munoz-Bonilla A, Fernandez-Garcia M (2012) Polymeric materials with antimicrobial activity. Prog Polym Sci 37(7):281–339

Murphin Kumar PS et al (2017) Synthesis of nano-cuboidal gold particles for effective antimicrobial property against clinical human pathogens. Microb Pathog 113:68–73

Narayanan KB, Sakthivel N (2011) Green synthesis of biogenic metal nanoparticles by terrestrial and aquatic phototrophic and heterotrophic eukaryotes and biocompatible agents. Adv Colloid Interf Sci 169:59–79

Ninganagouda S, Rathod V, Singh D, Hiremath J, Singh AK, Mathew J, Manzoor ul-Haq (2014) Growth kinetics and mechanistic action of reactive oxygen species released by silver nanoparticles from *Aspergillus niger* on of reactive oxygen species. J Phys Chem B 112(43):13608–13619

Oza G, Pandey S, Mewada A, Kalita G, Sharon M (2012) Facile biosynthesis of gold nanoparticles exploiting optimum pH and temperature of fresh water algae *Chlorella pyrenoidusa*. Adv Appl Sci Res 3:1405–1412

Palanikumar L, Ramasamy SN, Balachandran C (2014) Size dependent antimicrobial response of zinc oxide nanoparticles. IET Nanobiotechnol 8(2):111–117

Panacek A, Kvítek L, Prucek R, Kolar M, Vecerova R, Pizúrova N, Sharma VK, Nevecna T, Zboril R (2006) Silver colloid nanoparticles: synthesis, characterization, and their antibacterial activity. J Phys Chem B 110(33):16248–16253

Parashar V, Parashar R, Sharma B, Pandey AC (2009) Parthenium leaf extract mediated synthesis of silver nanoparticles: a novel approach towards weed utilization. Digest J Nanomater Biostruct 4:45–50

Parial D, Patra HK, Roychoudhury P, Dasgupta AK, Pal R (2012) Gold nanorod production by cyanobacteria-a green chemistry approach. J Appl Phycol 24:55–60

Pey P, Packiyaraj MS, Nigam H, Agarwal GS, Singh B, Patra MK (2014) Antimicrobial properties of CuO nanorods and multi-armed nanoparticles against *B. anthracis* vegetative cells and endospores. Beilstein J Nanotechnol 5:789–800

Philip D (2010) Green synthesis of gold and silver nano particles using *Hibiscus rosasinensis*. Phys E 42:1417–1424

Poulose S, Panda T, Nair PP, Th'eodore T (2014) Biosynthesis of silver nanoparticles. J Nanosci Nanotechnol 14(2):2038–2049

Prasad R (2014) Synthesis of silver nanoparticles in photosynthetic plants. J Nanopart 2014:963961. https://doi.org/10.1155/2014/963961

Prasad R, Pandey R, Barman I (2016) Engineering tailored nanoparticles with microbes: quo vadis. WIREs Nanomed Nanobiotechnol 8:316–330. https://doi.org/10.1002/wnan.1363

Prasad R, Jha A, Prasad K (2018a) Exploring the realms of nature for nanosynthesis. Springer International Publishing, New York. ISBN 978-3-319-99570-0. https://www.springer.com/978-3-319-99570-0

Prasad R, Kumar V, Kumar M, Wang S (2018b) Fungal Nanobionics: principles and applications. Springer Nature Singapore Pvt Ltd., Singapore. ISBN 978-981-10-8666-3. https://www.springer.com/gb/book/9789811086656

Rai M, Yadav A, Gade A (2009) Silver nanoparticles as a new generation of antimicrobials. Biotechnol Adv 27(1):76–83

Rajeshkumar S, Kannan C, Annadurai G (2012) Green synthesis of silver nanoparticles using marine brown algae *turbinaria conoides* and its antibacterial activity. Int J Pharm Bio Sci 3(4):502–510

Rajeshkumar S et al (2013) Antibacterial activity of algae mediated synthesis of gold nanoparticles from turbinaria conoides. Der Pharm Chem 5(2):224–229

Raji V, Kumar J, Rejiya CS, Vibin M, Shenoi VN, Abraham A (2011) Selective photothermal efficiency of citrate capped gold nanoparticles for destruction of cancer cells. Exp Cell Res 317(14):2052–2058

Raliya R, Biswas P, Tarafdar JC (2015) TiO$_2$ nanoparticle biosynthesis and its physiological effect on mung bean (*Vigna radiata* L.). Biotech Rep 5:22–26

Ramalingam N et al (2018) Green synthesis of silver nanoparticles using red marine algae and evaluation of its antibacterial activity. J Pharm Sci Res 10(10):2435–2438

Rauwel P, Kuunal S, Ferdov S, Rauwel E (2015) A review on the green synthesis of silver nanoparticles and their morphologies studied via TEM. Adv Mater Sci Eng 2015:682749

Ravinder Singh C, Kathiresan K, Anandhan S (2015) A review on marine based nanoparticles and their potential applications. Afr J Biotechnol 14:1525–1532

Reddy P, Urban S (2008) Linear and cyclic C18 terpenoids from the southern Australian marine brown alga *Cystophora moniliformis*. J Nat Prod 71(8):1441–1446

Richards DG, McMIllin DL, Mein EA (2002) Gold and its relationship to neurological/glandular conditions. Int J Neurosci 112:31–53

Ruparelia JP, Chatterjee AK, Duttagupta SP, Mukherji S (2008) Strain specificity in antimicrobial activity of silver and copper nanoparticles. Acta Biomater 4(3):707–716

Salari Z et al (2016) Sustainable synthesis of silver nanoparticles using macroalgae *Spirogyra varians* and analysis of their antibacterial activity. J Saudi Chem Soc 20:459–464

Schrofel A, Kratosova G, Bohunicka M, Dobrocka E, Vavra I (2011) Biosynthesis of gold nanoparticles using diatoms-silica-gold and EPS-gold bionanocomposite formation. J Nanopart Res 13:3207–3216

Shabnam N, Pardha-Saradhi P (2013) Photosynthetic electron transport system promotes synthesis of Au-nanoparticles. PLoS One 8:e71123

Shiny PJ, Mukherjee A, Chandrasekaran N (2013) Marine algae mediated synthesis of the silver nanoparticles and its antibacterial efficiency. Int J Pharm Pharm Sci 5(2):239–241

Silver S (2003) Bacterial silver resistance: molecular biology and uses and misuses of silver compounds. FEMS Microbiol Rev 27:341–353

Singaravelu G, Arokiamary JS, Kumar VG, Govindaraju K (2007) A novel extracellular synthesis of mondisperse gold nanoparticles using marine alga, *Sargassum wightti* Greville. Coll Surf B 57(1):97–101

Slomberg DL, Lu Y, Broadnax AD, Hunter RA, Carpenter AW, Schoenfisch MH (2013) Role of size and shape on biofilm eradication for nitric oxide-releasing silica nanoparticles. ACS Appl Mater Interfaces 5(19):9322–9329

Sondi I, Salopek-Sondi B (2004) Silver nanoparticles as antimicrobial agent: a case study on E. coli as a model for Gram negative bacteria. J Colloid Interface Sci 275(1):177–182

Sri Vishnu Priya R, Narendhrans S, Sivaraj R (2016) Potentiating effect of ecofriendly synthesis of copper oxide nanoparticles using brown alga: antimicrobial and anticancer activities. Bull Mater Sci 39(2):361–364

Stalin Dhas T et al (2014) Facile synthesis of silver chloride nanoparticles using marine alga and its antibacterial efficacy. Spectrochim Acta A Mol Biomol Spectrosc 120:416–420

Tavaria FK, Costa EM, Gens EJ, Malcata FX, Pintado ME (2013) Influence of abiotic factors on the antimicrobial activity of chitosan. J Dermatol 40(12):1014–1019

Thajuddin N, Subramanian G (1992) Survey of cyanobacterial flora of the southern east coast of India. Bot Mar 35:305–314

Thakkar KN, Mhatre SS, Parikh RY (2010) Biological synthesis of metallic nanoparticles. Nanomed Nanotechnol Biomed 6:257–262

Uma Suganya KS et al (2015) Blue green alga mediated synthesis of gold nanoparticles and its antibacterial efficacy against gram positive organisms. Mater Sci Eng C 47:351–356

Varisco M, Khanna N, Brunetto PS, Fromm KM (2014) New antimicrobial and biocompatible implant coating with synergic silver-vancomycin conjugate action. Chem Med Chem 9(6):1221–1230

Venkatpurwar V et al (2011) Green synthesis of silver nanoparticles using marine polysaccharide: study of in-vitro antibacterial activity. Mater Lett 65:999–1002

Yousefzadi M et al (2014) The green synthesis, characterization and antimicrobial activities of silver nanoparticles synthesized from green alga *Enteromorpha flexuosa* (wulfen). J Agardh Mater Lett 137:1–4

Zan L, Fa W, Peng T, Gong ZK (2007) Photocatalysis effect of nanometer TiO_2 and TiO_2-coated ceramic plate on Hepatitis B virus. J Photochem Photobiol B 86(2):165–169

Chapter 7
Green and Bio-Mechanochemical Approach to Silver Nanoparticles Synthesis, Characterization and Antibacterial Potential

Matej Baláž, Zdenka Bedlovičová, Mária Kováčová, Aneta Salayová, and Ľudmila Balážová

Contents

M. Baláž · M. Kováčová
Department of Mechanochemistry, Institute of Geotechnics, Slovak Academy of Sciences, Košice, Slovakia

Z. Bedlovičová (✉) · A. Salayová
Department of Chemistry, Biochemistry and Biophysics, Institute of Pharmaceutical Chemistry, University of Veterinary Medicine and Pharmacy, Košice, Slovakia
e-mail: zdenka.bedlovicova@uvlf.sk

Ľ. Balážova
Department of Pharmacognosy and Botany, University of Veterinary Medicine and Pharmacy, Košice, Slovakia

© Springer Nature Switzerland AG 2020
R. Prasad et al. (eds.), *Nanostructures for Antimicrobial and Antibiofilm Applications*, Nanotechnology in the Life Sciences,
https://doi.org/10.1007/978-3-030-40337-9_7

Abstract The area of nanotechnology has been enjoying enormous research interest recently because of the unique characteristics of nanoscale substances. Silver nanoparticles (AgNPs) are attractive for their broad spectrum of utilization in various sectors. Among the several ways of AgNPs preparation, we focused on green synthesis using phytochemicals present in plants without the use of chemicals and surfactants that are harmful for humans and environment. In this chapter, we provide a comprehensive review of the various options of green preparation of silver nanoparticles including bio-mechanochemical synthesis. Just few reports on the bio-mechanochemical synthesis combining tools of mechanochemistry (ball milling) and of green synthesis (plant material) were reported until now, and these studies are briefly reviewed. They show green and sustainable character of this method and underline the growing interest of scientists towards mechanochemistry. Silver and its compounds are known as strong antibacterial agents, so we have focused on description of the methods frequently applied to determine the antimicrobial properties of AgNPs.

Keyword Silver nanoparticles (AgNPs) · antibacterial activity · mechanochemistry · plant extract · green synthesis

7.1 Introduction

Due to small particle size, various shapes and high surface areas, silver nanoparticles show different chemical, physical and biological properties and lead to the broad spectrum of their applications in pharmaceutical field, catalysis, sensors, spectroscopy or electronics (Abbasi et al. 2016; Prasad et al. 2016). Silver provides activity against many bacterial species, so it is commonly used in medical industry, including creams with silver content to avoid local infections after open wounds or burns, soaps, catheters, pastes and textiles (Becker 1999; Catauro et al. 2004; Dallas et al. 2011; García-Barrasa et al. 2011).

Preparation of AgNPs can be realized by various ways. In general, we can classify these techniques as chemical, biological and physical. Some of these techniques are simple with good nanoparticle size control and can be achieved by affecting the process of reaction, but on the other hand, there are some difficulties and complications with stability of the products (Kowshik et al. 2002). Despite that, many of the reactants and surfactants engaged in these processes are dangerous to human health and the environment, and so, a "green" approach for metal nanoparticles is necessary.

Methods of classical chemical or physical approach to nanoparticles synthesis are environmentally toxic and economically demanding (Kalaiarasi et al. 2013). These objectives have forced the scientific community to search for new, easy and environmentally friendly choice that would not be harmful for humans, animals and environment. These facts shifted the attempts to synthesize Ag nanoparticles using green synthesis methods that have been indicated to be easy, economically advantageous and environmentally friendly (Prasad 2014). The green preparation of silver nanoparticles means preparing nanoparticles using biological routes for Ag(I) ion reduction by natural sources of reducing agents. These sources include microorganisms (algae, yeast, fungi, bacteria), plants and plant extracts or their products, such as small biomolecules including vitamins, amino acids or polysaccharides (Sharma et al. 2009; Ahmed et al. 2016; Mohammadlou et al. 2016; Prasad et al. 2016, 2018). Nanoparticles prepared by green approach are much superior to nanoparticles prepared by chemical and physical processes, as green methods exclude the use of toxic and consumptive chemicals and save the energy needed for synthesis (Clark et al. 2002; Kharissova et al. 2013; Prasad et al. 2016).

Nowadays, investigation on nanometals' biosynthesis by plant extracts has become immensely popular. This alternative green approach to the synthesis of AgNPs, which utilizes plant sources in the synthesis of metal nanoparticles, has developed in the new area of "phytonanotechnology" (Rajan et al. 2015; Prasad 2019). This method reflects the newest epoch in rapid and harmless methods for nanoparticles synthesis. For that reason, the present work has focused on silver nanoparticles synthesis using plant extracts. As already mentioned before, there are several ways of silver nanoparticles preparation. Among them, an environmentally friendly solvent-free approach called mechanochemistry has been used. However, the reports on bio-mechanochemical synthesis (co-milling of plant material and silver nitrate) are rather scarce (Rak et al. 2016; Arancon et al. 2017; Baláž et al. 2017a, b).

This chapter provides a brief introduction into the topic of green synthesis of silver NPs, outlining the most important group of compounds responsible for reduction, commonly used characterization techniques and methods used for antibacterial activity determination. The second part is devoted to the bio-mechanochemical preparation of silver NPs.

7.2 Green Synthesis of Silver Nanoparticles

The term "green chemistry" is generally understood as a principle to reduce chemistry-related impacts on human health and to eliminate the environmental pollution. Green chemistry represents a different way of thinking how chemistry can be done, for example, by reducing waste, saving energy, using non-toxic chemicals and solvents (de Marco et al. 2018; Joshi et al. 2018).

Nature is an endless source of compounds which can be used for silver nanoparticles synthesis by green chemistry approach. Synthetic methods using biomaterials are in line with the principles of "green" chemistry, as safe and non-toxic reagents are used to prepare nanoparticles. This part of the chapter is focused on green approach for the Ag nanoparticles synthesis using plant extracts. These procedures are performed outside of a living organism using an extract without cells producing reducing molecules with reducing capacity.

Green approach to nanoparticles synthesis is based on bottom-up approach, where the reducing agents contained in plants are liable for the Ag(I) ion bioreduction to prepare AgNPs. This process starts after incubation of the plant extracts with silver salts (silver nitrate is mostly used). In general, the biosynthesis of Ag nanoparticles involves three steps. The first is an induction step involving the reduction of Ag(I) ions followed by the second phase of growth, where larger aggregates are produced, and finally, the third, stabilizing step, in which the stabilization of prepared nanoparticles is carried out (Marchiol et al. 2014). The green approaches include the choice of suitable solvent used for the Ag nanoparticles synthesis and the choice of environmentally friendly reducing agent and non-toxic material as a capping agent for nanoparticle stabilization. The green reduction of silver ion is completed by natural material, mainly extracts obtained from plants, which contain reducing compounds, for example phenolics, flavonoids, terpenoids or saccharides (Prabhu and Poulose 2012), which also are responsible for their stabilization. In the plant-mediated nanotechnology field, a variety of plant parts (flowers, fruit, leaves, root, stem, seed, peel) are used (Kharissova et al. 2013; Mittal et al. 2013; Ahmed et al. 2016; Carmona et al. 2017).

In the concept of green approach of AgNPs synthesis, elimination of the use of toxic chemical is expected. Extractable phytochemicals therefore play a dual role as both reducing and stabilizing representatives for the production of metal nanoparticles containing AgNPs simultaneously without any involvement of chemicals. This approach is dependent on the amount of reductants contained in the extract, and therefore, the key step is to extract these *reductants* and *capping agents*.

7.2.1 Extraction

The ways of efficient extraction of the bioactive compounds were intensively studied in literature (Azwanida 2015). Several techniques have been studied to optimalize the extraction process of these compounds, such as drying methods, temperature of extraction and extraction solvent type. Extraction of different plant parts can be done by several extraction procedures. The initial stage in the plant extract production is the preparation of plant samples. Different parts of plants such as leaves (Ajitha et al. 2015; Khatoon et al. 2018), barks (Velayutham et al. 2013), roots (Behravan et al. 2019), fruits and flowers can be extracted from fresh or dried plant material prior to AgNPs synthesis. Generally, after the collection of plant/plant parts, the cleaning of plant material by washing it twice or more times with distilled water follows. Both, fresh and dried plant samples are used in plant extract preparation. After drying procedure, the plants are powdered or ground. Lowering the particle size allows better contact between sample and extraction solvent.

The solvent types used in the extraction importantly influence the quantity of the extracting reducing agents. The most frequently used solvent for extraction of bio-reducing compound is water. There are few reports regarding the use of organic solvents for extraction, like *Murraya koenigii* methanol leaf extract (Suganya et al. 2013), *Ziziphora tenuior* methanol extract (Sadeghi and Gholamhoseinpoor 2015), *Malva parviflora* ethanol/water extract (Zayed et al. 2012). In most cases, the solvent was evaporated and the residue diluted in water for silver nanoparticles synthesis.

Traditional solvent extractions of biologically active compounds from plant raw material are carried out in aqueous solution or organic solvents and by application of heat and mixing. The most frequently used traditional methods include soxhlet extraction (Sadeghi and Gholamhoseinpoor 2015), maceration (Zayed et al. 2012), percolation (Padalia et al. 2015) and sonication.

Instead of conventional extraction techniques, several modern extraction techniques were developed. Mechanochemical-assisted extraction (MCAE) belongs to the innovative pre-extraction techniques (Wu et al. 2017), the forming of which is nearly connected with the development of mechanochemistry (Baláž et al. 2013). This method has been established and put in an application for the bioactive compounds' extraction. In addition to the extraction method of heat reflux, the method of MCAE has higher yield of extraction, shortened time of extraction and lower temperature of extraction, so it can decrease or evade the reduction of thermolabile compounds' amount. MCAE's main goal is mechanical treatment of powder mixtures of plant materials and suitable reagents in special mill activators to solubilize the biologically active compounds (Lomovsky et al. 2017). In general, the MCAE procedures involve three steps, including sample preparation, mechanochemical treatment of plant material with solid reagent in ball mill and subsequent extraction procedure. There are several ways how MCAE provides improvement of extractable biomolecules: (a) the particle size reduction leading to the total contact surface area increase, so the probability of the reaction of biologically active compounds with the solid agents increases; (b) cell wall damage; and (c) chemical form transformation. The MCAE also allows the chemical conversion of compounds to enhance the solubility in water, so water can be used as an solvent of extraction as an alternative to organic solvents (Wu et al. 2017). This method was used for improved extraction, for example flavonoids from bamboo (Xie et al. 2013), chondroitin sulphate from shark cartilage (Wang and Tang 2009), polysaccharides extraction from *Ganoderma lucidum* spores (Zhu et al. 2012) or alkaloids from *Stephania tetrandra, Stephania moore* (Wang et al. 2019).

Plant extracts can be also obtained via microwave-assisted extraction (MAE). The MAE is a versatile technique which offers lower extraction period, higher extraction yield and less energy consumption. In addition, the microwave oven produces high temperature, which facilitates the breakage of the cell wall when the plant material is extracted, and as a result, active substances extraction is facilitated in the solvent. Water can be used in MAE under controlled temperatures. Different plants were subjected to MAE prior to AgNPs synthesis like red cabbage (Demirbas et al. 2016), green tea (Sökmen et al. 2017) and *Anthriscus cerefolium (L.)* Hoffm. (*Apiaceae*) (Ortan et al. 2015).

Safarpoor used ultrasound-assisted extraction (UAE) of 40 kHz frequency for preparation of *Thymus daenensis* and *Silybum marianum* ethanolic extract prior to AgNPs synthesis. (Safarpoor et al. 2018). Ultrasound-assisted extraction by effect of acoustic cavitation from the ultrasound facilitates inorganic and organic substances leaching from plant matrix. The extraction mechanism can be described by the diffusion through the cell wall and washing the contents of cell after destruction of the cell walls. The procedure is simple to use and relatively cheap that facilitates release of compounds. The UAE or sonification can be advantageously used for shortening the extraction time and extraction of termolabile biological compounds (Azmir et al. 2013; Azwanida 2015).

7.2.2 Biomolecules with Reducing Ability

The green synthesis of nanoparticles is based on the bioactive compounds extracted from plants that are able to reduce the metal ions to their elemental form. These metabolites are from group of terpenoids, flavonoids, tannins, phenolic acids, saponins, steroids, alkaloids, saccharides, proteins, amino acids, enzymes or vitamins (Kulkarni and Muddapur 2014). The mentioned substances containing reducing groups (OH, CH=O, NH_2, SH) are able to reduce silver ions to the elementary form Ag(0) and produce the Ag nanoparticles, but also play an important role in their stability. Metal nanoparticles prepared by plant extracts were found to be functionalized with biomolecules that have primary amino, carbonyl, hydroxyl or other stabilizing functional groups (Kulkarni and Muddapur 2014; Sadeghi and Gholamhoseinpoor 2015). Different concentrations and ratio of active compound in plant have effect on the morphological diversity of nanoparticles (triangles, cubes, spherical, ellipsoids and others) and their properties (Makarov et al. 2014).

7.2.2.1 Flavonoids

Flavonoids represent a wide group of plant secondary phenolic compounds based on the flavan nucleus and can be divided into isoflavonoids, flavonols, flavones, flavanones, chalcones and anthocyanins (Kelly et al. 2002). The number, positions and types of substitutions influence the preparation and stabilization of nanoparticles. Tautomeric transformation of enol-form of flavonoid molecule to its oxo-form is responsible for the synthesis of nanoparticles, because it may release hydrogen which can reduce metal ions to NPs. For example; it is predicted that luteolin and rosmarinic acid present in *Ocimum basilicum* are responsible for AgNPs formation from silver ions by changing their tautomeric forms (Ahmad et al. 2010). Dihydromyricetin-mediated silver nanoparticles were synthesized by green way from $AgNO_3$ and flavonoid dihydromyricetin (also known as ampelopsin) isolated from *Ampelopsis grossedentata*. Hydroxyl and carbonyl groups of dihydromyricetin were involved in formation of spherical Ag nanoparticles (Ameen et al. 2018).

7.2.2.2   Terpenoids

Terpenoids are secondary plant metabolites constructed from two or more isoprene units. Silver nanoparticles were prepared, for example, by rapid biosynthesis using leaf extract of *Thuja occidentalis* (Kanawaria et al. 2018). The nanoparticles were also encapsulated and stabilized by organic layer which contains mainly terpenes and terpenoids. These secondary metabolites take a part in preparation of nanoparticles and in their stabilization; it means prevention of agglomeration (Kanawaria et al. 2018). The main secondary metabolites from leaf extracts of *Azadirachta indica* (flavonoids and terpenoids) act as a reducing agent for the preparation of silver nanoparticles from $AgNO_3$ and also as a capping agent (Roy et al. 2017). Ag nanoparticles were prepared in a very short time using clove extract. The main constituent of clove extract is terpenoid eugenol which is responsible for bioreduction of metallic ions to nanoparticles. The mechanism of bioreduction via eugenol terpenoid found in the clove has been described. The induction effect is mediated by methoxy and allyl groups, which are located in the eugenol structure in the *para* and *ortho* positions to the -OH group capable of releasing the proton, thereby forming a resonance-stabilized anionic form. It is then able to release two electrons from one eugenol molecule (Singh et al. 2010).

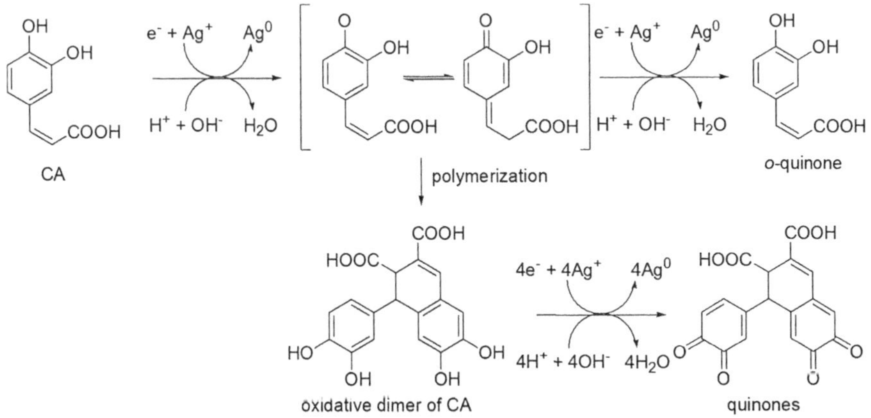

Fig. 7.1 Supposed mechanism of the AgNPs formation using caffeic acid in rice husk extract (Liu et al. 2018)

7.2.2.3 Phenolic Acids

High nucleophilicity of aromatic ring may support the chelating ability. Phenolic acids containing hydroxyl and carbonyl groups are considered as the major phytochemicals. For example, in rice husk, phenolic acids were investigated for their reducing ability to Ag(0) ion. An alkaline medium fits the transfer of the electron from the phenolic acids to reduce Ag(I) to create AgNPs, and in Fig. 7.1, the proposed mechanism for generating AgNPs prepared by caffeic acid (CA) in the rice husk extracts is shown. The reduction process of Ag(I) includes releasing of electrons to reduce Ag(I) to Ag(0), form caffeic acid-derived free radical, and consequently, another Ag(I) is reduced and CA is oxidized to *o*-quinone. Additionally, a caffeic acid dimer is produced which provides four electrons for Ag(I) reduction and is oxidized itself to quinone form (Fig. 7.1). The quinone derivatives can be combined to create a steric hindrance around the AgNPs, which avoids aggregation and enhances the stability of NPs (Liu et al. 2018).

7.2.2.4 Saccharides

One of the biomolecules used as bioreducing agents are saccharides. Saccharides represent a wide group of polyhydroxy derivatives. The reducing ability of saccharides depends on two factors. The first is the content of individual monosaccharides and the second is binding between them. Saccharides that are used for synthesis of nanoparticles have many advantages; (a) availability, (b) low price (c) environmental reasons (d) capping ability (Panigrahi et al. 2004).

Glucose is supposed to be a strong reducing agent than fructose. However, it was found that fructose is better for synthesis of smaller nanoparticles and hence utilized for the preparation of Au–AgNPs composite (Au-core, Ag-shell) of ~10 nm (Panigrahi et al. 2005).

It was shown that isolated polysaccharides and extract of biological material containing polysaccharides and other substances are suitable for biosynthesis of nanoparticles. Isolated polysaccharides such as starch or heparin serve as both a capping agent and a reducing agent (Raveendran et al. 2003; Huang and Yang 2004). However, also some additional substances, such as β-D-glucose, can serve as a reducing reagent and starch can serve as a capping agent (Raveendran et al. 2003; Vigneshwaran et al. 2006).

7.2.2.5 Proteins

Synthesis of nanoparticles is also connected with presence of amino acids and proteins in plants extracts. It was observed that proteins act as both reducing and stabilizing agents. The big advantage of such prepared nanoparticles is their biocompatibility (Tan et al. 2010). On the other hand, the disadvantage is the speed of reaction. The synthesis is relatively slow (hours, days) in comparison to the other plant metabolites (terpenoids, phenolic acids, flavonoids, saccharides), which took only few minutes (Ashraf et al. 2016; Baláž et al. 2017a). AgNPs synthesized by $AgNO_3$ using extract of *Malva parviflora* were stabilized in solution containing proteins produced by the biomass (Zayed et al. 2012). Nitrogen of the amide group and silver ion coordination confirm the fact that the NH_2 group of proteins has the strong capacity to capture the metal. It indicates that the proteins from plant extracts could cover the nanoparticles by layer, which has ability to avoid agglomeration and thereby to stabilize the solutions containing nanoparticles (Zayed et al. 2012). Although amino acids and proteins have the ability to reduce precursors of nanoparticles, they are very rarely used. Usually, different agents are used as reducers and proteins or amino acids serve only as capping agents, for example in bio-mechanochemical synthesis (Baláž et al. 2017a), simulated sunlight radiation (Huang et al. 2016) and sonochemical route (Ghiyasiyan-Arani et al. 2018).

7.2.3 Synthesis Procedure

After the extraction the solid residue is filtered off, the filtrate is used for nanoparticles synthesis. From literature survey, it follows that myriad of different plant extracts has already been used for the green synthesis of AgNPs (Ahmed et al. 2016; Ismail et al. 2016). We have selected just few recent interesting examples in our chapter. However, the principle of all the studies is always the same (Fig. 7.2).

The mostly used source of Ag(I) ions is silver nitrate aqueous solution in various concentrations. For example, aqueous extracts of *Adiantum capillus-veneris L.* were used for AgNPs biosynthesis by 1, 2 and 3 mM silver nitrate solution at room temperature via mechanical stirring (Omidi et al. 2018). The change of colour from yellow to dark brown was indication of silver nanoparticles formation. According to transmission electron microscopy (TEM) analysis, they synthesized nanoparticles of 18.4 nm size. Strong antibacterial activity was evidenced against both Gram-positive (*Staphylococcus aureus*) and Gram-negative (*Escherichia coli*) bacterial strains.

Green synthesis of AgNPs using a root extract of *Lepidium draba* weed was successfully implemented by Benakashani et al. (Benakashani et al. 2017). As source of Ag(I) ions, they used 0.01 M $AgNO_3$ with various amounts of root extract (2, 2.5,

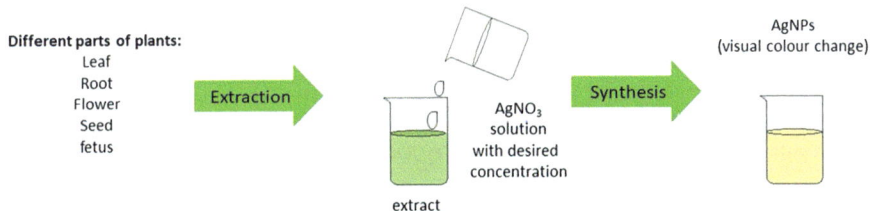

Fig. 7.2 Schematic representation of green synthesis of silver nanoparticles by plant extract

3, 3.5 mL). The evidence of nanoparticles generating was visually controlled by colour change from green to dark brown as is generally known. The tested antibacterial activity was determined as higher as Ag(I) ions.

Seeds used for silver NPs synthesis were used by Ping and his colleagues (Ping et al. 2018). As reducing and stabilizing agent, they used grape seed extract and 0.01 M AgNO$_3$ was reduced. The average size of prepared AgNPs was 54.8 nm when synthesized at room temperature. Antibacterial activities have not been detected yet.

The very interesting and improved green synthesis method by photoreduction was used by Viet and his working group using the *Buddleja globosa* leaf extracts (Van Viet et al. 2018). For Ag nanoparticles synthesis, they used low-power UV light in the presence of poly (vinyl pyrrolidone) (PVP) as the stabilizing agent. The procedure was optimized (ratio of ethanol/water/silver nitrate; pH value and time of reduction) to obtain spherical Ag nanoparticles with average size of 16 ± 2 nm. The prepared Ag nanoparticles were tested against *E. coli* strains with approximately 100% elimination.

7.3 Characterization Techniques

Characterization of Ag nanoparticles by various physical methods is very important in order to evaluate their physicochemical properties that can affect their behaviour, bio-distribution or safety. Synthesized silver nanoparticles are characterized by wide spectrum of methods including UV-Vis spectroscopy, infrared spectroscopy with Fourier transform (FTIR), X-ray photoelectron spectroscopy (XPS), dynamic light scattering (DLS), zeta potential, X-ray diffraction (XRD), scanning electron microscopy (SEM) and transmission electron microscopy (TEM). These techniques are very briefly reviewed below.

Very often, discrepancies in the papers regarding particle size determined by various methods occur. It is mainly because some methods focus on the individual nanoparticles, whereas the others on the large grains. This is explained by a very simplified figure (Fig. 7.3).

7.3.1 Ultraviolet and Visible Spectroscopy

Ultraviolet and visible spectroscopy (UV-Vis) is a broadly used method due to its availability, simplicity and sensitivity of measurement and is a relevant technique for the initial characterization of synthesized AgNPs by monitoring the synthesis during the process (Patil and Sastry 1997). In silver nanoparticles, the valence and conduction bands lie closely to each other, and electrons are able to move freely. This action leads to the formation of a surface plasmon resonance (SPR) absorption band occurring as a result of oscillation of AgNPs electrons with the wave of light

Fig. 7.3 Schematic illustration showing which techniques report the size of individual nanoparticles of few nanometres and those that of agglomerates/ grains of size larger than 100 nm

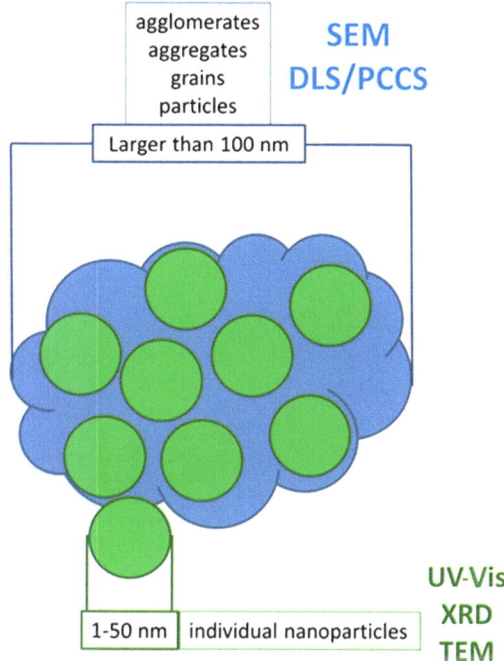

(Noginov et al. 2007; Van Der Merwe 2010). The absorption of Ag nanoparticle depends on the size of particles, dielectric medium and chemical environment (Noginov et al. 2007). The SPR band can serve as a simple detection of biosynthe-sized nanoparticles' stability with respect to unchanged wavelength of this absorption band (Zhang et al. 2016). As an example, the UV-Vis spectra showing surface plasmon resonance bands of AgNPs prepared by using the *Origanum vulgare L.* water extract (whole plant) and silver nitrate water solution (Baláž et al. 2017a) are shown in Fig. 7.4.

The time-dependent investigation of Vis spectra (Fig. 7.4) showed that soon after mixing *O. vulgare* L. water extract (showing two peaks in UV region) and silver nitrate water solution, the absorbance at 445 nm occurred, corresponding to the spherical silver nanoparticles. With reaction time, its intensity increased and after 7 min, the absorbance reached the maximum and after that, there were only slight changes (Baláž et al. 2017a).

7.3.2 Fourier Transform Infrared Spectroscopy

Fourier transform infrared spectroscopy (FTIR) represents accurate and reproducible possibility to detect small changes in functional groups by measuring absorbance. FTIR spectroscopy is relatively commonly technique used to identify which

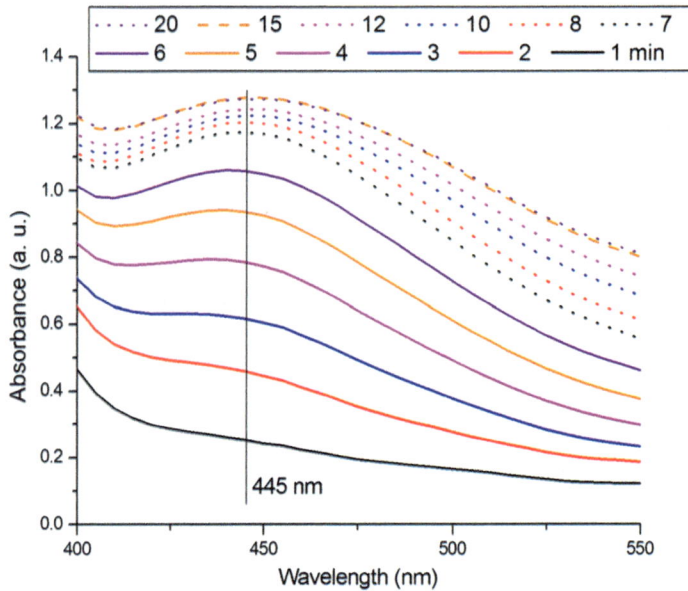

Fig. 7.4 Time-dependent absorption spectra during AgNPs synthesis and observation of SPR at 445 nm (Baláž et al. 2017a), copyright by Springer

molecules are responsible for bioreduction of Ag(I) ions in synthesis of AgNPs. In infrared spectra, the interactions between nanomaterials and biomolecules or the conformational states of bounded biomolecules (e.g. proteins) can be observed (Jiang et al. 2005; Shang et al. 2007; Lin et al. 2014). In addition to classical potassium bromide-based method, attenuated total reflection (ATR)-FTIR spectroscopy can also serve as a method for observation of the chemical changes on the surface (Lin et al. 2014). Sample preparation is much easier for conventional FTIR as there is no need to prepare pellet. ATR-FTIR method seems to be a good candidate for investigation of the surface characteristics of nanomaterial, but it is not a very sensitive technique at nanoscale due to its penetration depth and also because it has the same order of magnitude as the wavelength of IR (Liu and Webster 2007). As an example, ATR-FTIR spectra from the same study from where the UV-Vis spectrum, as shown, are provided (Fig. 7.5).

This complex, as shown in Fig. 7.5, provides a comparison of the FTIR spectra of all species included in the synthesis of AgNPs in the mentioned study. The spectra are described in detail in Fig. 7.5; however, the presence of –OH and the main peaks of the polymer in the spectrum of the final stabilized product can be emphasized. Also, no peaks of silver nitrate were observed in X-ray diffraction.

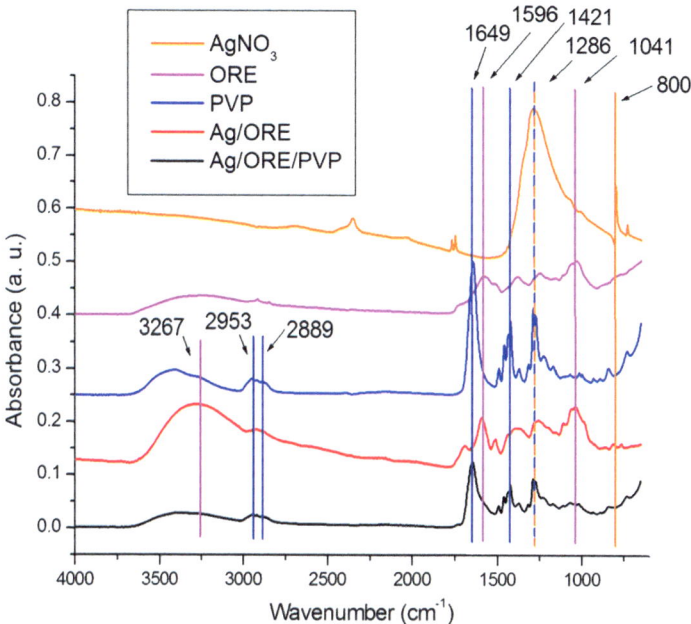

Fig. 7.5 ATR-FTIR spectra of silver nitrate, pure *Origanum vulgare L.* extract (ORE), polyvinyl pyrrolidone and both PVP-stabilized and non-stabilized AgNPs (Baláž et al. 2017a), copyright by Springer

7.3.3   X-Ray Diffraction

X-ray diffraction (XRD) is very extended analytical method which can be used to analyse the structure of molecules as well as the crystal structures, qualitative identification of studied structures and, last but not least, the particle sizes (Singh et al. 2013). The X-ray light reflects on crystal, and this reflection leads to the formation of diffraction peaks that reflect structural characteristics of studied crystal. By XRD technique, a broad spectrum of materials from biomolecules, inorganic catalysts to polymers and nanoparticles and nanostructures can be analysed (Vaia and Liu 2002; Macaluso 2009; Khan et al. 2013). A rich plethora of compounds can be easily detected, as almost every phase has a unique diffraction pattern. There are, of course, exceptions, and then other techniques are necessary. It is also possible to identify impurities; however, the detection limit of commercial X-ray diffractometers is around 5%. This method has also some disadvantages including problems with crystal growth and the ability to obtain data related to single conformation (Sapsford et al. 2011). With regard to silver NPs preparation, it is very good method, as silver nitrate and silver are well-distinguishable. However, it is necessary to dry obtained powder for analysis. XRD patterns showing the progress of the bio-mechanochemical synthesis of AgNPs when using *Origanum vulgare L.* plant and eggshell membrane are shown in Fig. 7.6.

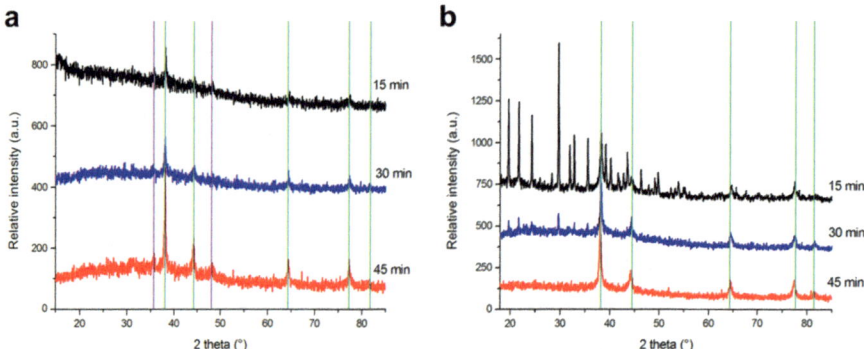

Fig. 7.6 XRD patterns of AgNPs synthesized using $AgNO_3$ and (**a**) ESM (eggshell membrane) mixture; (**b**) ORE *(Origanum vulgare)* mixture. Time of milling is included in figure, where green lines respond to the reflections of Ag (JCPDS 01–087-0717) and violet lines to WC (JCPDS 00–051-0939). All the non-marked peaks in (**b**) correspond to silver nitrate, $AgNO_3$ (JCPDS 01–074-2076) (Baláž et al. 2017b), copyright by Elsevier

It can be seen that after 15 min, the peaks corresponding to elemental silver were already present in both studied samples. The synthesis was completed a little bit faster for the eggshell membrane and took only 30 min. In case of *Origanum vulgare,* L. after 30 min of milling, some peaks of $AgNO_3$ were still visible, so the milling was conducted for 45 min in order to consume the remaining $AgNO_3$. In Fig. 7.6, also the reflections corresponding to qusongite (WC, material of milling chamber and balls) can be seen after 45 min, which is a proof of the wear from milling media. Therefore, it was concluded that for Ag/ESM system, 30 min of milling is enough.

7.3.4 Dynamic Light Scattering

The dynamic light scattering (DLS) method is also commonly used for physico-chemical characterization of synthesized nanoparticles and plays an important role in size distribution of small particles in suspension or solution (Sapsford et al. 2011; Lin et al. 2014; Zhang et al. 2016). The mentioned method measures the scattered laser light passing through a studied sample and is based on interaction of light with the studied particles. DLS is a technique used for determining the scattered light intensity as a function of time, and the average hydrodynamic size of prepared nanoparticles dispersed in liquids can be determined (Sapsford et al. 2011; Zhang et al. 2016). An example of particle size distribution determined by DLS and results of zeta potential measurement (described below) for AgNPs prepared by green synthesis using *Origanum vulgare L.* is provided in Fig. 7.7. The particle size is bimodal, with the average value of 136 nm (Sankar et al. 2013).

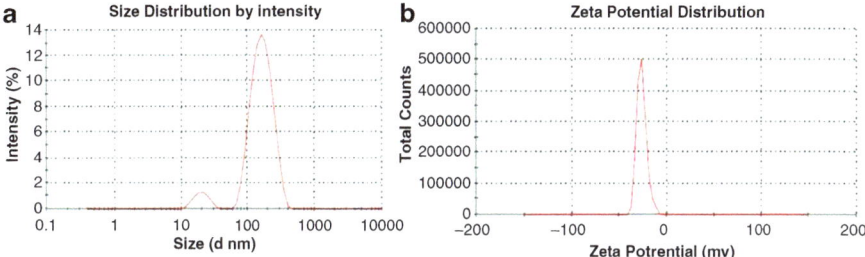

Fig. 7.7 Particle size distribution determined by dynamic light scattering (**a**) and zeta potential measurement (**b**) of the silver NPs prepared by green synthesis (Baláž et al. 2017a), copyright by Springer

7.3.5   Zeta Potential

Zeta (ξ) potential of nanomaterials in solution provides information about the charge of a nanoparticle-bioconjugate and its stability. Zeta potential is usually determined by application of an electric field across the prepared sample and measures the velocity at which charged species move to electrode (Pons et al. 2006). Zeta potential can also be applied to infer nanoparticles stability. Values with potential of 30 mV or more indicate stability, while values with potential lower than 30 mV predict particles with instability or agglomeration trend (Sapsford et al. 2011). However, this depends on the utilization of various stabilizers, as some of them act as steric ones, which do not affect the final zeta potential value then. The zeta potential measurement showing the detected value of −26 mV is shown in Fig. 7.7b.

7.3.6   Transmission Electron Microscopy

Transmission electron microscopy (TEM) is a very important, valuable and frequently applied method for nanoparticles characterization. The electrons penetrate through the sample in this case. TEM technique is used to quantitatively measure particle size, morphology and size distribution. The TEM analysis provides good spatial resolution and possibility for other analytical measurements, but on the other hand, it requires high vacuum and thin sample section, so the preparation of the sample is very important in order to get high-quality and representative pictures (Lin et al. 2014; Zhang et al. 2016). For high magnification, high-resolution TEM (HRTEM) is used. Moreover, selected area diffraction (SAD) indicating the exact phase composition of the particular area in the samples can be recorded and subsequently compared with the XRD measurements. This analysis results in the rings or points in the reciprocal space, from which the interplanar distance can be calculated and individual crystallographic planes can be indexed. TEM is often connected with the energy-dispersive X-ray spectroscopy (EDX or EDS) providing information

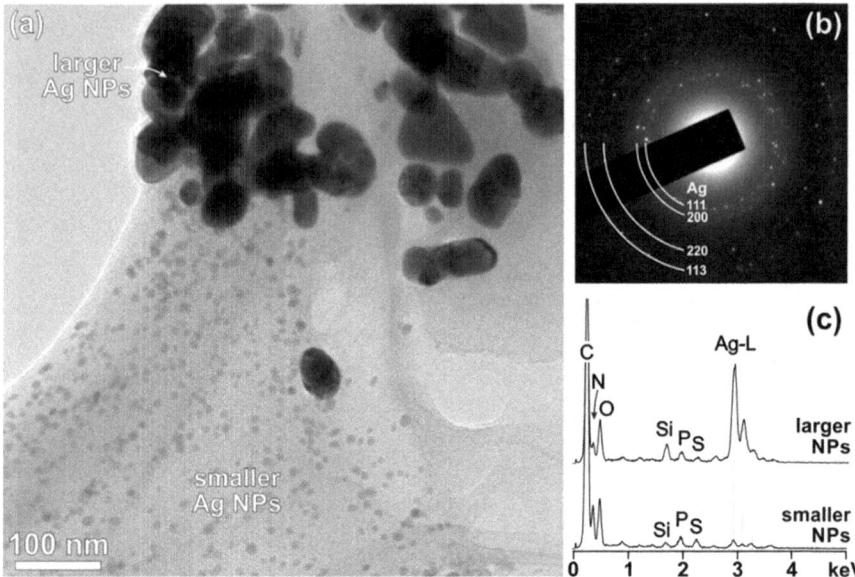

Fig. 7.8 (**a**) TEM images of the Ag/ORE/PVP indicating differentiation of the smaller and larger fractions of AgNPs. (**b**) SAD pattern of larger AgNPs fraction. (**c**) EDS spectrum of the larger and smaller AgNPs areas (Baláž et al. 2017a)

about individual elements present in the system. The example of TEM analysis of silver nanoparticles prepared by Baláž and his co-workers from water extract of *Origanum vulgare L.* after the stabilization with the polymer including SAD pattern and EDS spectra is provided in Fig. 7.8 (Baláž et al. 2017a).

Regarding particle size, two different populations can be detected from the samples, larger ones of about 38 nm and smaller ones of about 7 nm. The shape of the larger particles seems to be pseudooctahedral, whereas the smaller ones exhibit spherical shape. These groups of nanoparticles are separated. The SAD patterns presented in Fig. 7.8b show the presence of silver with cubic crystal structure. The EDS spectra of the sample after stabilization (Fig. 7.8c) did not show only silver, but also other elements presented in the plant matrix.

7.3.7 Scanning Electron Microscopy

Scanning electron microscopy (SEM) method represents a valuable technique in surface imaging used in nanomaterials studies. In contrast with TEM, the electron beam ends on the sample. This method is appropriate for investigating particle size, size distribution and shape of nanoparticles and surface of prepared particles. By SEM technique, we can study the morphology of prepared nanoparticles and obtain histogram from pictures by measuring and counting the particles manually or by software (Fissan et al. 2014; Lin et al. 2014; Zhang et al. 2016). The SEM method is also very often combined with EDX method; however, it is less accurate than in combination with TEM, as also bulk material (not seen on the photograph) is analysed (the beam goes deep into the material), Also, elemental mapping showing the distribution of individual elements is often used. The disadvantage of SEM is inability to resolve the internal structure of particles, as it is impossible to zoom so much to see individual NPs, but we can obtain valuable information about purity and aggregation of nanoparticles (Zhang et al. 2016).

7.3.8   X-Ray Photoelectron Spectroscopy

X-ray photoelectron microscopy (XPS) is an electron spectroscopy for chemical analysis (ESCA) and represents a valuable method to obtain qualitative and quantitative spectroscopic information of surface chemical analysis to obtain an estimated empirical formulae (Demathieu et al. 1999; Manna et al. 2001; Zhang et al. 2016). XPS technique is carried out under vacuum; X-ray irradiation of the nanoparticles leads to the emission of electrons, and the amount of escaped electrons from the surface of nanoparticles gives XPS spectra. The main advantage is that it can distinguish among different oxidation states of elements. An example of XPS spectra of bio-mechanochemically synthesized AgNPs showing only the peaks corresponding to elemental Ag(0) is shown in Fig. 7.9.

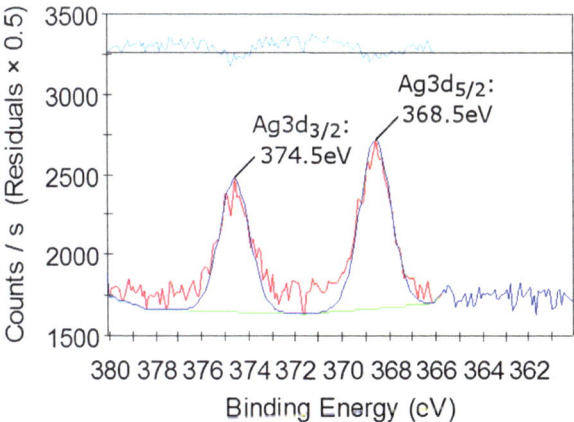

Fig. 7.9 XPS spectra of AgNPs containing composite (Rak et al. 2016), copyright by Royal Society of Chemistry

7.4 Antibacterial Activity

Due to development of bacterial resistance of classical antibiotics, scientists are forced to find out more effective antibacterial compounds. There are many studies dealing with the synthesis and antibacterial and toxicological properties' testing of NPs. The silver NPs make no exception. The interest of the scientific community on silver nanoparticles as antibacterial agents has been increased in the last few decades. Unusual properties of AgNPs have led to their utilization in medicinal field, and the main attention is focused on their antimicrobial properties.

Many authors confirmed that Ag nanoparticles show antimicrobial effects and eliminate the bacteria even at low concentrations (mg L^{-1}) (Panáček et al. 2006) without toxic influence on human cells (Carlson et al. 2008; AshaRani et al. 2009). As silver nanoparticles have shown potential as antibacterial substances, several methods are being used to evaluate their effect on different Gram-positive and Gram-negative bacterial strains. They will be briefly reviewed below.

7.4.1 Evaluation of Antibacterial Activity

Antibacterial activity studies are performed by different techniques, which lead to obtaining the results in different ways so the comparison of acquired data can be complicated. The mostly used methods of antibacterial activity evaluation are the determination of: (a) minimal inhibitory concentration (MIC), (b) half effective concentration (EC_{50}), (c) minimal bactericidal concentration (MBC) and then (d) time-kill test, (e) disc-diffusion method and (f) well-diffusion method.

The commonly used method of evaluation of antibacterial activity is the disc-diffusion method. This well-known method is based on preparation of agar plates inoculated by tested microorganism. After incubation, the discs are impregnated with silver nanoparticles at desired concentration. Prepared agar plates are incubated under appropriate conditions for examined bacterial strains, and sensitivity analysis to AgNPs is carried out by measurement of the inhibition zone diameter around the disc or well. These methods are advantageous for simplicity and economic reasons and are commonly used for antibacterial activity of silver nanoparticles determination (Balouiri et al. 2016).

Antibacterial characteristics of AgNPs are most often studied by dilution methods and quantitative assays for the determination of MIC. MIC values represent the lowest concentrations of silver nanoparticles that completely inhibit the growth of tested microorganisms. The less common method for quantification of antibacterial activity in comparison with MIC method is MBC (minimum bactericidal concentration) and is defined as the lowest concentration of tested antibacterial compound that kills 99.9% of microorganisms after 24 h of incubation. The most suitable technique for the antibacterial effect is its determination by the time-kill test (or time-kill curve) and can also be applied for establishing the synergistic effect for combination of two or more antimicrobial substances. These experiments offer the data about the dynamic interactions between various bacterial strains and tested antibacterial compounds. The time-kill test provides a time-dependent (usually 0, 4, 6, 8, 10, 12 and 24 h) or concentration-dependent antibacterial activity, and obtained data are typically presented in the form of graph (Balouiri et al. 2016). EC_{50} (effective concentration) and IC_{50} (inhibitory concentration) are used in discovery processes to determine suitability of the drugs. The EC_{50} is half maximal effective concentration reffering of drug, which induces half-maximal response. The IC_{50} is the half maximal inhibitory concentration corresponding to the 50% control (Sebaugh 2011; Mohammadlou et al. 2016).

7.4.2 Antibacterial Potential of AgNPs

Silver in Ag(0) oxidation state is well-known as antibacterial agent and has been applied as therapeutics of some diseases since ancient time (Russell and Hugo 1994). Silver nanoparticles prepared by different ways were broadly examined on a large number of pathogens including Gram-positive and Gram-negative bacteria. Some works have shown that the silver nanoparticles are more effective against Gram-negative bacteria (Chopade et al. 2013; Kokila et al. 2016), but on the other hand, the opposite results have also been reported (El Kassas and Attia 2014). These differences in sensitivity can be explained by the differentiation in the thickness and composition of biomolecules presented in the biomembranes as the cell wall of Gram-positive bacteria is composed by peptidoglycans and is thicker than Gram-negative bacteria cell wall (Franci et al. 2015; Dakal et al. 2016; de Jesús Ruíz-Baltazar et al. 2018). The importance of antibacterial activity determination in various bacterial species becomes the essence of the need to understand the mechanism, resistance and future applications of Ag nanoparticles. Ultimate reports on antimicrobial activity are summarized in Table 7.1.

Though the antibacterial activity of AgNPs has been widely considered and reported, some factors affecting the AgNPs antimicrobial properties are arising, e.g. shape, size and concentration of NPs and capping agents (Ahmed et al. 2016; Aziz et al. 2015, 2016). Nakkala et al. (2017) studied Ag nanoparticles with average size of 21 nm and the size distribution was observed to be 1–69 nm when synthesized using medicinal plant *Ficus religiosa*. These synthesized silver NPs resulted in strong antibacterial effect against *Pseudomonas fluorescens, Salmonella typhi, Escherichia coli and Bacillus subtilis*. Lower concentrations of silver nanoparticles indicated delayed effect on bacterial cell growth which may be caused by the bacteriostatic effect, while at the concentrations of 60 and 100 μg, the Ag nanoparticles showed bactericidal effect and no cell growth was detected. The green synthesized silver nanoparticles using the *Cynara cardunculus* leaf extract exhibited an antibacterial effect on *E. coli* and *S. aureus* at concentration of 20 mM; at lower concentration, the effect is minimal (de Jesús Ruíz-Baltazar et al. 2018). Antibacterial effect of AgNPs prepared by aqueous extracts from *Berberis vulgaris* using disc-diffusion

Table 7.1 Selected methods used in antibacterial evaluation of green synthesized Ag nanoparticles

Method	Tested bacteria	Plant	Precursor	References
Agar well-diffusion method	G−: *E. coli, P. aeruginosa, K. pneumoniae, Salmonella typhimurium, Shigella flexneri, Proteus mirabilis*	*Fritillaria* (flower)	10 mM AgNO$_3$	Hemmati et al. (2019)
	G+: *S. epidermidis, S. aureus, S. saprophyticus, S. pneumoniae, Streptococcus pyogenes, B. subtilis, Enterococcus faecalis, Listeria monocytogenes*			
MIC	G−: *E. coli*	*Berberis vulgaris* (root, leaf)	0.5, 1, 3, and 10 mM AgNO$_3$	Behravan et al. (2019)
	G+: *S. aureus*			
MIC	G−: *E. coli, P. aeruginosa*	*Annona reticulate* (leaf)	1 mM AgNO$_3$	Parthiban et al. (2018)
	G+: *S. aureus, B. cereus*			
Disc-diffusion	G−: *Enterobacter aerogenes, Pasteurella multocida*	*Coriandrum sativum L.* (leaf)	1 mM AgNO$_3$	Ashraf et al. (2018)
	G+: *S. aureus, B. subtilis*			
Agar well-diffusion method	G−: *E. coli, K. pneumoniae*	*Cleome viscosa L.* (fruit)	1 mM AgNO$_3$	Lakshmanan et al. (2018)
	G+: *S. aureus, B. subtilis*			
Disc-diffusion method	G−: *E. coli*	*Cynara cardunculus* (leaf)	50 mM AgNO$_3$	de Jesús Ruíz-Baltazar et al. (2018)
	G+: *S. aureus*			
Agar-well diffusion method	G−: *E. coli, P. aeruginosa, S. typhimurium*	*Origanum vulgare L.* (whole plant)	1, 2.5, 5, and 10 mM AgNO$_3$	Baláž et al. (2017a)
	G+: *S. aureus, B. cereus, Listeria monocytogenes*			
Microdilution method, MIC	G−: *E. coli, P. aeruginosa*	*Pelargonium endlicherianum* (root)	5 mM AgNO$_3$	Şeker Karatoprak et al. (2017)
	G+: *S. epidermidis*			
Agar well-diffusion method	G−: *E. coli, P. aeruginosa*	*Lantana camara* (leaf)	1 mM AgNO$_3$	Shriniwas and Subhash (2017)
	G+: *S. aureus*			

(continued)

Table 7.1   (continued)

Method	Tested bacteria	Plant	Precursor	References
Macrodilution broth method, MIC	G−: *E. coli, K. pneumoniae, P. aeruginosa, P. vulgaris*	*Diospyros sylvatica* (root)	10 mM silver acetate	Pethakamsetty et al. (2017)
	G+: *B. subtilis, B. pumilis, S. pyogenes, S. aureus*			
Disc diffusion method	G−: *E. coli, K. pneumoniae*	*Carica papaya* (peel)	0.25–1.25 mM AgNO₃	Kokila et al. (2016)
	G+: *S. aureus, B. subtilis*			
Agar well-diffusion method	G−: *E. coli, Pseudomonas sp., Klebsiella* sp.	*Alstonia scholar* (leaf)	1 mM AgNO₃	Ethiraj et al. (2016)
	G+: *Bacillus sp., Staphylococcus* sp.			
Agar well-diffusion method	G−: *E. coli, P. aeruginosa, K. pneumoniae*	*Ocimum tenuiflorum, Solanum trilobatum, Syzygium cumini, Centella asiatica, Citrus sinensis* (leaf, peel)	1 mM AgNO₃	Logeswari et al. (2015)
	G+: *S. aureus*			
Disc diffusion method	G−: *E. coli, P. fluorescens, S. typhi*	*Ficus religiosa* (leaf)	1 mM AgNO₃	Nakkala et al. (2017)
	G+: *B. subtilis*			
Agar diffusion method	G−: *P. aeruginosa, S. typhi, E. coli, K. pneumoniae*	*Tamarindus indica* (fruit)	5 mM AgNO₃	Jayaprakash et al. (2017)
	G+: *B. cereus, S. aureus, Micrococcus luteus, B. subtilis*			

MIC minimum inhibitory concentration, *G-* represents Gram-negative bacteria, *G+* represents Gram-positive bacteria

method at (1, 3, 5 mM) concentrations were evaluated on *S. aureus* and *E. coli*. As a negative control, water was used and as a positive control were chosen gentamicin and streptomycin. The MIC tests showed much higher antibacterial effect against Gram-negative bacteria, *E. coli*, than against Gram-positive, *S. aureus* (Behravan et al. 2019). Authors predict that silver nanoparticles have high affinity to sulphur and phosphorus-containing biomolecules, which is the main reason of AgNPs antibacterial properties. It means that silver nanoparticles influence the proteins containing sulphur which are concentrated inside and outside of the cell biomembrane and impact the possibilities of bacteria cell apoptosis (Yamanaka et al. 2005; Tamboli and Lee 2013). Antibacterial activity of silver nanoparticles is also affected by size and shape. It has been found that the smaller nanoparticles with larger surface-to-volume ratio exhibit higher antimicrobial effect than larger particles (Gurunathan 2015; Ethiraj et al. 2016; Deng et al. 2016). In general, higher concentration of antimicrobial agent results in more effective antimicrobial effect, but particles with small size can lead to apoptosis of bacterial cells at a lower concentration. As we mentioned before, the shape of nanoparticles also influences the interaction with Gram-negative bacteria, for example *E. coli* (Ge et al. 2014; Deng et al. 2016). The interaction between AgNPs with different shapes and *E. coli* has been studied (Pal et al. 2007). The interactions were similar, but inhibition results were different. Authors discussed that AgNPs with the same surface area, but different shapes, may have unequal effective surface area in considerations of active facets. At that time, authors were unable to give an estimation of how the surface area of different NPs influences their ability to kill bacteria (Pal et al. 2007).

Many authors and reviewers have reported the mechanism of action of AgNPs in detail, but the strict mechanism of the antibacterial effect of silver and silver nanoparticles is not fully elucidated. Most of studies consider multiple mechanisms of action, but the main three different mechanisms are expected: (a) irreversible damage of bacterial cell membrane through direct contact; (b) generation of reactive oxygen species (ROS); and (c) interaction with DNA and proteins (Sintubin et al. 2012; Durán et al. 2016; Sheng and Liu 2017; Rajeshkumar and Bharath 2017; Aziz et al. 2015, 2016, 2019) (Fig. 7.10).

7.5 Bio-mechanochemical Synthesis

Mechanochemistry has been in existence for a very long time, but its name was established much later. For example, creating fire by just banging two stones together in ancient times can be also considered mechanochemistry. Valuable works regarding history of this branch of chemistry were written by Takacs (Takacs 2004, 2013, 2018). Among many others, at least two important mechanochemists from the past can be mentioned, namely Carey Lea and Ostwald. The first one found out that the application of energy in the form of pressure and temperature yields different

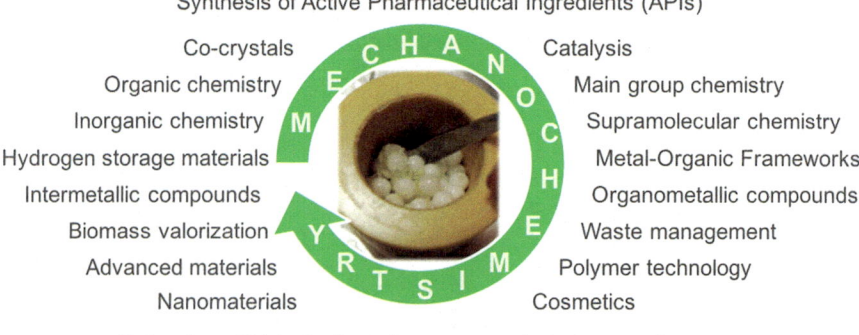

Fig. 7.10 Fields of applications of mechanochemistry (Baláž et al. 2019)

results (Lea 1893). Particularly, if temperature is applied on mercury or silver chloride, just sublimation occurs, whereas pressure effect results in the decomposition of these compounds into elements. Ostwald put the term "mechanochemistry" on the same level as thermochemistry, photochemistry and electrochemistry (Ostwald 1919). Since these preliminary works, mechanochemistry underwent a great development. Recently, it has penetrated into almost every branch of chemistry and materials science, for example pharmacy, organic synthesis, waste management, extraction of biologically active compounds from natural resources, nanotechnology, catalysis, etc. (Boldyreva 2013).

The term "mechanochemistry" is composed of two words: mechanics and chemistry. In simple words, by supplying the mechanical energy, it is possible to perform chemical reactions which would normally need higher temperatures or organic solvents under ambient conditions directly in solid state, which brings about very important environmental aspect. From the physical point of view, the energy is supplied to the powders by milling. The materials can be treated for two reasons, either to activate material or to perform reactions. The milling process can be conducted in dry or wet environment. There are many types of mills applying many different regimes (like pressing, shearing, impact, etc.). The principles of work of some basic high-energy mills are described in Fig. 7.11 (Baláž et al. 2019).

Most mechanochemical experiments are performed on a laboratory scale; however, the scale-up of this method is also gaining importance, namely, its wide usage in pharmaceutical industry (Al Shaal et al. 2010).

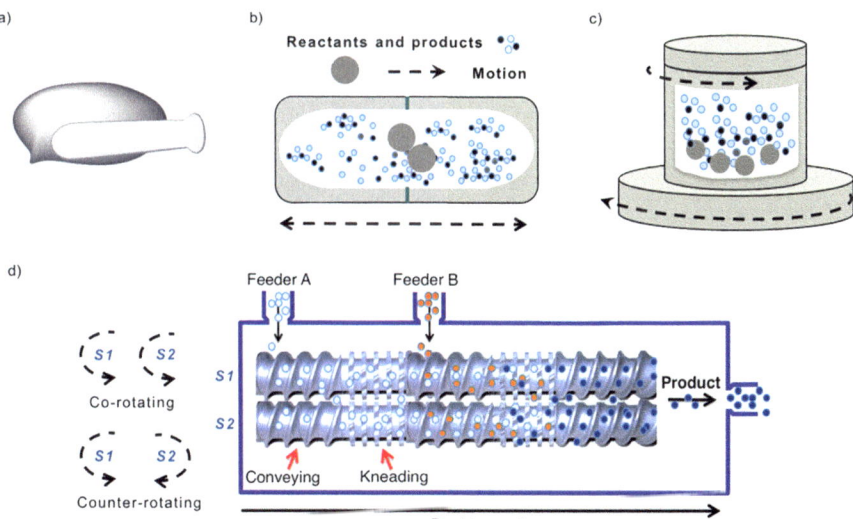

Fig. 7.11 Working principles of selected mechanochemical devices: (**a**) mortar and pestle, (**b**) shaker/vibratory mill (horizontal and vertical oscillations are possible), (**c**) planetary ball mill, (**d**) continuous twin-screw extruder. Adapted from ref. (Baláž et al. 2019)

The mechanochemical approach is also very important for nanomaterials synthesis (Baláž et al. 2013). Particularly, it represents a top-down approach by creating nanoparticles by comminution of coarse materials. The other strategy is to use the micro-sized elements for the preparation of the compounds with crystallite sizes in nano-range. This can be achieved by high-energy milling. Another possibility is the utilization of ultrafine milling (e.g. wet stirred media milling) (Romeis et al. 2016), where the size of the product is diminished into very small particles, which can be applied, for example, in biomedicine as nanosuspensions (Peltonen and Hirvonen 2010).

At the very beginning of the new millennium, nanoparticles synthesis by this method was just starting and a short review paper by McCormick and Tsuzuki was elaborated (Tsuzuki and McCormick 2004). Different metallic nanoparticles were successfully prepared by reduction of corresponding chloride with sodium metal. The authors also introduced a concept of dilution agent (NaCl in this case), because the reactions were exothermic, and it was necessary to avoid combustion. Also, the use of smaller grinding balls was beneficial as it also delayed the combustion. Many binary oxides and sulphides with particle size in the nanorange were successfully synthesized by the same research group from the corresponding chlorides. Since that time, the field broadened significantly, as pointed out earlier (Baláž et al. 2013).

There is a large number of papers on silver nanoparticles synthesis using mechanochemistry. One of the approaches is the utilization of AgCl as silver precursor, and the reduction is achieved via displacement reaction with another metal (Keskinen et al. 2001; Le et al. 2008; Khakan et al. 2017). In other works, composites of AgNPs with different compounds for various applications were prepared (Wang et al. 2010; Dai et al. 2016; Chambers et al. 2017). A cryomilling of micro-sized elemental Ag was also used (Kumar et al. 2016). The mechanochemical reduction of Ag_2O (Khayati and Janghorban 2012) or $AgNO_3$ was also used (Tiimob et al. 2017).

However, the reports utilizing natural materials for mechanochemical synthesis of AgNPs are very scarce. Namely, the utilization of materials of plant origin is an unexplored field. Such an approach can be called bio-mechanochemical synthesis and in further text, few works in which the combination of mechanochemistry and plant material for the production of AgNPs has been applied are discussed.

The first paper elaborated by Rak et al. in 2016 reports on the preparation of silver nanoparticles by ball milling method, in which lignin serves as a reducing agent and polyacrylamide (PAM) as a supporting polymer (Rak et al. 2016). During the synthesis, $AgNO_3$ was used as a source of silver. Regarding reducing agents, two types of lignin were used, namely commercially available one (Kraft lignin—KLig) and synthetic one (Thermomechanical pulp lignin—TMPLig). In the second case, lignin was prepared from TMP by oxidation prior to experiments. The difference between the two lignins is in the fibre size, as TMP lignin is larger and its structure is very similar to original lignin, so it can be also can called natural lignin. At first only solid $AgNO_3$ and lignin were milled for 90 min and AgNPs were obtained. However, as the authors wanted to use the product in the antibacterial filters, an introduction of the third component, namely polyacrylamide (PAM), was necessary.

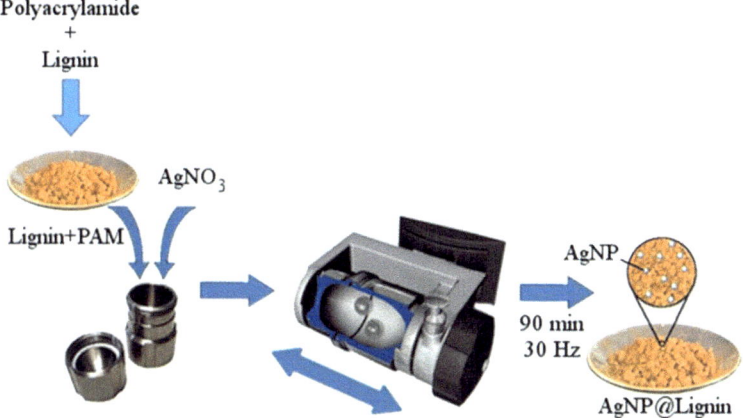

Fig. 7.12 Scheme of mechanochemical synthesis of AgNPs/lignin/polyacrylamide composites (Rak et al. 2016)

At first, lignin was left to form a composite with PAM, and then it was co-milled with $AgNO_3$ under similar conditions as in the case of PAM-free experiments. In Fig. 7.12, the scheme of the whole synthetic process is demonstrated.

Using TEM, the authors have shown that the prepared AgNPs were embedded in the organic lignin matrix. In the samples with PAM, AgNPs are encased in the polymer. The AgNPs were spherical and polydisperse with same range of sizes in all prepared samples so the PAM or the used type of lignin did not affect studied AgNPs in matter of size and shape.

The results of PXRD confirmed the presence of elemental silver. The authors also used XPS method (the spectrum was already shown in Fig. 7.9) and observed a peak located at 368.5 eV, which supports the presence of Ag in the zero-valence form. FTIR spectra have shown the peaks corresponding to lignin and PAM also in the AgNP-containing samples, thus showing that the structure of the materials was maintained.

The antibacterial activity was tested on two Gram-positive (*S. aureus, Enterococcus faecium*) and three Gram-negative (*Pseudomonas aeruginosa, Klebsiella pneumoniae, E. coli*) bacteria. For antimicrobial tests, the Kraft lignin was selected and two types of porous plugs were prepared. One consisted of AgNPs, Kraft lignin and PAM while another only Kraft lignin and PAM without any silver in it. Afterwards, the bacterial dispersion was passed through the plugs, and filtrates were diluted and inoculated onto agar plates. On the agar plates incubated with solutions that were passed through the silver-containing composites, no bacterial growth was noticed. On the contrary, no inhibition of bacterial growth was observed on the agar plates that were treated with the AgNPs-free solution. The scheme of this process and the photographs of the agar plates are presented in Fig. 7.13a and b, respectively.

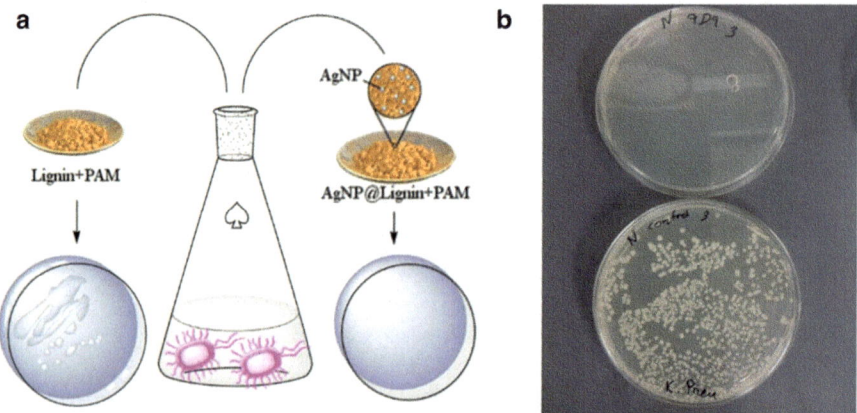

Fig. 7.13 (**a**) Schematic illustration showing the antibacterial activity of AgNP@KLig/PAM; (**b**) photos of agar plates incubated with a solution of *K. pneumonia* that were passed through the filter plug with a silver-containing composite AgNP@KLig/PAM (top), and the silver-free KLig/PAM control (bottom). Reprinted from (Rak et al. 2016) [17], copyright by Royal Society of Chemistry

The authors showed that mechanochemically prepared AgNPs@KLig/PAM possesses antibacterial activity and generated adequate amounts of Ag(I) to be able to eliminate all studied bacterial strains. The mechanism of antibacterial activity of silver nanoparticles most probably lies in the local oxidative generation of Ag(I) ions at aerobic conditions, which releases antimicrobially active silver ions in contact with water (Xiu et al. 2012).

The utilization of biomass-derived polysaccharides as a reducing agent for AgNPs synthesis was realized by Arancon et al. in 2017 (Arancon et al. 2017), namely, the polysaccharides were isolated from the fungi *Abortiporus biennis* (PS1), *Lentinus tigrinus* (PS2), *Rigidoporus microporus* (PS5) and *Ginkgo polypore* (PS10). The polysaccharides were then ball-milled with AgNO₃ in the ratio 1:2 for 30 min. After milling, the obtained powders were calcined at 600 °C to remove the unbound organic matrix. Finally, the toxicity of the products was investigated by MTT assay.

The first part of the paper is devoted to the characterization of the prepared AgNPs. The authors also focus on the interaction between polysaccharides and AgNPs by using FTIR, and they registered some bands corresponding to organics species, as some functional groups could still be detected in the spectra, despite the calcination step. This points to successful interaction. The calcined Ag-PS10 powders were further investigated by X-ray diffraction (Fig. 7.14).

A non-traditional result pointing to the preparation of silver in hexagonal crystal structures was observed. The authors claim that they have also cubic silver, but the intensity of the main peak, which should be located at 2 theta value at 38.1°, is really low, and the same situation is also with the third most intensive peak of cubic phase which should be located at 64.5°. The authors claim that they have the combination of both crystal structures based on the TEM analysis; however, the actual crystal

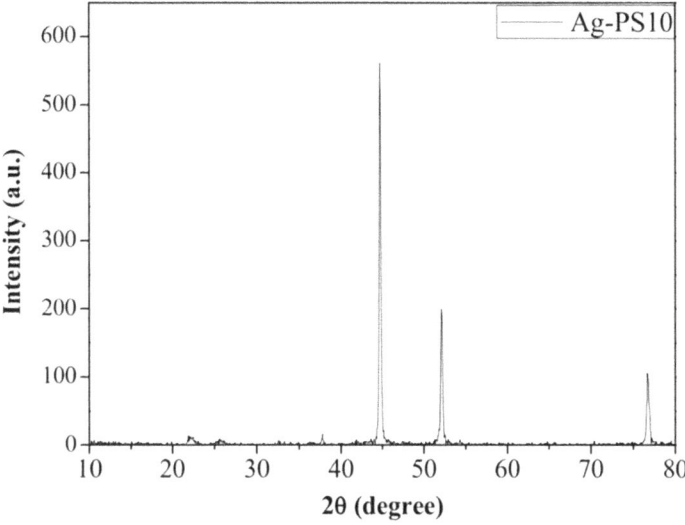

Fig. 7.14 XRD pattern of Ag-PS10 sample (Arancon et al. 2017), copyright by Elsevier

structure does not have to be the same as the actual shape of the particles. Namely, in most papers, spherical AgNPs are observed (mostly claimed based on UV-Vis measurements), but when XRD is done, the authors observe cubic reflections (Sankar et al. 2013). The authors have also used XPS method with the conclusion that they have Ag in elemental form, as the position of Ag peak was at 368.1 eV and it fits in the range between 368.0 and 368.3, which, according to the authors, corresponds to Ag in this oxidation state. Also, the thermal decomposition during calcination showing the evaporation of residual moisture and decomposition of unbound organics was observed.

The second part of paper is devoted to the MTT assay to show the toxicity on two human cell lines, namely A549 lung adenocarcinoma and SH-SY5Y neuroblastoma cell line. The products have shown almost no toxicity towards the studied cell lines; a statistically significant decrease in viability was observed only in one case when the highest concentration was used. The example is presented for Ag-PS10 sample (Fig. 7.15).

In the end, the authors also analysed the Ag content in the soluble silver supernatant of Ag–PS nanohybrid dispersions in supplemented cell culture media, and they detected higher amount of Ag in the media of A549 cell line.

The mechanochemical approach was also applied as a stabilization procedure after the classical green synthesis (Baláž et al. 2017a). Firstly, Ag nanoparticles have been prepared by a classical green synthesis in liquid medium by mixing silver nitrate water solution and aqueous extract of *Origanum vulgare, L.* (whole plant was used). Generation of AgNPs was monitored by measuring the UV-Vis spectra (Fig. 7.4) and also by visual colour change of the solution from light brownish

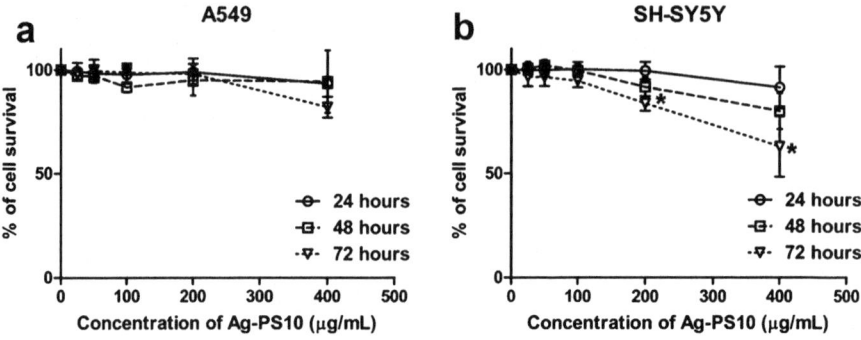

Fig. 7.15 The percentage cell viability estimated by MTT test on (**a**) A549 and (**b**) SH-SY5Y cell lines using increasing concentration of Ag-PS10 NPs (0, 25, 50, 100, 200 and 400 µg mL^{-1}) after exposition time of 24, 48 and 72 h. An OD value of control cells (unexposed cells) was taken as 100% viability (0% cytotoxicity). Results were displayed as mean ± SD of three independent experiments (*significantly different from control; $p < 0.05$). Reprinted from (Arancon et al. 2017), copyright by Elsevier

yellow at the beginning to dark brownish red at the end of the reaction. The reaction was completed after 15 min and subsequently, the stabilizing agent, polyvinylpyrrolidone (PVP) was added to the prepared nanosuspension. The effect of milling on the particle size distribution was studied. The authors found out that before the milling with PVP, two fractions of particles were present with sizes ranging between 110 and 330 nm for the larger one and between 15 and 80 nm for the smaller one. After the milling for 60 min, bimodal particle size distribution was also observed, but the milling decreased the average size of nanoparticles.

The increase in the intensity of the SPR peak with reaction time was already described earlier (Fig. 7.4). After the stabilization with PVP, this peak was red-shifted to 472 nm.

The final sample after green synthesis and the PVP-capped NPs were investigated in terms of photoluminescent properties. In the case of PVP-free AgNPs, just two weak emission peaks were recorded. On the other hand, sample of AgNPs with PVP exhibited one strong emission peak, which indicated that the addition of PVP enhances the photoluminescence emission intensity. Also, infrared spectroscopy (FTIR) was employed for the characterization (Fig. 7.5).

The transmission electron microscopy (TEM) was used to analyse the size, morphology and distribution of AgNPs both in the absence and presence of PVP. The results for the PVP-stabilized sample are shown in Fig. 7.8. The TEM analysis of the sample prior to stabilization is shown in Fig. 7.16.

As already mentioned, two different populations were found in both samples. In the sample without PVP, the populations were intermixed and dispersed in *O. vulgare* matrix, whereas after the stabilization, the separation of these groups occurred. The SAD pattern presented in Fig. 7.16b shows mainly the presence of silver with cubic crystal structure, but also the presence of AgCl was detected. Such particle is shown in Fig. 7.16c. Also, twinning phenomenon, common for silver NPs, was

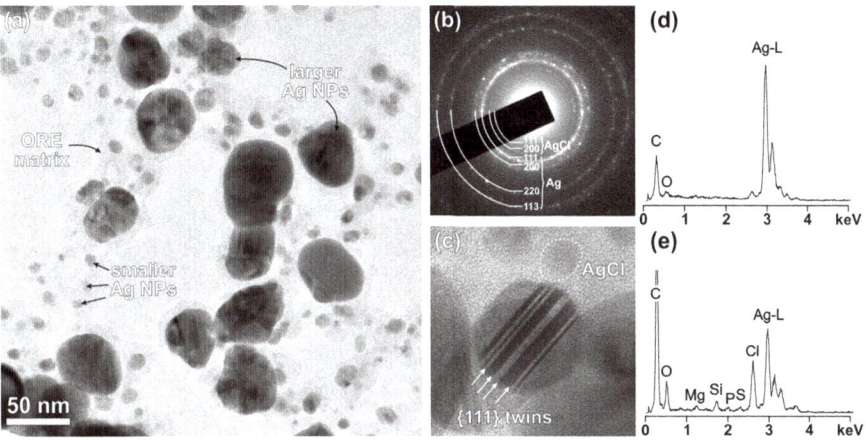

Fig. 7.16 Transmission electron microscopy (TEM) images of Ag/ORE sample. (**a**) Particle size distribution at low magnification (**b**) SAD pattern (**c**) larger AgNPs is the presence of parallel twins; a small particle of AgCl displaying weaker contrast due to the lower average density is encircled (**d**) EDS spectrum of larger AgNP and (**e**) EDS spectrum of small AgNPs in ORE matrix (Baláž et al. 2017a), copyright by Springer

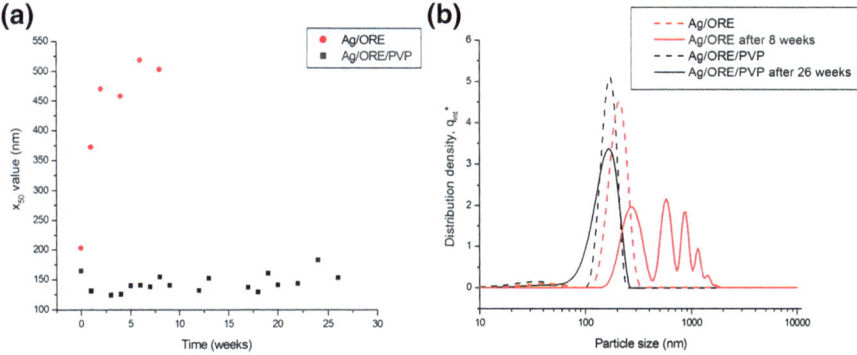

Fig. 7.17 Stability of the Ag/ORE/PVP and Ag/ORE nanosuspensions. (**a**) The hydrodynamic particle diameter, (x_{50}) change with the time of storage. (**b**) Particle size distribution (Baláž et al. 2017a), copyright by Springer

observed as it is shown in the same figure. From the EDS spectra in Fig. 7.16d and e, almost pure silver in the case of larger particle and considerable amount of chlorine in the case of smaller AgNPs with *Origanum* remains were detected.

The last part was devoted to the stability studies. Zeta potential and photon cross-correlation spectroscopy (PCCS) (quite similar to DLS method) were the techniques used for demonstrating the stability of prepared nanoparticles. After the addition of PVP, the zeta potential decreased, which is the proof of stabilization of the system. PCCS has shown the formation of agglomerates and a quick increase in average hydrodynamic particle diameter during storage of the PVP-free sample (this can be well seen for Ag/ORE sample after 8 weeks). As can be seen in Fig. 7.17, the

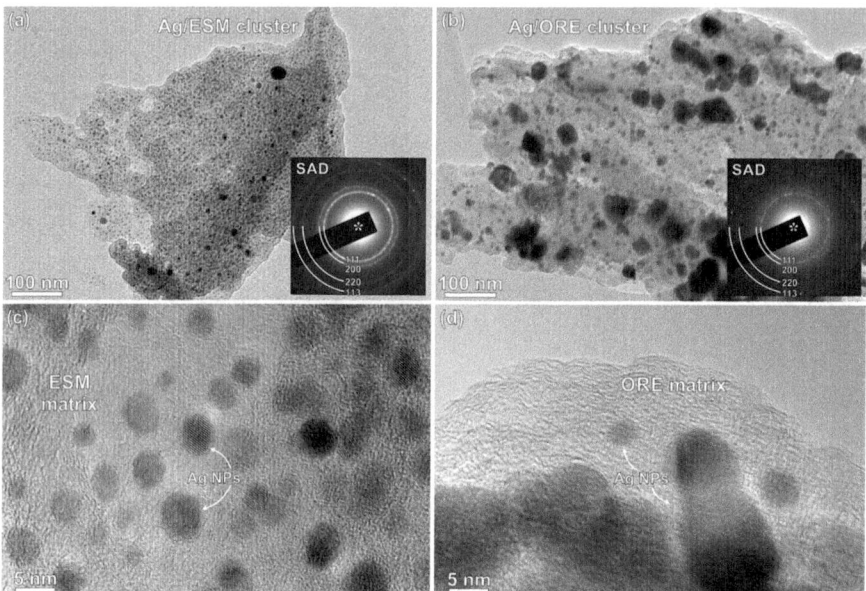

Fig. 7.18 (**a**) TEM images of Ag/ESM and (**b**) Ag/ORE at low magnification with inserted SAD patterns indicating the crystallinity of AgNPs in both samples. (**c**) HRTEM image of well-dispersed AgNPs in ESM matrix and (**d**) agglomeration of AgNPs in the ORE matrix. (Baláž et al. 2017b), copyright by Elsevier

introduction of PVP into the system led to a noticeable enhancement in the stability of the nanosuspensions. The average hydrodynamic particle diameter did not increase significantly during 26 weeks storage at 4 °C.

The main added value of this paper is the introduction of wet stirred media milling for the stabilization of silver NPs. The antibacterial activity was not investigated in this study.

The last paper to be discussed in this sub-chapter is the completely solid-state approach (Baláž et al. 2017b). An eggshell membrane (ESM) and *O. vulgare*, L. (ORE) plant were used as natural reducing agents. As a silver precursor, silver nitrate was used. The whole synthesis was accomplished in planetary ball mill with the ratio of $AgNO_3$ to eggshell membrane or *O. vulgare*, L. plant 1:1. The course of reaction was followed by X-ray diffraction, as already described in Fig. 7.6.

From the SEM images, the AgNPs embedded within the organic matrices of both materials could be identified. In the case of Ag/ORE sample, the NPs seemed to be denser. Besides silver, the EDS analysis has shown the presence of phosphorous and chlorine coming from the matrix of Ag/ORE sample. Also, zeta potential in water was measured, yielding more negative value for Ag/ORE sample, which suggests its better stability.

TEM analysis of both samples confirmed the presence of AgNPs dispersed in the organic matrix (Fig. 7.18). The AgNPs differ in size, namely Ag/eggshell membrane sample is comprised of smaller particles than Ag/*O. vulgare* sample. In the latter sample, the separate nanoparticles have tendency to merge and form larger clusters. Therefore, two different populations of nanoparticles are present in this sample. This phenomenon was not observed for Ag/eggshell membrane composite.

Both samples were also tested for antibacterial activity on six strains of bacteria (three Gram-positive and three Gram-negative) at four different concentrations (1, 2.5, 5 and 10 mM). The antibacterial activity is expressed as the percentage of relative inhibition zone diameter (RIZD) with respect to gentamicin standard antibiotic. In the case of pure biomaterials, no antibacterial activity was detected. In Ag-containing samples, the antibacterial activity was observed. The antibacterial activity was higher for the nanoparticles prepared with *Origanum vulgare, L.* plant. Also, the lowest concentration (1 mM) was not effective in inhibiting bacterial growth, whereas in the case of *L. monocytogenes* and *S. aureus*, small antibacterial activity of sample Ag/*O. vulgare* was observed. The Ag/*O. vulgare* nanoparticles at concentrations 2.5, 5 and 10 mM reduced the growth of all tested bacteria. The concentration 2.5 mM was quite interesting because in four of six studied bacteria, the antibacterial activity was higher than those with 5 mM concentration. It is possible that the silver ions had better distribution in *Origanum vulgare, L.* matrix and thus better effect. The antibacterial activity towards all studied bacteria for Ag/eggshell membrane sample was observed at 5 and 10 mMAg AgNO$_3$ concentration.

The last discussed study could ignite a new research field, which could be called bio-mechanochemical synthesis as it combines the tools of mechanochemistry and green synthesis (bio-approach) using plants. Together with the studies mentioned earlier, it definitely opens new possibilities and brings mechanochemical research closer to nature.

7.6   Conclusion

From the economical (low cost) and environmental point of view, the plant- and plant extract-based synthesis represents a green chemistry-based approach for silver nanoparticles synthesis and has opened new ways in treating and/or controlling various diseases of humans. This approach avoids the problem of using hazardous chemicals and subsequent addition of stabilization agents. Nature is the source of enormous number of extractable biomolecules (e.g. terpenoids, sugars, alkaloids, polyphenols, flavonoids, proteins) suitable for reduction of silver ions. Commonly used characterization techniques (UV-Vis, FTIR, XRD, SEM, TEM, etc.) can be also used to describe green-synthesized nanoparticles to evaluate and control the particle size, size distribution, shape of nanoparticles or to detect the degree of agglomeration. The last part was devoted to bio-mechanochemical synthesis of AgNPs, which represents a combination of mechanochemistry being a relatively old branch of chemistry, with biosynthesis using plants. This approach is still very rarely described in literature. The combination of mechanochemical-assisted extraction (MCAE) of plant materials and bio-reduction of silver ions in one-step procedure and could open a new research pathway.

Generally, nanoparticles with antimicrobial properties are qualified as a promising variant to antibiotics. They represent an appreciable future prospective in consideration in the development of pathogen-resistant solution. Ultimately, silver nanoparticles used in medical and pharmaceutical field are a rapidly developing field of science where the green synthesis provides easy and environmentally acceptable access to them.

Acknowledgments The authors wish to acknowledge the Slovak Research and Development Agency (VEGA 2/0044/18) and Slovak Research and Development Agency (APVV-18-0357) for financial support.

References

Abbasi E, Milani M, Fekri Aval S et al (2016) Silver nanoparticles: synthesis methods, bioapplications and properties. Crit Rev Microbiol 42:173–180. https://doi.org/10.310 9/1040841X.2014.912200

Ahmad N, Sharma S, Alam MK et al (2010) Rapid synthesis of silver nanoparticles using dried medicinal plant of basil. Colloid Surf B 81:81–86. https://doi.org/10.1016/j.colsurfb.2010.06.029

Ahmed S, Ahmad M, Swami BL, Ikram S (2016) A review on plants extract mediated synthesis of silver nanoparticles for antimicrobial applications: a green expertise. J Adv Res 7:17–28. https://doi.org/10.1016/j.jare.2015.02.007

Ajitha B, Ashok Kumar Reddy Y, Sreedhara Reddy P (2015) Green synthesis and characterization of silver nanoparticles using Lantana camara leaf extract. Mater Sci Eng C 49:373–381. https://doi.org/10.1016/j.msec.2015.01.035

Al Shaal L, Müller RH, Shegokar R (2010) SmartCrystal combination technology - scale up from lab to pilot scale and long term stability. Pharmazie 65:877–884. https://doi.org/10.1691/ph.2010.0181

Ameen F, AlYahya SA, Bakhrebah MA et al (2018) Flavonoid dihydromyricetin-mediated silver nanoparticles as potential nanomedicine for biomedical treatment of infections caused by opportunistic fungal pathogens. Res Chem Intermed 44:5063–5073. https://doi.org/10.1007/s11164-018-3409-x

Arancon RAD, Balu AM, Romero AA et al (2017) Mechanochemically synthesized Ag-based nanohybrids with unprecedented low toxicity in biomedical applications. Environ Res 154:204–211. https://doi.org/10.1016/j.envres.2017.01.010

AshaRani PV, Low Kah Mun G, Hande MP, Valiyaveettil S (2009) Cytotoxicity and genotoxicity of silver nanoparticles in human cells. ACS Nano 3:279–290. https://doi.org/10.1021/nn800596w

Ashraf JM, Ansari MA, Khan HM et al (2016) Green synthesis of silver nanoparticles and characterization of their inhibitory effects on AGEs formation using biophysical techniques. Sci Rep 6:20414. https://doi.org/10.1038/srep20414

Ashraf A, Zafar S, Zahid K et al (2018) Synthesis, characterization, and antibacterial potential of silver nanoparticles synthesized from Coriandrum sativum L. J Infect Public Health 12:275. https://doi.org/10.1016/j.jiph.2018.11.002

Aziz N, Faraz M, Pandey R, Sakir M, Fatma T, Varma A, Barman I, Prasad R (2015) Facile algae-derived route to biogenic silver nanoparticles: synthesis, antibacterial and photocatalytic properties. Langmuir 31:11605–11612. https://doi.org/10.1021/acs.langmuir.5b03081

Aziz N, Pandey R, Barman I, Prasad R (2016) Leveraging the attributes of Mucor hiemalis-derived silver nanoparticles for a synergistic broad-spectrum antimicrobial platform. Front Microbiol 7:1984. https://doi.org/10.3389/fmicb.2016.01984

Aziz N, Faraz M, Sherwani MA, Fatma T, Prasad R (2019) Illuminating the anticancerous efficacy of a new fungal chassis for silver nanoparticle synthesis. Front Chem 7:65. https://doi.org/10.3389/fchem.2019.00065

Azmir J, Zaidul ISM, Rahman MM et al (2013) Techniques for extraction of bioactive compounds from plant materials: a review. J Food Eng 117:426–436. https://doi.org/10.1016/j.jfoodeng.2013.01.014

Azwanida NN (2015) A review on the extraction methods use in medicinal plants, principle, strength and limitation. Med Aromat Plants 4:3–8. https://doi.org/10.4172/2167-0412.1000196

Baláž P, Achimovicová M, Baláž M et al (2013) Hallmarks of mechanochemistry: from nanoparticles to technology. Chem Soc Rev 42:7571–7637. https://doi.org/10.1039/c3cs35468g

Baláž M, Balážová Ľ, Daneu N et al (2017a) Plant-mediated synthesis of silver Nanoparticles and their stabilization by wet stirred media milling. Nanoscale Res Lett 12:83–91. https://doi.org/10.1186/s11671-017-1860-z

Baláž M, Daneu N, Balážová Ľ et al (2017b) Bio-mechanochemical synthesis of silver nanoparticles with antibacterial activity. Adv Powder Technol 28:3307–3312. https://doi.org/10.1016/j.apt.2017.09.028

Baláž M, Vella-Zarb L, Hernández JG et al (2019) Mechanochemistry: a disruptive innovation for the industry of the future. Chimica Oggi - Chemistry Today 37:32–34

Balouiri M, Sadiki M, Ibnsouda SK (2016) Methods for in vitro evaluating antimicrobial activity: a review. J Pharm Anal 6:71–79. https://doi.org/10.1016/j.jpha.2015.11.005

Becker RO (1999) Silver ions in the treatment of local infections. Met Based Drugs 6:311–314. https://doi.org/10.1155/MBD.1999.311

Behravan M, Hossein Panahi A, Naghizadeh A et al (2019) Facile green synthesis of silver nanoparticles using Berberis vulgaris leaf and root aqueous extract and its antibacterial activity. Int J Biol Macromol 124:148–154. https://doi.org/10.1016/j.ijbiomac.2018.11.101

Benakashani F, Allafchian A, Jalali SAH (2017) Green synthesis, characterization and antibacterial activity of silver nanoparticles from root extract of Lepidium draba weed. Green Chem Lett Rev 10:324–330. https://doi.org/10.1080/17518253.2017.1363297

Boldyreva E (2013) Mechanochemistry of inorganic and organic systems: what is similar, what is different? Chem Soc Rev 42:7719–7738. https://doi.org/10.1039/c3cs60052a

Carlson C, Hussein SM, Schrand AM et al (2008) Unique cellular interaction of silver nanoparticles: size-dependent generation of reactive oxygen species. J Phys Chem B 112:13608–13619. https://doi.org/10.1021/jp712087m

Carmona ER, Benito N, Plaza T, Recio-Sánchez G (2017) Green synthesis of silver nanoparticles by using leaf extracts from the endemic Buddleja globosa hope. Green Chem Lett Rev 10:250–256. https://doi.org/10.1080/17518253.2017.1360400

Catauro M, Raucci MG, Gaetano FDE, Marotta A (2004) Antibacterial and bioactive silver-containing Na2O.CaO.2SiO2 glass prepared by sol-gel method. J Mater Sci Mater Med 15:831–837. https://doi.org/10.1023/B:JMSM.0000032825.51052.00

Chambers C, Stewart SB, Su B et al (2017) Silver doped titanium dioxide nanoparticles as antimicrobial additives to dental polymers. Dent Mater 33:e115–e123. https://doi.org/10.1016/j.dental.2016.11.008

Chopade BA, Singh R, Wagh P et al (2013) Synthesis, optimization, and characterization of silver nanoparticles from *Acinetobacter calcoaceticus* and their enhanced antibacterial activity when combined with antibiotics. Int J Nanomedicine 8:4277. https://doi.org/10.2147/IJN.S48913

Clark JH, Lancaster M, Holder JV et al (2002) Handbook of green chemistry and technology. Blackwell Science Ltd, Oxford, UK

Dai Y, Yan XH, Wu X et al (2016) Facile self-assembly of AgNPs/WS2nanocomposites with enhanced electrochemical properties. Mater Lett 173:203–206. https://doi.org/10.1016/j.matlet.2016.03.050

Dakal TC, Kumar A, Majumdar RS, Yadav V (2016) Mechanistic basis of antimicrobial actions of silver nanoparticles. Front Microbiol 7:1–17. https://doi.org/10.3389/fmicb.2016.01831

Dallas P, Sharma VK, Zboril R (2011) Silver polymeric nanocomposites as advanced antimicrobial agents: classification, synthetic paths, applications, and perspectives. Adv Colloid Interf Sci 166:119–135. https://doi.org/10.1016/j.cis.2011.05.008

de Jesús Ruíz-Baltazar Á, Reyes-López SY, de Lourdes Mondragón-Sánchez M et al (2018) Biosynthesis of Ag nanoparticles using Cynara cardunculus leaf extract: evaluation of their antibacterial and electrochemical activity. Results Phys 11:1142–1149. https://doi.org/10.1016/j.rinp.2018.11.032

de Marco BA, Rechelo BS, Tótoli EG et al (2018) Evolution of green chemistry and its multidimensional impacts: a review. Saudi Pharm J 27:1–8. https://doi.org/10.1016/j.jsps.2018.07.011

Demathieu C, Chehimi MM, Lipskier JF et al (1999) Characterization of dendrimers by X-ray photoelectron spectroscopy. Appl Spectrosc 53:1277–1281. https://doi.org/10.1366/0003702991945524

Demirbas A, Welt BA, Ocsoy I (2016) Biosynthesis of red cabbage extract directed Ag NPs and their effect on the loss of antioxidant activity. Mater Lett 179:20–23. https://doi.org/10.1016/j.matlet.2016.05.056

Deng H, McShan D, Zhang Y et al (2016) Mechanistic study of the synergistic antibacterial activity of combined silver nanoparticles and common antibiotics. Environ Sci Technol 50:8840–8848. https://doi.org/10.1021/acs.est.6b00998

Durán N, Durán M, de Jesus MB et al (2016) Silver nanoparticles: a new view on mechanistic aspects on antimicrobial activity. Nanomedicine 12:789–799

El Kassas HY, Attia AA (2014) Bactericidal application and cytotoxic activity of biosynthesized silver Nanoparticles with an extract of the red seaweed Pterocladiella capillacea on the HepG 2 cell line. Asian Pacific J Cancer Prev 15:1299–1306. https://doi.org/10.7314/APJCP.2014.15.3.1299

Ethiraj AS, Jayanthi S, Ramalingam C, Banerjee C (2016) Control of size and antimicrobial activity of green synthesized silver nanoparticles. Mater Lett 185:526–529. https://doi.org/10.1016/j.matlet.2016.07.114

Fissan H, Ristig S, Kaminski H et al (2014) Comparison of different characterization methods for nanoparticle dispersions before and after aerosolization. Anal Methods 6:7324–7334. https://doi.org/10.1039/c4ay01203h

Franci G, Falanga A, Galdiero S et al (2015) Silver Nanoparticles as potential antibacterial agents. Molecules 20:8856–8874. https://doi.org/10.3390/molecules20058856

García-Barrasa J, López-de-Luzuriaga JM, Monge M (2011) Silver nanoparticles: synthesis through chemical methods in solution and biomedical applications. Cent Eur J Chem 9:7–19. https://doi.org/10.2478/s11532-010-0124-x

Ge L, Li Q, Wang M et al (2014) Nanosilver particles in medical applications: synthesis, performance, and toxicity. Int J Nanomedicine 9:2399–2407. https://doi.org/10.2147/IJN.S55015

Ghiyasiyan-Arani M, Salavati-Niasari M, Masjedi-Arani M, Mazloom F (2018) An easy sonochemical route for synthesis, characterization and photocatalytic performance of nanosized FeVO4 in the presence of aminoacids as green capping agents. J Mater Sci Mater Electron 29:474–485. https://doi.org/10.1007/s10854-017-7936-9

Gurunathan S (2015) Biologically synthesized silver nanoparticles enhances antibiotic activity against Gram-negative bacteria. J Ind Eng Chem 29:217–226. https://doi.org/10.1016/j.jiec.2015.04.005

Hemmati S, Rashtiani A, Zangeneh MM et al (2019) Green synthesis and characterization of silver nanoparticles using Fritillaria flower extract and their antibacterial activity against some human pathogens. Polyhedron 158:8–14. https://doi.org/10.1016/j.poly.2018.10.049

Huang H, Yang X (2004) Synthesis of polysaccharide-stabilized gold and silver nanoparticles: a green method. Carbohydr Res 339:2627–2631. https://doi.org/10.1016/j.carres.2004.08.005

Huang M, Du L, Feng J-X (2016) Photochemical synthesis of silver nanoparticles/eggshell membrane composite, its characterization and antibacterial activity. Sci Adv Mater 8:1641–1647. https://doi.org/10.1166/sam.2016.2777

Ismail M, Gul S, Khan MA, Khan MI (2016) Plant mediated green synthesis of anti- microbial silver nanoparticles—a review on recent trends. Rev Nanosci Nanotechnol 5:119–135. https://doi.org/10.1166/rnn.2016.1073

Jayaprakash N, Vijaya JJ, Kaviyarasu K et al (2017) Green synthesis of Ag nanoparticles using tamarind fruit extract for the antibacterial studies. J Photochem Photobiol B Biol 169:178–185. https://doi.org/10.1016/j.jphotobiol.2017.03.013

Jiang X, Jiang J, Jin Y et al (2005) Effect of colloidal gold size on the conformational changes of adsorbed cytochrome c: probing by circular dichroism, UV-visible, and infrared spectroscopy. Biomacromolecules 6:46–53. https://doi.org/10.1021/bm049744l

Joshi N, Jain N, Pathak A, Singh J, Prasad R, Upadhyaya CP (2018) Biosynthesis of silver nanoparticles using Carissa carandas berries and its potential antibacterial activities. J Sol-Gel Sci Techn 86(3):682–689. https://doi.org/10.1007/s10971-018-4666-2

Kalaiarasi R, Prasannaraj G, Venkatachalam P (2013) A rapid biological synthesis of silver nanoparticles using leaf broth of Rauvolfia Tetraphylla and their promising antibacterial activity. Indo Am J Pharm Res 3:8052–8062

Kanawaria SK, Sankhla A, Jatav PK et al (2018) Rapid biosynthesis and characterization of silver nanoparticles: an assessment of antibacterial and antimycotic activity. Appl Phys A Mater Sci Process 124:1–10. https://doi.org/10.1007/s00339-018-1701-7

Kelly EH, Dennis JB, Anthony RT (2002) Flavonoid antioxidants: chemistry, metabolism and structure-activity relationships. J Nutr Biochem 13:572–584. https://doi.org/10.1016/S0955-2863(02)00208-5

Keskinen J, Ruuskanen P, Karttunen M, Hannula SP (2001) Synthesis of silver powder using a mechanochemical process. Appl Organomet Chem 15:393–395. https://doi.org/10.1002/aoc.159

Khakan B, Miandashti AR, Ashjari M (2017) Mechanochemical synthesis of silver nanoparticles. J Powder Metall Min 6:2–5. https://doi.org/10.4172/2168-9806.1000174

Khan A, Asiri AM, Rub MA et al (2013) Synthesis, characterization of silver nanoparticle embedded polyaniline tungstophosphate-nanocomposite cation exchanger and its application for heavy metal selective membrane. Compos Part B Eng 45:1486–1492. https://doi.org/10.1016/j.compositesb.2012.09.023

Kharissova OV, Dias HVR, Kharisov BI et al (2013) The greener synthesis of nanoparticles. Trends Biotechnol 31:240–248. https://doi.org/10.1016/j.tibtech.2013.01.003

Khatoon A, Khan F, Ahmad N et al (2018) Silver nanoparticles from leaf extract of Mentha piperita: eco-friendly synthesis and effect on acetylcholinesterase activity. Life Sci 209:430–434. https://doi.org/10.1016/j.lfs.2018.08.046

Khayati GR, Janghorban K (2012) The nanostructure evolution of Ag powder synthesized by high energy ball milling. Adv Powder Technol 23:393–397. https://doi.org/10.1016/j.apt.2011.05.005

Kokila T, Ramesh PS, Geetha D (2016) Biosynthesis of AgNPs using Carica Papaya peel extract and evaluation of its antioxidant and antimicrobial activities. Ecotoxicol Environ Saf 134:467–473. https://doi.org/10.1016/j.ecoenv.2016.03.021

Kowshik M, Deshmukh N, Vogel W et al (2002) Microbial synthesis of semiconductor CdS nanoparticles, their characterization, and their use in the fabrication of an ideal diode. Biotechnol Bioeng 78:583–588. https://doi.org/10.1002/bit.10233

Kulkarni N, Muddapur U (2014) Biosynthesis of metal nanoparticles: a review. J Nanotechnol 2014:510246. https://doi.org/10.1155/2014/510246

Kumar N, Biswas K, Gupta RK (2016) Green synthesis of Ag nanoparticles in large quantity by cryomilling. RSC Adv 6:111380–111388. https://doi.org/10.1039/c6ra23120a

Lakshmanan G, Sathiyaseelan A, Kalaichelvan PT, Murugesan K (2018) Plant-mediated synthesis of silver nanoparticles using fruit extract of Cleome viscosa L.: assessment of their antibacterial and anticancer activity. Karbala Int J Mod Sci 4:61–68. https://doi.org/10.1016/j.kijoms.2017.10.007

Le MT, Kim DJ, Kim CG et al (2008) Nanoparticles of silver powder obtained by mechanochemical process. J Exp Nanosci 3:223–228. https://doi.org/10.1080/17458080802301997

Lea MC (1893) On endothermic decompositions obtained by pressure (second part.) transformations of energy by shearing stress. Am J Sci 46:413–420

Lin PC, Lin S, Wang PC, Sridhar R (2014) Techniques for physicochemical characterization of nanomaterials. Biotechnol Adv 32:711–726. https://doi.org/10.1016/j.biotechadv.2013.11.006

Liu H, Webster TJ (2007) Nanomedicine for implants: a review of studies and necessary experimental tools. Biomaterials 28:354–369. https://doi.org/10.1016/j.biomaterials.2006.08.049

Liu YS, Chang YC, Chen HH (2018) Silver nanoparticle biosynthesis by using phenolic acids in rice husk extract as reducing agents and dispersants. J Food Drug Anal 26:649–656. https://doi.org/10.1016/j.jfda.2017.07.005

Logeswari P, Silambarasan S, Abraham J (2015) Synthesis of silver nanoparticles using plants extract and analysis of their antimicrobial property. J Saudi Chem Soc 19:311–317. https://doi.org/10.1016/j.jscs.2012.04.007

Lomovsky OI, Lomovskiy IO, Orlov DV (2017) Mechanochemical solid acid/base reactions for obtaining biologically active preparations and extracting plant materials. Green Chem Lett Rev 10:171–185. https://doi.org/10.1080/17518253.2017.1339832

Macaluso RT (2009) Introduction to powder diffraction and its application to nanoscale and heterogeneous materials. ACS Symp Ser 1010:75–86. https://doi.org/10.1021/bk-2009-1010.ch006

Makarov VV, Love AJ, Sinitsyna OV et al (2014) "Green" nanotechnologies: synthesis of metal nanoparticles using plants. Acta Nat 6:35–44. https://doi.org/10.1039/c1gc15386b

Manna A, Imae T, Aoi K et al (2001) Synthesis of dendrimer-passivated noble metal nanoparticles in a polar medium: comparison of size between silver and gold particles. Chem Mater 13:1674–1681. https://doi.org/10.1021/cm000416b

Marchiol L, Mattiello A, Pošćić F et al (2014) In vivo synthesis of nanomaterials in plants: location of silver nanoparticles and plant metabolism. Nanoscale Res Lett 9:101. https://doi.org/10.1186/1556-276X-9-101

Mittal AK, Chisti Y, Banerjee UC (2013) Synthesis of metallic nanoparticles using plant extracts. Biotechnol Adv 31:346–356. https://doi.org/10.1016/j.biotechadv.2013.01.003

Mohammadlou M, Maghsoudi H, Jafarizadeh-Malmiri H (2016) A review on green silver nanoparticles based on plants: synthesis, potential applications and eco-friendly approach. Int Food Res J 23:446–463

Nakkala JR, Mata R, Sadras SR (2017) Green synthesized nano silver: synthesis, physicochemical profiling, antibacterial, anticancer activities and biological in vivo toxicity. J Colloid Interface Sci 499:33–45. https://doi.org/10.1016/j.jcis.2017.03.090

Noginov MA, Zhu G, Bahoura M et al (2007) The effect of gain and absorption on surface plasmons in metal nanoparticles. Appl Phys B Lasers Opt 86:455–460. https://doi.org/10.1007/s00340-006-2401-0

Omidi S, Sedaghat S, Tahvildari K et al (2018) Biosynthesis of silver nanoparticles with adiantum capillus-veneris L leaf extract in the batch process and assessment of antibacterial activity. Green Chem Lett Rev 11:544–551. https://doi.org/10.1080/17518253.2018.1546410

Ortan A, Fierascu I, Ungureanu C et al (2015) Innovative phytosynthesized silver nanoarchitectures with enhanced antifungal and antioxidant properties. Appl Surf Sci 358:540–548. https://doi.org/10.1016/j.apsusc.2015.07.160

Padalia H, Moteriya P, Chanda S (2015) Green synthesis of silver nanoparticles from marigold flower and its synergistic antimicrobial potential. Arab J Chem 8:732–741. https://doi.org/10.1016/j.arabjc.2014.11.015

Pal S, Tak YK, Song JM (2007) Does the antibacterial activity of silver nanoparticles depend on the shape of the nanoparticle? A study of the Gram-negative bacterium Escherichia coli. Appl Environ Microbiol 73:1712–1720. https://doi.org/10.1128/AEM.02218-06

Panáček A, Kvítek L, Prucek R et al (2006) Silver colloid nanoparticles: synthesis, characterization, and their antibacterial activity. J Phys Chem B 110:16248–16253. https://doi.org/10.1021/jp063826h

Panigrahi S, Kundu S, Ghosh SK et al (2004) General method of synthesis for metal nanoparticles. J Nanopart Res 6:411–414. https://doi.org/10.1007/s11051-004-6575-2

Panigrahi S, Kundu S, Ghosh SK et al (2005) Sugar assisted evolution of mono- and bimetallic nanoparticles. Colloid Surf A 264:133–138. https://doi.org/10.1016/j.colsurfa.2005.04.017

Parthiban E, Manivannan N, Ramanibai R, Mathivanan N (2018) Green synthesis of silver-nanoparticles from Annona reticulata leaves aqueous extract and its mosquito larvicidal and anti-microbial activity on human pathogens. Biotechnol Reports 20:e00297. https://doi.org/10.1016/j.btre.2018.e00297

Patil V, Sastry M (1997) Electrostatically controlled diffusion of carboxylic acid derivatized Q-state CdS nanoparticles in thermally evaporated fatty amine films. J Chem Soc 93:4347–4353. https://doi.org/10.1039/a704355d

Peltonen L, Hirvonen J (2010) Pharmaceutical nanocrystals by nanomilling: critical process parameters, particle fracturing and stabilization methods. J Pharm Pharmacol 62:1569–1579. https://doi.org/10.1111/j.2042-7158.2010.01022.x

Pethakamsetty L, Kothapenta K, Nammi HR et al (2017) Green synthesis, characterization and antimicrobial activity of silver nanoparticles using methanolic root extracts of Diospyros sylvatica. J Environ Sci 55:157–163. https://doi.org/10.1016/j.jes.2016.04.027

Ping Y, Zhang J, Xing T et al (2018) Green synthesis of silver nanoparticles using grape seed extract and their application for reductive catalysis of direct Orange 26. J Ind Eng Chem 58:74–79. https://doi.org/10.1016/j.jiec.2017.09.009

Pons T, Uyeda HT, Medintz IL, Mattoussi H (2006) Hydrodynamic dimensions, electrophoretic mobility, and stability of hydrophilic quantum dots. J Phys Chem B 110:20308–20316. https://doi.org/10.1021/jp065041h

Prabhu S, Poulose EK (2012) Silver nanoparticles: mechanism of antimicrobial action, synthesis, medical applications, and toxicity effects. Int Nano Lett 2:32. https://doi.org/10.1186/2228-5326-2-32

Prasad R (2014) Synthesis of silver nanoparticles in photosynthetic plants. J Nanopart 2014:963961. https://doi.org/10.1155/2014/963961

Prasad R, Pandey R, Barman I (2016) Engineering tailored nanoparticles with microbes: quo vadis. WIREs Nanomed Nanobiotechnol 8:316–330. https://doi.org/10.1002/wnan.1363

Prasad R, Jha A, Prasad K (2018) Exploring the realms of nature for nanosynthesis. Springer International Publishing, New York. ISBN 978-3-319-99570-0. https://www.springer.com/978-3-319-99570-0

Prasad R (2019) Plant Nanobionics: Approaches in Nanoparticles Biosynthesis and Toxicity. Springer International Publishing (ISBN 978-3-030-16379-2). https://www.springer.com/gp/book/9783030163785

Rajan R, Chandran K, Harper SL et al (2015) Plant extract synthesized silver nanoparticles: an ongoing source of novel biocompatible materials. Ind Crop Prod 70:356–373. https://doi.org/10.1016/j.indcrop.2015.03.015

Rajeshkumar S, Bharath LV (2017) Mechanism of plant-mediated synthesis of silver nanoparticles—a review on biomolecules involved, characterisation and antibacterial activity. Chem Biol Interact 273:219–227. https://doi.org/10.1016/j.cbi.2017.06.019

Rak MJ, Friščić T, Moores A (2016) One-step, solvent-free mechanosynthesis of silver nanoparticle-infused lignin composites for use as highly active multidrug resistant antibacterial filters. RSC Adv 6:58365–58370. https://doi.org/10.1039/c6ra03711a

Raveendran P, Fu J, Wallen SL (2003) Completely "green" synthesis and stabilization of metal nanoparticles. J Am Chem Soc 125:13940–13941. https://doi.org/10.1021/ja029267j

Romeis S, Schmidt J, Peukert W (2016) Mechanochemical aspects in wet stirred media milling. Int J Miner Process 156:24–31. https://doi.org/10.1016/j.minpro.2016.05.018

Roy P, Das B, Mohanty A, Mohapatra S (2017) Green synthesis of silver nanoparticles using Azadirachta indica leaf extract and its antimicrobial study. Appl Nanosci 7:843–850. https://doi.org/10.1007/s13204-017-0621-8

Russell AD, Hugo WB (1994) 7 antimicrobial activity and action of silver. Prog Med Chem 31:351–370. https://doi.org/10.1016/S0079-6468(08)70024-9

Sadeghi B, Gholamhoseinpoor F (2015) A study on the stability and green synthesis of silver nanoparticles using Ziziphora tenuior (Zt) extract at room temperature. Spectrochim Acta 134:310–315. https://doi.org/10.1016/j.saa.2014.06.046

Safarpoor M, Ghaedi M, Asfaram A et al (2018) Ultrasound-assisted extraction of antimicrobial compounds from Thymus daenensis and Silybum marianum: antimicrobial activity with and without the presence of natural silver nanoparticles. Ultrason Sonochem 42:76–83. https://doi.org/10.1016/j.ultsonch.2017.11.001

Sankar R, Karthik A, Prabu A et al (2013) Origanum vulgare mediated biosynthesis of silver nanoparticles for its antibacterial and anticancer activity. Colloid Surf B 108:80–84. https://doi.org/10.1016/j.colsurfb.2013.02.033

Sapsford KE, Tyner KM, Dair BJ et al (2011) Analyzing nanomaterial bioconjugates: a review of current and emerging purification and characterization techniques. Anal Chem 83:4453–4488. https://doi.org/10.1021/ac200853a

Sebaugh JL (2011) Guidelines for accurate EC50/IC50 estimation. Pharm Stat 10:128–134. https://doi.org/10.1002/pst.426

Şeker Karatoprak G, Aydin G, Altinsoy B et al (2017) The effect of *Pelargonium endlicherianum* Fenzl. Root extracts on formation of nanoparticles and their antimicrobial activities. Enzym Microb Technol 97:21–26. https://doi.org/10.1016/j.enzmictec.2016.10.019

Shang L, Wang Y, Jiang J, Dong S (2007) PH-dependent protein conformational changes in albumin:gold nanoparticle bioconjugates: a spectroscopic study. Langmuir 23:2714–2721. https://doi.org/10.1021/la062064e

Sharma VK, Yngard RA, Lin Y (2009) Silver nanoparticles: green synthesis and their antimicrobial activities. Adv Colloid Interf Sci 145:83–96. https://doi.org/10.1016/j.cis.2008.09.002

Sheng Z, Liu Y (2017) Potential impacts of silver nanoparticles on bacteria in the aquatic environment. J Environ Manag 191:290–296. https://doi.org/10.1016/j.jenvman.2017.01.028

Shriniwas PP, Subhash KT (2017) Antioxidant, antibacterial and cytotoxic potential of silver nanoparticles synthesized using terpenes rich extract of Lantana camara L. leaves. Biochem Biophys Reports 10:76–81. https://doi.org/10.1016/j.bbrep.2017.03.002

Singh AK, Talat M, Singh DP, Srivastava ON (2010) Biosynthesis of gold and silver nanoparticles by natural precursor clove and their functionalization with amine group. J Nanopart Res 12:1667–1675. https://doi.org/10.1007/s11051-009-9835-3

Singh DK, Pandey DK, Yadav RR, Singh D (2013) A study of ZnO nanoparticles and ZnO-EG nanofluid. J Exp Nanosci 8:567–577. https://doi.org/10.1080/17458080.2011.602369

Sintubin L, Verstraete W, Boon N (2012) Biologically produced nanosilver: current state and future perspectives. Biotechnol Bioeng 109:2422–2436. https://doi.org/10.1002/bit.24570

Sökmen M, Alomar SY, Albay C, Serdar G (2017) Microwave assisted production of silver nanoparticles using green tea extracts. J Alloys Compd 725:190–198. https://doi.org/10.1016/j.jallcom.2017.07.094

Suganya A, Murugan K, Kovendan K et al (2013) Green synthesis of silver nanoparticles using Murraya koenigii leaf extract against Anopheles stephensi and Aedes aegypti. Parasitol Res 112:1385–1397. https://doi.org/10.1007/s00436-012-3269-z

Takacs L (2004) M. Carey Lea, the first mechanochemist. J Mater Sci 39:4987–4993. https://doi.org/10.1023/B:JMSC.0000039175.73904.93

Takacs L (2013) The historical development of mechanochemistry. Chem Soc Rev 42:7649–7659. https://doi.org/10.1039/c2cs35442j

Takacs L (2018) Two important periods in the history of mechanochemistry. J Mater Sci 53:13324–13330. https://doi.org/10.1007/s10853-018-2198-3

Tamboli DP, Lee DS (2013) Mechanistic antimicrobial approach of extracellularly synthesized silver nanoparticles against Gram positive and Gram negative bacteria. J Hazard Mater 260:878–884. https://doi.org/10.1016/j.jhazmat.2013.06.003

Tan YN, Lee JY, Wang DIC (2010) Uncovering the design rules for peptide synthesis of metal nanoparticles. J Am Chem Soc 132:5677–5686. https://doi.org/10.1021/ja907454f

Tiimob BJ, Mwinyelle G, Abdela W et al (2017) Nanoengineered eggshell-silver tailored Copolyester polymer blend film with antimicrobial properties. J Agric Food Chem 65:1967–1976. https://doi.org/10.1021/acs.jafc.7b00133

Tsuzuki T, McCormick PG (2004) Mechanochemical synthesis of nanoparticles. J Mater Sci 39:5143–5146. https://doi.org/10.1023/B:JMSC.0000039199.56155.f9

Vaia RA, Liu W (2002) X-ray powder diffraction of polymer/layered silicate nanocomposites: model and practice. J Polym Sci Part B Polym Phys 40:1590–1600. https://doi.org/10.1002/polb.10214

Van Der Merwe PA (2010) Surface plasmon resonance. Physics (College Park Md) 627:1–50. https://doi.org/10.1007/978-1-60761-670-2

Van Viet P, Sang TT, Bich NHN, Thi CM (2018) An improved green synthesis method and Escherichia coli antibacterial activity of silver nanoparticles. J Photochem Photobiol B Biol 182:108–114. https://doi.org/10.1016/j.jphotobiol.2018.04.002

Velayutham K, Rahuman AA, Rajakumar G et al (2013) Larvicidal activity of green synthesized silver nanoparticles using bark aqueous extract of Ficus racemosa against Culex quinquefasciatus and Culex gelidus. Asian Pac J Trop Med 6:95–101. https://doi.org/10.1016/S1995-7645(13)60002-4

Vigneshwaran N, Nachane RP, Balasubramanya RH, Varadarajan PV (2006) A novel one-pot "green" synthesis of stable silver nanoparticles using soluble starch. Carbohydr Res 341:2012–2018. https://doi.org/10.1016/j.carres.2006.04.042

Wang P, Tang J (2009) Solvent-free mechanochemical extraction of chondroitin sulfate from shark cartilage. Chem Eng Process Process Intensif 48:1187–1191. https://doi.org/10.1016/j.cep.2009.04.003

Wang J, White WB, Adair JH (2010) Optical properties of core-shell structured Ag/SiO2nanocomposites. Mater Sci Eng B 166:235–238. https://doi.org/10.1016/j.mseb.2009.11.026

Wang S, Zhang R, Song X et al (2019) Mechanochemical-assisted extraction of active alkaloids from plant with solid acids. ACS Sustain Chem Eng 7:197–207. https://doi.org/10.1021/acssuschemeng.8b02902

Wu K, Ju T, Deng Y, Xi J (2017) Mechanochemical assisted extraction: a novel, efficient, eco-friendly technology. Trends Food Sci Technol 66:166–175. https://doi.org/10.1016/j.tifs.2017.06.011

Xie J, Lin YS, Shi XJ et al (2013) Mechanochemical-assisted extraction of flavonoids from bamboo (Phyllostachys edulis) leaves. Ind Crop Prod 43:276–282. https://doi.org/10.1016/j.indcrop.2012.07.041

Xiu Z, Zhang Q, Puppala HL et al (2012) Negligible particle-specific antibacterial activity of silver nanoparticles. Nano Lett 12:4271–4275. https://doi.org/10.1021/nl301934w

Yamanaka M, Hara K, Kudo J (2005) Bactericidal actions of a silver ion solution on Escherichia coli, studied by energy-filtering transmission electron microscopy and proteomic analysis. Appl Environ Microbiol 71:7589–7593. https://doi.org/10.1128/AEM.71.11.7589-7593.2005

Zayed MF, Eisa WH, Shabaka AA (2012) Malva parviflora extract assisted green synthesis of silver nanoparticles. Spectrochim Acta Part A Mol Biomol Spectrosc 98:423–428. https://doi.org/10.1016/j.saa.2012.08.072

Zhang X-F, Liu Z-G, Shen W, Gurunathan S (2016) Silver nanoparticles: synthesis, characterization, properties, applications, and therapeutic approaches. Int J Mol Sci 17:1534. https://doi.org/10.3390/ijms17091534

Zhu X, Chen X, Xie J et al (2012) Mechanochemical-assisted extraction and antioxidant activity of polysaccharides from Ganoderma lucidum spores. Int J Food Sci Technol 47:927–932. https://doi.org/10.1111/j.1365-2621.2011.02923.x

Chapter 8
Nanobiotechnology and Supramolecular Mechanistic Interactions on Approach for Silver Nanoparticles for Healthcare Materials

Bianca Pizzorno Backx

Contents

Abstract Nanobiotechnology is an essential area of science that relates advances in nanoscience associated with biotechnology. Nanostructures have very different potentials when compared to the material of origin on the macro scale, and the supramolecular interactions often potentiate these properties. For silver nanoparticles (AgNPs) synthesis, a stable medium is essential to establish the nanostructure. There are various routes to synthesize the AgNPs. This chapter involves the synthesis of AgNPs using the two approaches, chemical and green route. The choice of green synthesis becomes appropriate due to the efficiency of the dispersive medium and low environmental impact. The supramolecular chemistry approach justifies

B. P. Backx (✉)
Universidade Federal do Rio de Janeiro—Campus Duque de Caxias, Rio de Janeiro, Brazil
e-mail: biapizzorno@caxias.ufrj.br

© Springer Nature Switzerland AG 2020
R. Prasad et al. (eds.), *Nanostructures for Antimicrobial and Antibiofilm Applications*, Nanotechnology in the Life Sciences,
https://doi.org/10.1007/978-3-030-40337-9_8

185

different ways to synthesize these nanoparticles that have a significant antimicrobial effect and can be used for healthcare material production.

Keywords Nanobiotechnology · Supramolecular interaction · Silver nanoparticles · Healthcare materials · Antimicrobial activity

8.1   Introduction

In the field of nanotechnology, the definition of a particle is a piece of matter with definite physical boundaries. In order to be on the nanometer scale, a particle must be of size ranging between 1 and 100 nm. Thus, nanoparticles (NPs) are distinct particles with the three external dimensions at the nanometer scale (Linic et al. 2015).

The synthesis of silver nanoparticles (Ag NPs) involves primordial steps independent of the chosen route. The first step concerns the choice of a precursor, which supplies the silver to the system. Ag NPs can be obtained in two ways: the top-down approach and the bottom-up approach. The first one refers to the superficial attack to get the metal in the nanostructure, and the second refers to the chemical and green methodology (Fig. 8.1).

Supramolecular chemistry highlights the influence of the reaction medium on the stabilization of nanostructures through resonance and equilibrium relations in the intermolecular interactions that govern the system. Through the existence of an

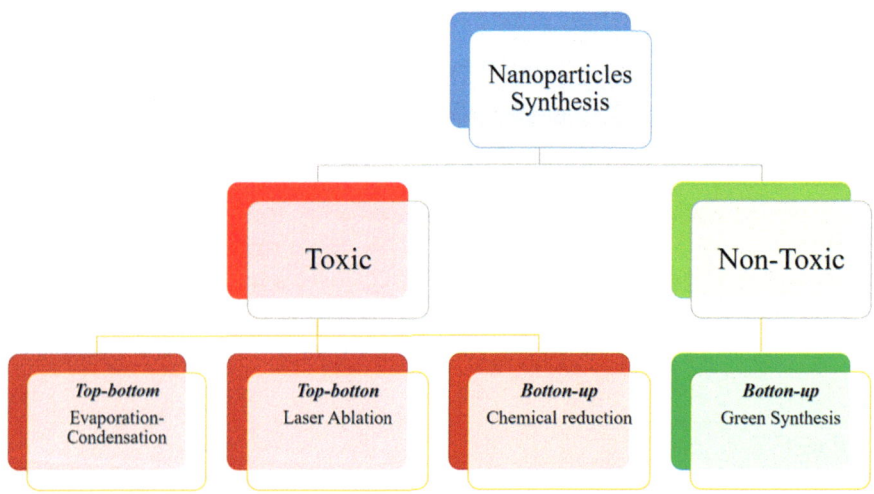

Fig. 8.1 Schematic summary of the approaches of silver nanoparticle synthesis from Backx et al. (2018) with permission

effective dispersive medium, silver nanoparticles undergo growth and nucleation and can stabilize without agglomerating (Reimers et al. 2017).

Particle agglomerations are caused by van der Waals forces that tend to decrease the surface energy of the system. When NPs are dispersed in a solvent, van der Waals forces and Brownian motions play essential roles, since the influence of gravity becomes negligible. Van der Waals forces are significant at short distances, and Brownian movement is the result of the collision between the particles and other atoms and molecules present in the solvent. The combination of these forces results in the formation of clusters and the establishment of nanoparticles (Li et al. 2018).

NPs can have entirely new or improved properties based on characteristics like size, morphology, and distribution, among others, when compared to the larger particles of the same source (bulk material) from where they were formed (Siegel et al. 2015). The size at the nanoscale influences NP reactivity because it increases the surface area-to-volume ratio. Other unique properties depend on several factors such as chemical composition, concentration, and the shape and the coating of the nanoparticle that is responsible for the interaction with the external environment (Grassian 2008).

Due to the growing need for ecologically correct synthesis routes, the interest in green nanoparticle synthesis studies is growing, which is the use of the biological method as an alternative to toxic chemical reagents in the synthesis of nanocomposites. The biological process consists of the use of plant extracts containing molecules capable of reducing metal ions of nanoparticles in only one step. In the case of silver, the reduction of the ion (Ag^+) to its metallic form (Ag°) occurs rapidly and under normal conditions of temperature and pressure and is called bioreduction (Thakkar et al. 2010; Akhtar et al. 2013; Lemire et al. 2013; Prasad et al. 2016).

The antimicrobial property of silver nanoparticles (AgNPs) makes these nanostructures applicable in different areas where it is necessary to combat the disordered proliferation of microorganisms. The antimicrobial effect against pathogens of medical relevance for humans, such as *Escherichia coli, Staphylococcus aureus, Pseudomonas aeruginosa*, among others, has been demonstrated. (Aziz et al. 2014; Lokina et al. 2014). Also, with the emergence of drug-resistant microorganisms used on the market today, AgNPs have arisen as a new way for the treatment of diseases caused by these pathogens (Rajpal et al. 2016; Khan et al. 2017). Thus, with their potential microbicide, they can be used in the coating of the curative, prosthesis, among others, reducing the risk of infection (Keat et al. 2015). Microorganisms were subjected to metals, creating a microorganism-metal coupling, which is very efficient in nanobiotechnology applications (Prasad 2014).

8.2 Obtaining Silver Nanoparticles

8.2.1 Top-Down

For the synthesis of AgNPs, there are two general approaches: top-down and bottom-up. In the top-down approach, the bulk material is broken into particles by physical methods such as evaporation-condensation or laser ablation. The synthesis by the first procedure, although having the advantage of uniformity in the distribution of NPs, requires a tubular furnace under atmospheric pressure that occupies a large area, consumes several kilowatts, and increases the ambient temperature around the material (Kruis et al. 2000).

The synthesis of AgNPs by laser ablation uses the bulk material, in this case, a silver metal plate submerged in an aqueous solution of sodium dodecyl sulfate (SDS) and irradiated with a laser, varying the concentration of SDS and the power of the laser (Mafune et al. 2000). The advantage of the synthesis by laser ablation is the absence of chemical reagents in the solutions, guaranteeing the purity of silver NPs.

8.2.2 Bottom-Up

The bottom-up technology is most commonly used in a chemical synthesis process that requires the major components such as metal precursors, agents responsible for the first step of the synthesis, and stabilizing agents. It is a fast and inexpensive technology for producing nanostructures and complementing the usual lithographic techniques. The real function of stabilizing agents is protecting the nanostructures by binding or absorption on the surface, avoiding the agglomeration, and inactivation of nanotechnology (Natsuki et al. 2015).

8.2.2.1 Chemical Reduction

Many methods of chemical synthesis have been developed over two simple ways: the Turkevich method consists of the chemical reduction of silver ions in salt form with sodium citrate, which also acts as a stabilizing agent in the resulting NPs (Pillai and Kamat 2004). The stability of the silver nanoparticles, in this case, is caused by the citruses whose carboxyl groups interact with the surface of the NPs that increases the electrostatic repulsion with the adjacent particles, making agglomeration difficult. In the Shiffrin-Brust method, sodium borohydride is used as the reducing agent and the acid replaced by a selected thiol, acting as a stabilizer (Perala and Kumar 2013; Brust et al. 1994). The borohydride ions are adsorbed on the surface, causing electrostatic repulsion, and because they are negative ions, this repulsion controls the size of the AgNPs, reducing the tendency to aggregate (Janardhanan et al. 2009).

The representative scheme demonstrates the action of the anions on the protective barrier of these AgNPs, stabilizing them (Fig. 8.2). The preparation of AgNPs, either by physical or chemical methods, is usually costly and can cause severe impacts on the environment due to the use of toxic chemicals and biological risks (Iravani et al. 2014; Toisawa et al. 2010).

There are several methods of synthesis of metallic nanoparticles. Their stability, size, and properties are dependent on the process of preparation. Some of the methods studied are an electrochemical reduction, thermal decomposition, vapor-metal deposition, and chemical reduction, among others. However, the most straightforward and most versatile technique is chemical reduction (Khanna et al. 2008).

As for the preparation of AgNPs, the simplest and most described method in the literature is the chemical reduction of metallic salts, such as silver nitrate. Soluble metal salts, reducing agents, and stabilizing agents are used for this (Muhammad et al. 2014; Pillai and Kamat 2004). Through the stabilizing agent, it is possible to obtain particles of various shapes and sizes, which, consequently, may result in different properties, thus allowing the diversification of the use of the nanoparticles. Chemical synthesis methods generally use toxic compounds such as sodium borohydride, sodium hydroxide, and ethylene glycol (Brust et al. 1994), which harm the environment and human health. They make it impossible to use nanoparticles in several areas, including medicine, because they can be adsorbed on the surfaces of synthesized nanoparticles. Still, the generation of hazardous by-products may occur, requiring special treatment, thus increasing production costs (Cauerhff and Castro 2013).

There is a growing intention to develop synthesis procedures that are less aggressive to the environment and human health, but with high yields and low cost.

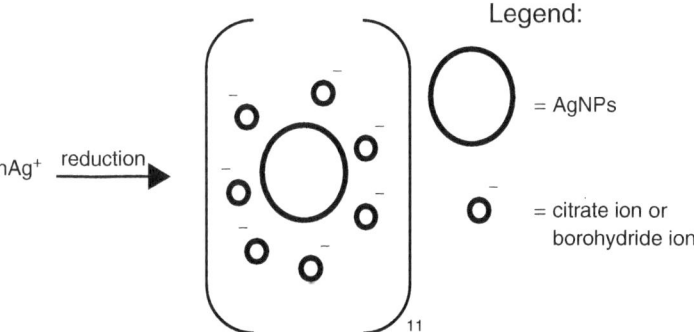

Fig. 8.2 The representative scheme demonstrates the action of the negative ions on the stabilization of AgNPs on the chemistry route

8.2.2.2   Green Synthesis

The preparation of AgNPs, either by physical or chemical methods, is usually costly and can cause severe impacts on the environment due to the use of toxic chemicals and the possibility of biological risks. Green chemistry focuses on reducing energy consumption; elimination of toxic solvents and organic substrates, substituting the least toxic or more natural; and reduction or elimination of by-products to be discarded (Syafiuddin et al. 2017; Prasad et al. 2018).

Medicinal plants are widely used to improve the symptoms of many diseases. Throughout many years, the role of "healer" was that who, using natural resources, mediated the cure of diseases. This culture still exists today, and the wide variety of medicinal plants supports this custom. There is a relevant contribution to the use of vegetables for therapeutic effects. They produce considering observations and available reports of the efficiency of plant extracts even without the real notion of the active principle of those extracts. The efficacy of many extracts is day by day increasing the interest of researchers in studies involving multiple areas of knowledge, which together enrich the knowledge about the inexhaustible natural medicinal source: the world flora (Ahmed et al. 2016; Prasad 2014; Aziz et al. 2015).

The use of plant extracts may be advantageous compared to other methods since the reaction rate for synthesis is very high; it allows large-scale production and does not use microorganisms, something that takes time for growing and generates costs with the maintenance of this growth (Prasad 2014; Patel et al. 2015; Keat et al. 2015). There is also the possibility of green synthesis occurring in different plant organs, such as flower, fruit, leaf, root, and fruit peels, which demonstrates the versatility of the method (Ahmed et al. 2016) (Fig. 8.3).

It is known that vegetables produce several antioxidant molecules, which could act in the production of nanoparticles. Biomolecules such as amino acids, proteins, polysaccharides, as well as secondary metabolites such as flavonoids, tannic acid,

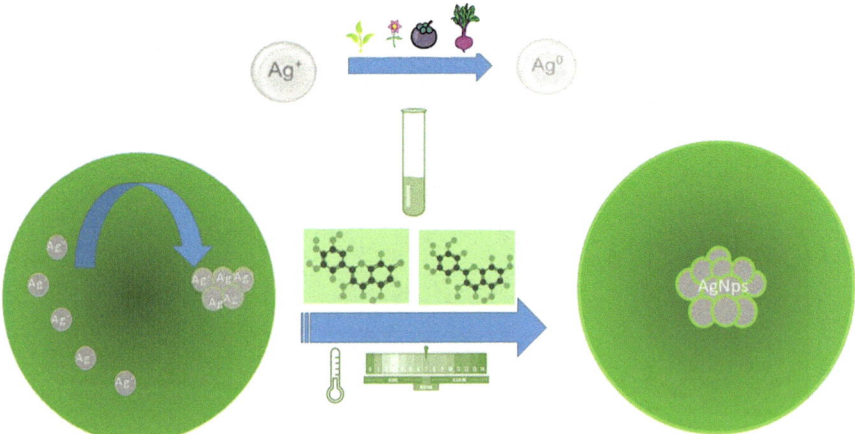

Fig. 8.3 Green synthesis of AgNPs

saponins, and terpenoids, can act in both reducing the metal precursor and stabilizing the nanoparticles and consequently, limiting their growth and preventing their agglomeration (Kong et al. 2003).

Generally, the precursor is an inorganic salt which in aqueous medium undergoes a solvation process, in which the water molecules make hydration layers, and repulsive forces allow the stabilization of the silver ions due to the repulsive forces. In this work, the precursor salt used was silver nitrate, which in an aqueous medium is dissociated to silver ions (monovalent cation, Ag^+). These ions are reduced by bioactive molecules that were present in the plant extract. In this way, the long and complex chains allow the reduction of the silver and make a shield that protects the silver in the fundamental state to oxidation, favoring the nucleation and growth essential for the formation of silver nanoparticles. The bioactive molecules, which are present in the dispersion medium, allow the efficient and stable environment to the AgNPs in the nanobiotechnology context (Zhang et al. 2016).

Supramolecular interactions allow the plant extract to be an effective medium because it provides stabilization and avoids oxidation through the forces of intermolecular interactions that are established. The aromatic rings present in these substances and the long chains provide a resonance effect that shields the silver in the ground state, favoring the process of growth and nucleation of AgNPs, as predicted by the Casimir effect, ceasing their growth by aggregation. Because they are very large molecules, they have a steric hindrance and promote a physical spacing that keeps them stable in the middle of the plant extract. Electrostatic stabilization occurs when repulsive electrostatic forces overpower the attractive forces of van der Waals (Reimers et al. 2017).

8.3 Colloidal System

A colloidal solution, in a simplified way, is understood as a dispersant matrix in which the size of the dispersed particulates is in a range of up to 100 nm. The essential objective of the present study is to understand a colloidal solution consisting of nanoparticles and the influence of the dispersive medium related to the stability of this colloidal dispersion (Phu et al. 2010).

Forces of intermolecular interactions influence the behavior of the substances entirely. Dispersive media, concentration, temperature, and the influence of pH, among other synthesis parameters, are directly related factors to NP morphology and stabilization. A peculiar relation between the medium and the nanostructures is observed in nanometric systems. Minimally interfering variations in molecular systems strongly influence the nanoscale (Moore et al. 2015).

For nanotechnology, the dispersion of nanoparticles is a study that must be thoroughly done. The stabilization is closely linked to the intermolecular forces, van der Waals, and hydrogen bonds, mainly supporting this dispersion in the colloidal system (Reimers et al. 2017).

It is crucial to take into account thermal agitation and electric repulsion of the surfaces in the influence of temperature in the route of synthesis since these parameters are essential to avoid the aggregation. Thus, a high critical micelle concentration should be favored by temperature increase, generating molecular vibrational energy increase, and, in the case of inorganic salts such as $AgNO_3$, a reduction of repulsive forces (Santos et al. 2016).

The first essential and straightforward protocol that ratifies the formation of a colloidal dispersion is the evaluation of the Tyndall effect by the passage of the beam of a laser, a simple pointer. At the moment that the path of the beam is revealed, it confirms that the route resulted in an effective colloidal dispersion. In this way, it is possible to proceed with the other characterization protocols. It is possible, through this simple test, to verify the efficiency of the synthesis (Yang et al, 2016).

The formation and stability of silver in the form of nanoparticles in a colloidal dispersion should be monitored. AgNPs exhibit completely new and optimized properties when relating the properties of silver on a larger scale, based on specific characteristics such as size, morphology, and distribution (Polte et al. 2012).

The nanometric scale has different properties when we compare the atom that originates the nanoparticles. Color is one of the modifications. Colloidal silver has a golden yellow color. A feature of noble metals in the form of nanoparticles is the intense coloration of their colloidal dispersions, which is caused by the absorption of surface plasmon resonance (SPR) (Yang et al. 2016).

When the particles aggregate, a new band is observed next, as a function of the interaction between the evanescent electric fields of the near nanoparticles, that is, Ag° aggregates interfere in the reading of the typical peak and in this way increase the width of the peak (Polte et al. 2012). At the nanometer scale, the particles are smaller than the wavelength of light. Electrons interact with the oscillating electric field of light, generating plasmonic waves. This phenomenon generates an electric dipole that resonates with electromagnetic radiation. Absorption and light scattering coincide (Scholl et al. 2012).

The condition of SPR is characteristic of the nature of metallic nanoparticles. Silver in the colloidal state has a peak of SPR revealed between 380 and 480 nm (Linic et al. 2015; Raveendran et al. 2006; Pal et al. 2007), which varies with the mean AgNPs size. The stability of the colloidal dispersion of AgNPs is provided by substances extracted from the plant extract, such as flavonoids and anthocyanins, which maintain the silver in the ground state, and interact with the surface of the particle, inhibiting aggregation. The SPR transition depends on the supramolecular interactions of the dispersion medium.

The typical peak is possible to be observed in UV-vis analysis when silver nanoparticles are in a colloidal state. Through the surface peak of resonance, it is likely to understand the behavior of light dispersion on AgNPs surface (Arvizo et al. 2012).

The specific surface area of the nanoparticles is increased, thereby increasing their biological efficiency due to the increase in surface energy. Silver has been recognized by its inhibitory effect on the action of bacteria and microorganisms commonly present in the industrial and medical industries (Pal et al. 2007). These include smart fabric coated with AgNPs with bactericidal action and curative coated with AgNPs to prevent the proliferation of microorganisms in burns and open wounds. As AgNPs are widely applied in humans, there is a growing need to develop green synthesis processes, which in this way become a win for humanity.

8.4   Plant Extract

Due to their large contact surface, particles at the nanoscale often clump together, forming secondary particles, which minimize the total area and thereby lower the surface energy of the system. When the nanoparticles are together, they form agglomerates called aggregates. Metal particles are also susceptible to oxidation because of their instability, making the use of a medium that delays oxidation and agglomeration necessary (Badawy et al. 2010).

Plant extracts act as antioxidants because of their various components. Among these, we can highlight phenolic compounds, flavonoids, anthocyanins, terpenoids, citric acid, functional groups (alcohols, aldehydes, amines), and heterocyclic compounds, among others. Some of these compounds may also act as stabilizing agents for NPs, covering the surface of the particles, limiting their growth, and preventing them from aggregating (Tripathi et al. 2013; Ingale and Chaudhari 2013; Saxena et al. 2012).

Phenolic acids are some of the substances that comprise the group of the phenolic compounds and are divided into three groups. They are characterized by having a benzene ring, a carboxylic group, and one or more hydroxyl and methoxyl groups in the molecule, imparting antioxidant properties. The first is composed of benzoic acids, which have seven carbon atoms and are the phenolic acids most easily found in nature. The second is formed by cinnamic acids, which have nine carbon atoms, seven of which are most commonly found in the plant kingdom, and the third, polymerization of the coniferyl, synapyl, and coumaryl alcohols. The content of these alcohols is an essential factor in the differentiation of the formed lignins (Walter and Marchesan 2011).

The flavonoids are polyphenols that function as sequesters of free radicals or as chelators of metals, acting on free radicals through two mechanisms: the first involving the inhibition of free radical formation and the second covering the elimination of free radicals (Svobodová et al. 2003). Because of this, flavonoids may represent important oxidation preventive agents. The leaf extract for the synthesis of nanoparticles is the most common choice, but there are also studies reporting the use of seeds, roots (Gnanadesigan et al. 2012), peels, and fruits (Backx et al. 2018). In this study, four types of plants were used: *Plinia Cauliflora* (fruit), *Melissa Officinalis* (leaves), *Beta vulgaris* (root) and *Citrus sinensis* (peel of fruit) (Table 8.1). To each plant and organ of the plant used, the green synthesis was efficient and presented the characteristic SPR peak of AgNPs (Fig. 8.4).

Green colloidal systems whose dispersive medium is derived from plant extracts have several physicochemical peculiarities, due to the presence of dozens of

Table 8.1 Bioactive extracted in each species of plant

Plant species	Active ingredients	Chemical structure
Plinia cauliflora	Cyanidin	
Melissa officinalis	Carvacrol	
Beta vulgaris	Betacyanin	
Citrus sinensis	Hesperidin	

chemical species with different reactivities and molecular interactions. Those inter-actions establish supramolecular relations involving oxidation, reduction, and the formation of new compounds with the variation of temperature and pressure or, even, influences of suprastructures in the colloidal dispersions by the power of supramolecular chemistry between the substances of the system (Liu et al. 2017).

The covalent bonds of global, organic chains with aromatic rings that justify the resonance and the electron-dislocation interaction generate highly ordered aggre-gates. Metallic bond in silver atoms is vulnerable. It is associated with the complex-ity of the orbital 4d and its stability with the electronic transition of the electron of the orbital 5s. In this context, the metallic bond between the atoms has the dynamics of decentralizing their electrostatics to this interaction (Toma et al. 2018).

The intermolecular interaction that occurs on the surface of these atoms and the molecules of the extract increases the repulsive energy and stabilizes the constituent atoms of these NPs. The resonance that occurs with the complex molecules of the bioactive extracted from the plants induces by the inductive effect, as predicted by the Casimir effect, the relocation of the electronic density along the chain of the molecules. Thus, the induced dipole formed favors the interaction of van der Waals and stabilization of AgNPs. This arrangement is highly corrected to produce a pat-tern without defects (Pinto 2018).

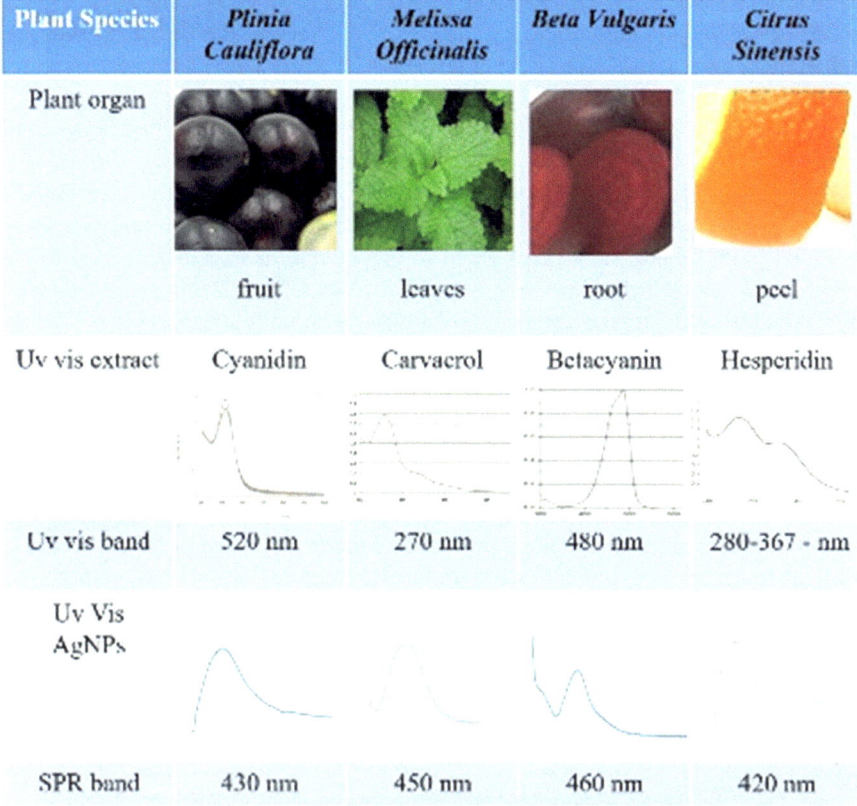

Plant Species	Plinia Cauliflora	Melissa Officinalis	Beta Vulgaris	Citrus Sinensis
Plant organ	fruit	leaves	root	peel
Uv vis extract	Cyanidin	Carvacrol	Betacyanin	Hesperidin
Uv vis band	520 nm	270 nm	480 nm	280-367 - nm
Uv Vis AgNPs				
SPR band	430 nm	450 nm	460 nm	420 nm

Fig. 8.4 Scheme of the plant organ, bioactive molecules, and SPR band for green synthesis

8.5 Mechanistic Approaches for Antimicrobial Activity

Recent studies have shown that some types of metallic nanoparticles, especially silver nanoparticles, have antimicrobial activity and are highly effective for fighting various kinds of microorganisms such as viruses, bacteria (such as *Staphylococcus aureus, Pseudomonas aeruginosa, Escherichia coli*), fungi that cause ringworm infections, and *Candida albicans* (Backx et al. 2018).

The antimicrobial property of AgNPs makes these nanostructures applicable in different areas of industry, such as the food industry, medicine, and other areas where it is necessary to fight the uncontrolled proliferation of microorganisms. Several studies have demonstrated the antimicrobial effect against pathogens of medical relevance for humans (Bumbudsanpharoke and Ko 2015; Nalawadea et al. 2014; Barua et al. 2013; Logeswari et al. 2015; Shrivastava et al. 2007). Also, with the emergence of microorganisms that are resistant to the drugs currently used in the market, AgNPs have emerged as an alternative for the treatment of diseases caused by microbial pathogens (Chaloupka et al. 2010; Rai et al. 2009). The antimicrobial action of AgNPs provides efficiency coating of catheters, prostheses, and bandages, reducing the risk of infection (Keat et al. 2015).

According to the literature, many plant extracts demonstrate significant antimicrobial activity and are known to be biosynthesizers of biologically active secondary metabolites, with emphasis on drugs that inhibit parasite growth; thus, it is possible to combine these properties with the antimicrobial activity of AgNPs (Prasad and Swamy 2013). The choice of the plant species to be assessed is an essential point of the study because, from this choice, a screening related to the origin of the selected plant species is made. Variables related to soil quality, the relative humidity of the region, temperature, atmospheric pollution, as well as the harvest and age of the plant may cause changes in the contents of the chemical constituents of these plant species (Ahmed et al. 2016).

Therefore, in addition to the medicinal function, the chosen extract is used as a vehicle of the AgNPs in an eco-friendly way, because it is all about green synthesis with more significant potential in the action of microorganisms (Sondi and Salopek-Sondi 2004).

In this way, nanobiotechnology presents solutions to several of these problems, with the lesser use of precursor materials, the use of non-toxic solvents, granting the reproducible reactive conditions, as well as obtaining the colloidal dispersion in a stable system.

Plant extracts are an excellent source of research for new antimicrobial agents since the molecular diversity of natural products is much higher than those derived from chemical synthesis processes. This study seeks the discovery of new bioactive molecules capable of optimizing the dispersion of the nanoparticles through the preparation of extracts to be tested for antimicrobial activity, making possible the pharmacological study of the flora, still entirely unexplored under this aspect (Huang and Lau 2016).

The mechanistic approaches to the antimicrobial activity of silver nanoparticles are still the aim of study for many research groups all over the world because it is still not fully understood. However, it is seen that AgNPs interact with a range of molecular processes that can lead to cell death. It is believed that silver nanoparticles interact directly with wall elements and the cell membrane, causing structural damage that alters its permeability, leading to the release of intracellular content (Joshi et al. 2018). This event is generated in the proton dissipation that affects the electrochemical gradient, essential for cellular respiration, and results in cell death (Aziz et al. 2016).

Another issue related to the mechanism of action of AgNPs is related to the entry of nanoparticles into the cell. Since the nucleotide of the DNA molecule consists of a phosphate group, a nitrogenous base, and a five-carbon atomic molecule, the silver, as AgNPs, forms complexes with nitrogen and phosphorus. Thus, there is the cancellation of DNA replication and fractionation due to chemical interaction disrupting protein synthesis and cellular metabolism (Singhal et al. 2017; Prasad et al. 2016). Antimicrobial action consists of inhibition of cellular functions, disruption of ATP production and DNA replication due to the entry of AgNPs, cell membrane

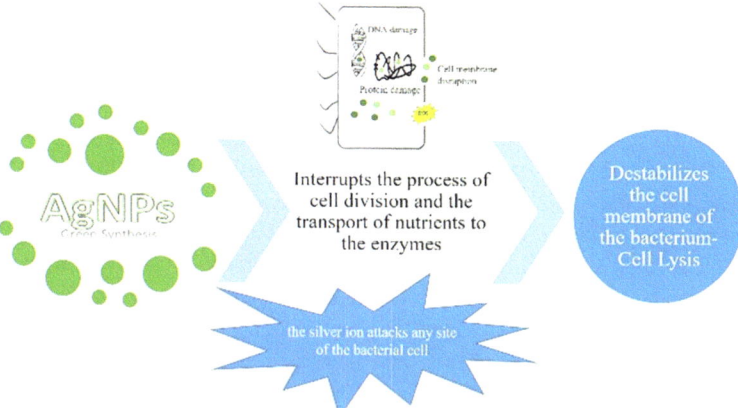

Fig. 8.5 Scheme of mechanistic approaches for antimicrobial activity

damage by the contact of AgNPs, and inducing the production of reactive oxygen species (ROS) and apoptosis (Fig. 8.5) (Joshi et al. 2018).

Another proposed mechanism is that AgNPs and their liberated silver ions (Ag$^+$) can act on the phosphate groups of the DNA molecules, resulting in damage and condensation, inhibiting the replication process. Also, AgNPs and silver ions (Ag$^+$) would form complexes with thiol groups (-SH) of cysteine residues in proteins inhibiting essential enzymatic functions, such as respiratory and protein translation, leading to cell death (Aziz et al. 2016; Patra and Baek 2017; Prasad et al. 2017). In this way, the destabilization of the plasma membrane potential occurs and results in the loss of the intracellular energetic binding of ATP, resulting in cell death. They act as catalysts for the production of reactive oxygen species, such as superoxide anions and hydrogen peroxide, generating oxidative stress leading to rupture of the cell membrane (Ramalingam et al. 2016). There is likely to be a synergistic action of different mechanisms, which makes the emergence of resistance more difficult. Also, the antimicrobial effect is dependent on the shape, size, coating, and concentration of the nanoparticles.

In this way, little concentration of AgNPs is dispersed on the surface of the fibers and textiles providing the inhibition of the growth of microorganisms, very common in hospital environments, which will have a better interaction with the bacterium, due to an increase in the surface area (Lee and Jeong 2004; Jeong et al. 2005). Thus, it is sought to affect the coating of textile surfaces such as towels, sheets, and materials present in hospital environments and used by health professionals. Those items may offer ways of contaminating the already debilitated individual. This coating can turn to be an essential pathway in patient protection and optimization of the time of rehabilitation, with a decrease in the incidence of diseases acquired by the prolonged stay of the patient in the hospital in full contact with those components.

Coatings modified with nanoparticles are used for various applications. Nanocoatings, as they are called, in the food industry, when in contact with food act as a barrier against the proliferation of microorganisms due to the antimicrobial properties. AgNPs are already successfully incorporated into plastic food storage containers and allow a considerable decrease in the proliferation of microorganisms (Prasad et al. 2017).

For curatives, the development of biodegradable polymer blends incorporated with AgNPs aims to decrease the proliferation of microorganisms in the wound and the impact of disposal of these materials on the environment. One of the mechanisms associated with the antimicrobial activity of biofilms may be associated with the chemical kinetics of stabilization and immobilization of silver in the polymer blends. In its metallic state, silver has low reactivity. The interaction of the biofilm, used as curative, made by the polymeric blend increased with the AgNPs, with the fluids of the wound, called exudate, and natural humidity of the skin creates an environment to the oxidation of silver. As the monovalent silver cation is highly reactive, it binds tissue proteins and causes structural damage to the membrane and bacterial cell wall, thereby inhibiting bacterial replication (Lara et al. 2011; Kim et al. 2007; Amro et al. 2000), thus rehydrating the injured site with the human humidity itself, decreasing the wound healing time and avoiding the proliferation of microorganisms in the wounds. Preliminary studies indicate dispersive efficiency of the nanoparticles in addition to mechanical properties optimized by the formation of polymer blends.

8.6   Silver Nanoparticles and Health Materials

8.6.1   Polymer Blends: Biofilms

A polymer blend is a mix of two polymers, and the interaction between the polymer phases provides the properties of a heterogeneous polymer blend (Jiang et al. 1991). The proximity between the two polymers characterizes the area between the polymer phases in a polymer blend, and so the mixture is entitled miscible blends. The miscible polymer blends are homogeneous down to the molecular level. The immiscible blends have a phase between two polymers, that is, there is no compatibility between the polymer phases, and this type of combination is a heterogeneous polymer blend. Most polymer blends are immiscible (George et al. 2015).

The separation area between two polymers in a blend can be reduced by the addition of interfacial agents known as compatibilizer agents (Chen and White 1993). These agents in polymer blends are essential because they are related to interfacial tension, phase morphology, and mechanical properties. These are generally molecules with polar and non-polar regions that can be aligned along the interface's area between the two polymers and the compatibility of the polymer blends to be increased, and this generates a reduction of the dispersed particle size, improved mechanical properties, and phase stability (Fakirov 2019). The presence of large and complex organic groups or polymers creates a physical shield that inhibits the growth of nanostructures and allows an efficient dispersion based on the efficiency of van der Waals interactions, steric stabilization, and electrostatic stabilization, which generates a stable colloidal matrix. Synthetic plastic is the primary material utilized for various applications (Cai et al. 2018).

An alternative to the use of synthetic plastics that remain in the environment is the use of biodegradable polymers made from renewable sources, the most targeted being the lipids, polysaccharides, and proteins (Nair and Laurencin 2007). Among the polysaccharides, the most used are the derivatives of starch, pectin, cellulose, and chitosan, among others. Biodegradable polymers, called biofilms, are characterized by being plastics where their degradation is made through microorganisms, fungi, bacteria, and algae (Backx and Santana 2018).

The physical, mechanical, and chemical properties of the biofilm can be optimized with the physical mixture of two polymers that interact through secondary forces, such as van der Waals. One way to produce the biofilm of this study is polymerizing the glycerol to provide polyglycerol to mix to the starch and make a polymer blend (Backx and Santana 2018).

The addition of silver nanoparticles in these polymer blends, such as in the biofilm application, is a powerful alternative as a bandage type, because of the antimicrobial potential of AgNPs, and the lower impact of these materials to the environment together became a powerful tool to other options curative to many lesions, such as burning skin. The polymer blend with the exudate can to rehydrate the skin and inhibit the microbial proliferation (Folestad et al. 2008).

8.6.2 Textile Industry

Another application of AgNPs is in the textile industry. The textiles, mainly in the hospital environment, can provide a favorable environment for the growth of microorganisms since they absorb humidity. With low immunity of a patient, the proliferation of fungi and bacteria present in tissues such as sheets, pillowcases, can cause infections and decrease the patient's life expectancy. Due to the antimicrobial efficacy of AgNPs reported in the literature, they have been studied for their implementation in antimicrobial textiles. Synthesis of AgNPs in media, which exhibit low or no toxicity but still show high efficiency, may make it feasible to apply them in hospital textiles to minimize diseases associated with the proliferation of these microorganisms to bedridden patients (Gopinath et al. 2017).

Studies to evaluate the efficiency of application of AgNPs in the textiles and effectiveness of washes are being developed to create appropriate methodologies to control the cytotoxicity and low or no environmental impact after the washings of the textiles treated with the AgNPs. Preliminary evaluations indicate impregnation of AgNPs in tissue fibers and antimicrobial efficiency even after washing (Backx et al. 2018).

In Fig. 8.6, it is possible to observe a graph obtained by the NanoSight equipment, which evaluates the dimensions of the synthesized AgNPs produced and their

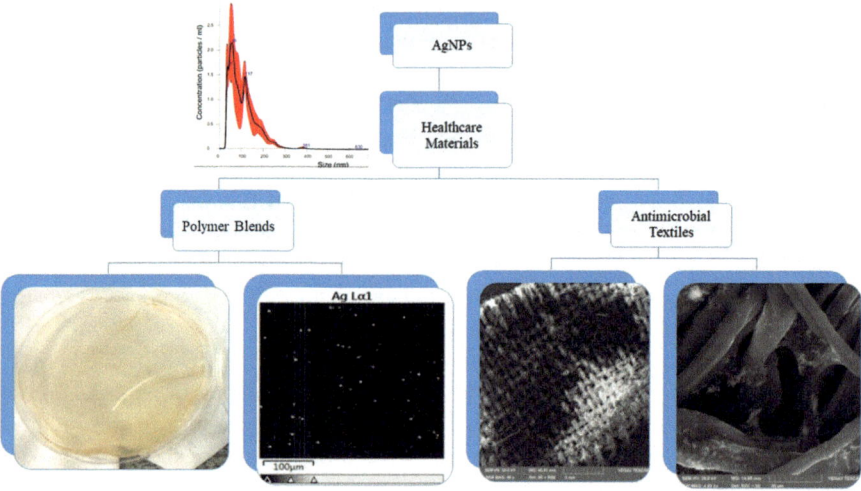

Fig. 8.6 This scheme shows the graph obtained by the NanoSight instrument, polymer blends (photo and the AgNPs X-ray map on polymer blend), and the textile (fibers of the textiles and the 4.89 kX zoom of fibers with AgNPs by SEM)

concentration. For the two health materials cited in this study, the nanoparticles were mostly of average size in the range between 1 and 50 nm. They increased the polymer blends, as can be seen in the X-ray mapping obtained by the scanning electron microscope (SEM). AgNPs are well dispersed on the surface of the biofilm and have spherical morphology. It is also possible to observe the efficiency of the dispersion of AgNPs in 100% cotton fabric fibers, widely used in hospital environments.

8.6.3 Antimicrobial Response

Bauer and Kirby described the disk diffusion test on agar. The concept of this method is based on the diffusion of an antimicrobial impregnated in a filter paper disk through the agar. The distribution of the antimicrobial leads to the formation of a bacterial growth inhibition halo that has a diameter. The test results are obtained by comparing the inhibition halo value with the criteria already published in references. The qualitative test is sufficient to guide the choice of antimicrobial agents. It is a practical, easy-to-execute, and idealized method for fast-growing bacteria. Reagents are relatively inexpensive, there is no need for special equipment, and there is excellent flexibility in choosing the number and type of antimicrobials to be tested.

The assays were initially performed against standard strains of *Staphylococcus aureus* and *E. coli* by the disk diffusion method. The resistance of pathogenic microorganisms to variable drugs has increased significantly due to the abusive use of

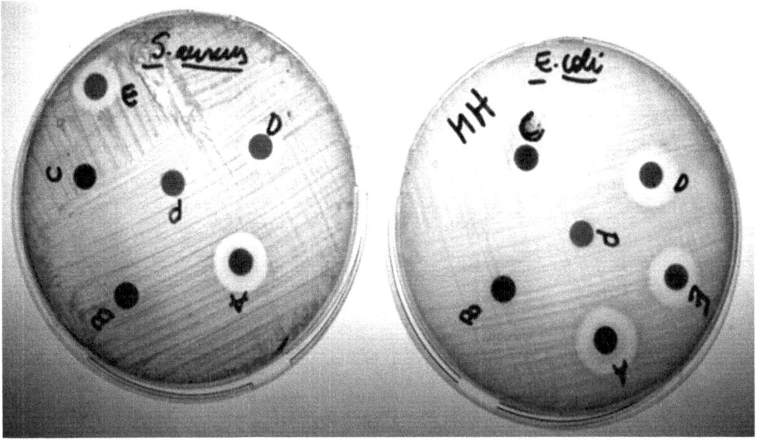

Fig. 8.7 Disk diffusion test. Legend: A—AgNO₃, B—Extract of *Melissa Officinalis leaves*, C—AgNP1 (0.05 mg/L), D—AgNP2 (0.5 mg/L), E—AgNP3 (1.2 mg/L), P—phosphate buffer saline (PBS)

antimicrobials that are easily obtained and end up undermining the treatment of infectious diseases. Thus, researchers have rising interests in the search for new means that provide ways of effectively acting in a careful approach to compensate for this resistance.

Seeding was done homogeneously so that microorganisms grow on the entire plate. Then, the white and sterile discs were placed. Each disk was added 10 μL of the solution to be tested at 37 °C for 24 h. The plate had six discs, one control, the white positive was silver nitrate (A), plant extract (B), and the other three with AgNPs following the preparation order of concentration and the negative white used was phosphate buffer saline (P). Samples of the AgNPs synthesized in the extracts were tested. For this test, the extract used was that of *Melissa Officinalis* leaves (Fig. 8.7).

It can be possible to visualize the zones of inhibition bacteria isolated, which appear as a clear area around the discs. The diameter of such zones of inhibition was measured using a meter ruler, and the mean value was recorded and expressed in millimeters, according to Table 8.2.

Through this experiment, it was possible to verify that the extracts of the plants studied do not present antimicrobial action, and that the AgNPs show greater microbicidal efficiency due to the more significant halo, related to the higher concentration of AgNPs. Researchers studied that the antimicrobial activity of AgNPs was dependent on the concentration of the AgNPs used and was strongly associated with the formation of holes in the cell wall of microorganisms. Following this, AgNPs cumulate in the microorganism membrane and cause permeability, resulting in cell death (Loo et al. 2015).

Table 8.2 The measurement of the diameter of zones of inhibition of the *S. aureus* and *E. coli* after 24 h of plating

Disks	*S. aureus* Zones of inhibition (mm)	*E. coli* Zones of inhibition (mm)
A (AgNO₃)	15.7	17.1
B (*Melissa Officinalis* leaf extract)	0	0
C (AgNP1 (0.05 mg/L)	0	0
D (AgNP1 (0.5 mg/L)	12.4	16.2
E (AgNP1 (1.2 mg/L)	13.9	15.3
P (PBS)	0	0

The treatment of pure extracts and spikes of different AgNPs concentrations

8.7 Conclusion and Future Outlook

Plant extracts can be widely used as a reducing and stabilizing medium for the green synthesis of silver nanoparticles. This observation demonstrates the potential of AgNPs as antimicrobial agents since they were efficiently dispersed and impregnated into the polymer blends and textiles. Thus, its use in hospital textiles and curative materials can be a preventive measure against the dissemination of hospital infection, which is a common cause of death within hospitals and has significant action against wound contamination.

In this way, it is possible to implement green routes for production of nanomaterials that have antimicrobial efficiency and low environmental impact and what effectively promotes these protocols as better synthetic routes, when compared to those already existent, that use toxic chemical substances and generate a much toxic waste that impacts directly on the environment. The possibilities are enormous when one takes into account the influence of the supramolecular interactions of plant bioactive, and that is related to the routes of synthesis.

The antimicrobial efficiency of AgNPs is increasingly being explored, and its potential in the application in various types of materials becomes a hope of a future combining health and environment.

The study of nanotoxicity and feasibility of use will be the next steps. If adjustments are necessary, new studies associating antimicrobial efficiency of AgNPs will be associated to pH, storage conditions, the influence of light, and comparison of extraction media for the isolation of biomolecules acting in the stabilization of AgNPs among others.

References

Ahmed S, Ahmad M, Swami BL, Ikram S (2016) A review on plants extract mediated synthesis of silver nanoparticles for antimicrobial applications: a green expertise. J Adv Res 7:17–28. https://doi.org/10.1016/j.jare.2015.02.007

Akhtar BS, Panwar J, Yun YS (2013) Biogenic synthesis of metallic nanoparticles by plant extracts. ACS Sustain Chem Eng 1:591–602. https://doi.org/10.1021/sc300118u

Amro NA, Kotra LP, Wadu-Mesthrige K, Bulychev A, Mobashery S, Liu G (2000) High-resolution atomic force microscopy studies of the *Escherichia coli* outer membrane: structural basis for permeability. Langmuir 16:2789–2796. https://doi.org/10.1021/la991013x

Arvizo RR, Bhattacharyya S, Kudgus RA, Giri K, Bhattacharya R, Mukherjee P (2012) Intrinsic therapeutic applications of noble metal nanoparticles: past, present, and future. Chem Soc Rev 4:2943–2970. https://doi.org/10.1039/c2cs15355f

Aziz N, Fatma T, Varma A, Prasad R (2014) Biogenic synthesis of silver nanoparticles using *Scenedesmus abundan*ce and evaluation of their antibacterial activity. J Nanoparticles 2014:689419. https://doi.org/10.1155/2014/689419

Aziz N, Faraz M, Pandey R et al (2015) Facile algae-derived route to biogenic silver nanoparticles: synthesis, antibacterial, and photocatalytic properties. Langmuir 31:11605–11612. https://doi.org/10.1021/acs.langmuir.5b03081

Aziz N, Pandey R, Barman I, Prasad R (2016) Leveraging the attributes of *Mucor hiemalis*-derived silver nanoparticles for a synergistic broad-Spectrum antimicrobial platform. Front Microbiol 7:1984. https://doi.org/10.3389/fmicb.2016.01984

Backx BP, Santana JC (2018) Green synthesis of polymer blend impregnated with silver nanoparticles in *Euterpe Oleracea* dispersive medium. IJGHC 7:424–429. https://doi.org/10.24214/IJGHC/GC/7/2/42429

Backx BP, Pedrosa BR, Delazare T et al (2018) Green synthesis of silver nanoparticles: a study of the dispersive efficiency and antimicrobial potential of the extracts of *Plinia cauliflora* for application in smart textiles materials for healthcare. J Nanomater Mol Nanotechnol 7:1–10. https://doi.org/10.4172/2324-8777.1000236

Badawy AME, Luxton TP, Silva RG et al (2010) Impact of environmental conditions (pH, ionic strength, and electrolyte type) on the surface charge and aggregation of silver nanoparticles suspensions. Environ Sci Technol 44:1260–1266. https://doi.org/10.1021/es902240k

Barua S, Yoo JW, Kolhar P et al (2013) Particle shape enhances the specificity of antibody-displaying nanoparticles. Proc Natl Acad Sci U S A 110(9):3270–3275. https://doi.org/10.1073/pnas.1216893110

Brust M, Bethell D, Walker M et al (1994) Synthesis of thiol-derivatized gold nanoparticles in a 2-phase liquid-liquid system. J Chem Soc Chem Commun 7(7):801–802. https://doi.org/10.1039/c39940000801

Bumbudsanpharoke N, Ko S (2015) Nano-food packaging: an overview of market, migration research, and safety regulations. J Food Sci 80:910–923. https://doi.org/10.1111/1750-3841.12861

Cai W, Xiao C, Qian L, Shuxun C (2018) Detecting van der Waals forces between a single polymer repeating unit and a solid surface in high vacuum. Nano Res 12:57–61. https://doi.org/10.1007/s12274-018-2176-8

Cauerhff A, Castro G (2013) Bionanoparticles, a green nanochemistry approach. Electron J Biotechnol 16(3):717–3458. https://doi.org/10.2225/vol16-issue3-fulltext-3

Chaloupka K, Malam Y, Seifalian AM (2010) Nanosilver as a new generation of nanoproduct in biomedical applications. Trends Biotechnol 28(11):580–588. https://doi.org/10.1016/j.tibtech.2010.07.006

Chen C, White JL (1993) Compatibilizing agents in polymer blends: interfacial tension, phase morphology, and mechanical properties. Polym Eng Sci 33:923–930. https://doi.org/10.1002/pen.760331409

Fakirov S (2019) Condensation polymers: their chemical peculiarities offer great opportunities. Prog Polym Sci 89:1–18. https://doi.org/10.1016/j.progpolymsci.2018.09.003

Folestad A, Gilchrist B, Harding K et al (2008) Wound exudate and the role of dressings. A consensus document. Int Wound J 5:3–12. https://doi.org/10.1111/j.1742-481X.2008.00439.x

George S, Sushama CM, Nair AB (2015) Efficiencies of dipolymer rubber blends (EPDM\ FKM) using common weight data envelopment analysis. Mater Res 20:6. https://doi. org/10.1590/1980-5373-mr-2016-1127

Gnanadesigan M, Anand M, Ravikumar S et al (2012) Antibacterial potential of biosynthesized silver nanoparticles using Avicennia marina mangrove plant. Appl Nanosci 2:143–147. https:// doi.org/10.1007/s13204-011-0048-6

Gopinath V, Priyadarshini S, Loke MF et al (2017) Biogenic synthesis, characterization of antibacterial silver nanoparticles and its cell cytotoxicity. Arab J Chem 10:1107–1117. https://doi. org/10.1016/j.arabjc.2015.11.011

Grassian VH (2008) When size really matters: size-dependent properties and surface chemistry of metal and metal oxide nanoparticles in gas and liquid phase environments. J Phys Chem C 112:18303–18313. https://doi.org/10.1021/jp806073t

Huang R, Lau BLT (2016) Biomolecule–nanoparticle interactions: elucidation of the thermodynamics by isothermal titration calorimetry. Biochim Biophys Acta 1860:945–956. https://doi. org/10.1016/j.bbagen.2016.01.027

Ingale AG, Chaudhari AN (2013) Biogenic synthesis of nanoparticles and potential applications: an eco-friendly approach. J Nanomed Nanotechol 4:165. https://doi. org/10.4172/2157-7439.1000165

Iravani S, Korbekandi H, Mirmohammadi SV et al (2014) Synthesis of silver nanoparticles: chemical, physical and biological methods. Res Pharm Sci 9:385–406

Janardhanan R, Karuppaiah M, Hebalkar N, Rao TN (2009) Synthesis and surface chemistry of nanosilver particles. Polyhedron 28(12):2522–2530. https://doi.org/10.1016/j.poly.2009.05.038

Jeong SH, Yeo SY, Yi SC (2005) The effect of filler particle size on the antibacterial properties of compounded polymer/silver fibers. J Mater Sci 40:5407–5411. https://doi.org/10.1007/ s10853-005-4339-8

Jiang R, Quirk RP, White JL, Min K (1991) Polycarbonate-polystyrene block copolymers and their application as compatibilizers agents in polymer blends. Polym Eng Sci 31:1545. https://doi. org/10.1002/pen.760312107

Joshi N, Jain N, Pathak A et al (2018) Biosynthesis of silver nanoparticles using Carissa carandas berries and its potential antibacterial activities. J Sol-Gel Sci-Tech 86:682–689. https://doi. org/10.1007/s10971-018-4666-2

Keat CL, Aziz A, Eid AM et al (2015) Biosynthesis of nanoparticles and silver nanoparticles. Bioresour Bioprocess 2:47. https://doi.org/10.1186/s40643-015-0076-2

Khan IM, Ahmad A, Miyan L, Ahmad M, Aziz N (2017) Synthesis of charge transfer complex of chloranilic acid as an acceptor with p-nitroaniline as a donor: crystallographic, UV–visible spectrophotometric and antimicrobials. J Mol Struct 1141:687–6697. https://doi.org/10.1016/j. molstruc.20117.03.050

Khanna PK, Kale TS, Koteshwarrao N et al (2008) Synthesis of oleic acid capped copper nano-particles via reduction of copper salt by SFS. Mater Chem Phys 110:21. https://doi. org/10.1016/j.matchemphys.2008.01.013

Kim JS, Kuk E, Yu KN et al (2007) Antimicrobial effects of silver nanoparticles. Nanomedicine 3:95–101. https://doi.org/10.1016/j.nano.2006.12.001

Kong JM, Chia LS, Goh NK, Chia TF, Brouillard R (2003) Analysis and biological activities of anthocyanins. Phytochemistry 64:923–933. https://doi.org/10.1016/S0031-9422(03)00438-2

Kruis F, Fissan H, Rellinghaus B (2000) Sintering and evaporation characteristics of gas - phase synthesis of size select PbS nanoparticles. Mater Sci Eng B 69:329. https://doi.org/10.1016/ S0921-5107(99)00298-6

Lara HH, Garza Trevino EN, Ixtepan-Turrent L et al (2011) Silver nanoparticles are broad-spectrum bactericidal and virucidal compounds. J Nanobiotech 9:30. https://doi. org/10.1186/1477-3155-9-30

Lee HJ, Jeong SH (2004) Bacteriostasis of Nanosized colloidal silver on polyester nonwovens. Text Res J 74:442–447. https://doi.org/10.1177/0040517512445331

Lemire JA, Harrison JJ, Turner RJ (2013) Antimicrobial activity of metals: mechanisms, molecular targets, and applications. Nat Rev Microbiol 11:371–384. https://doi.org/10.1038/nrmicro3028

Li B, Feng Z, He L et al (2018) Self-assembled supramolecular nanoparticles for targeted delivery and combination chemotherapy. Chem Med Chem 13(19):2037–2044. https://doi.org/10.1002/cmdc.201800291

Linic S, Aslam U, Boerigter C, Morabito M (2015) Photochemical transformations on plasmonic metal nanoparticles. Nat Mater 14:567–576. https://doi.org/10.1038/nmat4281

Liu Y, Chung Chang Y, Chen HH (2017) Silver nanoparticle biosynthesis by using phenolic acids in rice husk extract as reducing agents and dispersants. J Food Drug Anal 26:649–656. https://doi.org/10.1016/j.jfda.2017.07.005

Logeswari P, Silambarasan S, Abraham J (2015) Synthesis of silver nanoparticles using plants extract and analysis of their antimicrobial property. J Saudi Chem Soc 19:311–317. https://doi.org/10.1016/j.jscs.2012.04.007

Lokina S, Stephen A, Kaviyarasan V et al (2014) Cytotoxicity and antimicrobial activities of green synthesized silver nanoparticles. Eur J Med Chem 76:256–263. https://doi.org/10.1016/j.ejmech.2014.02.010

Loo SL, Krantz WB, Fane AG, Hu X, Lim TT (2015) Effect of synthesis routes on the properties and bactericidal activity of cryogels incorporated with silver nanoparticles. RSC Adv 5:44626–44635. https://doi.org/10.1016/j.jcis.2015.09.007

Mafune F, Kohno J, Takeda Y et al (2000) Structure and stability of silver nanoparticles in aqueous solution produced by laser ablation. J Phys Chem B 104:8333–8337. https://doi.org/10.1021/jp001803b

Moore TL, Rodriguez-Lorenzo L, Hirsch V et al (2015) Nanoparticle colloidal stability in cell culture media and impact on cellular interactions. Chem Soc Rev 44:6287–6305. https://doi.org/10.1039/C4CS00487F

Muhammad M, Khalid N, Aziz MD et al (2014) Synthesis of silver nanoparticles by silver salt reduction and its characterization. IOP Conf Ser Mater Sci Eng 60:012034. https://doi.org/10.1088/1757-899X/60/1/012034

Nair LS, Laurencin T (2007) Biodegradable polymers as biomaterials. Prog Polym Sci 32:762–798. https://doi.org/10.1016/j.progpolymsci.2007.05.017

Nalawadea P, Mukherjee S, Kapoor S (2014) Biosynthesis, characterization, and antibacterial studies of silver nanoparticles using pods extract of *Acacia auriculiformis*. Spectrochim Acta Part A 129:121–124. https://doi.org/10.1016/j.saa.2014.03.032

Natsuki J, Natsuki T, Hashimoto Y (2015) A review of silver nanoparticles: synthesis methods, properties, and applications. Int J Mater Sci Appl 4:325. https://doi.org/10.11648/j.ijmsa.20150405.17

Pal S, Tak YK, Song JM (2007) Does the antibacterial activity of silver nanoparticles depend on the shape of the nanoparticle? A study of the Gram-negative bacterium Escherichia coli. Appl Environ Microbiol 73:1712–1720. https://doi.org/10.1128/AEM.02218-06

Patel V, Berthold D, Puranik P, Gantar M (2015) Screening of cyanobacteria and microalgae for their ability to synthesize silver nanoparticles with antibacterial activity. Biotech Rep 5:112–119. https://doi.org/10.1016/j.btre.2014

Patra JK, Baek KH (2017) Antibacterial activity and synergistic antibacterial potential of biosynthesized silver nanoparticles against foodborne pathogenic bacteria along with its anticandidal and antioxidant effects. Front Microbiol 8:167. https://doi.org/10.3389/fmicb.2017.00167

Perala SRK, Kumar S (2013) On the mechanism of metal nanoparticle synthesis in the Brust–Schiffrin method. Langmuir 29(31):9863–9873. https://doi.org/10.1021/la401604q

Phu DV, Lang VTK, Lan NTK et al (2010) Synthesis and antimicrobial effects of colloidal silver nanoparticles in chitosan by γ-irradiation. J Exp Nanosci 5.169–179. https://doi.org/10.1080/17458080903383324

Pillai Z, Kamat P (2004) What factors control the size and shape of silver nanoparticles in the citrate ion reduction method? J Phys Chem B 108:945–951. https://doi.org/10.1021/jp037018r

Pinto F (2018) The Casimir effect and its role in nanotechnology applications. Mat Today Proc 5:15976–15982. https://doi.org/10.1016/j.matpr.2018.05.041

Polte J, Tuaev X, Wuithschick M et al (2012) Formation mechanism of colloidal silver nanoparticles: analogies and differences to the growth of gold nanoparticles. ACS Nano 6:5791–5802. https://doi.org/10.1021/nn301724z

Prasad R (2014) Synthesis of silver nanoparticles in photosynthetic plants. J Nanoparticles 2014:963961. https://doi.org/10.1155/2014/963961

Prasad R, Swamy VS (2013) Antibacterial activity of silver nanoparticles synthesized by bark extract of *Syzygium cumini*. J Nanoparticles 2013:431218. https://doi.org/10.1155/2013/431218

Prasad R, Pandey R, Barman I (2016) Engineering tailored nanoparticles with microbes: quovadis? WIREs Nanomed Nanobiotechnol 8:316–330. https://doi.org/10.1002/wnan.1363

Prasad R, Bhattacharyya A, Nguyen QD (2017) Nanotechnology in sustainable agriculture: recent developments, challenges, and perspectives. Front Microbiol 8:1014. https://doi.org/10.3389/fmicb.2017.01014

Prasad R, Jha A, Prasad K (2018) Exploring the Realms of Nature for Nanosynthesis. Springer International Publishing (ISBN 978-3-319-99570-0) https://www.springer.com/978-3-319-99570-0

Rai M, Yadav A, Gade A (2009) Silver nanoparticles as a new generation of antimicrobials. Biotechnol Adv 27(1):76–83. https://doi.org/10.1016/j.biotechadv.2008.09.002

Rajpal K, Aziz N, Prasad R, Varma RG, Varma A (2016) Extremophiles as biofactories of novel antimicrobials and cytotoxics—an assessment of bioactive properties of six fungal species inhabiting ran of Kutch, India. Indian J Sci Technol 9:1–9. https://doi.org/10.17485/ijst/2016/v9i24/89609

Ramalingam B, Parandhaman T, Das SK (2016) Antibacterial effects of biosynthesized silver nanoparticles on surface ultrastructure and nanomechanical properties of Gram-negative bacteria viz. *Escherichia coli* and *Pseudomonas aeruginosa*. ACS Appl Mater Interfaces 8:4963–4976. https://doi.org/10.1021/acsami.6b00161

Raveendran P, Fu J, Wallen SL (2006) A simple and green method for the synthesis of au, ag, and au–ag alloy nanoparticles. Green Chem 8:34–38. https://doi.org/10.1039/B512540E

Reimers JR, Ford MJ, Marcuccio SM et al (2017) Competition of van der Waals and chemical forces on gold-sulfur surfaces and nanoparticles. Nat Rev Chem 1:17. https://doi.org/10.1038/s41570-017-0017

Santos MS, Tavares FW, Biscaia Jr EC (2016) Molecular thermodynamics of micellization: micelle size distributions and geometry transitions. Braz J Chem Eng 33(3):515. https://doi.org/10.1590/0104-6632.20160333s20150129

Saxena A, Tripathi RM, Zafar F et al (2012) Green synthesis of silver nanoparticles using an aqueous solution of F*icus benghalensis* leaf extract and characterization of their antibacterial activity. Mater Lett 67:91–94. https://doi.org/10.1016/j.matlet.2011.09.038

Scholl JA, Koh AL, Dionne JA (2012) Quantum plasmon resonances of individual metallic nanoparticles. Nature 483:421–427. https://doi.org/10.1038/nature10904

Shrivastava S, Bera T, Roy A, Singh G et al (2007) Characterization of enhanced antibacterial effects of novel silver nanoparticles. Nanotechnology 18:225103. https://doi.org/10.1088/0957-4484/18/22/225103

Siegel J, Rezníčková A, Slepička P, Švorčík V (2015) Noble metal nanoparticles prepared by metal sputtering into glycerol and their grafting to the polymer surface. Nanopart Technol 5:31. https://doi.org/10.5772/61403

Singhal U, Khanuja M, Prasad R, Varma A (2017) Impact of synergistic association of ZnO-nanorods and symbiotic fungus *Piriformospora* indica DSM 11827 on *Brassica oleracea* var. botrytis (Broccoli). Front Microbiol 8:1909. https://doi.org/10.3389/fmicb.2017.01909

Sondi I, Salopek-Sondi B (2004) Silver nanoparticles as antimicrobial agent: a case study on *E. coli* as a model for Gram-negative bacteria. J Colloid Interface Sci 275:177–182. https://doi.org/10.1016/j.jcis.2004.02.012

Svobodová A, Psotová J, Walterová D (2003) Natural phenolics in the prevention of UV-induced skin damage. A review. Biomed Papers 147(2):137–145. https://doi.org/10.5507/bp.2003.019

Syafiuddin A, Salmiati S, Salim MR et al (2017) A review of silver nanoparticles researchers trends, global consumption, synthesis, properties, and future challenges. J Chin Chem Soc 64(7):732–756. https://doi.org/10.1002/jccs.201700067

Thakkar KN, Mhatre SS, Parikh RY (2010) Biological synthesis of metallic nanoparticles. Nanomedicine 6:257–262. https://doi.org/10.1016/j.nano.2009.07.002

Toisawa K, Hayashi Y, Takizawa H (2010) Synthesis of highly concentrated ag nanoparticles ina heterogeneous solid-liquid system under ultrasonic irradiation. Mater Trans 51:1764–1768. https://doi.org/10.2320/matertrans.MJ201005

Toma HE, Zamarion VM, Toma SH, Araki K (2018) The coordination chemist at gold nanoparticles. J Braz Chem Soc 21(7):1158. https://doi.org/10.1590/S0103-50532010000700003

Tripathi M, Rana D, Shrivastava A et al (2013) Biogenic synthesis of silver nanoparticles using Saraca indica leaf extract and evaluation of their antibacterial activity. Nano Biomed Eng 5(1):50–56. https://doi.org/10.5101/nbe.v5i1.p50-56

Walter M, Marchesan E (2011) Phenolic compounds and antioxidant activity of rice. Braz Arch Biol Technol 54(2):371. https://doi.org/10.1590/S1516-89132011000200020

Yang H, Wang Y, Chen X et al (2016) Plasmonic twinned silver nanoparticles with molecular precision. Nat Commun 7:12809. https://doi.org/10.1038/ncomms12809

Zhang F, Liu ZG, Shen W, Gurunathan S (2016) Silver nanoparticles: synthesis, characterization, properties, applications, and therapeutic approaches. Int J Mol Sci 17(9):1534. https://doi.org/10.3390/ijms17091534

Chapter 9
Polymer Macromolecules to Polymeric Nanostructures: Efficient Antibacterial Candidates

J. Lakshmipraba, Rupesh N. Prabhu, and V. Sivasankar

Contents

J. Lakshmipraba (✉) · R. N. Prabhu
Post Graduate and Research Department of Chemistry, Bishop Heber College,
Tiruchirappalli, Tamil Nadu, India

V. Sivasankar
Department of Civil Engineering, School of Engineering, Nagasaki University,
Nagasaki, Japan

Post Graduate and Research Department of Chemistry, Pachaiyappa's College,
Chennai, Tamil Nadu, India

© Springer Nature Switzerland AG 2020
R. Prasad et al. (eds.), *Nanostructures for Antimicrobial and Antibiofilm Applications*, Nanotechnology in the Life Sciences,
https://doi.org/10.1007/978-3-030-40337-9_9

Abstract Microbes pose a serious threat to life among human beings due to numerous infectious diseases. Though there is significant progress in the development of effective antimicrobial drugs, many infectious diseases are still difficult to treat. Polymers render a potential antimicrobial strategy to combat pathogens and gained a considerable attention in the recent past. Polyamine compounds are familiar about their vitality in many biological processes. Conspicuously, polyethyleneimine (PEI) as a polymeric chelating agent draws advantages such as high water solubility, modulation of functional groups, reliable molecular weight and physico-chemical stabilities. As reactive amino groups are abundant, a wide range of chemically modified cations with desirable properties make PEI remarkable. PEI offers an effective antimicrobial property, thanks to the hydrophobicity and positive charge density potentiated by alkylation. N-alkyl-substituted PEI immobilized over various knitted textiles showed evidence of strong bactericidal activity against a variety of airborne bacteria. It has also been realized that the molecular weight of polyethyleneimine and the antimicrobial activity are directly proportional to each other. The structure-activity relationship (SAR) plays a compelling role in comprehending the enhanced antimicrobial activity of linear and branched PEIs. In accordance with the above details, the present chapter focuses on the synthesis, characterization and antimicrobial applications of certain polymers, polymer metal complexes and nanopolymer materials.

Keywords Polyethyleneimine (PEI) · Antimicrobial property · Nanopolymer ·
Structure-activity relationship

9.1   Introduction about Polymer

A polymer can be a rationalized chemical compound with large molecules made up of copious repeating units. In other words, polymer is a long-chain macromolecule that consists of multiple small repeating units that are linked together to form an array. The small repeating units are known as monomers. In the early 1920s, chemists suggested that most of the molecules have molecular weights of a few thousands, and molecules with higher molecular weight cannot be prepared. Further, it was suggested that macromolecules are aggregates of small molecules. Hermann Staudinger, a German chemist working in organic chemistry, initially questioned this point of view. He studied various compounds existing in nature such as rubber, proteins, cellulose and starch. He proposed that they were macromolecules composed of more than 10,000 atoms. He came up with a polymeric structure for rubber where copious numbers of isoprene units were held together by strong covalent bonds (Staudinger 1920). In 1953, for "his discoveries in the field of macromolecular chemistry", he was awarded the Nobel Prize in Chemistry. The terms "monomer" and "polymer" were derived from Greek words, "mono (one)", "poly (many, numerous)" and "meros (part)".

Polymers are an integral part of the modern world. Many objects that we use in our day-to-day life have polymeric origin. Polymers can be characterized by uneven molecular weight (depending on its source or synthetic strategy or method of extraction). They have low specific gravity and generally resist corrosion. A polymer, based on the nature and number of functional groups present in the backbone, can form one-, two- or three-dimensional networks. Each repeating unit present in the polymer is known as -mer (or the basic unit); where the term "polymer" means a large number of repeating units. These repeating units contain mainly carbon and hydrogen, though sometimes other atoms such as oxygen, nitrogen, sulphur, chlorine, fluorine, silicon or phosphorous can also be present. A long chain is formed when many -mers are chemically attached together. Polymers differ from other molecules for the reason that due to their long-chain nature, the chains are intertwined in solution or in solid state or, for specific macromolecular structures, to become lined up as regular arrays in solid state. These molecular characteristics result in properties similar to those of solids, such as fibre-forming qualities, elasticity, strength and film-forming properties, which are normally not seen in small molecules. Many objects in daily life are composed of polymers; some of which are listed in Table 9.1.

Table 9.1 Polymers and applications

Monomer structure	Polymer structure	Name of the polymer	Applications/uses
$H_2C=CH_2$	$-(CH_2-CH_2)_n-$	Polyethylene	Plastic bags, bottles, toys, electrical insulation
$H_2C=CHCl$	$-(CH_2-CHCl)_n-$	Polyvinyl chloride (PVC)	Pipes, bags for intravenous solutions, tubing, floor coverings, non-food packaging
$F_2C=CF_2$	$-(CF_2-CF_2)_n-$	Polytetrafluoro-ethylene (PTFE)	Non-stick coatings, electrical insulation
$HClC=CHCl$	$-(CHCl-CHCl)_n-$	Polyvinylidene chloride (saran)	Food wrap
$H_2C=CHCN$	$-(CH_2-CHCN)_n-$	Polyacrylonitrile (Orlon or PAN)	Fibres for textiles, carpets, upholstery
$H_2C=CHOCOCH_3$	$-(CH_2-CHOCOCH_3)_n-$	Polyvinyl acetate	Carpenter's glue, paper adhesive
H_2C-CH_2 with N–H	$-(CH_2-CH_2-NH)_n-$	Polyethylenimine (PEI)	Detergents, CO_2 removal in space craft, adhesives, water treatment agents, cosmetics

9.2 Classification of Polymers

The following are some of the basic ways in which polymers can be classified (Young and Lovell 1991; Carreher Jr. 2003; Gupta 2010).

9.2.1 Classification of Polymers Based on Source

Depending on the source of origin, polymers can be grouped into three types.

1. *Natural Polymers*: Natural polymers are the polymeric materials which are found in nature in plants, animals and other natural sources. Some common examples of natural polymers include cellulose and starch (polymers of glucose, which are found in plants), proteins (polymers of amino acids, which are found in humans and animals) or natural rubber (polymer of isoprene, which is obtained from the latex of rubber trees).
2. *Synthetic Polymers*: These polymers can be synthesized in laboratories. They are produced commercially by industries for the day-to-day needs of mankind. Some commonly used polymers in our daily life include polyethylene (used in plastic for packaging), bakelite (used as electrical switches and many other electrical appliances), nylon (used in fishing nets, dress materials, etc.), polyvinyl chloride ([PVC] used in pipes, as insulation on electrical cables), Teflon (used as coating in non-stick pans), etc.
3. *Semi-Synthetic Polymers*: Semi-synthetic polymers are polymers procured from natural sources and then making further modification by physical or chemical treatment in a laboratory before obtaining their final form of commercial and practical importance. Vulcanized rubber, cellulose acetate, etc. belong to this category.

9.2.2 Classification of Polymers Based on Structure

Polymers can be classified into three different types based on their structure as shown in Fig. 9.1.

1. *Linear Polymers*: The monomer units in these are covalently attached together to form long continuous carbon-carbon chains, the remaining two valencies being satisfied by hydrogen or other small atoms or functional groups. They have high densities and high melting points due to greater packing efficiency. One example for linear polymers is polyvinyl chloride (PVC) which is largely used for the manufacture of electric cables, pipes, etc.

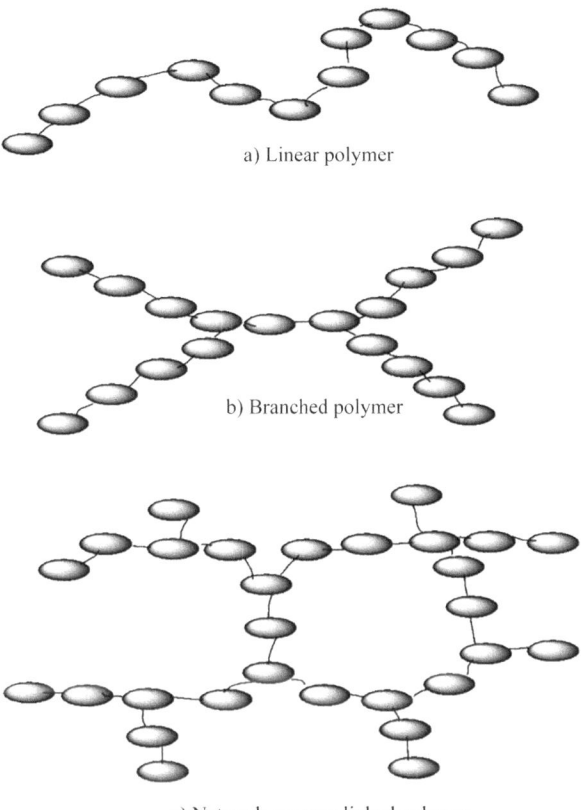

a) Linear polymer

b) Branched polymer

c) Network or cross-linked polymer

Fig. 9.1 Classification of polymers accroding to their structure

2. ***Branched Chain Polymers***: The structure of these polymers can be considered as branches emerging from random points on a single linear backbone. A small chain is generally covalently attached to the main linear backbone. As a result of branching, the polymers are less dense with low melting points. This is because, the polymers are not closely packed. Low-density polyethylene (LDPE) is generally used in manufacture of plastic bags and multipurpose containers. In addition, brush polymers, graft polymers, star-shaped polymers, comb polymers and dendrimers fall into this category.

3. ***Cross-linked or Network Polymers***: Here, the monomer units are linked together, resulting in the formation of a three-dimensional network. The monomers can have multiple functional groups and are connected randomly by covalent bonds between two or more linear chains. They are, however, brittle and hard. Common examples are bakelite (used as electrical insulators), vulcanized rubber (used in tyres), etc.

9.2.3 Classification of Polymers Based on Mode of Polymerization

Based on the manner of polymerization, polymers can be stratified into two as follows.

1. **Addition Polymers**: Addition polymers are obtained by the recurring addition of the monomer units. The monomer units generally possess unsaturated bonds (double bonds or triple bonds). In this type of polymerization, small molecules like water or alcohol, etc., are not expelled (i.e., no by-product is obtained). The empirical formulae of addition polymers are the same as their monomers. They are generally initiated by the presence of free radical (Free-radical addition polymerization), proton, or Lewis acid (cationic addition polymerization), anions such as amide or alkoxide (anionic addition polymerization). Polymerization of ethane to form polyethene or styrene to polystyrene are some examples (Fig. 9.2).
2. **Condensation Polymers**: The condensation reaction of monomers, accompanied by the elimination of small molecules such as water, HCl, ammonia, alcohol, etc., results in the fabrication of condensation polymers. Two or more functional groups are generally present in the monomers. The empirical formulae of condensation polymers and their monomers are different. A common example is the formation of nylon-66 (a polyamide) by the polymerization of adipic acid and hexamethylenediamine along with elimination of water molecules (Fig. 9.3).

9.2.4 Classification of Polymers Based on Molecular Forces

The properties manifested by polymers are a function of the extent of the force of attraction between the polymer molecules. Polymers can be categorized into four, based on the nature of the forces of attraction between them.

1. **Elastomers**: Elastomers have elasticity with high viscosity. They are rubbery-like solid polymeric materials. When a small force is applied, they are easily expandable. When the force is removed, they, generally, recover their original shape. These kinds of polymers are held together by weak intermolecular forces. They are examples of amorphous polymers. The most common examples of these are rubber bands, hair bands and vulcanized rubber.

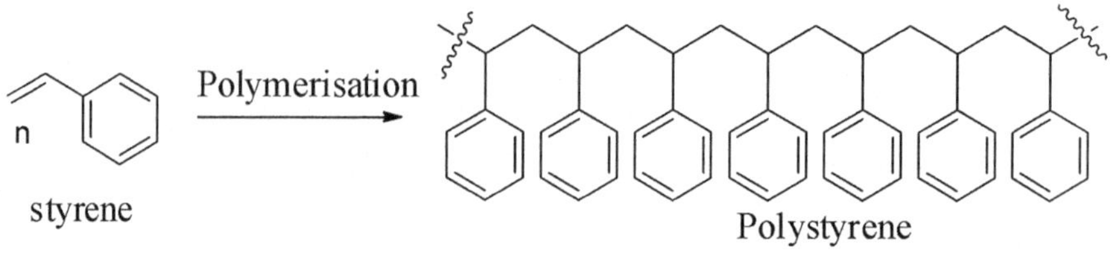

Fig. 9.2 Polymerization of styrene to polystyrene

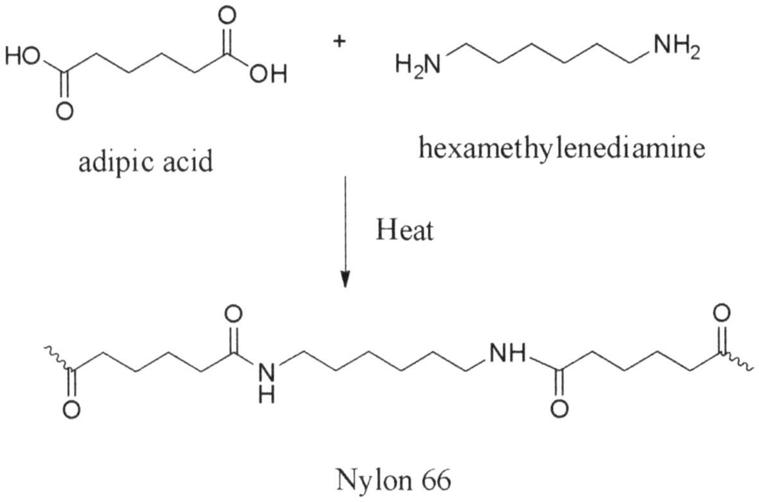

Fig. 9.3 Formation of nylon-66 by the polymerization reaction

2. ***Fibres***: These polymers are long and very thin in diameter. They can be easily entwined in fabric. They resemble a thread. The intermolecular forces between the chains are strong, which may be either dipole-dipole interaction or hydrogen bonds. The chains are thus closely packed and are inelastic with high tensile strength. Fibres are known to have abrupt and high melting points. Fibres can be natural or synthetic. Common examples are nylon-66, silk and dacron.
3. ***Thermoplastics***: Thermoplastics are long-chain polymers. The chains in the polymer are usually held together by van der Waal's forces. On the application of heat, these polymers become soft and upon cooling, they form a hard mass. They can be reheated, reshaped and frozen repeatedly and are thus mechanically recyclable. They do not possess any cross-linking bonds. Common examples are polyethylene, polystyrene, polycarbonate or polypropylene.
4. ***Thermosetting Plastics***: Thermosetting plastics are polymers with low molecular masses. Upon heating, cross-links are formed between the polymer chains; thereby, the polymers become hard and infusible. This reaction is irreversible, and once set, it is not possible to modify their shape. They are a three-dimensional network. They are resistant to consecutive mechanical deformation or softening due to heat or attack of solvents. Example: bakelite (a phenol-formaldehyde resin).

9.2.5 Classification of Polymers Based on the Type of Monomer Involved in the Structure

A polymer may possess identical or different monomers and can then be classified as homopolymers or copolymers as shown in Fig. 9.4 (Singh and Dubey 2009).

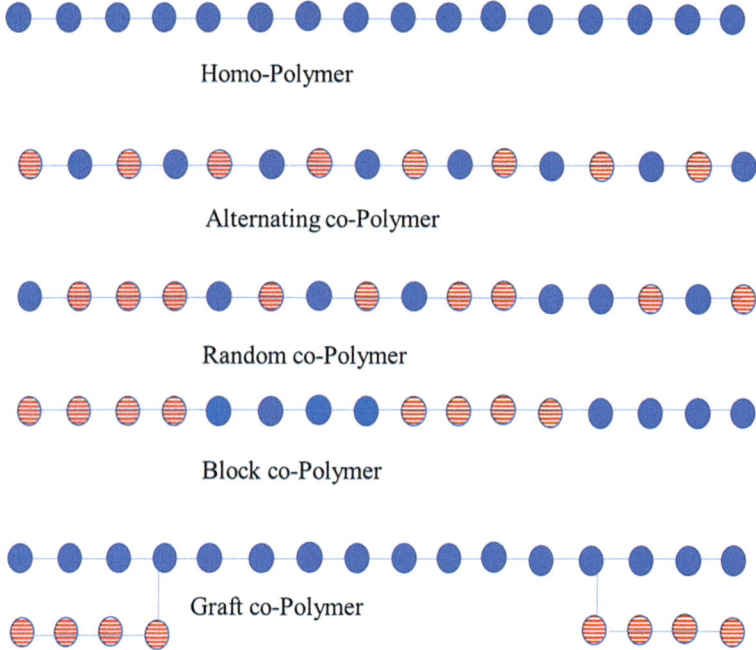

Fig. 9.4 Classification of polymer base on the monomer type

1. ***Homopolymers***: When identical monomer is repeated throughout the chain of the polymer, it is known as homopolymer. Or in other words, they have same monomeric units. Polyvinyl chloride, polyethylene, polystyrene and are some common examples.
2. ***Copolymers***: When monomers forming the polymers are different along the entire chain, it is known as copolymer. Copolymers resulted by the polymerization of two, three and four different monomeric species are also called bipolymers, terpolymers and quaterpolymers, respectively. Copolymers can be further divided into multiple types:

   (a) ***Alternating Copolymers***: In this type of copolymers, two monomeric units are arranged in an alternating fashion. The two monomers combine in 1:1 ratio. An example is copolymer of is nylon-66, which is synthesized by the polymerization of adipic acid and hexamethylenediamine.
   (b) ***Random Copolymers***: In this, the two monomers combine with each other in any order. The ratio of the monomers incorporated into the copolymer depends on variety of factors such as the properties of the individual monomers, the conditions in which the polymerization takes place, the conversion of the polymerization, etc.
   (c) ***Block Copolymers***: In this type of copolymer, all of similar types of monomer are assimilated in one part of the chain, and then all of the other is reacted in somehow. A block copolymer can be considered to be two homopolymers fused together at one of the ends.

(d) **Graft Copolymers**: These polymers are obtained, when a polymeric chain of one type of monomer gets grafted onto a polymeric chain of another monomer.

9.2.6 Classification of Polymers Based on Tacticity

Tacticity may be defined as the geometrical arrangement (orientation) of the pendant groups of the monomer units with respect to the polymeric backbone (main chain). In other words, it is the relative stereochemistry of adjacent chiral carbon atoms within a polymer. On this basis, polymers may be classified into three types as shown in Fig. 9.5.

1. **Isotactic Polymers**: In isotactic polymers, all the pendant groups or substituents are situated on the same side of the polymeric backbone. Isotactic polymers are generally semi-crystalline and often form a helical configuration. Polypropylene formed by Ziegler-Natta catalysis is an example of isotactic polymer.
2. **Syndiotactic Polymers**: In syndiotactic polymers, the substituents have alternate positions along the polymeric backbone. Syndiotactic polystyrene, which is prepared by metallocene catalysis polymerization, is crystalline with a melting point of 161 °C. Gutta percha, which is chemically same as natural rubber, is also an example for syndiotactic polymer.
3. **Atactic Polymers**: In atactic polymers, the substituent groups are randomly positioned along the polymeric backbone. Atactic polymers are usually amorphous, due to the random nature. Polyvinyl chloride, a polymer that is formed by free-radical mechanism, is usually atactic.

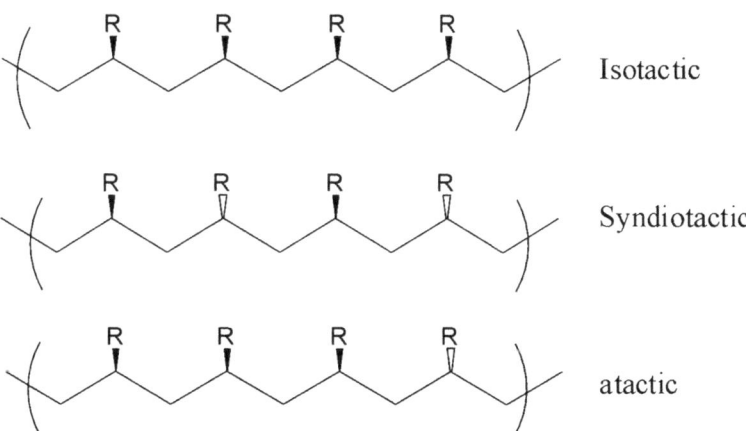

Fig. 9.5 Classification of polymer based on geometrical arragement

9.2.7 Biopolymers

Biopolymers are polymers that occur within living organisms. Some biopolymers can also be chemically synthesized these days from biological materials. Biological macromolecules or biopolymers can be grouped into four. They are lipids, carbohydrates, nucleic acids and proteins. These polymers are made up of different monomeric units and serve different functions.

1. *Lipids*: They are biomolecules that do not interact appreciably with water molecules, that is, they are insoluble in water. They can be further classified as waxes, fats, phospholipids, saccharolipids and sterol lipids. Fatty acids are lipid monomers with a carboxyl group attached to one end of a long-chain hydrocarbon. Fatty acids form complex polymers such as triglycerides, phospholipids and waxes. Steroids or sterol lipids are made up of four fused carbon ring-like structures. They behave as hormones and signalling molecules. In saccharolipids, the carboxylate group of the fatty acids are attached to the sugar backbone. Lipids form cell membranes and protect the organs. Further, they insulate the body and help to store energy.

2. *Carbohydrates*: They are biopolymers that are consisting of sugar monomers. They are prime requisite for the storage of energy. Carbohydrates are also known as saccharides or sugars and, their monomeric units are known as monosaccharides. Carbohydrates are obtained by the condensation polymerization of same or different monosaccharide units. The monomeric units are held together by glycosidic bonds. Some carbohydrates are linear (as in amylase) whereas some can be branched (as in amylopectin). Glucose is an important monosaccharide. It acts as the energy source during cellular respiration by breaking down into smaller molecules. Cellulose and starch are other examples of polysaccharide made up of glucose monomers.

3. *Nucleic Acids*: They contain polynucleotide chain made up of nucleotide monomers. Deoxyribonucleic acid (DNA) and ribonucleic acid (RNA) are examples. For the protein synthesis, nucleic acids contain instructions. They are carriers of genetic information from one generation to the next.

4. *Proteins*: They are large biomolecules which form complex structures. Amino acid monomers are fused together to form proteins. They play a vital role in many biological functions such as muscle movement, transportation of molecules, etc. Haemoglobin, collagen, antibodies and enzymes are examples.

9.2.8 Inorganic Polymers

There are four main classifications of inorganic polymers as discussed below (Archer 2001; Rahimi 2004).

1. *Wholly Inorganic Polymers*: They are the major components present in soil, sand and mountains. They find applications as abrasives, fibres, coating materi-

als, cutting materials, building and construction materials, flame retardants, lubricants, catalysts, etc. Common examples are carborundum (silicon carbide), fibrous glass, Portland cement, boron nitride, silica, antimony(III) oxide, poly(sulphur nitride), etc.

2. *Inorganic-Organic Polymers*: They contain organic moieties attached to inorganic units in the chain. The area of this class of polymers is voluminous, the most important being polysilanes, poly(carbosilane), polysiloxanes, polyphosphazenes, polycarbophosphazenes, etc. They generally possess strong electronic absorption, high conductivity, photoconductivity, photosensitivity and find applications in semiconductors, non-linear optical materials, electro-optical materials, carrier species for biologically active agents, chemotherapeutic models, etc.

3. *Organometallic Polymers*: The organometallic polymers are made of more than 40 elements consisting of main group of metals (such as Si or Ge), transition metals or rare earth elements along with the elements (C, H, N, O, B, P, halides) which are found in organic polymers. The variations of organometallic polymers appear to be endless. These new types of materials have salient features of both organic and inorganic materials such as low density, different structural and functional properties, high electrical conductivity and high temperature stability. Metals may be pendent or may be present in the main chain.

4. *Hybrid Organic-Inorganic Polymers*: Countless different combinations of the organic and inorganic moieties are possible, and hence, a library of hybrid organic-inorganic polymers of fascinating properties can be synthesized by the incorporation of appropriate building blocks. They are multifunctional with a wide range of interesting properties and applications. Some of these polymers display enhanced mechanical strength, resistance to abrasion, improved weathering and outstanding corrosion resistance, excellent adhesion on metal surfaces, reduced effect on health and environment, etc. These new materials are also sought for improved optical properties, electrical properties, luminescence, ionic conductivity, as well as chemical or biochemical activity. Some examples are epoxy polysiloxanes, perfluorosulfonic acid polymers, polyaryletherketones, polyarylethersulfones, sulfonated polyimides, polybenzimidazoles, etc.

9.2.9 Polymer-Anchored Metal Complexes

The polymer-anchored metal complexes are of immense interest in industry and academia due to their extensive and interesting applications. This technique of immobilization on an inert support has ground much care due to their simple separation from the reaction mixture, resulting in ease of handling, operational flexibility, greater thermal stability, higher efficiency, better product selectivity, and economy in various industrial processes. Transition metal complexes on polymer support have a wide range of applications in organic synthesis, as curing agent for epoxy resin, as ion exchanger, as catalyst, as oxidizing agents, as reducing agents, as photosensitizers, as agriculturally active reagents, as pharmacologically active reagents, etc.

9.3   Analysis of Polymers

Average molecular weight concept, number average concept (determined by cryoscopy, ebulliometry and osmometry methods), weight average concept (measured by light scattering and ultracentrifuge techniques), viscosity average concept (determined using viscometer), etc., are the various ways molecular weights of polymers can be expressed.

   Various techniques can be used for the chemical analysis of polymers. Some of them are listed: nuclear magnetic resonance spectroscopy, elemental analysis, powder x-ray diffraction studies, infrared spectroscopy, mass spectrometry, gas chromatography, electron magnetic resonance spectroscopy, light microscopy, scanning electron microscopy, differential calorimetric analysis, differential thermal analysis, thermogravimetric analysis, etc. The important test methods for measuring the physical properties of polymers are as follows: tensile strength, fatigue, brittleness, ductility, hardness, tear resistance, abrasion resistance, etc.

9.4   Factors Affecting Activity of Antimicrobial Polymers

Antimicrobial activity of polymers is generally reflected as function of counterbalance between multiple factors. Various factors such as molecular weight, hydrophilicity, pH, charge density, temperature, presence of moisture, alkyl chain length, etc., contribute to the antimicrobial activity. Some of the notable factors which affect the antimicrobial activity are portrayed below.

9.4.1   Molecular Weight (MW) Versus Alkyl Chain Length of Polymer

Molecular weight (MW) plays a role in regulating the physical as well as chemical properties of polymers. For polyacrylates and polymethyl acrylates comprising biguanide as sidechain, molecular weight is the deciding factor in governing the bactericidal action. The gilt-edge range reported in the literature is in between 5×10^4 and 1.2×10^5 Da; significant decrease in activity arises beyond the range of the above molecular weight (Palermo and Kuroda 2009).

   In the same way, poly(tributyl-4-vinylbenzyl phosphonium chloride) also exhibits the most favourable antimicrobial activity in the range of 1.6×10^4–9.4×10^4 Da (Epand et al. 2004). Further, the role of molecular weight is paradoxical for chitosan, and different reports findings are reported from different research groups. Gram-negative bacteria endured larger challenge than Gram-positive bacteria (Svenson and Tomalia 2005; Lichter and Rubner 2009). Meanwhile, ε-PL, with varying length of the alkyl chain, displayed compelling effects on activity. Polymers of L-lysine residues possessing the chain length of nine displayed excellent inhibition of microbial growth (Park et al. 2004).

9.4.2 Charge on the Polymers

When the positively charged polymers are used, they interact with the bacterial cells in a better way via electrostatic interaction. In chitosan system, as the degree of deacetylation increases, the charge density also increases. Further, this will increase the electrostatic interaction and consequently enhance the antimicrobial activity (Takahashi et al. 2013). A better antibacterial inhibition towards *Staphylococcus aureus* at larger degree of deacetylation by chitosan system was also described (Porter et al. 2000). Furthermore, the modification in chitosan structures like guanidinylated chitosan (Schmitt et al. 2006) and asparagine N-conjugated chitosan oligosaccharide (Schmitt et al. 2004) produced very high antimicrobial activity. But, in the case of N-carboxyethyl chitosan, there is no amino group. So, this compound fails to show any antimicrobial activity (Tossi et al. 2000).

9.4.3 Hydrophilic Interaction of Polymers

One of the most indispensable prerequisites for the compound to demonstrate antimicrobial activity is hydrophilicity. By changing the MW and hydrophobic groups in the amphiphilic polymethacrylate derivatives, the antimicrobial activities were enhanced (Venkataraman et al. 2010). In addition, derivatization of chitosan such as saccharization, acylation, alkylation, metallization and quaternization showed high water solubility and also higher antimicrobial activity than the underivatized original form (Mowery et al. 2007; Dohm et al. 2010).

9.4.4 Role of Counter Ions

In the case of quaternary ammonium/phosphonium compounds, counter ion effect played a crucial role. Due to the strong interaction affinity of counter ions with quaternary compounds, less antimicrobial activity was sensed for these, resulting in sluggish release of free ions in the reaction medium. When compared to free quaternary ammonium compounds, higher antimicrobial activity was observed for quaternary ammonium compounds having chloride and bromide as counter anions (Zhang et al. 1999).

9.4.5 Effect of pH

In the case of chitosan and other polymers having amphoteric nature, a variation in activity with the change in pH has been reported. The antimicrobial activity was controlled by pH for chitosan. Due to increased water solubility and formation of polycations at acidic pH, the activity was maximum in this condition. Yet, at basic pH, there are no reports on antimicrobial activity (Epand et al. 2008).

9.5   Antibacterial Activity

Table 9.2 lists the microorganisms used in this chapter. The terminology used in this chapter consists of: minimum inhibitory concentration (MIC), which is the lowest concentration of the compound or drug that prevents visible microorganism growth after overnight incubation; minimal bactericidal concentration (MBC), which is the lowest concentration of an antibacterial compound/drug mandatory to kill a bacterium over a fixed, somewhat extended period, under a specific set of conditions; and IC_{50} which is the half-maximal inhibitory concentration.

Table 9.2 Microorganisms described in the text

Gram-negative bacteria	*Bacteroides forsythus, Chlamydia pneumoniae, Cellulophaga lítica, Chlamydia trachomatis, Escherichia coli, Enterobacter aeruginosa, Haemophilus influenzae, Klebsiella pneumoniae, Mycoplasma gallisepticum, Neisseria gonorrhoeae, Pseudomonas fluorescens, Proteus vulgaris, Pseudomonas putida, Porphyromonas gingivalis, Pseudomonas aeruginosa, Pseudoalteromonas haloplanktis, Proteus mirabilis, Serratia liquefaciens, Stenotrophomonas maltophilia, Salmonella typhi, Spiroplasma citri, Salmonella typhimurium, Spiroplasma floricola, Salmonella enteritidis, Shigella boydii, Shigella dysenteriae, Shigella sonnei, Spiroplasma melliferum, Yersinia pseudotuberculosis, Yersinia enterocolitica*
Gram-positive bacteria	*Actinomyces viscosus, Acholeplasma laidlawii, Bacillus subtilis, Bacillus cereus, Bifidobacterium bifidum, Brochothrix thermosphacta, Bacillus macroides, Bacillus coagulans, Bacillus megaterium, Bacillus thuringiensis, Bifidobacterium breve, Clostridium difficile, Enterococcus faecium, Enterococcus faecalis, enterococcus hirae, lactobacillus salivarius, Listeria monocytogenes, lactobacillus casei, Mycobacterium tuberculosis, mycobacterium smegmatis, Micrococcus luteus, Pediococcus pentosaceus, Streptococcus pneumonia, Staphylococcus hominis, Streptococcus pyogenes, Staphylococcus epidermidis, Streptococcus mutans, Staphylococcus aureus, Staphylococcus haemolyticus, Staphylococcus citreus, Staphylococcus saprophyticus*
Fungi	*Alternaria alternata, Aspergillus terreus, Aspergillus fumigatus, Aspergillus Niger, Aspergillus flavus, Botrytis cinerea, Byssochlamys fulva, Cladosporium cladosporioides, Chaetosphaeridiaceae globosum, Eurotium tonophilum, Fusarium oxysporum, Fusarium moniliforme, Mucor circinelloides, Microsporum gypseum, Pyrobaculum islandicum, Penicillium funiculosum, Penicillium digitatum, Penicillium citrinum, Penicillium pinophilum, Rhizopus oryzae, Rhizoctonia bataticola, Rhizopus stolonifer, Stachybotrys chartarum, Sporotrichum pulverulentum, Trichophyton mentagrophytes, Trichoderma virens, Trichoderma lignorum, Trichoderma viridis, Trichophyton rubrum*
Yeasts	*Aureobasidium pullulans, Candida glabrata, Candida albicans, Cryptococcus neoformans, Candida tropicalis, Candida parapsilosis, Candiada utilis, Debaryomyces hansenii, Hanseniaspora guilliermondii, Kluyveromyces fragilis or marxianus, Pichia stipitis, Pichia jadinii, Rhodotorula rubra, Saccharomyces cerevisiae*
Algae	*Amphora coffeaeformis, Dunaliella tertiolecta, Navicula incerta*
Viruses	*Herpes simplex, Human Immunology, Influenza A, Simian 40, Varicella zoster*

9.5.1   Polyethyleneimines

A variety of polymers such as polyamines, polyamino acids, polystyrene, polyvinyl chloride, etc., can be used to anchor metal complexes. This chapter focuses only on polyethyleneimine (PEI) as the polymeric backbone, as it is beyond the scope of this chapter to cover all types of polymer backbones.

Polyethyleneimines (PEIs) are synthetic, non-biodegradable, cationic molecules. The polymer molecule contains amine function as repeating units between two aliphatic carbon chains. A branched polyethyleneimine (BPEI), on the other hand, contains all kinds of 1°, 2°, and 3° amino functional groups in its structure. But, in the case of linear polyethyleneimine (LPEI), only 1° and 2° amine groups are present in its structure. Ring-opening polymerization of aziridine route was generally adopted for the synthesis of BPEI. When the reaction condition is altered, different degrees of branching can be achieved (Fig. 9.6).

LPEI synthesis was made effective by ring-opening polymerization reaction of 2-ethyl-2-oxazoline and followed by hydrolysis. LPEI has a melting point ~75 °C and is solid at room temperature. It is soluble in hot water at low pH. They are readily soluble in organic solvents. Irrespective of the molecular weight, BPEI is liquid. PEI have enormous range of applications in diverse fields, because of their polycationic character. Some of applications of PEI are discussed here.

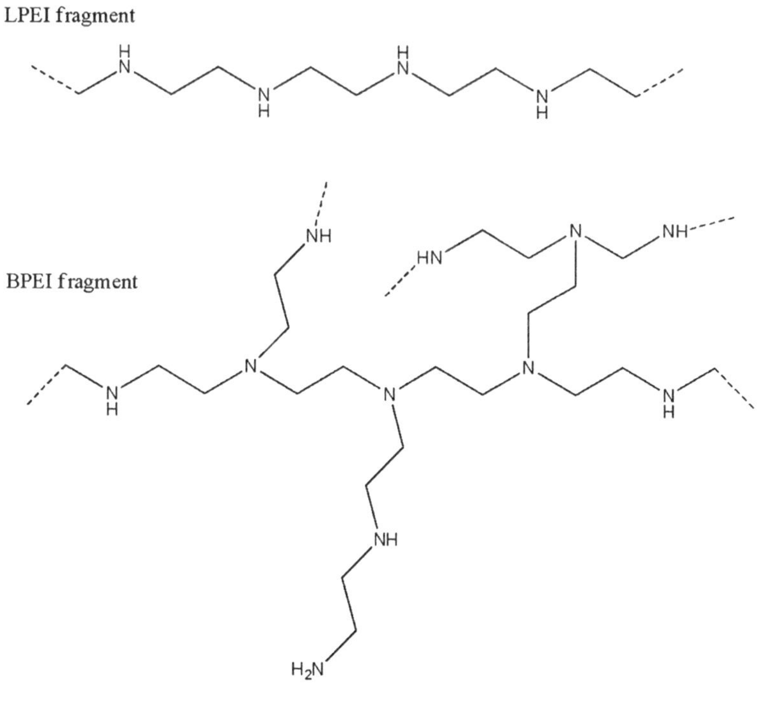

Fig. 9.6 Structure of PEI

PEI has a large number of reactive amino groups. To attain enticing desirable physicochemical properties, modification in PEI is generally carried out. First, for the unsubstituted PEI, antimicrobial activity was scrutinized by the covalent interaction of PEI on the surface of glass material. But, in this case, when compared with an untreated surface of glass, no reduction in microbial count was seen. The primary significant requirements of antimicrobial activities are hydrophobicity and positive charge density. Incorporation of alkyl groups of varying lengths improves both these effects. Many efforts were done towards the incorporation of N-alkyl-PEI to numerous macroscopic and nano-scaled particles, commercial plastics, textiles, glass, different natural and synthetic materials, organic and inorganic compounds and monolithic and porous surface materials. All these above-mentioned derivatives stemmed in near 100% refrainment of air- and water-borne microbes, together with antibiotic- and pathogenic-resistant strains without any report of emergence of resistance. The foremost criterion reported for the mechanism of antibacterial activity was cell membrane rupture. For the mammalian cells (monkey kidney), these surfaces are nontoxic. A significant bacterial activity against several air-borne bacteria was exhibited in the case of N-alkylated PEI immobilized over a wide range of woven textiles such as wool, cotton and polyester. There is a compelling correlation between MW of the PEI and its activity. When MW of the PEI is high, excellent antimicrobial activity is generally observed, whereas negligible activity is observed for PEI polymers having low molecular weight. Substituted PEIs were also employed as contrary to *Candida albicans*. Dimethylaminoethyl methacrylate anchored to PEI produces a reduction in bacterial growth of up to 92% in the prosthetics used by laryngectomized patients. Substituted PEIs are encouraging coating materials of various other medical accessories.

The minimum inhibitory concentration (MIC) was analysed for the PEI to understand the antimicrobial activities. It is the least strength of PEI needed to entirely suppress the bacterial growth under standard experimental conditions. Both LPEI and BPEI are having different water solubilities, and hence, various procedures were followed for the preparation of solutions required for the antibacterial assay. BPEIs are soluble in hot water; LPEIs have poor water solubility even at low concentrations. But, they are willingly soluble in ethanol. Hence, in order to prepare the assay stock solution with a variety of polymer concentrations, LPEI was dissolved in ethanol initially, and then, consecutively, twofold of 0.01% acetic acid was added. The BPEIs are highly soluble in water, and the bulk solutions were thus made in tris-buffered saline. Bacteria solution was taken in Mueller-Hinton broth, and then, this was treated with the bulk solutions in order to figure out the MIC. Five percent ethanol was the highest concentration of alcohol in the solution to be estimated. Further, the control experiments using different solvents did not show any notable variation in the bacterial upsurge even after incubating for 18 h, which was evaluated by turbidity of the solution as well as OD_{600}.

Various BPEIs were examined; it was inferred that B-PEI$_{1.1}$ showed the minimum MIC value of 250 mg mL^{-1} against *Escherichia coli*, whereas BPEI$_{12}$ did not show any significant activity (MIC > 1000 mg mL^{-1}). These results indicated that increasing molecular weight does not increase the antibacterial activity. For comparison, under the same assay conditions, natural antimicrobial peptide magainin 2 (MW = 2300) showed MIC of 125 mg mL^{-1}. In addition, from the titration results, it was registered that BPEI$_{12}$ had 68 mol% of positively charged ammonium groups, whereas other BPEIs had greater content of positively charged ammonium groups. The results of reversed-phase high-performance liquid chromatography suggested that the intrinsic hydrophobicity of BPEI$_{1.1}$ was lower than that of BPEI$_{12}$. Though it was described earlier that in the case of cationic amphiphilic PEIs (Pasquier et al. 2008) and poly-(propylene imine) dendrimers (Chen et al. 2000; Chen and Cooper 2002) increasing the hydrophobicity increases their antimicrobial activity, PEI$_{12}$ exhibited a curtailed activity against E. coli when compared with remaining PEIs. MIC value of LPEIs was found to be 31 mg mL^{-1} (E. coli), which was most likely due to the similar molecular weight (MW = 4400 and 6500). Additionally, most active BPEI$_{1.1}$ displayed MIC value eight times higher than that of the LPEIs. The observation suggested that the LPEIs inhibit the growth of E. coli to a greater extent. BPEIs quaternized with long alkyl groups were reported to have reduced antibacterial activity due to excess hydrophobicity. This could be due to the formation of aggregates, which results in the lowering of the number of polymer chains available for interaction with the cell membranes of the bacteria (Pasquier et al. 2008). Based on the results of dynamic light scattering studies, it was inferred that unmodified BPEI and LPEI did not show any light scattering in MH broth assay solution or in the phosphate buffer solution. This implies that, under the studied experimental conditions, they did not form any aggregates that can reduce the interaction with the bacterial cell membranes, thereby decreasing the efficiency.

The hydrophilic nature of unmodified BPEIs was found to be effective to magnify the number of active polymer chains against bacterial cells. The MICs of BPEIs were found to be 16–31 mg mL^{-1}, when the activity was screed against the Grampositive bacteria S. aureus; however, under similar conditions, magainin-2 showed no potent activity (MIC > 250 mg mL^{-1}). As inferred from the results, the MIC values obtained were considerably lower than those for E. coli. BPEI$_{12}$ did not produce any activity against E. coli (MIC > 1000 mg mL^{-1}); however, S. aureus showed MIC value of 16 mg mL^{-1}, yielding a MIC selectivity index (i.e., the ratio of MIC E. coli to that of MIC S. aureus) of greater than 64. Other BPEIs were also found to exhibit the MIC selectivity index higher than 8. From these results, it was understood that the BPEIs were more selective active against S. aureus than E. coli. Similarly, both LPEIs showcased an MIC value of 8 mg mL^{-1} for S. aureus, which was fourfold lesser when compared to that of E. coli. Thus, the LBEIs showcased greater selectivity towards S. aureus when compared to that of E. coli (having MIC selective index of 4). It was further described that antibacterial activity can be altered by modifying the counter anions used for ammonium groups, although the mechanism and mode of interaction is still unclear (Chen et al 2000; Kanazawa et al. 1993; Lienkamp et al. 2009). So it was believed that negatively charged ions, if present, may influence the antibacterial activity of PEIs.

9.5.2 Hyperbranched and Dendritic Polymers

Highest positive charge density on BPEI plays a vital role in its antimicrobial activity. Modification of substituents in the amine groups to change the hydrophobic interaction in the BPEI polymeric backbone was found to reinforce the antimicrobial activity. Water solubility is generally preserved in the case of quaternized BPEI, but they have competence to absorb the bacterial membrane also. As such, numerous reports are there to improvise the degree of quaternization and the substituents on the alkyl chain in the ammonium groups. By using tertiary amination reaction protocol, quaternized BPEI was synthesized (Gao et al. 2007). *E. coli* bacterium was used to the study the antibacterial properties of these quaternized BPEI by colony counting method and by using enzyme activity method. The effect of pH on the biocidal activity was also evaluated. The outcome of the study revealed magnificent results due to the presence of cationic groups that influence the activity of the polymer. For the *E. coli* protein, the isoelectric point is 4.5, signalling a minimum response in antibacterial activity with change in pH. In the case of quaternized BPEI, for pH more than this isoelectric point value, increase in the antimicrobial activity was described and reaching a constant value in weakly acidic to basic pH.

Sterilization is the key factor that results in the activity of polymer. This was confirmed by enzyme activity studies. Small quaternary ammonium salts are used to imitate this work. Cell death in polymer was as a result of eruption of cell membranes. The contents between the cells are thus released, similar to that reported in the case of small molecular quaternary ammonium compounds. BPEI was further prepared by functionalization of the 1° amine groups with tetrasubstituted ammonium groups, varying the length of the alkyl chains and introduction of allylic and benzylic groups (Pasquier et al. 2007, 2008). This strategy fine-tunes the length of the hydrophobic groups and modulates the hydrophilic to hydrophobic balance in the BPEIs and also the molecular weight of the functionalized BPEIs. Their antimicrobial activity was then evaluated against *Bacillus subtilis* and *E. coli*. The MICs of *B. subtilis* in the range of 0.03–0.04 mg mL^{-1} were reported, whereas the MICs for *E. coli* were ten times higher. It was inferred that, in this case, the activity decreases with increasing molecular weight. Moreover, increase in the ratio of hydrophobic moiety when compared to the cationic groups increases the activity (Pasquier et al. 2008).

9.5.3 Polymer-Metal Complexes

Arunachalam and co-workers reported a variety of polymer metal complexes containing copper(II), in which the percentage of Cu(II) in the polymer was varied. Complexes of the formulation [Cu(phen)(L-tyr)BPEI]ClO$_4$, [Cu(phen)(L-thr) (BPEI)]ClO$_4$·2H$_2$O,   [Cu(phen)(L-arg)BPEI]Cl,   [Cu(phen)(L-phe)(BPEI)] ClO$_4$·4H$_2$O, and [Cu(phen)$_2$(BPEI)]Cl$_2$·4H$_2$O, (where L-Thr = L-theronine, L-arg = L-arginine, L-tyr = L-tyrosine, L-phe = L-phenylalanine, phen = 1,10-phenanthroline) were synthesized (Kumar and Arunachalam 2007, 2009; Kumar et al. 2008; Lakshmipraba et al. 2013, 2017). Disc diffusion method was employed against various bacteria and fungi (*C. albicans, Pseudomonas aeruginosa, E. coli, B. subtilis, S. aureus*) to evaluate the antimicrobial activity, and the results were compared with control drugs such as clotrimazole and ciprofloxacin. The studies implied that the antimicrobial activity of the polymer metal complexes was higher in the case of bacteria. Further, the antimicrobial studies and cytotoxicity studies were performed using MTT assays and other in vitro studies. The results revealed that the polymer metal complexes possessing a greater degree of coordination exhibited higher activity and were specific for cell rupturing.

9.5.4 Nanoparticles

Numerous advantages of nanoparticles arise from their small size and unique properties (Prasad et al. 2016). So, many studies were carried out by incorporating metallic nanoparticles (such as titanium oxide, silver, copper) into polymeric material so as to obtain pivotal antibacterial activity. It is an added advantage of nanosized materials. Moreover, antimicrobial activities in many health care areas using polymer-nanoparticles have gained significant attention.

Dental research is one of the areas where, in the last few years, dental restoration materials has attained much attraction due to their clinical survival, which was considered as an important criteria and timely replacement (Beyth et al. 2014a, b). To answer this situation, nanoparticles from cross-linked quaternary PEI incorporated into resin were synthesized and were then optimized for charge, size, thermal stability and antibacterial action (Farah et al. 2013). The above-mentioned nanoparticles were effective against S. *aureus, Staphylococcus epidermidis, P. aeruginosa* and *E. coli*. One percent loading of nanoparticles on resin shows complete inhibition of *E. faecalis* and S. *aureus*, whereas a 2% loading of nanoparticles was required for other bacteria. Nanoparticles on resin offer the advantage of being biocompatible. But, they are not able to diffuse in agar plate (Beyth et al. 2006, 2008).

Antimicrobial action of PEI nanoparticle-containing composite resins was investigated. They removed the formation of bacterial plaque. Further, they protect the surface of composite resin from roughness, which then inhibits the formation of secondary carriers (Beyth et al. 2010). Quaternary ammonium PEI nanoparticles possessing *N*-octyldimethyl residues were reported to have good antibacterial action (Domb et al. 2013). Silica nanoparticles with functionalized quaternary ammonium moieties were also reported by addition of PEI into silica nanoparticle. This was further cross-linked with diiodopentane, followed by alkylation using octyl iodide and quaternization using methyl iodide (S-QAPEI). The size of particles ranges from 2 to 3 μm with the zeta potential of +50–60 mV. They were found to have strong antibacterial activity (Farah et al. 2014; Yudovin-Farber et al. 2010).

9.5.5 Nanopolymers

PEI showed MICs values of 195.31 mg^{-1} for bacteria and 48.83 mg^{-1} for yeast strains. MLCs corresponded to 326 MIC for *S. aureus* ATCC 29213, *S. aureus* (Sa1) and *Acinetobacter baumannii* (Ab1). For *S. epidermidis* ATCC 155, *S. epidermidis* (Se1) and *A. baumannii* ATCC 19606, the minimal lethal concentration (MLC) value was equivalent to 8 × MIC. For *C. albicans*, the MIC and MLC value was reported as 48.83 mg L^{-1}. The MIC of nanoPEI was 1250 mg L^{-1} for *S. epidermidis* (Se1), *C. albicans* ATCC 90028, *S. aureus* ATCC 29213, *C. albicans* (Ca1), S. aureus (Sa1), *S. epidermidis* ATCC 155, and 2500 mg L^{-1} for *A. baumannii* ATCC 19606 and *A. baumannii* (Ab1).

9.5.6 Biofilm Activity

When PEIs of concentrations equivalent to MLC, MIC, 2 × MIC and 0.56 MIC were used, an appreciable decrease in the biofilm metabolic activity was observed. It showed a dose-dependent activity of both clinical and ATCC strains of *A. baumannii* and of *S. aureus* ($P < 0.05$). Also, for *S. epidermidis*, produced inhibition, at all the concentrations, which were not dose-dependent ($P < 0.05$) for the concentration studied, remarkably inhibited the biofilm metabolic activity. However, in the case of yeasts, significant inhibition of the metabolic activity of the biofilm was observed at the concentration of ε48.83 mg^{-1} (MIC, 2 × MIC and 4 × MIC). The lowest concentration of 24.42 mg^{-1} (0.5 × MIC) did not tarnish the metabolic activity of the biofilm. Biofilm metabolic activity of all *S. aureus, A. baumannii* and *S. epidermidis* ($P > 0.05$) strains showed curtailed activity when nanoPEI with the concentration of 0.56 MIC and 0.5 × MIC were used. In the case of *C. albicans*, appreciable metabolic activity of biofilm of both clinical and ATCC strains was observed only at concentration 2 × MIC (2500 mg^{-1}). Thirty-four percent of biofilm inhibition and 57% of ATCC stain inhibition were observed for the clinical yeast. For concentrations equivalent to the MIC and 0.5 × MIC, no inhibition of biofilm was observed. Some Gram-negative bacteria, Gram-positive bacteria and yeast were chosen for the antimicrobial action of PEI and nanoPEI and also effect on biofilm formation on PUR-Catheters. The antimicrobial activity of both PEI and nanoPEI was understood by the analysis of the obtained results. These results were similar for all the strains studied. Surprisingly, PEI with low concentration was found to inhibit yeast growth predominantly when compared to bacterial strains. From the literature reports, it was inferred that strong antibacterial effect was exhibited by nanoPEI (Beyth et al. 2006, 2008). This is due to the strong interaction between the polycationic nature of polymer and negatively charged surfaces of bacteria (Beyth et al. 2008). This interaction may be responsible for the high activity of polymers in cell permeability and rupturing of cell membranes. Beyth et al. (2006) reported substantial antibacterial effect of PEI nanoparticles against *Streptoccocus mutans*. The same antimicrobial properties were maintained over 1 month for the composite resin materials containing PEI nanoparticles. In addition, quaternary ammonium-PEI (QAPEI) nanoparticles were reported to comprehensively inhibit the growth of Gram-positive (*Staphylococcus aureus*) bacteria as well as Gram-negative (*E. coli*) bacteria. QAPEI particle- composite resin produced complete inhibition of growth of *Streptococcus mutans*. The above studies imply that quaternary ammonium moieties attached to polymer play an important role in the inhibition of bacterial growth in vitro. The result further revealed that they can possibility act as additives in medical equipment (Yudovin-Farber et al. 2010). The results were further supported by Beyth et al. (2008) by differential bacterial sensitivity towards nanoPEI. The most effective QAPEI derivative was QAPEI alkylated with octyl halide at 1:1 mole ratio (primary amine of PEI monomer units/alkylating agent), which inhibited $>10^6$ *S. aureus* growth at a concentration of 80 µg mL^{-1}. There was a strict correlation between the activity of the compounds and their overall octyl content. Although several structural parameters influenced the antibacterial potency of the QAPEI nanoparticles, the most important features appear to be the degree of alkylation and degree of methylation, whereas other parameters had less significant impact (Beyth et al. 2008; Yudovin-Farber et al. 2008). Incorporation of QAPEI nanoparticles in bone cements produced durable antibacterial effect without changing the biocompatibility of the cement (Beyth et al. 2014a, b). NanoPEI produced cell membrane disruption and in cell depolarization except in the case of *A. baumannii* strains. When *C. albicans* were used, cell depolarization was detected. Interestingly, in biofilm, covalently bounded quaternized polydimethylaminoethylmethacrylate (poly-DMAEMA) and PEI reduced biofilm activity up to 92%, which was observed for *C. albicans* (De Prijck et al. 2010). The high bacterial activity against both bacteria was also observed when alkylated PEI attached to flat macroscopic surfaces and to surfaces of nanoparticles was used. The overall studies implied that at sub-inhibitory concentration was prominent for all bacteria tested for the anti-biofilm effect of nanoPEI. Further, biofilm metabolic activity was observed only at 2 MIC for the inhibition of *C. albicans*. In the special case, the MLC was not analysed.

9.6   Conclusion

The antimicrobial action of polymer macromolecules was examined by employing different bacteria, fungi and yeast. PEI has high molecular weight and polydispersity. Under these conditions, the interaction of PEI with cell membranes is strenuous to rationalize. But, development of variation of end group and monodisperse molecular weight is shown as a path to improve antimicrobial activity. Incorporation of metal complexes to BPEI generally showed a potential activity towards bacteria. The antimicrobial activity of PEI nanoparticles and resins showed biocompatibility and also good antimicrobial activity. NanoPEI were studied with different bacteria, fungi and yeast strains, and they resulted in excellent MIC values. NanoPEI showed significant metabolic activity, many of which are under clinical trial. Numerous PEI and nano-PEI reported in the literature exhibited interesting activity against bacteria and fungi.

9.7 Future Outlook

The variation of counter anions on in the PEI polymer on the antibacterial activity would be of considerable interest for further investigation. PEI and nanoPEI appear to be promising candidates for the augmentation of novel and indwelling catheters. By using the above materials, construction of novel medical devices that have little susceptibility to bacteria, fungi and yeast infection can be produced on a large scale. In vitro analysis of nanoparticles also creates new scope for the production of effective devices for clinical trials. The in vivo studies of PEI and nanoPEI require animal testing and can be only speculated for the time being. The fascinating antimicrobial results on the surface of polyurethane (PUR) catheters and their derivatives signify the need for probing their possible application in the field of medicine. Further research has to be explored in the field of water-soluble polymer metal complexes in order to enhance the biocompatibility in device formation.

References

Archer RD (2001) Inorganic and organometallic polymers. Wiley-VCH, Inc., New York, USA

Beyth N, Yudovin Farber I, Bahir R, Domb AJ, Weiss EI (2006) Antibacterial activity of dental composites containing quaternary ammonium polyethyleneimine nanoparticles against Streptococcus mutans. Biomaterials 27:3995–4002

Beyth N, Houri Haddad Y, Baraness Hadar L, Yudovin Farber I, Domb AJ, Weiss EI (2008) Surface antimicrobial activity and biocompatibility of incorporated polyethylenimine nanoparticles. Biomaterials 29:4157–4163

Beyth N, Yudovin Farber I, Bahir R, Domb AJ, Weiss EI (2010) Long-term antibacterial surface properties of composite resin incorporating polyethyleneimine nanoparticles. Quintessence Int 41:827–835

Beyth N, Farah S, Domb AJ, Weiss EI (2014a) Antibacterial dental resin composites. React Funct Polym 75:81–88

Beyth S, Polak D, Milgrom C, Weiss EI, Matanis S, Beyth N (2014b) Antibacterial activity of bone cement containing quaternary ammonium polyethyleneimine nanoparticles. J Antimicrob Chemother 69:854–855

Carreher CE Jr (2003) Polymer chemistry, 6th edn. Marcel Dekker, Inc., Florida, USA

Chen CZS, Cooper SL (2002) Interactions between dendrimer biocides and bacterial membranes. Biomaterials 23:3359–3368

Chen CZS, Beck-Tan NC, Dhurjati P, van Dyk TK, LaRossa RA, Cooper SL (2000) Quaternary ammonium functionalized poly(propylene imine) dendrimers as effective antimicrobials: structure-activity studies. Biomacromolecules 1:473–480

De Prijck K, De Smet N, Coenye T, Schacht E, Nelis HJ (2010) Prevention of Candida albicans biofilm formation by covalently bound dimethylaminoethylmethacrylate and polyethylenimine. Mycopathologia 170:213–221

Dohm MT, Mowery BP, Czyzewski AM, Stahl SS, Gellman SH, Barron AE (2010) Biophysical mimicry of lung surfactant protein B by random nylon-3 copolymers. J Am Chem Soc 132:7957–7967

Domb AJ, Beyth N, Farah S (2013) Quaternary ammonium antimicrobial polymers. MRS Online Proc Lib Arch 1569:97–107

Epand RF, Raguse TL, Gellman SH, Epand RM (2004) Antimicrobial 14- helical β-peptides: potent bilayer disrupting agents. Biochemistry (Mosc) 43:9527–9535

Epand RF, Mowery BP, Lee SE, Stahl SS, Lehrer RI, Gellman SH, Epand RM (2008) Dual mechanism of bacterial lethality for a cationic sequence-random copolymer that mimics host-defense antimicrobial peptides. J Mol Biol 379:38–50

Farah S, Khan W, Farber I, Kesler Shvero D, Beyth N, Weiss EI, Domb AJ (2013) Crosslinked QA-PEI nanoparticles: synthesis reproducibility, chemical modifications, and stability study. Polym Adv Technol 24:446–452

Farah S, Aviv O, Laout N, Ratner S, Beyth N, Domb A (2014) Antimicrobial silica particles loaded with quaternary ammonium polyethyleneimine network. J Polym Adv Technol 25:689–692

Gao BJ, Zhang X, Zhu Y (2007) Studies on the preparation and antibacterial properties of quaternized polyethyleneimine. J Biomater Sci Polym Ed 18:531–544

Gupta AL (2010) Polymer chemistry. Revised Edition. Pragati Prakashan Educational Publishers, Meerut, India

Kanazawa A, Ikeda T, Endo T (1993) Polymeric phosphonium salts as a novel class of cationic biocides. II. Effects of counter anion and molecular weight on antibacterial activity of polymeric phosphonium salts. J Polym Sci A Polym Chem 31:1441–1447

Kumar RS, Arunachalam S (2007) DNA binding and antimicrobial studies of some polyethyleneimine–copper(II) complex samples containing 1,10-phenanthroline and l-theronine as co-ligands. Polyhedron 26:3255–3262

Kumar RS, Arunachalam S (2009) DNA binding and antimicrobial studies of polymer–copper(II) complexes containing 1,10-phenanthroline and l-phenylalanine ligands. Eur J Med Chem 44:1878–1883

Kumar RS, Arunachalam S, Periasamy VS, Preethy CP, Riyasdeen A, Akbarsha MA (2008) DNA binding and biological studies of some novel water-soluble polymer–copper(II)–phenanthroline complexes. Eur J Med Chem 43:2082–2091

Lakshmipraba J, Arunachalam S, Riyasdeen A, Dhivya R, Vignesh S, Akbarsha MA, James RA (2013) DNA/RNA binding and anticancer/antimicrobial activities of polymer-copper(II) complexes. Spectrochim Acta A 109:23–31

Lakshmipraba J, Arunachalam S, Gandi DA, Thirunalasundari T, Vignesh S, James RA (2017) Interaction of polymer-anchored copper(II) complexes with tRNA studied by spectroscopy methods and biological activities. Luminescence 32:309–316

Lichter JA, Rubner MF (2009) Polyelectrolyte multilayers with intrinsic antimicrobial functionality: the importance of mobile polycations. Langmuir 25:7686–7694

Lienkamp K, Madkour AE, Kumar KN, Nusslein K, Tew GN (2009) Antimicrobial polymers prepared by ring-opening metathesis polymerization: manipulating antimicrobial properties by organic counterion and charge density variation. Chem Eur J 15:11715–11722

Mowery BP, Lee SE, Kissounko DA, Epand RF, Epand RM, Weisblum B, Stahl SS, Gellman SH (2007) Mimicry of antimicrobial host-defense peptides by random copolymers. J Am Chem Soc 129:15474–15476

Palermo EF, Kuroda K (2009) Chemical structure of cationic groups in amphiphilic polymethacrylates modulates the antimicrobial and hemolytic activities. Biomacromolecules 10:1416–1428

Park ES, Kim HS, Kim MN, Yoon JS (2004) Antibacterial activities of polystyrene-block-poly(4-vinyl pyridine) and poly(styrenerandom-4-vinyl pyridine). Eur Polym J 40:2819–2822

Pasquier N, Keul H, Heine E, Moeller M (2007) From multifunctionalized poly(ethylene imine)s toward antimicrobial coatings. Biomacromolecules 8:2874–2882

Pasquier N, Keul H, Heine E, Moeller M, Angelov B, Linser S, Willumeit R (2008) Amphiphilic branched polymers as antimicrobial agents. Macromol Biosci 8:903–915

Porter EA, Wang X, Lee HS, Weisblum B, Gellman SH (2000) Nonhaemolytic β-amino-acid oligomers. Nature 404:565

Prasad R, Pandey R, Barman I (2016) Engineering tailored nanoparticles with microbes: quo vadis. WIREs Nanomed Nanobiotechnol 8:316–330. https://doi.org/10.1002/wnan.1363

Rahimi A (2004) Inorganic and organometallic polymers: a review. Iran Polym J 13:149–164

Schmitt MA, Weisblum B, Gellman SH (2004) Unexpected relationships between structure and function in α/β-peptides: antimicrobial foldamers with heterogeneous backbones. J Am Chem Soc 126:6848–6849

Schmitt MA, Weisblum B, Gellman SH (2006) Interplay among folding, sequence, and lipophilicity in the antibacterial and hemolytic activities of α/β-peptides. J Am Chem Soc 129:417–428

Singh J, Dubey RC (2009) Organic polymer chemistry. Pragati Prakashan Educational Publishers, Meerut, India

Staudinger H (1920) Über polymerisation. Ber Dtsch Chem Ges 53:1073–1085

Svenson S, Tomalia DA (2005) Dendrimers in biomedical applications reflections on the field. Adv Drug Deliv Rev 57:2106–2129

Takahashi H, Palermo EF, Yasuhara K, Caputo GA, Kuroda K (2013) Macromol Biosci 13:1285–1299

Tossi A, Sandri L, Giangaspero A (2000) Amphipathic α-helical antimicrobial peptides. Biopolymers (Peptide Sci) 55:4–30

Venkataraman S, Zhang Y, Liu LH, Yang YY (2010) Design, syntheses and evaluation of hemocompatible pegylated-antimicrobial polymers with well-controlled molecular structures. Biomaterials 31:1751–1756

Young RJ, Lovell PA (1991) Introduction to polymers, 2nd edn. Chapman & Hall, London, UK

Yudovin-Farber I, Beyth N, Nyska A, Weiss EI, Golenser J, Domb AJ (2008) Surface characterization and biocompatibility of restorative resin containing nanoparticles. Biomacromolecules 9:3044–3050

Yudovin-Farber I, Golenser J, Beyth N, Weiss EI, Domb AJ (2010) Quaternary ammonium polyethyleneimine: antibacterial activity. J Nanomater 2010:1–11

Zhang Y, Jiang J, Chen Y (1999) Synthesis and antimicrobial activity of polymeric guanidine and biguanidine salts. Polymer 40:6189–6198

Chapter 10
Antibiofilm, Antifouling, and Anticorrosive Biomaterials and Nanomaterials for Marine Applications

Mani Jayaprakashvel, Mnif Sami, and Ramesh Subramani

Contents

M. Jayaprakashvel (✉)
Department of Marine Biotechnology, AMET Deemed to be University,
Chennai, Tamil Nadu, India

M. Sami
Laboratory of Molecular and Cellular Screening Processes, Centre of Biotechnology of Sfax,
Sfax, Tunisia

R. Subramani
School of Biological and Chemical Sciences, The University of the South Pacific,
Suva, Republic of Fiji

© Springer Nature Switzerland AG 2020
R. Prasad et al. (eds.), *Nanostructures for Antimicrobial and Antibiofilm Applications*, Nanotechnology in the Life Sciences,
https://doi.org/10.1007/978-3-030-40337-9_10

Abstract Formation of biofilms is one of the most serious problems affecting the integrity of marine structures both onshore and offshore. These biofilms are the key reasons for fouling of marine structures. Biofilm and biofouling cause severe economic loss to the marine industry. It has been estimated that around 10% of fuel is additionally spent when the hull of ship is affected by fouling. However, the prevention and control treatments for biofilms and biofouling of marine structures often involve toxic materials which pose severe threat to the marine environment and are strictly regulated by international maritime conventions. In this context, biomaterials for the treatment of biofilms, fouling, and corrosion of marine structures assume much significance. In recent years, due to the technological advancements, various nanomaterials and nanostructures have revolutionized many of the biological applications including antibiofilm, antifouling, and anticorrosive applications in marine environment. Many of the biomaterials such as furanones and some polypeptides are found to have antibiofilm, antifouling, and anticorrosive potentials. Many of the nanomaterials such as metal (titanium, silver, zinc, copper, etc.) nanoparticles, nanocomposites, bioinspired nanomaterials, and metallic nanotubes were found to exhibit antifouling and anticorrosive applications in marine environment. Both biomaterials and nanomaterials have been used in the control and prevention of biofilms, biofouling, and corrosion in marine structures. In recent years, the biomaterials and nanomaterials were also characterized to have the ability to inhibit bacterial quorum sensing and thereby control biofilm formation, biofouling, and corrosion in marine structures. This chapter would provide an overview of the biomaterials from diverse sources and various category of nanomaterials for their use in antibiofilm, antifouling, and anticorrosion treatments with special reference to marine applications.

Keywords Antibiofilm · Antifouling · Anticorrosion · Biomaterials · Nanomaterials · Marine structures

10.1   Introduction

Bacteria are unicellular microorganisms present in almost all ecosystems in our planet. They usually exist in two different formats such as planktonic forms and biofilms. Planktonic forms are free-living bacteria which exist either individually or collectively as a colony, whereas in biofilms, bacteria live in communities within an extracellular matrix made of exopolysaccharides produced by the same bacteria itself. According to Flemming et al. (2016), bacterial biofilms are the emergent form of bacterial life. The life of biofilms is entirely different from that of planktonic forms. Biofilms are demonstrating various emergent properties such as communal mutualism, capture of space, and nutrition. The biofilms also demonstrate extended survival as a community by exhibiting resistance to antibiotics. By forming biofilms, bacteria evade the environmental challenges such as drying, washing action of water, and air (Roberts et al. 2013).

The biofilms, upon maturation, can cause serious changes in the metallic properties of substrates in which they form microcolonies. Thus, beyond the electrochemical influences on metal corrosion, biofilm-induced corrosion is also found naturally. Moreover, biofilms on solid surfaces in aquatic environments make the attachment of macrofouling organisms easier. This makes all structures in aquatic environment more prone to biofouling. Thus, biofilms can cause serious environmental, medical, industrial, and economic impacts by inciting various diseases, biocorrosion, and biofouling (Videla and Characklis 1992). Three sectors, viz., medical, food industry, and shipping industry, are the most affected by the biofilms. The biofilms in human and animals, both on internal organs and medical implants, cause serious challenge to the medical sector. In various industries, more specifically in food, beverage, and dairy industries, biofilms are of serious concern. In ships, the biofilms on external surfaces, propellers and piping systems, cause serious problems such as lowering fuel efficiency and weakening the strength of metals and also form the foundation for the biofouling. Thus, the economic losses due to biofilms in shipping are many fold higher than any other industry.

Marine environment is a diverse environment wherein different ecosystems are existing. Each of the ecosystems in marine environment is unique. Marine environment has all extremities of temperatures, pH, atmospheric pressure, tidal action, wind, salinity, nutrients, etc. Hence, the organisms living in the marine environment are so dynamic in producing new and potential biomolecules to face all these extremities as adaptation. The centuries-long continued research on the terrestrial environment has yielded millions of metabolites and thousands of bioactivities. However, due to the fast-evolving microbes, especially in the fields of control and treatment of biofilms, biocorrosion, and biofouling, we are in salient need of alternative and new molecules to develop alternative strategies. Since the terrestrial bioresources are having the disadvantage of finding previously reported molecules or bioactivities, we need alternative bioresources for the exploration of new molecules. Marine bioresources, especially, marine microorganisms, are so diverse and hence could offer themselves as potential choice for the exploration of novel biomolecules with potent biological functions. Thus, this chapter would emphasize on the role of marine microorganisms for the production of antibiofilm, antifouling, and anticorrosion metabolites. Besides, the nanotechnology is revolutionizing almost all branches of scientific research. The role of nanotechnology in improving the prevention and control strategies for the successful management of biofilms, biocorrosion, and biofouling is also emphasized in this chapter.

10.2 Antibiofilm Compounds

10.2.1 The Biology of Biofilms

Biofilms are multicellular communities of bacteria colonized together in a self-produced extracellular matrix, usually made of exopolysaccharides (López et al. 2020). Biofilms are architecturally unique arrangement of bacteria to evade environmental stresses. Researchers have concluded that in natural environment, bacteria survive in solid surfaces majority through the formation of biofilms (Costerton and Lewandowski 1995). In an ecological perspective, biofilms are communities of bacteria, which live in an interdependent mutualistic relationship among multiple bacterial genera as well. Their association as a biofilm community is more complex. However, they exhibit coordinated functions to survive as a biofilm on solid surfaces (Davey and O'Toole 2000).

Biofilms largely consist of water (more than 90%) only, like any other living system. Most of the biofilms contain microbial cells, nucleic acids (DNA and RNA), polysaccharides, proteins, and water. Biofilm forming bacteria exhibit different strategies for formation of biofilms. The strategies are different due to difference in environment and species involved. Formation of bacterial biofilms usually comprises four steps (Parsek and Singh 2003; Jamal et al. 2015; Jamal et al. 2017).

10.2.1.1   Bacteria Adhere to Solid Surfaces

The bacterial cells which are planktonic make initial adhesion with solid surfaces. The locomotion appendages of bacteria such as flagella, pili, and fimbriae are helpful in establishing initial cell adhesion with solid surfaces. A solid-liquid interface promotes the biofilm formation. This attachment is found to be reversible

10.2.1.2   Formation of Microcolonies

After successful initial attachment with solid surfaces, bacteria regulate their physiological and metabolic activities to form microcolonies. The microcolonies thus formed may have single species or also have multiple species depending on the nature of environment.

10.2.1.3   Formation of 3D Matrix with Exopolymers

All the biofilms are encompassed inside an extracellular matrix. The extracellular matrix is largely composed of exopolysaccharides. However, proteins and DNA are also present in minor quantities.

10.2.1.4   Maturation of Biofilms

Once bacterial biofilms are protected with extracellular matrix, they form water-filled channels for the distribution of nutrients. These channels act as circulatory systems to transport nutrients in and dispose wastes out. The biofilm forming bacteria are having programmed mechanisms for stopping the exopolysaccharide protection at certain point of time, after which the planktonic bacterial cells are released into the environment.

The different processes involved in the biofilm formation in general are depicted in Fig. 10.1.

10.3   Biofilm Forming Bacteria: Medical Implications

Though many genera of bacteria are capable of producing biofilms, the following few bacterial genera are studied extensively for biofilm formation. These bacteria are also highly significant in view of industrial, medical, and environmental safety.

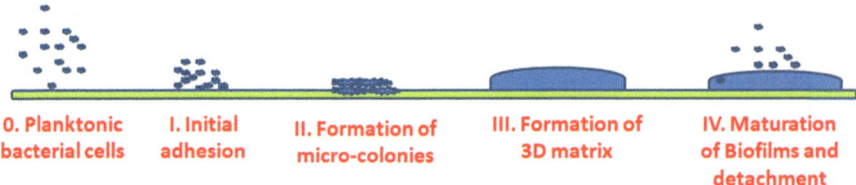

Fig. 10.1 Summary of basic processes involved in microbial biofilm formation

10.3.1 *Escherichia coli*

E. coli is one of the most widely studied facultatively anaerobic bacterium. Most of the strains of the *E. coli* are active colonizers in the gastrointestinal tract of human and animals. Because of their ability to form infectious biofilms in both internal systems and environmental systems, *E. coli* is assumed to be one of the model organisms in biofilm studies (Beloin et al. 2008). *E. coli* was also found associated as a normal flora in gut as a commensal. It was also reported to cause serious infectious biofilms in the intestine. They also form biofilms in medical devises and implants. Because of their high versatility and resistant mechanisms, antibiotic treatment for the control of *E. coli* is still a hurdle.

10.3.2 *Pseudomonas aeruginosa*

Pseudomonas aeruginosa is a Gram-negative ubiquitous bacterium. It is also regarded as most notorious opportunistic pathogen in animals and human. *P. aeruginosa* are reported to form biofilms in internal organs, medical devices, and medical implants. They are also involved in the nosocomial infections. The biofilm forming *P. aeruginosa* are quite difficult to be managed by the immune system and antibiotic therapy (Mulcahy et al. 2014). Biofilm forming *P. aeruginosa* are causing cystic fibrosis, a most dangerous lung biofilm infection in humans (Rasamiravaka et al. 2015). Besides, *Pseudomonas* was also found associated with biofilms of food industry (Amina and Bensoltane 2015).

10.3.3 *Staphylococcus aureus*

Staphylococcus aureus is a Gram-positive nonmotile bacterium yet forms biofilms. Strains of *S. aureus* are known for their biofilms in wounds and medical implants and can cause multiple chronic infections. There are several highly antibiotic-resistant strains reported (Lister and Horswill 2014) Because of their antibiotic resistance and sturdy biofilms, *S. aureus* produce stable and recurrent internal infections with huge mortality rate. The biofilms of *S. aureus* are associated with several infections in many internal organs such as the bones, teeth, gums, sinus, and eyes; chronic wound infections; etc. (Archer et al. 2011).

10.3.4   Streptococcus epidermidis

Streptococcus epidermidis is also found associated with infectious biofilms in humans. They cause biofilms in medical devices also.

10.3.5   Streptococcal Biofilms

Streptococcus is a versatile genera of Gram-positive bacteria which are found to establish biofilms in internal tissues and tooth surfaces of humans (Suntharalingam and Cvitkovitch 2005). *Streptococcus pyogenes* (group A streptococci, GAS) is a serious pathogen which causes biofilm-mediated infections in the skin and mucosal membranes. Several antibiotic-resistant GAS were also reported to cause bacterial pharyngitis in children by the formation of biofilms (Conley et al. 2003; Fiedler et al. 2015).

10.3.6   Enterobacter cloacae

Enterobacter cloacae is a Gram-negative bacterium found in the intestines of humans and animals. They cause severe infections such as nosocomial infections, respiratory tract infection, urinary tract infections, endocarditis, skin and soft tissue infections, osteomyelitis, and ophthalmic infections by forming biofilms. They also cause biofilms in food industries and also cause food spoilage (Jamal et al. 2015; Cai et al. 2018; Zurob et al. 2019).

10.3.7   Klebsiella pneumoniae

Klebsiella pneumoniae is a Gram-negative pathogenic bacteria. It causes biofilms and results in urinary tract infections (UTI), pneumonia, and soft tissue infections. It also causes meningitis (Vuotto et al. 2017).

10.4 Biofilms in Industrial Sector

Bacteria such as *Salmonella* spp., *Listeria* spp., pathogenic *E. coli*, *Campylobacter* spp., *Bacillus cereus*, and *S. aureus* are found to produce biofilms in various facets of the food industry. Since they are also pathogenic, they cause food contamination. They cause huge loss to the food industry especially in the dairy, seafood processing, meat processing, beverages, and ready-to-eat food industries (Laxmi and Bhat 2018; Giaouris and Simoes 2018; Srey et al. 2013). Most of the biofilm forming bacteria are highly resistant to the mechanical washing and chemical disinfection processes followed in food industries (Shi and Zhu 2009a, b). The prevention and control of biofilms in food industries needs careful understanding of the exact steps involved in biofilm formation. The identification of biofilm site is the foremost need to design prevention or control strategy: use of proper sanitation methods, appropriate food preservatives, antibiofilm enzymes, organosilanes, probiotics, phage therapy, extracts from aromatic plants, inhibitors of quorum-sensing system, bacteriocins, and nanomaterials (Kregiel 2014, Meireles et al. 2016; Merino et al. 2019; Lamas et al. 2018). Besides bacteria, yeasts such as *Rhodotorula mucilaginosa*, *Candida krusei*, *Candida kefyr*, *and Candida tropicalis* were also reported to have formed biofilms in beverage industries and are best controlled by the usage of effective disinfectants such as sodium hypochlorite on the ultrafiltration membranes (Tarifa et al. 2018).

Bacteria such as *Pseudomonas*, *Campylobacter*, *Salmonella*, *Enterobacter*, *Proteus*, *Citrobacter*, *E. coli*, *Klebsiella*, *Staphylococcus*, and *Listeria* were found associated with infectious biofilms that cause huge contamination, spoilage, and economic losses to the food (seafood, meat, poultry) processing industries (Wang et al. 2016a, b; Wang et al. 2017a, b, c; Li et al. 2017; Pang et al. 2019; Wang et al. 2019). Biofilm formation in dairy industry is one of the major concern of quality dairy products. Biofilm forming bacteria such as *Bacillus* spp., *Pseudomonas* spp., *Klebsiella* spp., *Listeria* spp., and *Staphylococcus* spp. were reported to form biofilms in different parts of the dairy manufacturing plants (Grutsch et al. 2018; Bremer et al. 2018). While it is envisaged that controlling biofilms through frequent cleaning would cost more and also limit the amount of product that can be manufactured (Bremer et al. 2018), alternate strategies such as use of sanitizers (Tang et al. 2010), phytochemicals (De Oliveira et al. 2018), and biopolymers (Felipe et al. 2019) were attempted for the control of biofilms in dairy industries.

Biofilms are also formed in unusual entities such as hemodialysis units (Pasmore and Marion 2008), spent nuclear fuel pools (Sarro et al. 2005, 2007), paper and board machines (Kolari et al. 2003), fluidized bed reactors (Jordening et al. 1992), and gas industry pipelines (Zhu et al. 2003). There are some new approaches such as use of ultrasound and electroporation, use of chemicals to hydrolyze exopolymers, physical disruption of biofilm adhesion, and use of phytochemicals that are tested globally. However, these attempts are mostly in experimentation level and needed much precision to be upscaled to suit the large industrial systems (Murthy and Venkatesan 2008; Xu et al. 2017; Wang et al. 2019).

10.5   Biofilms in Aquatic Ecosystems

In natural aquatic ecosystems, bacteria, diatoms, and protozoa form biofilms. These biofilms are complex and dynamic. They are usually developed on all exposed interfaces of the aquatic ecosystems. Their diversity, composition, and effects vary from nature of the ecosystems (Neagu et al. 2017). The complexities of microbial biofilms in aquatic environment are very large yet poorly understood. These biofilms have important consequences for the performance of aquatic environments and the ecological responses (Besemer 2015). Metagenomic analysis of biofilm forming bacteria revealed that the genes responsible for biofilm formation in aquatic environment are conserved across wide range of bacteria and most of these genes are identical to that of pseudomonads (Anupama et al. 2018). The major consequence of biofilms in aquatic ecosystems is that they could accumulate toxic pollutants such as pesticide residues. Through a recent study, Fernandes et al. (2019) have concluded that some of the biofilms formed in the aquatic ecosystems tend to accumulate the herbicide glyphosate in the polluted waters. Antibiotic resistance genes (ARG) were also found conserved in biofilms of aquatic ecosystem. This is very serious that these ARG can be transferred to other bacteria in the same ecosystem by horizontal gene transfer; thereby it can increase the risk of overpopulation of antibiotic-resistant pathogens (Proia et al. 2016). *Legionella pneumophila*, an upcoming waterborne pathogen, also develops as biofilms in industry and domestic appliances and spread through aerosols that arise from reclaimed water sources. *Legionella* are found associated with infectious biofilms in reclaimed water from wastewater treatment plants, cooling towers, spray irrigation, and toilet flushing and pose serious health risks (Caicedo et al. 2019; De Giglio et al. 2019). Biofilms in the river beds are also reported to influence the river bed structure and function. They were reported to reduce the river bed erosion and maintain the flow of stream channels (Piqué et al. 2016). However, in most cases, river biofilms and sediments are reported to be the main reservoirs for infectious pathogens. This appears to be a serious risk for the public health (Mackowiak et al. 2018).

10.6   Biofilms in Marine Environment

Marine environment is a dynamic environment where the living organisms face a constant stress. Besides heavy salinity, temperature fluctuation, strong tidal and wave actions, dynamic pressure across water columns, and nutritional selectivity, the microorganisms especially marine bacteria have conquered the sea and established their survival on solid marine surfaces through biofilms. Biofilm formation not only is advantageous for the biofilm forming bacteria but also favors ecological and biogeochemical functions in the changing marine environment. Among various marine ecosystems, intertidal systems are highly fluctuating ecosystems. The microorganisms that form biofilms in these intertidal ecosystems are forming a protective microenvironment to alleviate the physical and chemical stress faced by the other organisms. Especially, the exopolysaccharide matrix of the marine biofilms functions as a solid surface for cells to carry out diverse physiological functions such as biogeochemical cycle, nutrient mobilization, and evading environmental stresses such as salinity, temperature, UV irradiation, and desiccation (Decho 2000). However, bacterial biofilms in marine environment result in the formation of biofouling and biocorrosion, which are deleterious to the environment and also causes huge economic loses to the marine industry. They also serve as reservoirs for pathogenic microbes and resistant genes (Wahl et al. 2012; Salta et al. 2013a, b, c; Dang and Lovell 2016). Biofilm forming bacteria colonize almost all possible marine structures in the seawater. Biofilms, also termed as microfouling in marine environment, are affecting both natural and artificial marine structures. Marine structures such as nets, offshore oil platforms and pipelines, ship hulls, ballast tank, and piping systems are greatly affected by biofilms which eventually allow higher organisms such as barnacles and mussels (termed as macrofoulers) to cause biofouling (Salta et al. 2013a, b, c). Marine biofilms have direct deleterious effects in various marine-related economic activities such as shipping, offshore oil rigs, desalination plants, aquaculture, and other industries (De Carvalho Carla 2018). The detailed review by de Carvalho Carla (2018) emphasizes that biofilms in marine environment could be deleterious to the marine environment in addition to the huge economic losses it causes to various industrial sectors (aquaculture, maritime transport, oil and gas industry, desalination plants) associated with the marine environment. It is rightly envisaged by many authors that the biofilms in any environment can be prevented effectively than be controlled totally. Nurioglu et al. (2015) have given an overview of various methods used for the prevention of biofilms and biofouling. They have summarized that paints and coatings such as copper, arsenic, tin, and mercury are used to prevent microbial adhesion. However, these biocides cause more environmental damage than the benefits. Now, the preventive paints and coatings are now done with fluorine and silicon-based materials or their combinations. Natural compounds, enzymes, nanomaterials, graphene, and ultrasonic waves have also been used for the prevention of marine biofilms (Armstrong et al. 2000a, b; Kristensen et al. 2008; Fabrega et al. 2011; Silva et al. 2019; Kurzbaum et al. 2019). However, considering the complexity in controlling the marine biofilms, alternative strategies or works on improvising the existing methods are still warranted.

10.7 Antibiofilm Metabolites from Marine Organisms

Marine environment covers almost 70% of the Earth's surface. The earlier life on Earth has also originated from the seas. So, the oceanic environment is having enormous diversity than its terrestrial counterpart. The biological diversity in the World Oceans accounts for about 95% of the total biodiversity (Qasim 1999; Subramani and Aalbersberg 2012). It appears that the emphasis has been shifted to marine environment for the search of novel metabolites with newer bioactivities. Because, it has been envisaged that research over centuries has completely studied the terrestrial bioresources, and hence, chances for getting newer metabolites from these bioresources are very limited (Subramani and Aalbersberg 2012). The marine environment is now becoming the focus of researchers worldwide as it harbors diverse bioresources. These bioresources produce a battery of novel metabolites, which often have unique structure, function, and applications with highest efficiency. Hence, marine organisms are collected, extracted, and fractionated to develop compounds for novel therapeutic and industrial applications (Gallimore 2017). Many of the marine organisms such as microorganisms, sponges, and seaweeds have been experimented for getting novel bioactive metabolites with antibiofilm potential. Among these organisms, marine microorganisms are becoming the promising source of effective metabolites with antibiofilm activity. The microbial metabolites are found to inhibit the biochemicals involved in biofilm formation. They also inhibit the initial attachment, biofilm development, and the quorum sensing among the biofilm forming bacteria (Adnan et al. 2018).

The published literature on the production of antibiofilm compounds from marine organisms is so enormous. Marine organisms such as seaweeds (red seaweeds and brown seaweeds), microalgae, cyanobacteria, crab shells, sponges, soft corals, fungi, bacteria, and actinomycetes have produced an array of antibiofilm metabolites. Metabolites such as sesquiterpenes, fucoidan and other polysaccharides, chitosan, ianthellin, brominated tyrosine, lectins, surfactants, enzymes, culture filtrates, and other uncharacterized compounds have been produced by these marine organisms against various biofilm forming bacteria. These organisms and antibiofilm metabolites have exhibited antibiofilm activities against biofilm forming bacteria by various mechanisms such as antimicrobial activity, destabilization of plasma membrane, inhibition of initial attachment, affecting virulence factors, inhibition of swarming motility, inhibition of quorum sensing, affecting cell surface hydrophobicity, inhibition of resistance characters, and cell wall degradation. The complete details such as the name of marine organisms that exhibited antibiofilm activity, the compounds they produced, the biofilm forming bacteria they inhibited, and the mechanisms of biofilm inhibition are summarized in Table 10.1.

10.8   Nanotechnology and Antibiofilm Strategies

Biofilms are complex microbial communities formed in diverse systems including marine structures. These biofilm forming bacteria are highly evolving in nature so that they could be able to develop resistance against traditionally used biocides. The biofilm forming bacteria are also developing resistance to the antibiotic metabolites synthesized for the control and prevention of biofilms. Hence, alternative strategies are always in search for the prevention and control of biofilms. Several new approaches such as inhibition of quorum sensing (QS), enzymatic disruption, use of natural products, paints coasted with bactericides, nanotechnology, and bioelectric approach have been attempted globally as an alternative strategy for the prevention

Table 10.1 Marine organisms, which produced antibiofilm compounds

Marine organism	Antibiofilm compound	Target biofilm organism	Possible mechanism/implication	References
Marine brown alga *Halidrys siliquosa*	Algal methanolic extract	*Staphylococcus, Streptococcus, Enterococcus, Pseudomonas, Stenotrophomonas,* and *Chromobacterium*	Antimicrobial nature	Busetti et al. (2015)
Red seaweed, *Laurencia dendroidea*	Elatol, sesquiterpene	*Leishmania amazonensis*	Destabilization of the plasma membrane	Santos et al. (2010)
Seaweed, *Chondrus crispus*	Extract of the seaweed	*Cobetia marina* and *Marinobacter hydrocarbonoclasticus*	Affected bacterial attachment	Salta et al. (2013a, b, c)
Fucus vesiculosus, seaweed	Fucoidan	*Streptococcus mutans* and *S. sobrinus*	Antimicrobial activity	Jun et al. (2018)
Marine brown alga *Halidrys siliquosa*	Partially fractionated methanolic extract	*Staphylococcus aureus* MRSA	Antimicrobial nature	Busetti et al. (2015)
Cyanobacteria *Westiellopsis prolifca*	Acetone extract	*Bacillus subtilis, Shigella* sp., *Proteus* sp.	Antimicrobial activity	Al-TmimiG et al. (2018)
Portunus sanguinolentus, crab shells (biowastes)	Chitosan	*Staphylococcus aureus*	Reducing the staphyloxanthin pigment, a characteristic virulence feature of the pathogen	Rubini et al. (2018)
Aiolochroia crassa, sponge	Ianthellin	Six different marine biofilm forming bacteria	Inhibition of initial attachment	Kelly-Quintos et al. (2005)
Soft coral, *Eunicea* sp.	Batyl alcohol (1) and fuscoside E peracetate (6)	*Pseudomonas aeruginosa* ATCC 27853 and *Staphylococcus aureus* ATCC 25923	Specific biofilm inhibition with lower antimicrobial effect	Díaz et al. (2015)
Aiolochroia crassa, sponge	Brominated tyrosine	Six different marine biofilm forming bacteria	Inhibition of swarming	Kelly-Quintos et al. (2005)
Aplysina fulva, marine sponge	Mucin-binding lectin	*Staphylococcus aureus, S. epidermidis,* and *Escherichia coli*	Reduce the biomass biofilm	Carneiro et al. (2019)

Marine organism	Antibiofilm compound	Target biofilm organism	Possible mechanism/implication	References
Marine-associated fungi	Mevalonolactone	Staphylococcus epidermidis	Interfere with the adherence and biofilm formation	Scopel et al. (2014)
Marine-derived fungus of Emericella variicolor	Sesterterpenes	Mycobacterium smegmatis	Antimicrobial activity	Arai et al. (2013)
Coral-associated bacterium Bacillus horikoshii	Cell-free extracts	Streptococcus pyogenes	Quorum-sensing inhibition properties and cell surface hydrophobicity reduction properties	Thenmozhi et al. (2009)
Marine bacterium Pseudoalteromonas sp. strain 3J6	Culture filtrates	Paracoccus sp. 4M6 and Vibrio sp. D01	Impaired biofilm formation	Dheilly et al. (2010)
Marine bacterium Vibrio sp. QY101	Bacterial exopolysaccharide (A101)	Pseudomonas aeruginosa and S. aureus	Inhibits initial adhesion and the development of complex architecture	Jiang et al. (2011)
Bacillus cereus, deep-sea bacterium	Amylase enzyme	Pseudomonas aeruginosa and Staphylococcus aureus	Inhibition of complete biofilm formation	Vaikundamoorthy et al. (2018)
Sponge-associated strain of Bacillus licheriformis	Polysaccharide; α-D-galactopyranosyl-(1→2)-glycerol-phosphate	Escherichia coli PHL628 and Pseudomonas fluorescens	Reduced the initial adhesion and biofilm development	Sayem et al. (2011)
Marine Actinomycetes Streptomyces, Nocardiopsis, Micromonospora	Supernatants	E. coli and S. aureus	Prevented bacteria to develop biofilms and resistance	Leetanasaksakul and Thamchaipenet (2018)
Actinomycetes Nocardiopsis sp.	Pyrrolo[1,2-a] pyrazine-1,4-dione, hexahydro-3-(2-methylpropyl)	Proteus mirabilis and E. coli.	Cell wall degradation in treated cells of the bacteria	Rajivgandhi et al. (2018)

and control of biofilms (Sadekuzzaman et al. 2015). Use of microbial metabolites from the marine bioresources, which are certainly new and novel, is one of the effective alternative strategy for prevention and control of biofilms. In addition to that, nanomaterials are also found as an efficient alternative (De Souza et al. 2014). Because of the highest heterogeneous nature of the biofilm, the biofilm forming bacteria develop multiple resistance mechanisms (Mah and O'Toole 2001). Hence, multiple and combination of control and prevention strategies are the present need.

The nanomaterials such as nanospheres, liposomes, dendrimers solid lipid nanoparticles, polymeric nanoparticles, and polymeric micelle have been used for the control of biofilms (Kasimanickam et al. 2013; De Souza et al. 2014; Mohankandhasamy and Lee 2016). Singh et al. (2017) have reviewed recent articles published on the biofilm control using nanomaterials. They have emphasized the role of nanomaterials and nanostructures in treating biofilms by interfering with quorum sensing. Biologically synthesized silver nanoparticles were reported to reduce biofilm formation by marine bacteria (Inbakandan et al. 2013). The bacterial biofilms in a natural marine environment were negatively affected by the silver nanoparticles (Fabrega et al. 2011). Biosurfactants such as rhamnolipid are coated with metal nanoparticles such as silver and iron oxide nanoparticles to enhance their antiadhesive and antibacterial activity against *S. aureus* (Khalid et al. 2019). Gold nanoparticles functionalized with proteolytic enzyme were found to have dual role such as antimicrobial and biofilm disruption in *Pseudomonas fluorescens* biofilms (Habimana et al. 2018). Similarly, chitosan nanoparticles were also functionalized by adding antibiotic (oxacillin) and enzyme (DNase) to reduce the thickness of exopolysaccharide matrix and number of viable cells in *Staphylococcus aureus* biofilms (Tan et al. 2018). Studies done across the globe have indicated that nanotechnology coupled with antibiofilm metabolites from marine bioresources could be an effective strategy for the prevention and control of biofilms.

10.9 Antifouling Compounds

The aggregation of undesired organic molecules and marine fouling organisms such as microorganisms, algae, plants, animals, and their by-products on human-made structures in the marine industry causes serious technical and economic complications globally (Callow and Callow 2011; Schultz et al. 2011). The accumulation of biofoulers on marine vessels can increase vessel's hydrodynamic volume and friction up to 60% (Vietti 2009a) resulting in drag increase affecting the vessel's speed by up to 10%, which could also elevate fuel consumption and thereby greenhouse gas emissions (Vietti 2009a, b). Several antifouling methods have been applied to combat biofouling worldwide (Magin et al. 2010; Shan et al. 2011). Metal-based compounds such as tributyltin and cuprous oxide and synthetic organic biocides Igarol and diuron have widely been used to control marine biofouling (Yebra et al.

2004). However, most of these antifoulants showed toxicity on nontarget species and possibly contaminate the marine environment, leading to bans and stringent rules on their use in antifouling coatings (van Wezel and van Wlaardingen 2004; Yebra et al. 2004; Guerin et al. 2007; Wang et al. 2016a, b). Therefore, effective and eco-friendly antifouling agents are desperately needed.

10.10 Fouling Organisms

Biofouling causes serious problems to marine industry products and aquaculture development leading to severe economic loss and also creates multiple ecological issues worldwide (Wang et al. 2017a, b, c). An estimate has found that over 4000 different biofouling organisms may be present in the marine environment (Shan et al. 2011). Among them, majorities are primarily living in shallow water along the coast and harbors, those that afford rich nutrients (Wang et al. 2017a, b, c). Generally, in biofouling, marine adhesion organisms are of two types. The first of these micro-foulers or biofilm organisms are marine bacteria, algae, and protozoa (Shan et al. 2011). Biofilms are ubiquitous in nature, as long as the natural and artificial surfaces are in contact with the water. The second category is macrofoulers such as barnacles, bryozoans, and tubeworms (Shan et al. 2011). Among these two, macrofoulers such as barnacles, mussels, polychaete worms, bryozoans, and seaweed are associated with more severe forms of biofouling in marine environments (Yebra et al. 2004).

10.11 Mechanism of Biofouling Process

Biofouling is accumulation of different organisms, a complex process, and common problem worldwide on man-made objects immersed in the waters (Shan et al. 2011). The series of phases involved in the biofouling process are as follows:

1. The initial short-term phase is when the surface submerged in the water adsorbs organic molecules such as protein, polysaccharide, and proteoglycan, forming the primary film which provides adhesive surface aiding attachment of microorganisms; those are sessile in this conditioned layer (Shan et al. 2011; Maureen and Robert 1994).
2. Microbial attachment on the substratum leads to the next phase in the formation of biofilm which is development of microfoulers (includes microalga and bacteria) that adhere to the surface (Shan et al. 2011).
3. Colonization of microorganisms in biofilm involves adsorption and adhesion. In this process, the adsorption is reversible while adhesion is irreversible. The reversible adsorption is generally ruled by physical effects such as Brownian

motion, electrostatic interaction, gravity, water flow, and van der Waals forces (Fletcher and Loeb 1979; Walt et al. 1985; Per et al. 2004; Chambers et al. 2006; Shan et al. 2011). However, irreversible adhesion is primarily governed through biochemical effects such as secretion of extracellular polymeric substances (EPS). Diatoms are the most important contributors for biofilm formation in marine ecosystems (Shan et al. 2011). Lewin (1984) reported that exclusively microfouling can increase up to 18% fuel consumption and reduce at least 20% sailing speed.

4. In this phase, larvae or spores of macrofoulers attach to the surface of the biofilm and then develop into a complex biological community. Generally, larvae of bryozoan members (Maki et al. 1989), polychaetes (Lau et al. 2003), and some other biofoulers (Hung et al. 2005) involve in the development of complex biological community (Shan et al. 2011).

5. In the last stage, attachment of marine invertebrates such as barnacles, mussels, and macroalgae develops a more complex community.

Focusing on inhibition of physical reactions may be easier to control biofouling rather than the biochemical reactions (Shan et al. 2011). However, it is very difficult to prevent biofilm formation and eventual biofouling as adhesion of diatoms and bacteria is very hard to disrupt (Yebra et al. 2004; Shan et al. 2011).

10.12   Natural Products as Antifouling Agents

In past years, several physical, mechanical, chemical, and biological methods were applied for the prevention of marine biofouling (Abarzua et al. 1999). Antifouling coatings containing heavy metals such as Cu, Pb, Hg, and As and organotins such as tributyltin (TBT) were generally used to control biofouling (Omae 2003;

Qian and Fusetani 2010; Qi and Ma 2017). However, owing to marine ecological protection and high toxicity of these antifouling coatings, they were banned globally (Rittschof et al. 2001; Faÿ et al. 2007; Qi and Ma 2017). Therefore, nontoxic or less toxic and environment-friendly antifouling coating is urgently needed. Marine natural products can be a better alternative for the chemicals commonly used in antifouling coatings. Marine organisms such as sponges, soft corals, ascidians, seaweeds, and microorganisms especially symbiotic microbes produce metabolites that demonstrate potential antifouling properties (Satheesh et al. 2016; Qi and Ma 2017; Wang et al. 2017a, b, c; Dahms and Dobretsov 2017). The secondary metabolites derived from marine organisms are nontoxic or low toxic to the marine ecological environment and degradable and showed high efficiency against biofouling (Qi and Ma 2017).

10.13 Antifouling Natural Products from Marine Microorganisms

10.13.1 Bacteria

Recently, natural products from the marine realm are in spotlight for the discovery of antifouling compounds (Wang et al. 2017a, b, c). Many antifouling compounds have been reported so far from seaweeds and marine invertebrates (Qian et al. 2015). However, the major issue for the commercialization of these natural products from macroorganisms in the marine sectors was found to be limited supply as coating material (Wang et al. 2017a, b, c). Interestingly, microbial-derived extracts and natural products are the particular target in the spotlight of marine natural products because of the probability of supplying perpetual quantities of antifouling compounds through fermentation and genetic engineering of the producing organisms and the potential of source renewal (Dobretsov et al. 2006, 2013; Satheesh et al. 2016; Wang et al. 2017a, b, c). The metabolites of marine organisms especially microorganisms with antifouling properties have been proved as effective inhibitors against fouling organisms with low and nontoxic properties (Qian and Fusetani 2010).

Marine microorganisms produced various natural products belonging to polyketides, lactones, nucleosides, peptides, phenyl ethers, fatty acids, steroids, benzenoids, alkaloids, and terpene antifouling agents (Wang et al. 2017a, b, c). Table 10.2 summarizes some marine bacteria reported for producing antifouling compounds. Members of *Bacillus* and *Streptomyces* are the top antifouling metabolite producers. Importantly, symbiotic bacteria associated with various marine organisms showed rich potential for searching for the antifouling compounds. To mention a few, two fatty acids reported from marine *Shewanella oneidensis* showed effective inhibition of germination of green alga *Ulva pertusa* spores (Bhattarai et al. 2007). A promising nontoxic antifouling diterpene isolated from *Streptomyces cinnabarinus* exhibited significant activity against the diatom *N. annexa* and the marine macroalga *U. pertusa* (Cho and Kim 2012). The steroids obtained from the filamentous bacterium *Leucothrix mucor* significantly inhibit the attachment of biofilm forming bacteria *Pseudomonas aeruginosa* and *Alteromonas* sp. (Cho 2012). Ubiquinones from sponge-derived *Alteromonas* sp. showed effective inhibition on larval settlement of *B. amphitrite* at lower concentrations (Kon-ya et al. 1995). The seaweed-derived *Streptomyces praecox* afforded environment-friendly diketopiperazines displaying strong inhibition of zoospore settlement of the seaweed *U. pertusa* and growth of the diatom *N. annexa* (Cho et al. 2012).

10.13.2 Fungi

Several marine fungi produced diverse chemical groups of antifouling metabolites (Table 10.3). Among the various fungi, *Aspergillus* species produced large amount of antifouling metabolites (Table 10.3). Bisabolane-type sesquiterpenoids from

Table 10.2 List of few marine bacterial strains reported for antifouling activity

Bacterial strain	Source	Antifouling compound	Target fouling organism	References
Pseudovibrio denitrificans	Ascidian	Diindol-3-ylmethanes	Larval settlement of *B. amphitrite* and the bryozoan *Bugula neritina*	Wang et al. (2015)
Bacillus sp.	Sponge, *Aplysina gerardogreeni*	Culture extracts	Prevent biofilm formation	Aguila-Ramírez et al. (2014)
Bacillus sp.	Sponge	Culture extracts	Anti-diatom activity	Jin et al. (2013)
Streptomyces violaceoruber	Seaweed	Butenolides	*Ulva pertusa*	Hong and Cho (2013)
Streptomyces praecox	Macroalga	Diketopiperazines	Active against the marine seaweed *Ulva pertusa* and fouling diatom	Cho et al. (2012)
Bacillus cereus	Sponge, *Sigmadocia* sp.	Culture extracts	Strong microalgal settlement inhibitory activity	Satheesh et al. (2012)
Streptomyces cinnabarinus	Seaweed rhizosphere sediment	Lobocompactol	Macroalga *U. pertusa* and the diatom *N. annexa*	Cho and Kim (2012)
Leucothrix mucor	Red algae	Steroids	Larval settlement of *B. amphitrite* and diatoms	Cho et al. (2012)
Bacillus sp.	Seagrass	Culture extracts	Activity against biofilm forming activity	Marhaeni et al. (2011)
Pseudoalteromonas haloplanktis	Ascidian	Culture extracts	Spore germination of *Ulva pertusa*	Ma et al. (2010)
Streptomyces albidoflavus	Deep-sea sediments	Butenolides	Larval settlement of *Balanus amphitrite*	Xu et al. (2010); Dickschat et al. (2005; 2010)
Streptomyces sp.	Deep-sea sediment	12-Methyltetradecanoic acid	*Hydroides elegans* larvae	Xu et al. (2009)
Shewanella oneidensis	Seawater	2-Hydroxymyristic, cis-9-oleic acid	*Ulva pertusa* spores	Bhattarai et al. (2007)

(continued)

Table 10.2 (continued)

Bacterial strain	Source	Antifouling compound	Target fouling organism	References
Pseudoalteromonas tunicata	Macroalga Ulva australis	Culture extracts	Settlement of Bugula neritina	Rao et al. (2007)
Vibrio sp.	Macroalga	Water-soluble macromolecules	Larval attachment of the polychaete Hydroides elegans	Harder et al. (2004)
Marine cyanobacterium Kyrtuthrix maculans	Exposed to sheltered rocky shores	Maculalactone A	Naupliar larvae of the barnacles B. amphitrite, Tetraclita japonica, and Ibla cumingii	Brown et al. (2004)
Pseudomonas sp.	Nudibranchs	Culture extracts	Larval settlement of B. amphitrite	Burgess et al. (2003)
Acinetobacter sp.	Ascidian, Stomozoa murrayi	6-Bromoindole-3-carbaldehyde	Larval settlement of B. amphitrite	Olguin-Uribe et al. (1997)
Alteromonas sp.	Sponge, Halichondria okadai	Ubiquinones	Larval settlement of B. amphitrite	Konya et al. (1995)

Aspergillus sp. (Li et al. 2012), a novel benzenoids from *Ampelomyces* sp. (Kwong et al. 2006), aspergilone A from *Aspergillus* sp. (Shao et al. 2011a, b), and eurotiumides A–D from *Eurotium* sp. (Chen et al. 2014) are the antifouling compounds that are of particular interest as those displayed activity against the larval settlement of *B. amphitrite* at minimum concentrations. Pestalachlorides obtained from marine-derived *Pestalotiopsis* sp. exhibited high antifouling potential against larval settlement of *B. amphitrite* and demonstrated no toxicity (Xing et al. 2016). The marine *Xylariaceae* sp. produced two antifouling compounds such as dicitrinin A and phenol A acid. Dicitrinin A showing significant effect against the attachment of *B. neritina* larvae and less toxicity than phenol A acid (Nong et al. 2013). Sterigmatocystin and methoxy sterigmatocystin isolated from marine *Aspergillus* sp. strongly inhibit the larval settlement of *B. amphitrite* with the EC_{50} values <0.125 μg/mL and are able to paralyze the larvae at effective concentration (Li et al. 2013; Wang et al. 2017a, b, c). Alkaloids reported from *Scopulariopsis* sp. demonstrated potent antilarval settlement activity at lower concentrations, and they were nontoxic, therefore suggesting that alkaloids are promising antifoulants (Shao et al. 2015).

Table 10.3 List of few marine fungal strains reported for antifouling activity

Fungal strain	Source	Antifouling compound	Target fouling organism	References
Sarcophyton sp.	Soft coral	Amibromdole	Larvae of *B. amphitrite*	Xing et al. (2016)
Pestalotiopsis sp.	Marine-derived	Pestalachlorides E and F	Larval settlement of *B. amphitrite*	Xing et al. (2016)
Scopulariopsis sp.	Gorgonian coral	Alkaloids	Larval settlement of barnacle *B. amphitrite*	Shao et al. (2015)
Eurotium sp.	Gorgonian coral	Eurotiumides A–D	Larval settlement of *B. amphitrite*	Chen et al. (2014)
Aspergillus terreus	Gorgonian coral *Echinogorgia aurantiaca*	Territrem and butyrolactone derivatives	Larval settlement of *B. amphitrite*	Nong et al. (2013)
Cochliobolus lunatus	Sea anemone	Cochliomycins D–F	Larval settlement of *B. amphitrite*	Liu et al. (2014)
Xylariaceae sp.	Sea gorgonian corals *Melitodes squamata*	Dicitrinin A and phenol A acid	Larval settlement of *B. neritina*	Nong et al. (2013)
Aspergillus sp.	Marine-derived	Sterigmatocystin and methoxy sterigmatocystin	Larval settlement of *B. amphitrite*	Li et al. (2013)
Aspergillus elegans	Soft coral *Sarcophyton* sp.	Phenylalanine derivatives and cytochalasins	Larval settlement of *B. amphitrite*	Zheng et al. (2013)
Aspergillus sp.	Gorgonian coral	Polyketides	Antifouling activity	Bao et al. (2017)
Aspergillus sp.	Sponge *Xestospongia testudinaria*	Sesquiterpenoids	Larval settlement of *B. amphitrite*	Li et al. (2012)
Aspergillus sp.	Gorgonian coral	Aspergilone A	Larval settlement of *B. amphitrite*	Shao et al. (2011a, b)
Cochliobolus lunatus	Gorgonian coral	Cochliomycins	Larval settlement of *B. amphitrite*	Shao et al. (2011a, b)
Letendraea helminthicola	Sponge	3-Methyl-*N*-(2-phenylethyl) butanamide and cyclo(D-Pro-D-Phe)	Antifouling activity	Yang et al. (2007)
Ampelomyces sp.	Biofilm developed on the glass slides submerged in Hong Kong waters	3-Chloro-2,5-dihydroxybenzyl alcohol	Larval settlement of *B. amphitrite* and *Hydroides elegans*	Kwong et al. (2006)

10.14   Antifouling Compounds from Marine Macroorganisms

10.14.1   Invertebrates

Marine invertebrates especially sponges, gorgonians, and soft corals are the prominent resources for producing antifouling compounds. Qi and Ma (2017) summarized a total of 198 antifouling compounds obtained from sponges, gorgonians, and soft corals which mostly belong to chemical classes such as diterpenoids, sesquiterpenoids, prostanoids, alkaloids, and steroids. To summarize a few, kalihinenes and kalihipyrans were obtained from the marine sponge *Acanthella cavernosa*, showing strong antifouling activity against *B. amphitrite* larvae (Okino et al. 1996a, b). Sesquiterpenes and phenol derivatives from the sponge *Myrmekioderma dendyi* showed activity against *B. amphitrite* larvae at nontoxic concentrations (Tsukamoto et al. 1997). Antifouling sesquiterpenoids and subergorgic acid isolated from a gorgonian *Anthogorgia* sp. and *Subergorgia suberosa* showed strong inhibition against the larval settlement of *B. amphitrite* larvae at $EC_{50} < 7.0$ μg/mL and 1.2 μg/mL, respectively (Qi et al. 2008; Chen et al. 2012). Cembranoid epimers obtained from the Caribbean gorgonian *Pseudoplexaura flagellosa* inhibited the bacterial biofilm maturation of *Pseudomonas aeruginosa*, *Vibrio harveyi*, and *Staphylococcus aureus* without inhibiting the bacterial growth (Tello et al. 2011). Steroids from sponge *Topsentia* sp. exhibited antifouling activity without toxicity against *B. amphitrite* larvae (Tsukamoto et al. 1997). However, antifouling pyrrole-derived compounds oroidin and mauritiamine that were isolated from the sponge *Agelas mauritiana* showed moderate activity against larval metamorphosis of *B. amphitrite* (Feng et al. 2013). Interestingly, aaptamine and isoaaptamine alkaloids derived from the sponge *Aaptos aaptos* showed antifouling activity against zebra mussel attachment (Diers et al. 2006). Number of antifoulants are reported from marine invertebrates; however, their mode of action is still unknown. However, it is interesting to note that antifoulants from marine invertebrates inhibit both microfoulers such as bacteria and diatoms and macrofoulers such as *B. amphitrite*, *B. albicostatus*, *B. improvises*, *B. neritina*, *Mytilus edulis*, *Psychotria viridis*, and *Halocynthia roretzi* (Qi and Ma 2017).

10.14.2   Macroalgae

Marine macroalgae are well-known for producing various commercial products such as cosmetics, various groups of antibiotics, and cytotoxic agents. Notably, several marine macroalgae constantly clean over decades of time suggesting their potential in antifouling properties (Dahms and Dobretsov 2017). Both the three major algal groups (green, brown, and red algae) were reported for antifouling compound production; however, green algae incurred less attention, while brown algae are being explored globally for antifouling compounds. In contrast, research on green algae for discovery of antifoulants has recently increased (Dahms and Dobretsov 2017). Organic extracts of seaweed *Ulva reticulata* showed moderate antifouling activity, and later chemical characterization of those extracts revealed that those metabolites have potential antibiofilm molecules (Prabhakaran et al. 2012). The compounds 3-bromo-5-(diphenylene)-2(5*H*)-furanone and β-carotene isolated from green algae *U. rigida* and *Ulva* sp., respectively, exhibited significant antifouling activity (Grosser et al. 2012; Chapman et al. 2014). Methanolic extracts of a brown alga *Padina tetrastromatica* displayed strong antibacterial, anti-diatom, and anti-mussel properties (Surest et al. 2014). Similarly, nonpolar extracts of native and invasive *Sargassum* spp. (Schwartz et al. 2017) and ethanol and dichloromethane extracts of *Sargassum muticum* (Silkina et al. 2012) inhibited the growth and attachment of diatoms. A total of six chromanol compounds were isolated from *Sargassum horneri*. These compounds effectively inhibited the settlement of the larvae of *M. edulis* (mussel) with an EC_{50} of 0.11–3.34 μg mL^{-1}. They also inhibited the zoosporic settlement of *U. pertusa* with an EC_{50} of 0.01–0.43 μg mL^{-1} and the diatom *N. annexa* with an EC_{50} of 0.008–0.19 μg mL^{-1} (Cho 2013). Besides, fatty acid derivatives especially docosane, hexadecanoic acid, and cholesterol trimethyl-silyl ether from red algae showed rich antifouling activities (Dahms and Dobretsov 2017). The ethanol and dichloromethane extracts from a red alga *Ceramium botryocarpum* were studied for growth inhibition of marine diatoms. The ethanol fraction of *C. botryocarpum* extracts was most efficient with growth of an EC_{50} 5.3 μg mL^{-1} with reversible diatom growth effect (Silkina et al. 2012). Polyether triterpenoids such as dehydrothyrsiferol and saiyacenols B and C isolated from the *Laurencia viridis* showed high inhibitory activity against marine biofouling organisms (Cen-Pacheco et al. 2015). Furthermore, omaezallenes, a newly discovered natural product from the red alga *Laurencia* sp., effectively inhibit the fouling marine organism (Umezawa et al. 2014). It is no doubt that marine macroalgae harbor rich amount of biofouling compounds; however, extraction, isolation, and commercialization are critical, expensive, and laborious.

10.15 Development of Marine Natural Product-Based Antifouling Treatments

The consequence of metal-based antifouling coatings in the marine sectors has been the subject of serious controvert on the ecology of the marine environment (Foster 1994). Tributyltin (TBT) is a chemical antifoulant, most widely used in the antifouling coatings especially in paints, which accumulates in marine sediments contaminating numerous marine species (Stewart and Thompson 1997; Hashimoto et al. 1998; Fisher et al. 1999) due to which some countries have banned the use of TBT (Burgess et al. 2003). Currently, copper and herbicide and 2-methylthio-4-tert-butyl-amino-6-cyclopropylamino-*s*-triazine (Irgarol®)-based paints are widely used as an alternative to TBT; however, copper and Irgarol also showed toxicity to several marine species (Claisse and Alzieu 1993; Batley et al. 1994; Thomas 2001; Gibbon 1995). Marine natural products can be an alternative for the chemicals as it is environment-friendly and effective in antifouling coatings (Willemsen and Ferrari 1993; Armstrong et al. 2000a, b; Burgess et al. 2003). Price et al. (1994) added sea pansy extracts into commercially available paint and found to be effective for only a short duration in the ocean. Later on, researchers used sponge extracts (Willemsen and Ferrari 1993) and gorgonian extracts (Bakus et al. 1994) in paints which successfully inhibited the barnacle settlements and tube worms. A paint added with culture extract of marine *Pseudomonas* sp. strain NUDMB50-11 exhibited excellent activity to inhibit the settlement of barnacle larvae *B. amphitrite* and algal spores of *Ulva lactuca* (Burgess et al. 2003). The extracellular compounds produced by a tunicate-derived *Pseudoalteromonas tunicata* exhibited strong inhibition against settlement of invertebrates and algal spores, growth of microbes, and surface colonization by diatoms (Holmstrom and Kjelleberg 1999; Burgess et al. 2003). Therefore, it is evident that number of epibiotic bacteria have the potential of producing biomolecules which can be incorporated into paints that retain their antifouling activity (Burgess et al. 2003).

10.16 Biological Synthesis of Nanoparticles

Nanotechnology is now playing crucial role in the contemporary technical world and applied science for designing, synthesizing, and exploiting small structures for a number of applications in medicine and life sciences (Dahms and Dobretsov 2017; Prasad et al. 2016, 2018a, b). Besides, nanoparticles have been applied in several processes such as wastewater treatment, industrial catalysis, chemical and biological sensors, agriculture and food industries, and modern electronic components such as wireless electronic logic and memory systems (Pugazhendhi et al. 2015; Dahms and Dobretsov 2017; Prasad et al. 2017a, b). Metal nanoparticles including silver, gold, and platinum have been applied in various fields of medicine, pharmaceuticals, and bioelectronics (Shankar et al. 2016; Aziz et al. 2014, 2015, 2016). Recently, metal and metal oxide nanoparticles and nanostructures have also been used in the antifouling activities (Yang et al. 2016a, b; Sathe et al. 2016; Al Naamani et al. 2017). Among the other metal oxide nanoparticles, zinc oxide nanoparticles are of particular interest due to it being inexpensive and colorless and their UV blocking properties (Al Naamani et al. 2017). Zinc oxide is a natural product employed in food processing and agriculture (Reddy et al. 2007) and recently used as a potential source as antimicrobial, antibiofilm (Dhillon et al. 2014), and antifouling agents with enhanced physical properties (Malini et al. 2015; Abiraman et al. 2016; Al Naamani et al. 2017). However, the most widely used nanoparticle is silver due to its size, shape, and broad applications including antifouling activity (Muthukumar et al. 2015; Yang et al. 2016a). Biological synthesis of nanoparticles using seaweeds is getting momentum in recent years which is environment-friendly, less cost-intensive, and less energy-consuming with reliable production method and antifouling applications. Al Naamani et al. (2017) developed nanocomposite chitosan-zinc oxide nanoparticle hybrid coatings that showed antibiofilm and anti-diatom activity against *Navicula* sp. and antibacterial activity against *Pseudoalteromonas nigrifaciens*. Biocompatible and functionalized silver nanoparticles synthesized from an aqueous extract of a macroalga *Enteromorpha compressa* are employed as a reducing and stabilizing agent displayed strong antimicrobial and anticancer activity which may potentially be applied in antifouling approach (Ramkumar et al. 2017). Green synthesis of metal nanoparticles could be a promising source in antifouling applications.

10.17 Anticorrosion Metabolites

Seawater covers more than 70% of the globe's surface. It contains many mineral salts, dissolved gases including oxygen (O_2), bacteria and other unicellular or multicellular organisms, and suspended solids and sediments that sometimes give it a high degree of turbidity. Seawater is therefore not simply a solution of sodium chloride. Thus, the sulfate ions arrive, in order of importance, after the Na^+ and Cl^- ions and are strongly involved in the mechanisms associated with the marine corrosion of steels and other metals. The chemical and biological specificities of seawater make it a particularly aggressive environment with regard to many materials, steels and others.

10.18 The Problem of Corrosion of Marine Structures

Corrosion is the set of phenomena of destruction or alteration of a solid body in contact with a fluid external medium. Such a definition brings together, of course, the most diverse phenomena, materials, and environments (Phull and Abdullahi 2017). The disorders observed on the metallic structures of harbor, fluvial, and seaside infrastructures are mainly caused by the corrosion phenomenon which manifests itself differently on the metallic parts according to the zones of exposure (splashing, tidal, immersion, etc.).

In general, the calculation of the lifetime of a metallic structure in an aquatic site takes into account a loss of thickness due to the uniform corrosion of the order of 0.1 mm/year; localized corrosion rates of the order of cm/year were recorded at some sites. The areas considered to be the site of this degradation deserve special attention and protection (Phull and Abdullahi 2017).

10.19 Corrosion Types and Damages

In desalination system, where seawater is used, corrosion of metals is of two types: general and localized corrosion. Both corrosions will bring great harm to the service life of equipment and safety of operation systems. Corrosion in the seawater primes to even significant economic losses, such as productivity loss, loss of end product, loss of efficiency, and product contamination. Even more serious, corrosion in the seawater leads to catastrophic major accidents, such as contamination of potentially toxic materials, causing pollution to the environment, and may also pose severe threat to the public health (Hou et al. 2018).

10.20 Corrosion Main Characteristics

According to seawater characteristics and metal corrosion law, the behavior of corrosion in sea is shown as four corrosion characteristics:

1. When dissimilar metals contact, anode metal may cause significant galvanic corrosion damage, which ascribed to the seawater has good conductivity and small corrosion resistance.
2. Since seawater has vast amount of chlorine, passive metals are prone to suffer localized corrosion in seawater, such as "pitting corrosion, crevice corrosion, and stress corrosion," and prone to suffer "erosion corrosion" in the high-velocity seawater.
3. Any factor of increasing limiting diffusion current density could aggravate metal corrosion.
4. According to the contact style of metal and seawater, the sea could be categorized into five zones: atmospheric zone, splash zone, tidal zone, immersion zone, and sea area. The metal corrosion in these areas is quite different, and the most serious corrosion appears in splash zone (Hou et al. 2018).

10.21 Living Organisms that Contribute to Corrosion in Marine Environment

During the corrosion process, the biological factor is likely to intervene to induce or accelerate the phenomenon. The corrosion is then called biocorrosion, biodeterioration of materials, or corrosion influenced (or induced) by microorganisms (MIC, microbially influenced corrosion). The ISO 8044 (1999) defined the terms "microbial corrosion" and "bacterial corrosion" as the interactions between living microorganisms and the material. In general, microorganisms would not directly use the materials as a source of nutrients. However, the drastic modification of the conditions on the surface thereof, under the influence of microbial metabolism, is likely to induce or accelerate its degradation.

10.22 Concept of Bacterial Metabolism

The prokaryotes, especially bacteria, that are generally associated with the biocorrosion processes of steels or other metals are mainly involved in sulfur and iron cycles: they use as terminal acceptor of electron a compound derived from sulfur or iron. The main groups from these metabolisms such as those of sulfate-reducing, thiosulfate-reducing, and sulfo-oxidant bacteria and more concisely bacteria using iron for their metabolism are also presented in the following sections.

10.23 The Microflora Linked to the Sulfur Cycle

The sulfur cycle is a major and complex life cycle. In nature, both parts of the sulfur cycle, aerobic and anaerobic, are usually overlaid and complement each other. In anaerobiosis, the sulfur cycle is entirely microbial. Sulfide comes from the reduction of sulfates by sulfate-reducing bacteria (SRB), the decline of primary sulfur by sulfate-reducing bacteria, the decline of thiosulfates by thiosulfate-reducing bacteria, and bacterial decomposition of sulfur-containing proteins. The sulfide is subsequently oxidized by the anoxygenic phototrophic bacteria that use it as an electron donor for their photosynthesis. Under aerobic conditions, the cycle is only partly biological, and the sulfide can be oxidized by aerobic chemilithotrophic bacteria (Caumette et al. 1986). The group of sulfate-reducing bacteria is slightly detailed below.

10.24 Sulfate-Reducing Bacteria (SRB)

Sulfate-reducing bacteria (SRB) are those that have been and are still the most studied in the field of the biocorrosion of steels and other metals (Beech and Gaylarde 1999). They are considered among the most harmful and are present in many cases of accelerated corrosion of metal structures. They are particularly involved in the accelerated corrosion of low-water port infrastructure (Pedersen and Hermanson 1991). Most recent studies, however, agree that bacterial consortia should be given a key role, particularly between microorganisms that are sulfurogenic and sulfo-oxidant (Pedersen and Hermanson 1991). From a morphological and physiological point of view, sulfate-reducing bacteria represent a complex and varied group of anaerobic bacteria (Rabus et al. 2002.) Within sulfurous microorganisms, sulfate-reducing bacteria were found to form the most important microbial group. They generally belong to the domain of *Bacteria* except of three species of the genus *Archaeoglobus* that belong to the domain of *Archaea*. Among *Bacteria*, the most described are *Desulfovibrio* with forty-one species and *Desulfotomaculum* with twenty species (Ralf et al. 2002).

10.25   The Microflora Linked to the Iron Cycle

The iron cycle consists of the transformation of iron (II) into iron (III) and vice versa. These operations take place under different physicochemical conditions (pH, temperature, etc.) in an abiotic or biotic way. The reduction or oxidation of iron by microorganisms plays a significant role in this cycle. Microorganisms associated with iron metabolism are the second most frequently cited group in biocorrosion phenomena of steels. It contains iron-reducing bacteria (IRB) and iron-oxidative bacteria (IOB). The IRB group is described below.

10.26   Iron-Reducing Bacteria (IRB)

IRBs form a metabolic group in which species can be phylogenetically distant (Lonergan et al. 1996). The described strains belong in their majority with δ-Proteobacteria and more precisely with the genera *Geobacter*, *Desulfuromonas*, and *Pelobacter*. However, some of the γ-Proteobacteria such as *Shewanella* spp. and *Geovibrio ferrireducens* also belong to the IRB (Caccavo et al. 1996). IRBs are divided into two groups according to their ability to completely oxidize organic matter to CO_2. This oxidation is coupled with the reduction of iron (III) contained in minerals (ferrihydrite, goethite, hematite). Some of these minerals are also produced by iron corrosion in the marine environment. Therefore, the presence of IRB could locally modify the rust layer, which would initiate the creation of anodic zones and induce a difference in electrochemical potential conducive to the development of a localized corrosion process (Lee and Newman 2003).

10.27   Compounds or Metabolites from Marine Organisms with Anticorrosive Potential

Recently, it has been demonstrated that incorporating bacteria into a protective coating of metals against marine corrosion is the solution devised at Sheffield Hallam University (Great Britain). It has been shown to be effective in preventing biocorrosion on an aluminum alloy. Moreover, some extracellular enzymes from marine organisms showed anticorrosive activity. This is the case of protease from the marine bacterium *Bacillus vietnamensis*. In fact, it has been projected that Cu ions coordinated with proteolytic enzymes and bonded with H_2O molecules which results in reduction in the oxygen availability in the environment, thereby preventing the corrosion of the copper-based alloys (Moradi et al. 2019).

Steel corrosion is a global problem in the marine environment. Many inhibitory treatments have been applied to alleviate the degradation of metallic constituents. Many of those methods are not cost-effective and not environment-friendly as well. Liu et al. (2018) presented a novel and "green" method which uses marine bacterium *Pseudoalteromonas lipolytica* for the prevention of steel corrosion in seawater. In this method, the marine bacterium would produce a biofilm over the steel and forms a hybrid form of film over the steel; thereby the steel is getting prevented from corrosion. The bacterium used in the study is capable of transitioning from a biofilm status to biomineralized film state which is crucial for its enduring anticorrosion potential. By forming a biomineralized film, the bacteria overpowers the volatility of biofilm defense on corrosion (Liu et al. 2018).

10.28 Anticorrosive Biomaterials in Industrial Applications

Biomaterial, one of the most interesting fields of modern science, deals with the biologically derived materials or substances that are used within a biological system. Although according to the sources two types are there yet in terms of several advantageous properties, natural biomaterials are far more important than synthetic biomaterials. Among the natural sources, the newest one and also the most potent one identified to be is the marine environment. The most undiscovered part of the Earth, the marine environment is the powerhouse of millions of undiscovered species generating the greatest biodiversity zone. Biomaterials from various marine organisms like sponges, ascidians, crustaceans, sessile organisms, corals, actinobacteria, seaweeds, and fungi have been reported. Development of new strategies coupled with chemical synthesis method could pave the way for future discovery in this actively growing field (Aritra Saha et al. 2014). Many studies were done to develop biomaterials with anticorrosive activities. As examples, a hybrid pigment based on acetylacetonate was tested for its effect on the protection of corrosion in an epoxy-ester polymeric coating. The results obtained exhibited that the epoxy-ester coating protection performance was significantly improved by adding zinc acetylacetonate (Palimi et al. 2018). On the other hand, the incorporation of plant extract in epoxy paint could increase its anticorrosive properties. This is the case of the *Gracilaria edulis* extract (Rajan et al. 2016).

10.29   Nanotechnology and Corrosion Control

Studies across the globe over the years have concluded that the nanotechnology has chiefly contributed for the management of metal corrosion through recent advancements in the cutting-edge technology. Nanotechnology, wherein materials of nanosize are employed, is having the scope for reinventing other and fresher technologies that are more effective than those used nowadays (Wansah et al. Wansah et al. 2014). The processes of nanocrystallization and modification of nanoscale chemical composition have revolutionized the field of nanotechnology. This has prompted the nanotechnology to be used in corrosion management (Shi et al. 2010). Nanoparticles are much smaller than the micro-sized particles and are not covered in the spectrum of visible light (400–700 nm) which are transparent to the human eye. They may contain very few aroms and molecules and may be arranged in two or three dimensions (Liu and Shi 2009). In addition to many applications, nanotechnology is also applied in the management of metal corrosion because of the formation of metal nanocomposites in thin-film coatings of metallic surfaces. Nickel-, silica-, and aluminum-based nanocomposites/nanomaterials and nanostructures were used in the management of metallic corrosions (Lekka et al. 2005). Titanium-based technologies are also developed for the prevention of corrosion in metallic steel through the formation of multilayer nanofilms (Shen et al. 2005). There have been many review articles published on the recent trends in the use of nanotechnology and about various nanomaterials for the management of metallic corrosion (Saji and Thomas 2007). Thus, the immensely effective nanotechnology can be effectively used for the prevention and control of corrosion of metallic particles in marine environment with cost-effectiveness and ecological friendliness.

10.30   Conclusion

Biofilms, biocorrosion, and biofouling are important biological processes that cause severe economic losses in many industries. The prevention and control strategies used conventionally for the management of the bioprocesses are becoming ineffective as the organisms operating these bioprocesses are fast evolving and develop resistance rapidly. Marine bioresources are potential sources for the exploration of novel bioactive metabolites with antibiofilm, anticorrosion, and antifouling properties. The nanoparticles functionalized with biomaterials from marine and other bioresources are considered as most desired alternative strategy for the eco-friendly and sustained prevention and control strategy for the management of biofilms, biocorrosion, and biofouling.

Acknowledgments Author MJ thanks the AMET deemed to be university for facilities and encouragement. Author MS thanks Centre of Biotechnology of Sfax for support. Author RS thanks University of South Pacific for support and facilities.

References

Abarzua S, Jakubowski S, Eckert S, Fuchs P (1999) Biotechnological investigation for the prevention of marine biofouling II. Blue-green algae as potential producers of biogenic agents for the growth inhibition of microfouling organisms. Bot Mar 42:459–465

Abiraman T, Kavitha G, Rengasamy R, Balasubramanian S (2016) Antifouling behavior of chitosan adorned zinc oxide nanorods. RSC Adv 6:69206–69217

Adnan M, Alshammari E, Patel M, Amir Ashraf S, Khan S, Hadi S (2018) Significance and potential of marine microbial natural bioactive compounds against biofilms/biofouling: necessity for green chemistry. Peer J 6:5049

Al-TmimiG SL, Al-Rrubaai G, Zaki NH (2018) Antibiofilm activity of intracellular extracts of *Westiellopsis prolifica* isolated from local environment in Baghdad. JGPT 10:281–288

Al Naamani L, Dobretsov S, Dutta J, Burgess G (2017) Chitosan-ZnO nanocomposite coatings for the prevention of marine fouling. Chemosphere 168:408–417

Amina M, Bensoltane A (2015) Review of pseudomonas attachment and biofilm formation in food industry. Poult Fish Wildl Sci 2:126

Anupama R, Mukherjee A, Babu S (2018) Gene-centric metegenome analysis reveals diversity of Pseudomonas aeruginosa biofilm gene orthologs in fresh water ecosystem. Genomics 110:89–97

Arai M, Niikawa H, Kobayashi M (2013) Marine-derived fungal sesterterpenes, ophiobolins, inhibit biofilm formation of Mycobacterium species. J Nat Med 67(2):271–275

Archer NK, Mazaitis MJ, Costerton JW, Leid JG, Powers ME, Shirtliff ME (2011) Staphylococcus aureus biofilms: properties, regulation, and roles in human disease. Virulence 2(5):445–459

Armstrong E, Boyd KG, Pisacane A, Peppiatt CJ, Burgess JG (2000a) Marine microbial natural products in antifouling coatings. Biofouling 16:215–224

Armstrong E, Boyd KG, Burgess JG (2000b) Prevention of marine biofouling using natural compounds from marine organisms. Biotechnol Annu Rev 6:221–241

Aziz N, Fatma T, Varma A, Prasad R (2014) Biogenic synthesis of silver nanoparticles using Scenedesmus abundans and evaluation of their antibacterial activity. J Nanopart 2014:689419. http://dx.doi.org/10.1155/2014/689419

Aziz N, Faraz M, Pandey R, Sakir M, Fatma T, Varma A, Barman I, Prasad R (2015) Facile algae-derived route to biogenic silver nanoparticles: Synthesis, antibacterial and photocatalytic properties. Langmuir 31:11605–11612. https://doi.org/10.1021/acs.langmuir.5b03081

Aziz N, Pandey R, Barman I, Prasad R (2016) Leveraging the attributes of Mucor hiemalis-derived silver nanoparticles for a synergistic broad-spectrum antimicrobial platform. Front Microbiol 7:1984. doi: 10.3389/fmicb.2016.01984

Bakus GJ, Wright M, Khan AK, Ormsby B, Gulko D, Licuanan W, Carriazo E, Ortiz A, Chan DB, Lorenzana D, Huxley MP (1994) Experiments seeking marine natural antifouling compounds. In: Thompson MF, Nagabhushanam R, Sarojini R, Fingerman M (eds) Recent developments in biofouling control. AABalkema, Rotterdam, pp 373–381

Bao K, Bostanci N, Thurnheer T, Belibasakis GN (2017) Proteomic shifts in multi-species oral biofilms caused by Anaeroglobus geminatus. Sci Rep 7(1):4409

Batley GE, Chapman JC, Wilson SP (1994) Environmental impacts of antifouling technology. In: Kjelleberg S, Steinberg P (eds) Biofouling: problems and solutions. UNSW, Sydney, pp 49–53

Beech IB, Gaylarde CC (1999) Recent advances in the study of biocorrosion—an overview. Microbiology 30:177–190

Beloin C, Roux A, Ghigo JM (2008) Escherichia coli biofilms. Curr Top Microbiol Immunol 322:249–289

Besemer K (2015) Biodiversity, community structure and function of bio films in stream ecosystems. Res Microbiol 166:774–781

Bhattarai HD, Ganti VS, Paudel B, Lee YK, Lee HK, Hong YK, Shin HW (2007) Isolation of antifouling compounds from the marine bacterium, Shewanella oneidensis SCH0402. World J Microbiol Biotechnol 23:243–249

Bremer P.J, Brooks S, Flint J, Palmer S, Burgess D, Lindsay D and Seale B (2018) Biofilm formation and control in the dairy industry, reference module in food science, Elsevier, Amsterdam

Brown GD, Wong HF, Hutchinson N, Lee SC, Chan BKK, Williams GA (2004) Chemistry and biology of maculalactone A from the marine cyanobacterium Kyrtuthrix maculans. Phytochem Rev 3:381–400

Burgess JG, Boyd KG, Armstrong E, Jiang Z, Yan L, Berggren M (2003) The development of a marine natural product-based antifouling paint. Biofouling 19:197–205

Busetti A, Thompson TP, Tegazzini D, Megaw J, Maggs CA, Gilmore BF (2015) Antibiofilm activity of the brown alga Halidrys siliquosa against clinically relevant human pathogens. Mar Drugs 13(6):3581–3605

Caccavo F, Frolund Jr, Van Ommen F, Kloekeand Nielsen PH (1996) Deflocculation of activated sludge by the dissimilatory Fe(III)-reducing bacterium *Shewanella alga BrY*. Appl Environ Microbiol 62:1487–1490

Cai S, Song Y, Chen C, Shi J, Gan L (2018) Natural chromatin is heterogeneous and self-associates in vitro. Mol Biol Cell 29(13):1652–1663

Caicedo K-H, Rosenwinkel, Exner M, Verstraete W, Suchenwirth R, Hartemann P, Nogueira R (2019) Legionella occurrence in municipal and industrial wastewater treatment plants and risks of reclaimed wastewater reuse: review. Water Res 149:21–34

Callow JA, Callow ME (2011) Trends in the development of environmentally friendly fouling-resistant marine coatings. Nat Commun 2:244–253

Caumette P (1986) Phototrophic sulfur bacteria and sulfate reducing bacteria causing red waters in a shallow brackish coastal lagoon (Prévost Lagoon, France). FEMS Microbiol Ecol 38:113–124

Cen-Pacheco F, Santiago-Benítez AJ, García C, Álvarez-Méndez SJ, Martín-Rodríguez AJ, Norte M, Martín VS, Gavín JA, Fernández JJ, Daranas AH (2015) Oxasqualenoids from laurencia viridis: combined spectroscopic–computational analysis and antifouling potential. J Nat Prod 78:712–721

Chambers LD, Stokes KR, Walsh FC (2006) Modern approaches to marine antifouling coatings. Surf Coat Technol 201:3642–3652

Chapman J, Hellio C, Sullivan T, Brown R, Russell S, Kiterringham E, Le Nor L, Regan F (2014) Bioinspired synthetic macroalgae: Examples from nature for antifouling applications. Int Biodeterior Biodegrad 86:6–13

Chen D, Yu S, Van Ofwegen L, Proksch P, Lin W (2012) Anthogorgienes A-O, new guaiazulene-derived terpenoids from a Chinese gorgonian Anthogorgia species, and their antifouling and antibiotic activities. J Agric Food Chem 60:112–123

Chen M, Shao CL, Wang KL, Xu Y, She ZG, Wang CY (2014) Dihydroisocoumarin derivatives with antifouling activities from a gorgonian-derived Eurotium sp. fungus. Tetrahedron 70:9132–9138

Cho JY (2012) Antifouling steroids isolated from red alga epiphyte filamentous bacterium Leucothrix mucor. Fish Sci 78:683–689

Cho JY (2013) Antifouling chromanols isolated from brown alga Sargassum horneri. J Appl Phycol 25:299–309

Cho JY, Kim MS (2012) Induction of antifouling diterpene production by Streptomyces cinnabarinus PK209 in co-culture with marine-derived Alteromonas sp. KNS-16. Biosci Biotechnol Biochem 76:1849–1854

Cho JY, Kang JY, Hong YK, Baek HH, Shin HW, Kim MS (2012) Isolation and structural determination of the antifouling diketopiperazines from marine-derived Streptomyces praecox 291-11. Biosci Biotechnol Biochem 76:1116–1121

Claisse D, Alzieu C (1993) Copper contamination as a result of antifouling paint regulations? Mar Pollut Bull 26:395–397

Conley J, Olson ME, Cook I S, Ceri II, Phan V, Davies HD (2003) Biofilm formation by group a streptococci: is there a relationship with treatment failure? J Clin Microbiol 41(9):4043–4048

Costerton J, Lewandowski Z (1995) Microbial biofilms. Annu Rev Microbial 49:71–45

Dahms HW, Dobretsov S (2017) Antifouling compounds from marine macroalgae. Mar Drugs 15:E265

Dang H, Lovell CR (2016) Microbial surface colonization and biofilm development in marine environments. Microbiol Mol Biol Rev 80(1):91–138

Davey ME, O'Toole GA (2000) Microbial biofilms: from ecology to molecular genetics. Microbiol Mol Biol Rev 4:847–867

De Carvalho Carla CCR (2018) Marine Biofilms: A Successful Microbial Strategy with Economic Implications. Front Mar Sci 5:126

De Giglio O, Napoli C, Apollonio F, Brigida S, Marzella A, Diella G, Calia C, Scrascia M, Pacifico C, Carlo Pazzani C, Vito Felice Uricchio VF, Montagna MT (2019) Occurrence of legionella in groundwater used for sprinkler irrigation in Southern Italy. Environ Res 170:215–221

De Oliveira AM, Férnandes MDS, Filho BAA, Gomes RG, Bergamasco R (2018) Inhibition and removal of staphylococcal biofilms using Moringa oleifera Lam. aqueous and saline extracts. J Environ Chem Eng 6:2011–2016

De Souza ME, Lopes LQS, Vaucher RA, Santos RCV (2014) Antibiofilm applications of nano-technology. Fungal Genom Biol 4:117

Decho AW (2000) Microbial biofilms in intertidal systems: an overview. Cont Shelf Res 20:1257–1273

Dhillon GS, Kaur S, Brar SK (2014) Facile fabrication and characterization of chitosan-based zinc oxide nanoparticles and evaluation of their antimicrobial and antibiofilm activity. Int Nano Lett 4:107

Diers JA, Bowling JJ, Duke SO, Wahyuono S, Kelly M, Hamann MT (2006) Zebra mussel antifouling activity of the marine natural product aaptamine and analogs. Mar Biotechnol 8:366–372

Dickschat JS, Wagner-Döbler I, Schulz S (2005) The chafer pheromone buibuilactone and ant pyrazines are also produced by marine bacteria. J Chem Ecol 31(4):925–947

Dickschat JS (2010) Quorum sensing and bacterial biofilms. Nat Prod Rep 27(3):343–69

Dobretsov S, Abed RM, Voolstra CR (2013) The effect of surface colour on the formation of marine micro and macrofouling communities. Biofouling 29(6):617–627

Dobretsov S, Dahms HU, Qian PY (2006) Inhibition of biofouling by marine microorganisms and their metabolites. Biofouling 22:43–54

Fabrega J, Zhang R, Renshaw JC, Liu WT, Jamie R (2011) Impact of silver nanoparticles on natural marine biofilm bacteria. Chemosphere 85:961–966

Faÿ F, Linossier I, Peron JJ, Langlois V, Vallée-Rehel K (2007) Antifouling activity of marine paints: study of erosion. Prog Org Coat 60:194–206

Fiedler T, Köller T, Kreikemeyer B (2015) Streptococcus pyogenes biofilms—formation, biology, and clinical relevance. Front Cell Infect Microbiol 5:15

Felipe V, Breser ML, Bohl LP, Da Silva ER, Morgante CA, Correa SG, Porporatto C (2019) Chitosan disrupts biofilm formation and promotes biofilm eradication in Staphylococcus species isolated from bovine mastitis. Int J Biol Macromol 126:60–67

Feng D, Qiu Y, Wang W, Wang X, Ouyang P, Ke C (2013) Antifouling activities of hymenialdisine and debromohymenialdisine from the sponge Axinella sp. Int Biodeterior Biodegrad 85:359–364

Fernandes G, Aparicio VC, Bastos MC, Gerónimo ED, Labanowski J, Prestes OD, Zanella R, dos Santos DR (2019) Indiscriminate use of glyphosate impregnates river epilithic biofilms in southern Brazil. Sci Total Environ 651:1377–1387

Fisher WS, Oliver LM, Walker WW, Manning CS, Lytle TF (1999) Decreased resistance of eastern oysters (Crassostrea virginica) to a protozoal pathogen (Perkinsus marinus) after sublethal exposure to tributyltin oxide. Mar Environ Res 47:185–201

Flemming HC, Jost Wingender Szewzyk U, Steinberg P, Scott A (2016) Rice biofilms: an emergent form of bacterial life. Nat Rev Microbiol 14:563–575

Fletcher M, Loeb GI (1979) Influence of substratum characteristics on the attachment of a marine pseudomonad to solid surfaces. Appl Environ Microbiol 37:67–72

Foster AC (1994) Current antifouling technologies. In: Kjelleberg S, Steinberg P (eds) Biofouling: problems and solutions. UNSW, Sydney, pp 44–48

Gallimore W (2017) Marine metabolites: oceans of opportunity. In: Badal S, Delgoda R (eds) Pharmacognosy. Academic Press, New York, pp 377–400

Giaouris EE, Simoes MV (2018) Pathogenic biofilm formation in the food industry and alternative control strategies. In: Holban AM, Grumezescu AM (eds) Handbook of food bioengineering, foodborne diseases, vol 11. Academic Press, New York, pp 309–377

Gibbon P (1995) Molluscan shellfisheries in estuaries. In: Earl RC (ed) Marine environmental management: review of events in 1994 and future trends. Candle Cottage, Kempley, Gloucestershire, UK, p 83

Grosser K, Zedler L, Schmitt M, Dietzek B, Popp J, Pohnert G (2012) Disruption-free imaging by Raman spectroscopy reveals a chemical sphere with antifouling metabolites around macroalgae. Biofouling 28:687–696

Grutsch AA, Nimmer PS, Pittsley RH, McKillip JL (2018) Bacillus spp. as pathogens in the dairy industry. In: Foodborne diseases. Academic Press, Cambridge, MA, 193–211

Guerin T, Sirot V, Volatier JL, Leblanc JC (2007) Organotin levels in seafood and its implications for health risk in high seafood consumers. Sci Total Environ 388:66–77

Habimana O, Zanoni M, Vitale S, O'Neill T, Xu DSB, Casey E (2018) One particle, two targets: a combined action of functionalised gold nanoparticles, against Pseudomonas fluorescens biofilms. J Colloid Interface Sci 526:419–428

Harder T, Dobretsov S, Qian PY (2004) Waterborne polar macromolecules act as algal antifoulants in the seaweed Ulva reticulata. Mar Ecol Prog Ser 274:133–141

Hashimoto S, Watanabe M, Noda Y, Hayashi T, Kurita Y, Takasu Y, Otsuki A (1998) Concentration and distribution of butyltin compounds in a heavy tanker route in the Strait of Malacca and in Tokyo Bay. Mar Environ Res 45:169–177

Holmstrom C, Kjelleberg S (1999) Marine Pseudoalteromonas species are associated with higher organisms and produce biologically active extracellular agents. FEMS Microbiol Ecol 30:285–293

Hong YK, Cho JY (2013) Effect of seaweed epibiotic bacterium Streptomyces violaceoruber SCH-09 on marine fouling organisms. Fish Sci 79:469–475

Hou X, Gao L, Zhendong C, Yin J (2018) Corrosion and protection of metal in the seawater desalination. Earth Environ Sci 108:022037

Hung OS, Thiyagarajan V, Wu R (2005) Effects of ultraviolet radiation on films and subsequent settlement of Hydroides elegans. Mar Ecol Prog Ser 304:155–166

Inbakandan D, Kumar C, Abraham LS, Kirubagaran R, Venkatesan R, Khan SA (2013) Silver nanoparticles with anti microfouling effect: a study against marine biofilm forming bacteria. Colloid Surf B 111:636–643

Jamal M, Tasneem U, Hussain T, Andleeb S (2015) Bacterial biofilm: its composition, formation and role in human infections. Res Rev J Microbiol Biotechnol 4:3

Jamal M, Ahmad W, Andleeb S, Jalil F, Imran M, Nawaz MA, Hussain T, Ali M, Rafiq M, Kamil MA (2017) Bacterial biofilm and associated infections. J Chin Med Assoc 81:7–11

Jin C, Xin X, Yu S, Qiu J, Miao L, Feng K (2013) Antidiatom activity of marine bacteria associated with sponges from San Juan Island, Washington. World J Microbiol Biotechnol 30:1325–1334

Jordening HJ, Melo LF, Bott TR, Fletcher M, Capdeville B (1992) Anaerobic biofilms in fluidized bed reactors. biofilms—science and technology, NATO ASI Series (Series E: Applied Sciences), vol 223. Springer, Dordrecht

Jun JY, Jung MJ, Jeong IH, Yamazaki K, Kawai Y, Kim BM (2018) Antimicrobial and antibiofilm activities of sulfated polysaccharides from marine algae against dental plaque bacteria. Mar Drugs 16:301

Kasimanickam RK, Ranjan A, Asokan GV, Kasimanickam VR, Kastelic JP (2013) Prevention and treatment of biofilms by hybrid and nanotechnologies. Int J Nanomedicine 8:2809–2819

Khalid HF, Tehseen B, Sarwar Y, Hussain SZ, Khan WS, Raza ZA, Sadia Bajwa Z, Kanaras AG, Hussain I, Rehman A (2019) Biosurfactant coated silver and iron oxide nanoparticles with enhanced anti-biofilm and anti-adhesive properties. J Hazard Mater 364:441–448

Kelly-Quintos C, Kropec A, Briggs S, Ordonez CL, Goldmann DA, Pier GB (2005) The role of epitope specificity in the human opsonic antibody response to the staphylococcal surface polysaccharide poly N-acetyl glucosamine. J Infect Dis 192(11):2012–2019

Kolari M, Nuutinen J, Rainey FA (2003) Salkinoja-Salonen Colored moderately thermophilic bacteria in paper-machine biofilms. J Int Microbiol Biotechnol 30:225

Kon-ya K, Shimizu N, Otaki N, Yokoyama A, Adachi K, Miki W (1995) Inhibitory effect of bacterial ubiquinones on the settling of barnacle, Balanus amphitrite. Experientia 51:153–155

Kregiel D (2014) Advances in biofilm control for food and beverage industry using organo-silane technology: a review. Food Control 40:32–40

Kristensen JB, Meyer RL, Laursen BS, Shipovskov S, Besenbacher F, Poulsen CH (2008) Antifouling enzymes and the biochemistry of marine settlement. Biotechnol Adv 26(5):471–481

Kurzbaum E, Iliasafov L, Kolik L, Jeana Starosvetsky J, Bilanovic D, Butnariu M, Armon R (2019) The titanic and other shipwrecks to biofilm prevention: the interesting role of polyphenol-protein complexes in biofilm inhibition. Sci Total Environ 658:1098–1105

Kwong TFN, Miao L, Li X, Qian PY (2006) Novel antifouling and antimicrobial compound from a marine-derived fungus Ampelomyces sp. Mar Biotechnol 8:634–640

Lamas A, Paz-Mendez AM, Regal P, Vazquez B, Miranda JM, Cepeda A, Franco CM (2018) Food preservatives influence biofilm formation, gene expression and small RNAs in Salmonella enterica. LWT 97:1–8

Lau SCK, Harder T, Qian PY (2003) Induction of larval settlement in the serpulid polychaete Hydroides elegans (Haswell): Role of bacterial extracellular polymers. Biofouling 19:197–204

Leetanasaksakul K, Thamchaipenet A (2018) Potential anti-biofilm producing marine actinomycetes isolated from sea sediments in Thailand. Agric Nat Resour 52(3):228–233

Laxmi M, Bhat SG (2018) Mitigation using bacteriophage. In: Holban AM, Grumezescu AM (eds) Handbook of food bioengineering, advances in biotechnology for food industry. Academic Press, New York, pp 393–423

Lee AK, Newman DK (2003) Microbial iron respiration: Impact on corrosion processes. Appl Microbiol Biotechnol 62:134–139

Lekka M, Kouloumbi N, Gajo M, Bonora PL (2005) Corrosion and wear resistant electrodeposited composite coatings. Electrochim Acta 50:4551–4556

Lewin R (1984) Microbial adhesion is a sticky problem. Science 224:375–377

Li D, Xu Y, Shao CL, Yang RY, Zheng CJ, Chen YY, Fu XM, Qian PY, She ZG, de Voogd NJ (2012) Antibacterial bisabolane-type sesquiterpenoids from the sponge-derived fungus Aspergillus sp. Mar Drugs 10:234–241

Li YX, Wu HX, Xu Y, Shao CL, Wang C, Qian PY (2013) Antifouling activity of secondary metabolites isolated from Chinese marine organisms. Mar Biotechnol 15:552–558

Li K, Yang G, Debru AB, Li P, Zong L, Li P, Bao Q (2017) SuhB regulates the motile-sessile switch in Pseudomonas aeruginosa through the Gac/Rsm pathway and c-di-GMP signaling. Front Microbiol 8:1045

Lister JL, Horswill AR (2014) Staphylococcus aureus biofilms: recent developments in biofilm dispersal. Front Cell Infect Microbiol 4:178

Liu Y, Shi X (2009) Electrochemical chloride extraction and electrochemical injection of corrosion inhibitor in concrete: state of the knowledge. Corros Rev 27:53–82

Liu QA, Shao CL, Gu YC, Blum M, Gan LS, Wang KL, Chen M, Wang CY (2014) Antifouling and fungicidal resorcylic acid lactones from the Sea Anemone-derived fungus Cochliobolus lunatus. J Agric Food Chem 62:3183–3191

Liu T, Guo Z, Zeng Z, Guo N, Lei Y, Liu T, Sun S, Chang X, Yin Y, Wang X (2018) Marine bacteria provide lasting anticorrosion activity for steel via biofilm-induced mineralization. ACS Appl Mater Interfaces 10:40317–40327

López D, Vlamakis H, Kolter R (2020) Biofilms. Cold Spring Harb Perspect Biol 2(7):a000398

Lonergan DJ, Jenter HL, Coates JD, Phillips EJ, Schmidt TM, Lovley DR (1996) Phylogenetic analysis of dissimilatory Fe(III)-reducing bacteria. J Bacteriol 178:2402–2408

Ma Y, Liu P, Zhang Y, Cao S, Li D, Chen W (2010) Inhibition of spore germination of Ulva pertusa by the marine bacterium Pseudoalteromonas haloplanktis CI4. Acta Oceanol Sin 29:69–78

Mackowiak M, Leifels M, Hamza IA, Jurzik L, Wingender J (2018) Distribution of Escherichia coli, coliphages and enteric viruses in water, epilithic biofilms and sediments of an urban river in Germany. Sci Total Environ 626:650–659

Magin CM, Cooper SP, Brennan AB (2010) Non-toxic antifouling strategies. Dent Mater 13:36–44

Mah TC, O'Toole GA (2001) Mechanisms of biofilm resistance to antimicrobial agents. Trends Microbiol 9:34–39

Maki JS, Rittschof D, Schmidt AR (1989) Factors controlling adhesion of bryozoan larvae: a comparison of bacterial films and unfilmed surfaces. Biol Bull 177:295–302

Malini M, Thirumavalavan M, Yang WY, Lee JF, Annadurai G (2015) A versatile chitosan/ZnO nanocomposite with enhanced antimicrobial properties. Int J Biol Macromol 80:121–129

Marhaeni B, Radjasa OK, Khoeri M, Sabdono A, Bengen DG, Sudoyo H (2011) Antifouling activity of bacterial symbionts of seagrasses against marine biofilm-forming bacteria. J Environ Prot 2:1245–1249

Maureen EC, Robert F (1994) The influence of low surface energy materials on bioadhesion—a review. Int Biodeterior Biodegrad 34:333–348

Meireles A, Borges A, Giaouris E, Simões M (2016) The current knowledge on the application of anti-biofilm enzymes in the food industry. Food Res Int 86:140–146

Merino L, Procura F, Trejo FM, Bueno DJ, Golowczyc MA (2019) Biofilm formation by Salmonella sp. in the poultry industry: detection, control and eradication strategies. Food Res Int 119:530–540

Mohankandhasamy R, Lee J (2016) Recent nanotechnology approaches for prevention and treatment of biofilm-associated infections on medical devices. Bio Med Res Int 2016:1851242

Moradi M, Sun Z, Song Z, Hu Z (2019) Effect of proteases secreted from a marine isolated bacterium *Bacillus vietnamensis* on the corrosion behaviour of different alloys. Bioelectrochemistry 126:64–71

Mulcahy LR, Isabella VM, Lewis K (2014) Pseudomonas aeruginosa biofilms in disease. Microb Ecol 68(1):1–12

Murthy SP, Venkatesan R (2008) Industrial biofilms and their control. In: Springer series on biofilms. Springer, Berlin, Heidelberg

Muthukumar K, Vignesh S, Dahms HU, Gokul MS, Palanichamy X, Subramian G, James RA (2015) Antifouling assessments on biogenic nanoparticles: A field study from polluted offshore platform. Mar Pollut Bull 101:816–825

Neagu L, Cirstea DM, Curutiu C, Mitache MM, Lazăr V, Chifiriuc MC (2017) Microbial biofilms from the aquatic ecosystems and water quality. In: Grumezescu AM (ed) Water purification. Academic Press, New York, pp 621–642

Nong XH, Zheng ZH, Zhang XY, Lu XH, Qi SH (2013) Polyketides from amarine-derived fungus Xylariaceae sp. Mar Drugs 11:1718–1727

Nurioglu AG, Esteves AC, de With G (2015) Non-toxic, non-biocide-release antifouling coatings based on molecular structure design for marine applications. J Mater Chem B 3:6547–6570

Okino T, Yoshimura E, Hirota H, Fusetani N (1996a) Antifouling kalihinenes from the marine sponge Acanthella cavernosa. Tetrahedron Lett 36:8637–8640

Okino T, Yoshimura E, Hirota H, Fusetani N (1996b) New antifouling kalihipyrans from the marine sponge Acanthella cavernosa. J Nat Prod 59:1081–1083

Olguin-Uribe G, Abou-Mansour E, Boulander A, Debard H, Francisco C, Combaut G (1997) 6-bromoindole-3-carbaldehyde, from an Acinetobacter sp. bacterium associated with the ascidian Stomozoa murrayi. J Chem Ecol 23(11):2507–2521

Omae I (2003) Organotin antifouling paints and their alternatives. Appl Organomet Chem 17:81–105

Parsck MR, Singh PK (2003) Bacterial biofilms: an emerging link to disease pathogenesis. Annu Rev Microbiol 57:677–701

Palimi MJ, Alibakhshi E, Ramezanzadeh B, Bahlakeh G, Mahdavian M (2018) Screening the anti-corrosion effect of a hybrid pigment based on zinc acetyl acetonate on the corrosion protection performance of an epoxy-ester polymeric coating. J Taiwan Inst Chem Eng 82:261–272

Pang X, Wong C, Chung H, Yuk H (2019) Biofilm formation of Listeria monocytogenes and its resistance to quaternary ammonium compounds in a simulated salmon processing environment. Food Control 98:200–208

Pasmore M, Marion K (2008) Biofilms in hemodialysis. In: The role of biofilms in device-related infections. Springer, Berlin/Heidelberg, pp 167–192

Pedersen A, Hermanson M (1991) Bacterial corrosion of iron in seawater in situ and in aerobic and anaerobic model systems. FEMS Microbiol Ecol 86:139–148

Per RJ, Kent MB, Ann IL (2004) Linking larval supply to recruitment: flow-mediated control of initial adhesion of barnacle larvae. Ecology 85:2850–2859

Phull B, Abdullahi AA (2017) Marine corrosion, Reference module in materials sciences and material engineering

Piqué G, Vericat D, Sabater S, Batalla RJ (2016) Effects of biofilm on river-bed scour. Sci Total Environ 572:1033–1046

Prabhakaran S, Rajaram R, Balasubramanian V, Mathivanan K (2012) Antifouling potentials of extracts from seaweeds, seagrasses and mangroves against primary biofilm forming bacteria. Asian Pac J Trop Biomed 2:316–322

Prasad R, Pandey R, Barman I (2016) Engineering tailored nanoparticles with microbes: quo vadis. WIREs Nanomed Nanobiotechnol 8:316–330. https://doi.org/10.1002/wnan.1363

Prasad R, Kumar M, Kumar V (2017a) Nanotechnology: an agriculture paradigm. Springer Nature Singapore Pte Ltd., Singapore. ISBN: 978-981-10-4573-8

Prasad R, Kumar V, Kumar M (2017b) Nanotechnology: food and environmental paradigm. Springer Nature Singapore Pte Ltd., Singapore. ISBN 978-981-10-4678-0

Prasad R, Jha A, Prasad K (2018a) Exploring the realms of nature for nanosynthesis. Springer International Publishing, New York. ISBN 978-3-319-99570-0) https://www.springer.com/978-3-319-99570-0

Prasad R, Kumar V, Kumar M, Wang S (2018b) Fungal nanobionics: principles and applications. Springer Nature Singapore Pte Ltd., Singapore. ISBN 978-981-10-8666-3 https://www.springer.com/gb/book/9789811086656

Price RR, Patchan K, Clare AS, Rittschof D, Bonaventura J (1994) Performance enhancement of natural antifouling compounds and their analogs through microencapsulation and controlled release. In: Thompson M-F, Nagabhushanam R, Sarojini R, Fingerman M (eds) Recent developments in biofouling control. A ABalkema, Rotterdam, pp 321–334

Proia L, Schiller DV, Melsio AS, Sabater S, Borrego CM, Mozaz SR, Balcázar JL (2016) Occurrence and persistence of antibiotic resistance genes in river biofilms after wastewater inputs in small rivers. Environ Pollut 210:121–128

Pugazhendhi E, Kirubha P, Palanisamy K, Gopalakrishnan R (2015) Synthesis and characterization of silver nanoparticles from Alpinia calcarata by Green approach and its applications in bactericidal and nonlinear optics. Appl Surf Sci 357:1801–1808

Qasim (1999) The Indian ocean: images and realities. Oxford and IBH, New Delhi, pp 57–90

Qi SH, Ma X (2017) Antifouling compounds from marine invertebrates. Mar Drugs 15:263

Qi SH, Zhang S, Yang LH, Qian PY (2008) Antifouling and antibacterial compounds from the gorgonians Subergorgia suberosa and Scripearia gracilis. Nat Prod Res 22:154–166

Qian PYXY, Fusetani N (2010) Natural products as antifouling compounds: recent progress and future perspectives. Biofouling 26:223–234

Qian PY, Li Z, Xu Y, Li Y, Fusetani N (2015) Mini-review: marine natural products and their synthetic analogs as antifouling compounds. Biofouling 31:101–122

Rabus R, Bruchert V, Amann J, Konneke M (2002) Physiological response to temperature changes of the marine, sulfate-reducing bacterium Desulfobacterium autotrophicum. FEMS Microbiol Ecol 2:409–417

Rajan R, Selvaraj M, Subramanian G (2016) Studies on the anticorrosive and antifouling properties of the *Gracilaria edulis* extract incorporated epoxy paint in the Gulf of Mannar Coast,
Mandapam. India. Prog Org Coat 90:448–454

Rajivgandhi G, Vijayan R, Maruthupandy M, Vaseeharan B, Manoharan N (2018) Antibiofilm
effect of Nocardiopsis sp. GRG 1 (KT235640) compound against biofilm forming Gram negative bacteria on UTIs. Microb Pathog 118:190–198

Ramkumar VR, Pugazhendhi A, Gopalakrishnan K, Sivagurunathan P, Saratale GD (2017)
Biofabrication and characterization of silver nanoparticles using aqueous extracts of seaweed
Enteromorpha compressa and its biomedical properties. Biotechnol Rep (Amst) 14:1–7

Rao D, Webb JS, Holmstrom C, Case R, Low A, Steinberg P (2007) Low densities of epiphytic
bacteria from the marine alga Ulva australis Inhibit settlement of fouling organisms. Appl
Environ Microbiol 73:7844–7852

Rasamiravaka T, Labtani Q, Duez P, El Jaziri M (2015) The formation of biofilms by pseudomonas
aeruginosa: A review of the natural and synthetic compounds interfering with control mechanisms. BMRI 2015: Article ID 759348

Reddy KM, Feris K, Bell J, Wingett DG, Hanley C, Punnoose A (2007) Selective toxicity of zinc
oxide nanoparticles to prokaryotic and eukaryotic systems. Appl Phys Lett 90:2139021–2139023

Rittschof D, Clintock JB, Baker BJ (eds) (2001) Natural product antifoulants and coatings development. In Marine chemical ecology. Taylor and Francis, Abingdon, UK, pp 543–566

Roberts JL, Khan S, Emanuel C, Powell LC, Pritchard MF, Onsoyen E, Myrvold R, Thomas DW,
Hill KE (2013) An in vitro study of alginate oligomer therapies on oral biofilms. J Dent 41:892

Rubini D, Banu SF, Nisha P, Murugan R, Thamotharan S, Percino MJ et al (2018) Essential oils
from unexplored aromatic plants quench biofilm formation and virulence of Methicillin resistant Staphylococcus aureus. Microb Pathog 122:162–173

Sadekuzzaman M, Yang S, Mizan M, Ha S (2015) Current and recent advanced strategies for combating biofilms. Compr Rev Food Sci Food Saf 14:491–509

Saha A, Yadav R, Rajendran N (2014) Biomaterials from sponges, ascidians and other marine
organisms. Int J Pharm Sci Rev Res 27:100–109

Saji V, Thomas J (2007) Nano-materials for corrosion control. Curr Sci 92:51–55

Salta M, Wharton JA, Blache Y, Stokes KR, Francois J (2013a) Briand marine biofilms on artificial
surfaces: structure and dynamics. Environ Microbiol 15:2879–2893

Salta M, Wharton JA, Blache Y, Stokes KR, Briand JF (2013b) Marine biofilms on artificial surfaces: structure and dynamics. Environ Microbiol 15:2879–2893

Salta M, Wharton JA, Dennington SP, Stoodley P, Stoke KR (2013c) Anti-biofilm performance
of three natural products against initial bacterial attachment. Int J Mol Sci 14:21757–22178

Sarro MI, Garcia AM, Moreno DA (2005) Biofilm formation in spent nuclear fuel pools and bioremediation of radioactive water. Int Microbiol 8:223–230

Sarro MI, García AM, Moreno DA (2007) Development and characterization of biofilms on stainless steel and titanium in spent nuclear fuel pools. J Ind Microbiol Biotechnol 34:433–441

Sathe P, Myint MTZ, Dobretsov S, Dutta J (2016) Self-decontaminating photocatalytic zinc
oxide nanorod coatings for prevention of marine microfouling: a mesocosm study. Biofouling
32:383–395

Satheesh S, Soniyamby AR, Sunjaiy Shankar CV, Punitha SMJ (2012) Antifouling activities of
marine bacteria associated with the sponge (Sigmodocia sp). J Ocean Univ China 11:354–360

Satheesh S, Ba-akdah MA, Al-Sofyani A (2016) Natural antifouling compound production by
microbes associated with marine macroorganisms—a review. Electron J Biotechnol 21:26–35

Santos ALSD, Galdino ACM, Mello TPD, Ramos LDS, Branquinha MH, Bolognese AM et al
(2010) What are the advantages of living in a community? A microbial biofilm perspective!
Mem Inst Oswaldo Cruz 113(9)

Sayem SA, Manzo E, Ciavatta L, Tramice A, Cordone A, Zanfardino A et al (2011) Anti-biofilm
activity of an exopolysaccharide from a sponge-associated strain of Bacillus licheniformis.
Microb Cell Factories 10(1):74

Schultz M, Holm BJ, Hertel W (2011) Economic impact of biofouling on a naval surface ship. Biofouling 27:87–98

Schwartz N, Dobretsov S, Rohde S, Schupp PJ (2017) Comparison of antifouling properties of native and invasive Sargassum (Fucales, Phaeophyceae) species. Eur J Phycol 52:116–131

Scopel M, Abraham WR, Antunes AL, Henriques AT, Macedo AJ (2014) Mevalonolactone: an inhibitor of staphylococcus epidermidis adherence and biofilm formation. Med Chem 10:246–251

Shan C, Jia Dao W, Hao Sheng C, DaRong C (2011) Progress of marine biofouling and antifouling technologies. Chin Sci Bull 56:598–612

Shankar PD, Shobana S, Karuppusamy I, Pugazhendhi A, Ramkumar VC, Arvindnarayan S, Kumar G (2016) A review on the biosynthesis of metallic nanoparticles (gold and silver) using bio-components of microalgae: Formation mechanism and applications. Enzym Microb Technol 95:28–44

Shao CL, Wang CY, Wei MY, Gu YC, She ZG, Qian PY, Lin YC (2011a) Aspergilones A and B, two benzylazaphilones with an unprecedented carbon skeleton from the gorgonian-derived fungus Aspergillus sp. Bioorg Med Chem Lett 21:690–693

Shao CL, Wu HX, Wang CY, Liu QA, Xu Y, Wei MY, Qian PY, Gu YC, Zheng CJ, She ZG, Lin YC (2011b) Potent antifouling resorcylic acid lactones from the gorgonian-derived fungus Cochliobolus lunatus. J Nat Prod 74:629–633

Shao CH, Xu RF, Wang CY, Qian PY, Wang KL, Wei MY (2015) Potent antifouling marine dihydroquinolin-2(1H)-one-containing alkaloids from the gorgonian coral-derived fungus Scopulariopsis sp. Mar Biotechnol 17:408–415

Shen GX, Chen YC, Lin CJ (2005) Corrosion protection of 316 L stainless steel by a TiO₂ nanoparticle coating prepared by sol–gel method. Thin Solid Films 489:130–136

Shi X, Zhu X (2009a) Biofilm formation and food safety in food industries. Trends Food Sci Technol 20:407–413

Shi X, Zhu X (2009b) Biofilm formation and food safety in food industry. Trends Food Sci Technol 20:407–413

Shi J, Votruba AR, Farokhzad OC, Langer R (2010) Nanotechnology in drug delivery and tissue engineering: from discovery to applications. Nano Lett 10:3223–3230

Silkina A, Bazes A, Mouget JL, Bourgougnon N (2012) Comparative efficiency of macroalgal extracts and booster biocides as antifouling agents to control growth of three diatom species. Mar Pollut Bull 64:2039–2046

Simon SF, Duprilot M, Mayer N, García V, Alonso MP, Blanco J, Nicolas-Chanoine MH (2019) Association between kinetic of early biofilm formation and clonal lineage in Escherichia Coli. Front Microbiol 10:1183

Silva ER, Ferreira O, Ramalho PA, Azevedo NF, Bayón R, Igartua A, Bordado JC, Calhorda MJ (2019) Eco-friendly non-biocide-release coatings for marine biofouling prevention. Sci Total Environ 650:2499–2511

Singh BN, Prateeksha, Upreti DK, Singh BR, Defoirdt T, Gupta VK, De Souza AO, Singh HB, Barreira JC, Ferreira IC, Vahabi K (2017) Bactericidal, quorum quenching and anti-biofilm nanofactories: a new niche for nanotechnologists. Crit Rev Biotechnol 37:525–540

Srey S, Jahid IJ, Ha SD (2013) Biofilm formation in food industries: a food safety concern. Food Control 31:572–585

Stewart C, Thompson JAJ (1997) Vertical distribution of butyltin residues in sediments of British Columbia harbours. Environ Technol 18:175–202

Subramani R, Aalbersberg W (2012) Marine actinomycetes: an ongoing source of novel bioactive metabolites. Microbiol Res 167:571–580

Suntharalingam P, Cvitkovitch DG (2005) Quorum sensing in streptococcal biofilm formation. Trends Microbiol 13:3–6

Tan Y, Su M, Leonhard M, Moser D, Greta M, Haselmann Wang J, Eder D, Stickler BS (2018) Enhancing antibiofilm activity with functional chitosan nanoparticles targeting biofilm cells and biofilm matrix. Carbohydr Polym 200:35–42

Tang X, Flint BSH, Brooks JD (2010) The efficacy of different cleaners and sanitisers in cleaning biofilms on UF membranes used in the dairy industry. J Membr Sci 352:71–75

Tarifa MC, Jorge E, Brugnoni LLI (2018) Disinfection efficacy over yeast biofilms of juice processing industries. Food Res Int 105:473–481

Tello E, Castellanos L, Arevalo-Ferro C, Rodríguez J, Jiménez C, Duque C (2011) Absolute stereochemistry of antifouling cembranoid epimers at C-8 from the Caribbean octocoral Pseudoplexaura flagellosa. Revised structures of plexaurolones. Tetrahedron 67:9112–9121

Thenmozhi R, Nithyanand P, Rathna J, Karutha Pandian S (2009) Antibiofilm activity of coral-associated bacteria against different clinical M serotypes of Streptococcus pyogenes. FEMS Immunol Med Mic 57(3):284–294

Tsukamoto S, Kato H, Hirota H, Fusetani N (1997) Antifouling terpenes and steroids against barnacle larvae from marine sponges. Biofouling 11:283–291

Umezawa T, Oguri Y, Matsuura H, Yamazaki S, Suzuki M, Yoshimura E, Furuta T, Nogata Y, Serisawa Y, Matsuyama-Serisawa K (2014) Omaezallene from red alga Laurencia sp.: structure elucidation, total synthesis, and antifouling activity. Angew Chem Int Ed 53: 3909–3912

Van Wezel AP, van Wlaardingen P (2004) Environmental risk limits for antifouling substances. Aquat Toxicol 66:427–444

Vaikundamoorthy R, Rajendran R, Selvaraju A, Moorthy K, Perumal S (2018) Development of thermostable amylase enzyme from Bacillus cereus for potential antibiofilm activity. Bioorg Chem 77:494–506

Videla HA, Characklis WG (1992) Biofouling and microbially influenced corrosion. Int Biodeterior Biodegrad 29:195–212

Vietti P (2009a) New hull coatings for navy ships cut fuel use, protect environment. Office of Naval Research, Accessed 6 Feb 2019

Vietti P (2009b) New hull coatings cut fuel use, protect environment (PDF). Currents: 36–38. _New_Hull_Coatings.pdf. Accessed 6 Feb 2019

Vuotto C, Longo F, Pascolini C, Donelli G, Libori MF, Tiracchia V, Salvia A, Varaldo PE (2017) Biofilm formation and antibiotic resistance in Klebsiella pneumoniae urinary strains. J Appl Microbiol 123:1003–1018

Wahl M, Goecke F, Labes A, Dobretsov S, Weinberger F (2012) The second skin: Ecological role of epibiotic biofilms on marine organisms. Front Microbiol 3:2

Walt DR, Smulow JB, Turesky SS et al (1985) The effect of gravity on initial microbial adhesion. J Colloid Interface Sci 107:334–336

Wang J, Lin J, Zhang Y, Zhang J, Feng T, Li H, Wang Y (2019) Activity improvement and vital amino acid identification on the marine-derived quorum quenching enzyme MomL by protein engineering. Mar Drugs 17(5):300

Wang KL, Xu Y, Lu L, Li Y, Han Z, Zhang J (2015) Low-toxicity diindol-3-ylmethanes as potent antifouling compounds. Mar Biotechnol 17:624–632

Wang H, Wang H, Xing T, Wu N, Xu X, Zhou G (2016a) Removal of Salmonella biofilm formed under meat processing environment by surfactant in combination with bio-enzyme. LWT-Food Sci Technol 66:298–304

Wang X, Huang Y, Sheng Y, Su P, Qiu Y, Ke C, Feng D (2016b) Antifouling activity towards mussel by small-molecule compounds from a strain of Vibrio alginolyticus bacterium associated with sea anemone Haliplanella sp. J Microbiol Biotechnol 27:460–470

Wang H, Qi J, Dong Y, Li Y, Xu X, Zhou G (2017a) Characterization of attachment and biofilm formation by meat-borne Enterobacteriaceae strains associated with spoilage. LWT 86:399–340

Wang KL, Wu ZH, Wang Y, Wang CY, Xu Y (2017b) Mini-review: antifouling natural products from marine microorganisms and their synthetic analogs. Mar Drugs 15:266

Wang H, Teng F, Yang X, Guo X, Tu J, Zhang C (2017c) Colonization and biofilm development in marine environments. Microbiol Mol Biol Rcv 80.91–138

Wansah JF, Udounwa AE, Ahmed AD, Essiett AA, Jackson EU (2014) Application of nanotechnology in the corrosion protection of steel oil pipes. Proceedings of the 1st African International Conference/Workshop on Applications of Page 103 Nanotechnology to Energy, Health and Environment, UNN, March

Willemsen PR, Ferrari GM (1993) The use of anti-fouling compounds from sponges in anti-fouling paints. Surface Coat Int 10:423–427

Xing Q, Gan LS, Mou XF, Wang W, Wang CY, Wei MY, Shao CL (2016) Isolation, resolution and biological evaluation of pestalachlorides E and F containing both point and axial chirality. RSC Adv 6:22653–22658

Xu Y, Li H, Li X, Xiao X, Qian PY (2009) Inhibitory effects of a branched-chain fatty acid on larval settlement of the polychaete hydroides elegans. Mar Biotechnol 11:495–504

Xu Y, He H, Schulz S, Liu X, Fusetani N, Xiong H, Xiao X, Qian PY (2010) Potent antifouling compounds produced by marine Streptomyces. Bioresour Technol 101:1331–1336

Xu D, Jia R, Li Y (2017) Advances in the treatment of problematic industrial biofilms World. J Microbiol Biotechnol 33:97

Yang LH, Miao L, Lee OO, Li X, Xiong H, Pang KL (2007) Effect of culture conditions on antifouling compound production of a sponge-associated fungus. Appl Microbiol Biotechnol 74:1221–1231

Yang JL, Li YF, Guo XP, Liang X, Xu YF, Ding DW, Bao WY, Dobretsov S (2016a) The effect of carbon nanotubes and titanium dioxide incorporated in PDMS on biofilm community composition and subsequent mussel plantigrade settlement. Biofouling 32:763–777

Yang JL, Li YF, Liang X, Guo XP, Ding DW, Zhang D, Zhou S, Bao WY, Bellou N, Dobretsov S (2016b) Silver nanoparticles impact biofilm communities and mussel settlement. Sci Rep 6:37406

Yebra DM, Kiil S, Dam-Johansen K (2004) Antifouling technology- past, present and future steps towards efficient and environmentally friendly antifouling coatings. Prog Org Coat 50:75–104

Zheng CJ, Shao CL, Wu LY, Chen M, Wang KL, Zhao DL (2013) Bioactive phenylalanine derivatives and cytochalasins from the soft coral-derived fungus, *Aspergillus elegans*. Mar Drugs 11:2054–2068

Zhu XY, Lubeck J, Kilbane JJ (2003) Characterization of microbial communities in gas industry pipelines. Appl Environ Microbiol 69:5354–5363

Zurob E, Dennett G, Gentil D, Montero-Silva F, Gerber U, Naulín P, Ramírez C (2019) Inhibition of wild Enterobacter cloacae biofilm formation by nanostructured graphene-and hexagonal boron nitride-coated surfaces. Nano 9(1):49

Chapter 11
Effect of Biotic and Abiotic Factors on the Biofilm Formation in Gram-Negative Non-fermenting Bacteria

Anna V. Aleshukina, Elena V. Goloshva, Iraida S. Aleshukina, Zinaida M. Kostoeva, Andrey V. Gorovtsov, and Elena A. Vereshak

Contents

Abstract Biofilm microorganisms are protected from the adverse factors of a physical, chemical, and biological nature, which include temperature effects, desiccation, ultraviolet radiation, antibiotics, disinfectants, and humoral and cellular factors of the immune system. Therefore, in recent years, a new strategy in the treatment of infections associated with biofilms is the search for the means of destroying the biofilms of bacteria. The chapter highlights the current understanding of the biofilm-forming ability of bacteria with an emphasis on this property in non-fermentative bacteria. Long-term monitoring of the circulation of these microorganisms indicates their stable detection among the dominant potential pathogens associated with the provision of medical care. The results of studies of biofilm formation among the

A. V. Aleshukina · E. V. Goloshva · I. S. Aleshukina
Laboratory of Virology, Microbiology and Molecular Biological Research Methods, Rostov Research Institute of Microbiology and Parasitology, Rostov-on-Don, Russia

Z. M. Kostoeva · A. V. Gorovtsov (✉) · E. A. Vereshak
Southern Federal University, Rostov-on-Don, Russia
e-mail: zmkostoyeva@sfedu.ru

© Springer Nature Switzerland AG 2020   273
R. Prasad et al. (eds.), *Nanostructures for Antimicrobial and Antibiofilm Applications*, Nanotechnology in the Life Sciences,
https://doi.org/10.1007/978-3-030-40337-9_11

detected non-fermenting bacteria are presented. The possibilities of mass spectrometric analysis in the selection of effective disinfectants in hospitals are shown.

Keywords Biofilm · Quorum sensing · Microbiota · Mass spectrometry

11.1 Introduction

Gram-negative bacteria are the most common cause of bacterial infections in many countries. Infections caused by non-fermentative Gram-negative bacteria (NFGNB) pose a serious problem under current healthcare conditions, especially for immuno-compromised patients. The difficulties of treating, controlling, and preventing infections caused by NFGNB are associated with their widespread distribution in the environment and multiple, natural or acquired antimicrobial resistance. Surveillance and Control of Pathogens of Epidemiological Importance (SCOPE) studies have shown that approximately one-fourth of Gram-negative bacteria belongs to NFGNB (Enoch et al. 2007; Rattanaumpawan et al. 2013).

The group of non-fermenting bacteria includes Gram-negative bacilli, which oxidize, but not ferment carbohydrates. The NFGNB that cause infections most frequently include representatives of several genera, which are conventionally divided into two groups: oxidase-positive, the genera *Pseudomonas* (except for the species *P. luteola* and *P. oryzihabitans*), *Burkholderia, Moraxella, Chryseobacterium, Alcaligenes, Eikenella, Flavobacterium,* and *Kingella,* and oxidase-negative, genera *Stenotrophomonas, Acinetobacter,* and *Bordetella* (except *B. pertussis, B. avium, B. bronchiseptica, B. hinzii*). Many of these bacteria grow well on simple nutrient media and normally inhabit the environment (soil, water, plants) and are also found in the human body. At the same time, they are the causative agents of opportunistic infections of humans and animals (Shaginyan and Chernukha 2005).

NFGNB are medium and small rods (turning into cocci); the size of cells may vary depending on the composition of the nutrient medium and the growth phase of the microbial population. Some representatives (*Moraxella* spp.) are highly resistant to alcohol discoloration. They do not form spores; some species are mobile due to polarly located or peritrichous flagella. For certain species (*Acinetobacter* spp.), the so-called twitching movement is characteristic, which is observed more frequently with increasing pH of the medium to 9.0. Many species form capsules (Zubkov 2003).

With the exception of *Burkholderia pseudomallei,* the causative agent of a specific nosological form—melioidosis, all other NFGNB cause opportunistic infections without a certain localization and pathognomonic symptoms (Demikhovskaya 2012).

The clinical forms of nosocomial infections associated with these microorganisms are largely of the same type. *Pseudomonas aeruginosa* and *Acinetobacter* spp. most often act as the etiological factor of nosocomial and ventilator-associated pneumonia; less often, they cause catheter-associated angiogenic sepsis. *Pseudomonas aeruginosa* also causes chronic lung infections in individuals with cystic fibrosis. The role of NFGNB should be considered for tertiary peritonitis, traumatic and implant-related osteomyelitis, arthritis, and meningitis and abscess of the brain after neurosurgical interventions. *Pseudomonas aeruginosa* is considered to be one of the most common causes of infection of burn wounds, and *Acinetobacter* may have etiological significance in peritonitis associated with dialysis, liver abscesses, and eye infections (Maurice et al. 2018).

Stenotrophomonas maltophilia, formerly known as *Pseudomonas maltophilia*, and then *Xanthomonas maltophilia*, is a common commensal that is often isolated from water, soil, and wastewater. However, it has also become an important agent causing opportunistic infections in immunocompromised patients who have undergone transplantation or suffer from oncological and hematological diseases (Gales et al. 2001).

Common risk factors for infections caused by these NFGNB include three groups of predictors interconnected with the severity of the patient's condition: duration of use of invasive treatment and monitoring methods (mechanical ventilation, prolonged bladder catheterization, central venous catheterization), ICU duration, and duration of antibacterial therapy.

11.2   Biofilm Bacteria: The Problem of Our Time

According to modern concepts, a biofilm is a continuous multilayer of bacterial cells attached to the interface of the phases and to each other and enclosed in a biopolymer matrix (Smirnova et al. 2010). Such bacterial communities can be formed by bacteria of one or several species and consist both of actively functioning and resting or uncultivated forms. Morphological differentiation of the community occurs in response to the action of such environmental factors as temperature, pH of the medium, and osmolarity and is accompanied by a change in metabolism. Thus, biofilms represent communities of microorganisms, the components of which perform various functions and have a single goal—maximum adaptation to a given ecological niche and the most effective distribution of the population. The intercellular interaction in the biofilm is carried out with the participation of a regulatory mechanism called "quorum sensation" (QS) (Ilyina et al. 2004; Chernukha et al. 2009; Romanova and Ginzburg 2011). Information is exchanged with the help of specialized chemical signaling molecules, helping the microbial community to act as a single organism, demonstrating a change in the phenotype, expressed in differences in growth and expression of specific genes. Details on this will be discussed below. Currently, QS regulation is found in more than 50 species of bacteria.

Different types of bacteria use QS for a coordinated response consistent with their local population density. QS-type gene regulation systems play a key role in the interaction of bacteria with higher organisms, animals, and plants, both in pathogenesis and in symbiosis. For the first time, the phenomenon of gene regulation by

QS type was discovered in the 1970s in *Aliivibrio* sea bacteria. However, the term "quorum sensing" was first introduced in 1994 by Greenberg E.P. with co-authors. As a result of active research, it was discovered that the QS phenomenon is widespread among bacteria. It turned out that a number of genes that determine the virulence of pathogenic microorganisms and drug resistance are regulated according to the QS principle (Manukhov 2011).

In the formation of biofilms there are several stages: adhesion, colonization, maturation, and dispersion:

Stage 1: Reversible adhesion to the surface. Most often, microorganisms exist in the form of free-floating masses or single (plankton) colonies. However, under normal conditions, most microorganisms tend to attach themselves to the surface and, ultimately, form a biofilm.

Stage 2: Permanent adhesion to the surface. As the bacteria multiply, they adhere more firmly to the surface, differentiate, and exchange genes, which ensures survival.

Stage 3: Formation of a mucous protective biofilm matrix. After firmly adhering to the surface, bacteria begin to form an exopolysaccharide surrounding matrix, known as the extracellular polymeric substance (EPS matrix). Small colonies of bacteria then form the initial biofilm. The composition of the mucus that the matrix consists of varies according to the species of microorganisms present in it, but the main components are polysaccharides, proteins, glycolipids, and bacterial DNA.

Stage 4: Dispersion is characterized by a periodic separation of individual cells from the biofilm that can eventually attach themselves to the surface and form a new colony (Afinogenova and Darovskaya 2011).

With the help of confocal laser scanning microscopy, it was shown that biofilms are heterogeneous in time and space. The cells in the mucous matrix are arranged in mushroomlike, column-like formations enclosed in an exopolysaccharide layer, which allows maintaining the concentration of nutrients necessary for population growth and also serves to protect cells from dehydration, as well as from humoral and cellular factors of the macroorganism resistance (Gostev and Sidorenko 2014). The matrix is divided by channels, cavities, and voids filled with water, ensuring the distribution of nutrients and the exchange of metabolic products with the environment. In the biofilm community, different bacteria are often closely related, since they perform different but mutually useful functions. The metabolic products (in fact, waste) of some bacteria can be used by others as the main source of food, and sometimes the reaction carried out by the first group of organisms can proceed only if the second group removes the final product from the environment. All biofilms have their own microenvironment, differing in pH levels, absorption of nutrients, and oxygen concentrations.

The process of biofilm formation in microbes is influenced by abiotic (temperature, desiccation, ultraviolet radiation, chemicals of various nature, etc.) and biotic (interaction with other biological objects) factors that have both a stimulating and depressing effect (Goloshva 2017).

Biofilms as a "fixed" form of the existence of microorganisms in the environment have long been known. Pathogenic and opportunistic bacteria, such as *Vibrio cholerae, Legionella, Listeria, Campylobacter, Klebsiella, Salmonella,* and *Pseudomonas*, inhabiting mainly soils and water bodies, attach to soil particles or exist in various associations with algae, rhizosphere of plants, protozoa, crustaceans, and other organisms. The study of the environmental patterns of biofilm formation makes it possible to identify the role of abiotic and biotic factors in the implementation of this existence strategy in nature.

In the works of various researchers (Dronina et al. 2012; Mayansky et al. 2011), it was shown that biofilms consisting of microorganisms of different taxa are stronger and thicker than biofilms formed by one species. The interaction probably occurs at the stage of formation of the extracellular matrix. In "mature" biofilms, unlike planktonic structures, competition between species is rarely found. Even in the case when one of the species, due to the high growth rate, occupies a leading position, the second retains its viability and high numbers. Similar relationships were found in biofilms formed by populations of rapidly growing *Klebsiella pneumoniae* and *P. aeruginosa*. A rare kind of competition can be observed when one species produces substances that inhibit the other species within the biofilm community. An example of such interaction can be found in a biofilm formed by two species of *Ruminococcus*, one of which forms a bacteriocin that is active against another. The most common types of relationships between the components of biofilms are commensalism and protocooperation. Commensalism is expressed in the one-sided influence of one of the biofilm components on the vital activity of another component. An example is oxygen consumption by an aerobic microorganism that promotes the growth of microaerophilic or anaerobic microorganisms. This type of relationship plays a role in microbial corrosion involving sulfate reducers. Protocooperation is expressed in the mutual positive influence of the components of biofilms on each other. Such relationships exist in biofilms formed by phototrophic and heterotrophic microorganisms. A typical example of the protocooperation is the interaction of cellulose-fermenting bacteria and methanogens. The latter, utilizing molecular hydrogen and formate formed during the fermentation, shift the thermodynamic equilibrium, preventing the accumulation of reduced coenzymes in fermenting cells and stimulating their synthesis of ATP. The previous case is an example of the transition of protocooperation to synergy, in which both components of a biofilm benefit from cooperation, and the formation or consumption of any product in a biofilm exceeds the value characteristic of individual populations. A typical example is the hydrolysis of cellulose in a biofilm containing both cellulolytic bacteria and microorganisms that are incapable of cellulose cleavage. The latter stimulate cellulose hydrolysis and cellulolytic bacteria growth by consuming low-molecular-weight hydrolysis products that repress cellulase biosynthesis (Nikolaev and Plakunov 2007),

An important factor influencing the formation of biofilms is the ability of microorganisms to produce mucous polymeric compounds, allowing them to adhere to various surfaces, maintain functional stability, and also survive in various ecological niches. Microorganisms inhabiting aquatic ecosystems are exposed to various organic substances that are products of human industrial activities. In the work of Litvinenko (2015), it is shown that organic substances are important regulators of biofilm formation in the underground hydrosphere. The presence of $Fe(OH)_3$ in the aquifer together with traces of such organic substances as phenanthrene and naphthalene can have a decisive influence on the adhesive activity of microorganisms and their ability to form biofilms. It was experimentally shown that the intensive formation of mucous polymeric compounds by microorganisms occurred in the presence of yeast extract. The intake of readily available nitrogen-containing organic substances increases the proportion of bacteria involved in the nitrogen cycle (ammonifiers, nitrifiers, denitrifiers) and provokes their development of biofilms. Under the limited amount of organic substances, bacterial cells were attached directly to insoluble $Fe(OH)_3$ particles. Electron microscopy showed the presence of large calcified globules in the composition of bacterial films. Their formation can be associated with the accumulation of Ca ions on the surface of bacterial cells, which increase the viscosity and elasticity of the biopolymer matrix of microbial films.

Unlike calcium ions, copper and zinc cations are able to inhibit biofilm formation in *S. aureus* strains (Cheknev et al. 2014).

Another example of mutually beneficial biofilm community is the coexistence of biofilm-forming bacteria, such as *P. aeruginosa, E. coli, Enterococcus* sp., *Klebsiella* sp. and their moderate phages. Studies in various hospitals (Zueva et al. 2014) showed that, for example, for the *Pseudomonas aeruginosa*, the frequency of joint isolation with the *Pseudomonas* bacteriophage was approximately 30%. At the same time, more than 60% of *P. aeruginosa* were not lysed by phage isolated from the material, taken from the same site. Moreover, the phages present in such biofilm ecosystems, not possessing any virulence of their own, having neither clinical nor epidemiological effects, can contribute to the formation of pathogenic properties in bacteria due to their acquisition of phage-mediated virulence genes. The mechanism involved in bacteriophages mediating horizontal genetic exchange between strains is lysogenic conversion or transduction. The interaction between moderate bacteriophage and bacteria, leading to the emergence of virulence genes in bacteria, enhances the microbial ability to persist in the host organism, increasing pathogenicity.

11.3 The Quorum-Sensing System from the Standpoint of Biochemistry and Genetics

Extracellular polymeric substance (EPS, exopolysaccharide) can represent from 50 to 90% of the total organic matter of biofilms and can be considered the primary material of the biofilm matrix (Sreeremya 2017). EPS, synthesized by bacterial cells, vary greatly in their composition and, consequently, in their chemical and physical properties. Some of them are neutral macromolecules, but most of them are polyanionic due to the presence of uronic acids (the most common is D-glucuronic acid, although D-galacturonic and D-mannuronic acids are also found) or pyruvate ketal groups. Inorganic residues, such as phosphate or less commonly sulfate, can also give polyanionic status (Sutherland 2001). The biopolymer matrix can also include metal ions, proteins (up to 60%), nucleic acids (up to 20%), lipids (up to 40%), and even humic substances. It is known that the synthesis of EPS is influenced by the nutrient content—an excess of available carbon and the restriction of nitrogen, potassium, and phosphate contribute to the synthesis of EPS (Shlepotina et al. 2014; Sreeremya 2017).

The formation and maintenance of structured multicellular bacterial communities depend entirely on the production and quantity of EPS. The concentration, adhesion, charge, sorption capacity, specificity, and character of the individual components of the EPS, as well as the three-dimensional structure of the matrix (dense areas, pores, and channels), determine the lifestyle of microorganisms in this biofilm. In summary, the morphology of a biofilm can be smooth and flat, rough, loose, or fibrous and also vary depending on the degree of porosity, the presence of voids, and channels filled with water (Flemming and Wingender 2010).

EPS provides mechanical stability of biofilms, contributes to their adhesion to surfaces, and forms a mucous polymer that binds and immobilizes the cells of the biofilm. In addition, the biofilm matrix acts as an external digestive system—by keeping extracellular enzymes close to cells, it allows them to metabolize dissolved, colloidal, and solid biopolymers (Flemming and Wingender 2010). It is the presence of the matrix that determines the specific properties of biofilms: resistance to the action of antibiotics, disinfectants, high temperatures, desiccation, ultraviolet radiation, factors of the humoral and cellular immunity of the microorganism, the ability to coordinate gene expression, and the exchange of genetic information between the biofilm microorganisms (Shlepotina et al. 2014).

The resistance of biofilm bacteria to the listed stress effects is due to a number of factors:

1. Reducing the possibility of entrance of antibacterial drugs into the biofilm. The biofilm matrix, which surrounds the bacterial cells, can directly bind or block the diffusion, and/or inactivate antibiotics. Inactivation of antibiotics was demonstrated on biofilm-forming *P. aeruginosa*, which have an increased ability to produce β-lactamases. The resistance is also associated with the filtering ability of the matrix. For example, the difficulty of ciprofloxacin entrance into the biofilm formed by *P. aeruginosa* was described. In addition, the elements of the matrix are not only a passive filter. Glycerol-phosphorylated beta-glucans produced by *P. aeruginosa* not only slow down the diffusion of aminoglycosides through the biofilm but also actively bind the antibiotics. Alginate mucus produced by *P. aeruginosa*, filling the extracellular space in biofilms, may also have an antibiotic binding effect (Chebotar et al. 2012; Chernyavsky 2013).

2. The slow rate of division and growth of bacteria in the biofilms compared to planktonic cultures. Within the biofilm, there is a gradient in the content of nutrients and oxygen. As a result, most of the metabolically active cells are located on

the periphery of biofilms, while the inner parts of the biofilms are dominated by metabolically inactive and dormant cells. It is known that actively growing cells are targets for the action of antibiotics; therefore, bacteria inside biofilms are insensitive to their effects.

3. The presence of the so-called persisting cells in the biofilms, constituting 1–10% of biofilms and resistant to the effects of both aggressive environmental factors and antimicrobial agents. The phenomenon of persistence is manifested in the fact that in the presence of "bacteriostatic" concentrations of antibiotics-inhibitors of protein synthesis, the formation of some proteins, especially those included into the cytoplasmic membrane, continues, albeit at a reduced rate. At the same time, there is a slow increase in the optical density of the bacterial population due to the growth of a part of the cells (Plakunov et al. 2010).

The genetic factors responsible for the level of resistance of biofilm bacteria to stress factors are also known. For example, the ndvB gene is expressed by *P. aeruginosa* only in biofilm state, and mutations in this gene make the biofilm much more sensitive to antibiotics compared to wild strains (Chernyavsky 2013).

As already mentioned, bacterial communities in biofilms are social systems characterized by certain cooperation and functional organization, implemented by the QS system. Global regulation provides both the formation of a biofilm and increased bacterial adhesion, the beginning of the synthesis of factors associated with the antagonistic activity (Rybalchenko and Bondarenko 2013).

From a genetic perspective, quorum sensing (QS) is the process of collective coordination of gene expression in a bacterial population that mediates the specific behavior of cells. The mechanism of QS is based on a complex hierarchical regulation of the target loci of the bacterial cell genome. In this case, regulation is carried out at different levels of exposure: transcriptional, translational, and posttranslational. For a specific cellular signal, the cells in the population respond with a specific response. To date, it has been established that cell-to-cell interactions affect cell differentiation, expression of virulence genes, growth processes, the nature, and direction of mobility (taxis), as well as bacterial apoptosis and toxin formation (Gostev and Sidorenko 2014).

The operation of the QS system can be considered using the example of *P. aeruginosa*. The central components of quorum sensing are the las and rhl systems, which activate gene expression depending on the density of microorganism cells. Each system is represented by two genes: one encodes an enzyme which synthesizes a specific autoinducer—an acylated homoserine lactone (lasI/rhlI); the other encodes a transcription activator with which the corresponding autoinducer binds (lasR/rhlR). The autoinducers of the las and rhl systems are N-(3-oxododecanoyl)-L-homoserine lactone (3-oxo-C_{12}-HSL) and N-butyryl-homoserine lactone (C_4-HSL), respectively. There is a special system, called MexEF-OprN pump, for the export of 3-oxo-C12-HSL from the cell. The las system controls the expression of genes encoding virulence factors such as elastases A and B, as well as an alkaline protease; rhl system controls the enzymes of rhamnolipid biosynthesis and pyocyanin. Recently, a third signaling molecule has been discovered that participates in quorum-sensing reactions in *P. aeruginosa*—2-heptyl-3-hydroxy-4-quinolone (PQS). This signaling molecule can control the expression level of las B, which encodes las B elastase, as well as the expression level of rhll, which encodes C_4-HSL synthase.

Mutations in the lasl gene disrupt the maturation of the biofilm, since the LasI protein does not synthesize 3-oxo-C_{12}-HSL, and the formation of the microfilm does not continue after the microcolony stage. The role of C_4-HSL in the formation of a biofilm remains unknown. Biofilms formed by LasI protein mutants are susceptible to detergents, while normal biofilms are stable (Gruzina 2003).

As part of the biofilm, microbes have a high resistance to effectors of the immune system, antibiotics, and disinfectants. Biofilm bacteria are able to survive when exposed to antibiotics in so high concentrations that they cannot be achieved in the human body at standard therapeutic dosages (Chebotar et al. 2012). Paradoxical facts of enhancing biofilm growth of *Staphylococcus capitis* in the presence of maximum therapeutic concentrations of oxacillin (Cui et al. 2012) and enhancement of the biofilm formation process by strains of *Staphylococcus aureus* under the influence of ciprofloxacin (Horovits et al. 2016) have been reported.

11.4 Abiotic and Biotic Factors That Inhibit Biofilm Formation

One of the ways to combat biofilms on biotic surfaces may be the use of substances that can interfere with the adhesion of pathogens and affect the disorganization of the extracellular matrix. In experimental and clinical studies, it was found that the drug *N*-acetylcysteine, which has pronounced mucolytic properties, is able to reduce the adhesion of certain pathogens to the mucous membranes of the respiratory tract and also have a direct destructive effect on the extracellular matrix. The combination of antimicrobial agents (thiamphenicol) with *N*-acetylcysteine has been shown to have high clinical efficacy in the eradication of biofilms from mucous membranes (Riise et al. 2000; Macchi et al. 2006).

Improving the delivery of antimicrobial drugs can help improve the penetration of biofilms. For example, the liposomal complex of amphotericin B has a more pronounced activity against resistant biofilms formed by *Candida* spp. (Lisovskaya et al. 2016).

Nanoparticles of various bactericidal substances, in particular nano-silver preparations (NSPs), pretend to be an alternative to antibiotics. Antimicrobial properties of silver have been known for a very long time. Silver ions exhibit a wide spectrum of action; they inhibit the growth of pathogenic bacteria, viruses, and fungi. Regarding the mechanism of action of silver on bacteria, it is known that silver compounds interact with proteins, various enzymes, and nucleic acids; cause structural changes in the cell wall and membranes of bacteria; and inhibit the cell's electron transfer chains, which ultimately leads to the destruction and death of bacteria.

However, many issues related to the action of silver nanoparticles (SNPs) and ions on bacteria, including the mechanisms of transport of NSPs in a bacterial cell, genetic control of the sensitivity, and resistance of bacterial cells to silver compounds, are little studied. Nano-silver prevents the adhesion of bacteria to the cell membranes, preventing the first stage of film formation. The use of nano-silver preparations is also promising in that, together with a pronounced antibacterial effect and a broad spectrum of action, the use of these drugs on bacteria does not form resistance (Chestnova et al. 2017). According to M.A. Radzig, the concentrations of $AgNO_3$, which suppress the growth of planktonic cells and the formation of biofilms, were significantly lower than in the case of the effect of NSPs. The almost simultaneous decline of planktonic growth and the formation of biofilms suggest that the decrease in the formation of biofilms is due to the suppression of bacterial growth in the presence of silver compounds and is not associated with the specific action of silver ions and nanoparticles directly on the formation of biofilms (Radzig 2013).

It is proposed to use nano-silver preparations as a disinfectant. Such drugs have filled the market and are actively advertised by companies as the safest for the environment and humans (AgBion-2, Sumerian Silver, etc.). Along with nano-silver, nano-variants of disinfectants based on active chlorine ("Nanoclo-2") of active oxygen ("Nika-peroxam") and combinations of nano-silver and active oxygen ("Silversil") are proposed for use.

Reactive oxygen species (ROS) are able to have a regulating influence on the interaction of bacteria with the surfaces that are colonized, determining the structural and functional state of microbial biocenoses (Bukharin and Sgibnev 2012). The study of the influence of low concentrations of ROS on the ability of bacteria to form biofilms and their adhesive characteristics showed a decrease in the adhesive properties of the studied microorganisms under the influence of hydrogen peroxide and hydroxyl radicals in concentrations of 0.5 and 0.05 mM. The stimulation of all studied concentrations of ROS on the biofilm formation was noted for *E. coli*, *S. aureus*, and *K. pneumoniae*, which can be explained by the stimulating effect of ROS on the production of extracellular polymers by bacteria, forming the matrix of biofilms. The obtained data correlate with the results of the study of the electromagnetic radiation influence on the biofilm formation in *P. aeruginosa*. The electromagnetic radiation at the frequencies of the molecular emission and absorption spectra of atmospheric oxygen stimulated the ability of bacteria to form biofilms (Pronina et al. 2010). It is possible that the change in the adhesive ability of bacteria and their ability to form biofilms is an adaptive response of microorganisms under conditions of oxidative stress.

One of the proposed methods of abiotic influence may be the use of electromagnetic radiation (Pronina et al. 2010). The effect of electromagnetic radiation at the frequencies of the molecular emission and absorption spectra of atmospheric oxygen and nitric oxide on adhesion, population development, and the formation of *P. aeruginosa* biofilms was studied. It was shown that irradiation at the frequency of the molecular emission and absorption spectra of nitric oxide reduced the ability of bacteria to form biofilms, which may be explained by the influence of electromagnetic radiation on the induction of ROS generated by both bacterial cells and immunocompetent cells of the microorganism.

When studying the effect of a pulse-periodic corona discharge on the viability of *E. coli* M-17 cells in biofilms, the disruption of the integrity in the surface and deep structures of biofilms was detected as well as changes in the morphological properties of *E. coli* M-17 cell's characteristic of sublethal thermal effects. The bactericidal effect of a pulse-periodic corona discharge on *E. coli* M-17 cells as part of biofilms has been shown (Rybalchenko et al. 2015).

In addition to the classical methods of antibacterial therapy, effective suppression of bacterial biofilms formed in wounds by ultrasound cavitation in case of purulent-septic complications of diabetic foot syndrome has been shown (Risman et al. 2011).

The most promising direction in finding ways to prevent the formation of bacterial biofilms and their destruction is the use of biotic substances, in view of their potential safety for the host organism. Bactericidal factors of cells and tissues perform an important function in the protection of the microorganism. A special place is occupied by low-molecular-weight proteins (peptides) with pronounced cationic properties, which include, in particular, platelet cationic protein. During the experiments, it was established that platelet cationic proteins are able to reduce the biofilm formation of *S. aureus* by almost 50% and increase the hydrophobicity of the planktonic and biofilm fractions of the microbial population (Zhurlov et al. 2012).

The value of lysozyme as a factor of the immune system capable of inhibiting coagulase-negative staphylococci biofilm formation with direct exposure to strains isolated from healthy people was shown (Gordina et al. 2016).

The anti-staphylococcal antibiotic batumin isolated from a strain of soil bacteria of the genus *Pseudomonas* was shown to effectively inhibit the biofilm formation of *Staphylococcus aureus*, *Klebsiella*, and *Candida*, isolated from humans (Bukharin et al. 2012). The obtained data on the inhibition of biofilm formation by low-molecular-weight substances of biotic origin opens up the prospect for their further use as drugs suitable for the fight against biofilms of persistent microorganisms.

Of particular importance is the interaction of the host's immune system with bacteria that are in the composition of biofilms (Bekhalo et al. 2010). It has been proven that neutrophils are able to actively destroy biofilms due to phagocytosis, elastase and lactoferrin secretion, and opsonin-dependent interaction with biofilms. In the study of the effects arising from the interaction of neutrophils with biofilms of *Staphylococcus aureus* in vitro, it was shown that human neutrophils have a destructive effect on biofilms, destroying the extracellular matrix by exocytosis and direct phagocytosis (Chebotar et al. 2012). The inhibition of biofilm formation of *S. aureus* and *E. coli* under the influence of secretory products of vaginal epithelial cells isolated from biotopes with normocenosis was also shown (Bukharin et al. 2012). A study of the combined effect of lactoferrin and ciprofloxacin antibiotic on growth and biofilm formation by *P. aeruginosa* showed that using a combination of lactoferrin and the doses of antibiotic, reduced compared with conventional therapeutic ones, inhibits the ability of bacteria *P. aeruginosa* to form biofilms.

Microorganisms of various species are able to form a biofilm together. The genetic diversity of microbes in biofilm communities makes them more resistant to stress factors. Community composition and interactions within it can have a profound effect on the behavior of bacteria. The interactions between bacteria are complex and highly dependent on the context. They can range from fierce competition for nutrients and niches, manifested by antagonistic behavior, to highly developed cooperation between different species that support their mutual growth under specific conditions (Gabrilska and Rumbaugh 2015).

To understand whether the interaction between two species of *S. aureus* and *P. aeruginosa* in cystic fibrosis can affect their ability to produce biofilm, Baldan et al. (2014) quantified the biomass of biofilm during individual and joint culture in a 1:1 ratio of *S. aureus* and *P. aeruginosa* by crystal violet staining. The obtained data proved the inhibitory effect of *P. aeruginosa* on the formation of *S. aureus* biofilm.

The ways to combat biofilms on biotic surfaces include the use of substances that influence the mechanisms of primary adhesion of bacteria to the surface, blocking the synthesis or destruction of the polymer matrix, and disruption of intercellular information exchange (Khrenov and Chestnova 2013).

Recently, the phenomenon of stimulating the formation of biofilms in pathogenic microorganisms with low concentrations of antibiotics has attracted increasing attention. A number of studies have revealed the effect of low concentrations of antibiotics on the adhesion of bacteria to solid surfaces. Subinhibitory concentrations of imipenem, which suppresses the synthesis of the bacterial cell wall, have been shown to increase adhesion of *A. baumannii* cells to polystyrene, and the antibiotic tobramycin at subinhibitory concentrations enhances the formation of biofilms by *P. aeruginosa* PAO1. However, other authors have shown that tobramycin subinhibitory concentrations inhibit the *N*-acetylhomoserine lactone-dependent quorum-sensing system in the *P. aeruginosa* PUPa3 strain, leading to a decrease in biofilm formation (Martyanov et al. 2015).

11.5 Biofilms of Human Microbiota

The question of biofilms formed by representatives of normal flora in the body of humans and warm-blooded animals deserves special mentioning.

At the turn of the twenty-first century, a new concept has emerged, viewing the microflora of the human body as an extracorporeal organ that performs multiple functions to maintain normal homeostasis (Shenderov 2001). While remaining invisible to the naked eye, this "organ" weighs about two kilograms and has approximately 10^{14} cells of microorganisms, which is ten times more than the number of the human body's own cells. Representatives of the normal microflora, organized in the biofilm community, form a "stocking" covering the intestinal wall, other mucous membranes, and human skin. The thickness of the biofilm ranges from 0.1 to 0.5 μm. The biofilm consists of a polysaccharide skeleton, including microbial polysaccharides together with mucin, which is produced by the cells of the microorganism. Microcolonies of bacteria that can be arranged in several layers are immobilized in the framework structure. For example, in a biofilm that covers the skin, microcolonies are 1–2 layers and, in the biofilm of the large intestine, 500–1000 layers.

The organization of microorganisms in a biofilm, balanced in species composition and functional distribution of community members, greatly increases the efficiency of the normal human microflora in performing its functions, the most important of which is to ensure colonization resistance. By colonization resistance, we mean a combination of mechanisms that impart stability to the normal microflora and prevent the host from foreign microorganisms entering the body. The decrease in colonization resistance due to various reasons leads to an increase in the number and expansion of the spectrum of opportunistic microorganisms and their translocation from primary localization sites with the possible development of inflammatory diseases (Bukharin et al. 2002; Bondarenko et al. 2003). Colonization resistance is determined by many factors. An important place is occupied by an adhesive and antagonistic activity of microbes that make up the normal microflora, competition for nutritional factors, and others (Goloshva 2005).

According to molecular studies, the composition of the microflora is genetically related within the biofilm community and is specific at the strain level for the individual. Such a microbial community persists throughout life, with a maximum variation in the concentration of individual microbes. Based on known facts, a model of the biofilm of the microbial community of the intestinal wall layer was proposed. Microorganisms in the amount of 10^{11} cells/cm^3 are distributed in the mucin layer consisting of a glycopeptide produced by the goblet cells of the epithelium of the intestinal mucosa, which is similar in chemical nature to the polysaccharide protective capsule, which surrounds many microbes. In a thin layer of mucin, microorganisms are distributed at a fairly close distance from each other (of the microbial cell size order). Such an arrangement provides contact between the cells and the chyme diffusing into the mucin for the rapid exchange of metabolic products, which corresponds to the concept of biofilm as a pseudocytological structure. In a sufficiently thick layer of mucin, the microbial biofilm may look like a layer of epithelium attached to cells or separately located conglomerates of cells (Osipov and Rodionov 2013).

The resistance of normal microflora to the effects of adverse factors of biotic and abiotic nature inside the biofilm is tens and hundreds of times higher than that of non-immobilized cells (El-Azizi et al. 2005). This fact forces us to take a fresh look at the mechanism of the origin and causes of the development of intestinal dysbiosis, as well as the ways of its correction. From this point of view, dysbacteriosis is not considered as fluctuations in the relative content of certain types of microorganisms. Intestinal dysbiosis is a fundamental violation of the structure of the biofilm of the mucous colon, and the collective immunity of the pathological biofilm negates the possibility of correcting dysbacteriosis with the help of probiotics. Biotechnological probiotics are foreign and are not embedded inside the intestinal biofilm, but are rejected due to biological incompatibility.

The widespread pollution of the environment, the accumulation of xenobiotics of a different spectrum in it, and the widespread use of antibacterial drugs lead to microecological disturbances of the evolutionarily established equilibrium between the organism and the microflora inhabiting it. A number of research works were devoted to changes occurring as a result of the irrational impact of broad-spectrum antibiotics on the luminal and parietal (biofilm) microflora of the intestine (Patrusheva 2000; Goloshva 2005; Aleshukina 2012).

Vorobev et al. (2001) showed that with normal microbiocenosis, indicators of the "specific content" of microorganisms in the parietal part of the intestine remain almost unchanged and are regulated by various factors, including the thickness of the mucin layer. Mucin, with its constituent microorganisms, is exfoliated into the intestinal lumen, and thus, the parietal microflora is actively involved in the formation of luminal microflora. Evaluation of the mucinase activity of the microbiota of the luminal part of the intestine can be used as indirect evidence of migration disorders in intestinal biotopes, since the reverse process is not excluded, especially in case of quantitative violations of the microbiota ratios.

The biofilm of the "extracorporeal organ" of the human intestine consists of a complex of microorganisms' biofilms and intestinal mucin. As already noted, the "building material" of this shell or film is predominantly the mucin glycopeptide, the carbohydrate portion of which is mainly represented by galactose. Increased lactose levels, often causing diarrhea in intestinal dysbiosis, lead to an increase in galactose and increased mucin synthesis. According to Aleshukina (2012), an increase in mucin production, in turn, leads to the activation of mucinolytic activity in intestinal bacteria (both resident and transient). It turns out a vicious circle in the substrate-enzyme relations of bacteria in the mucosal layer: an increase in galactose → an increase in the synthesis of mucin (substrate) → an increase in the synthesis of mucinases (enzymes) by bacteria → destruction of the mucin → an increase in the content of galactose. The vicious circle is terminated with the complete depletion of the substrate or a decrease in the number of mucinase-producing bacteria. The combined increase in the proteolytic and mucinase activities of eubionts in dysbiotic disorders contributes to the violation of the peptide and carbohydrate parts of the mucin. As a result, the biofilm becomes thinner; the colonization resistance of the macroorganism is reduced. And along with an increase in the factors of persistence and pathogenicity occurrence in symbionts (*E. coli*) and optionally present enterobacteria (*Klebsiella* sp., *Enterobacter* sp., *Citrobacter* sp., *Proteus* sp.), there is an increase in translocation of bacterial metabolites and the microorganisms themselves into other biotopes (intestine, gallbladder, blood). At the same time, the proteolytic activity and the mucinolytic ability of bifidobacteria and lactobacilli are recorded less frequently, and the degree of manifestation is much lower even during normobiosis. In dysbacteriosis, these indicators are reduced, which also contributes to the reduction of the colonization resistance of the organism in the biotope.

Experimental evaluation of biofilm-forming capacity has shown that symbiotic bifidobacteria have reduced biofilm-forming activity compared with optionally present epidermal staphylococci in the intestinal microbiota. Pseudomonads and

Staphylococcus aureus possessed the most pronounced ability to produce biofilms (Aleshukina et al. 2016). Such behavior of microorganisms of the intestinal microbiota can be regarded as a criterion of pathogenicity.

11.6 MALDI-TOF Mass Spectrometry in the Selection of Disinfectants to Neutralize Biofilm-Forming NFGNB

In this study, the properties of 200 strains of non-fermenting bacteria isolated from patients from various departments of a medical facility were studied.

11.6.1 Mass Spectrometry Studies

For the identification of bacterial species by the MALDI-TOF MS method, a microflex benchtop mass spectrometer with the MALDI Biotyper database (Bruker Daltonics, Germany) was used. Preparation of pure 24-h cultures of strains on solid nutrient media for the study and analysis was carried out according to the instructions for the device. Strains of NGFNB were applied to the steel plate (Bruker Daltonics) in two replications. The target was dried for several minutes (5–15) at room temperature. Then, 1 μl of the CHCA matrix (α-cyano-4-hydroxycinnamic acid) was applied to each sample, dried, and placed in a mass spectrometer for analysis (initial background).

The level of identification of bacteria was interpreted according to the criteria specified in the instructions: 2.300–3.000 high probability of identification of the species; 2.000–2.299 reliable identification of the genus, probable identification of the species; 1.700–1.999 probable identification of the genus; and 0.00–1.699 unreliable identification.

The obtained mass spectra were studied with an estimate of the variation of peaks of different length of the span. In accordance with this parameter, the peaks were divided into groups: high-span "major" peaks (the ratio of ion mass to charge (m/z) is more than or equal to 5000), medium-span peaks (from 5000 to 2500), low-span "minor" peaks (less than or equal to 2500).

Twenty-four-hour broth cultures of NGFNB were exposed to various disinfectants. The strains were exposed to a 0.25% Ultrason solution (DonDe LLC)—and a 0.3% Nika-peroxam solution (Genix NPF LLC). As a control, MS profiles were used in the presence of saline.

Three milliliter of the prepared disinfectant solution was added to polyester Petri dishes (in the number of strains) and incubated at 37 °C for 60 min. This procedure was performed to model the treatment of surfaces with disinfectant solutions. According to the manufacturer, the bactericidal substances are attached to the surfaces and retain their protective effect against the microbes.

The strains were grown on nutrient agar at 37 °C for 24 h. One billion CFU suspension was prepared in the 24-h cultures of microorganisms according to the turbidity standard for 10 McFarland units.

After the incubation, the disinfectant was removed with a pipette, and the plates were washed with saline, which was then also removed by aspiration. The suspensions of microorganisms were added to the treated Petri dishes and incubated at 37 °C for 24 h. At the expiration of the incubation period, the sensitivity to disinfectants was determined by using a microflex MALDI Biotyper mass spectrometer. As a control, MS profiles were used, which were obtained when exposed to saline.

After the application of all tested disinfectants, the proteomic profiles of cultures were characterized by an increase in minor and major MS peaks by a factor of 2–3 compared with the initial profiles and profiles in the presence of an isotonic sodium chloride solution, which indicates the denaturation of proteins and the effectiveness of the disinfectant. In some strains, this effect led to a complete profile change and nonrecognition of cultures during biotyping.

Twenty-seven strains were selected to study their ability to form biofilms, as a factor in the protection of microorganisms from various antimicrobial agents. According to the photometric analysis, all the studied strains had a high ability to form biofilms (biofilm index >1.1). At the same time, 85.7% among *Acinetobacter* and 42.1% among *P. aeruginosa* strains had a very high biofilm formation capacity—more than 2.

The study of *P. aeruginosa* resistance (19 strains) included an assessment of their properties in the biofilm state: therefore, the effect of Ultrason solutions belonging to the group "quaternary ammonium salts + amine + guanidine" and "Nika-peroxam" of the group "quaternary ammonium salts + active oxygen-containing (hydrogen peroxide)," on planktonic and fixed forms. The effectiveness of disinfectants can be estimated by changing the spectra of bacterial proteins. As a control, MS profiles obtained by exposure to a saline solution were used.

When exposed to the Nika-peroxam and Ultrason disinfectants, the planktonic forms of *P. aeruginosa* showed an increase in the frequency of occurrence of proteins with an amplitude of peaks 5681, 7490, 8576, and 9625 Da - Daltons on average by a factor of 2–3. In the fixed forms of pseudomonads, under the action of disinfectants, an increase in the average frequency of the occurrence of proteins with an amplitude of 5692, 7505, 8572, 9713, and 14,705 was observed (Fig. 11.1).

The increase in MS peaks compared to baseline profiles and with the profiles in the presence of an isotonic solution of sodium chloride indicates the denaturation of proteins and the effectiveness of these disinfectants. A more pronounced effect was indicated for the disinfectant "Nika-peroxam" including hydrogen peroxide, which was characterized by an increase in the number of "heavier" major patterns both in the planktonic form and in the fixed one. Such denaturation led in some strains to a complete profile change and nonrecognition of cultures during biotyping. The study of changes in the mass spectrometry profiles of pseudomonads showed the effectiveness of the use of the tested disinfectants for the inactivation of microorganisms at different stages of biofilm formation.

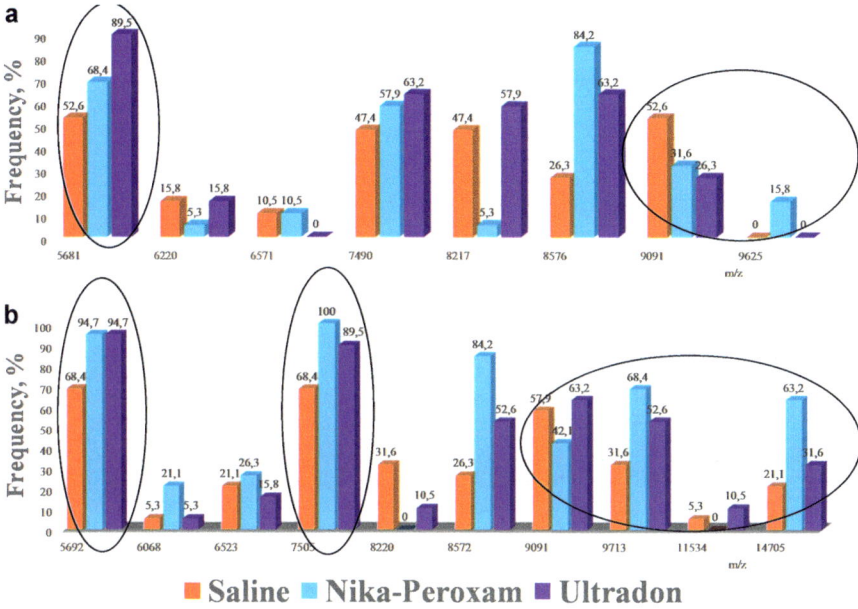

Fig. 11.1 The effect of disinfectants on *P. aeruginosa*: (**a**) in planktonic form; (**b**) in fixed form. Oval shapes mark the loci of proteomic profiles differing in planktonic and fixed forms of *P. aeruginosa* under the action of disinfectants

11.7   Conclusions

The relevance of the study is due to the fact that more than 80% of human infections are associated with the formation of biofilms. They play a special role in the development of infections caused by the use of medical devices and equipment. Biofilms can colonize almost any tissue and surface. They make a great contribution to the development of infectious processes associated with the provision of medical care. According to WHO criteria (February 28, 2017), non-fermentative Gram-negative bacteria *A. baumannii* and *P. aeruginosa* belong to the group of critical priority superbugs with pan-antibiotic resistance leading to complicated infectious diseases that cannot be treated with conventional therapy and frequently lead to a lethal outcome.

We have considered the influence of factors of abiotic and biotic nature on the process of biofilm formation in microorganisms. The examples of mutually beneficial existence of microorganisms belonging to different taxa in the biofilm community have been shown. For non-fermenting bacteria, an analysis of promising areas for the control of microbial biofilms on biotic surfaces has been proposed, including using nanoparticles that are alternative or complementary to traditional methods of antimicrobial prophylaxis and therapy.

The development of methodological approaches for the determination of biofilm formation by non-fermenting bacteria and their sensitivity to disinfectants based on mass spectrometric analysis is presented.

References

Afinogenova AG, Darovskaya EN (2011) Microbial biofilm wounds: state of the issue. Traumatol Orthop Russia 3(61):119–125. (In Russian)

Aleshukina AV (2012) Microbial-host relations in colon biotopes with dysbacteriosis. Dissertation of the doctor of medical sciences, Moscow, 289 p. (In Russian)

Aleshukina AV, Goloshva EV, Yagovkin EA, Tverdokhlebova TI (2016) Biofilm formation of different types of bacteria. In: Materials of the interregional scientific-practical conference with international participation "Actual issues of diagnosis and prevention of infectious and parasitic diseases in southern Russia", Rostov-on-Don, October 13–14, pp 169–174. (In Russian)

Baldan R., Cigana C., Testa F., Bianconi I., De Simone M., Pellin D., Di Serio C, Bragonzi A., Cirillo DM, Palaniyar N (2014) Adaptation of Pseudomonas aeruginosa in Cystic Fibrosis Airways Influences Virulence of Staphylococcus aureus In Vitro and Murine Models of Co-Infection. PLoS ONE 9 (3):e89614

Bekhalo VA, Bondarenko VM, Sysolyatina EV, Nagurskaya EV (2010) Immunobiological features of bacterial cells that make up the "medical biofilms". Microbiology 4:97–107. (In Russian)

Bondarenko VM, Gracheva NM, Matsulevich TV (2003) Intestinal dysbacteriosis in adults. KMK Press, Moscow, p 224. (In Russian)

Bukharin OV, Sgibnev AV (2012) The effect of reactive oxygen species on adhesive characteristics and the production of biofilms by bacteria. J Microbiol Epidemiol Immunobiol 3:70–73. (In Russian)

Bukharin OV, Valyshev AV, Perunova NB, Chelpachenko OE, Mironova AR, Trasevich AV (2002) Bacterial-fungal associations of the intestine in terms of colonization by yeast-like fungi of the genus Candida. J Microbiol Epidemiol Immunobiol 5:45–48. In Russian

Bukharin OV, Churkina LN, Perunova NB, Ivanova EV, Novikova IV, Avdeeva LV, Yaroshenko LV (2012) The effect of anti-staphylococcal antibiotic batumin on the biofilm formation of microorganisms. J Microbiol Epidemiol Immunobiol 2:8–12. (In Russian)

Chebotar IV, Konchakova ED, Evteeva NI (2012) Neutrophil-dependent destruction of biofilms formed by Staphylococcus aureus. J Microbiol Epidemiol Immunobiol 1:10–15. (In Russian)

Cheknev SB, Vostrova EI, Piskovskaya LS, Vostrov AV (2014) Effect of copper and zinc cations linked by gamma globulin proteins in the culture of Staphylococcus aureus. J Microbiol Epidemiol Immunobiol 3:4–9. (In Russian)

Chernukha MY, Danilina GA, Alekseeva GV, Shaginyan IA, Gintsburg AL (2009) The role of the quorum sensing regulatory system in the formation of biofilms by the bacteria Burkholderia cepacia and Pseudomonas aeruginosa. J Microbiol Epidemiol Immunobiol 4:39–43. (In Russian)

Chernyavsky VI (2013) Bacterial biofilms and infections (lecture). Ann Mechnikov Inst 1:86–90. (In Russian)

Chestnova TV, Gladkikh PG, Korotkova AS (2017) The combined effect of silver nanoparticles in combination with methyluracil and antibiotics on recovery processes in infectious peritonitis. Bulletin of new medical technologies. Electronic Edition, 11 (3). (In Russian)

Cui B, Smooker PM, Rouch DA, Daley AJ, Deighton MA (2012) Differences between two clinical Staphylococcus capitis subspecies revealed by biofilm, antibiotic resistance and PFGE profiling. J Clin Microbiol 51(1):9–14. https://doi.org/10.1128/JCM.05124-11

Demikhovskaya EV (2012) Non-fermenting bacteria in the aspect of multiple antibiotic resistance of pathogens of nosocomial infections. Dis Antibiot 1:89–95. (In Russian)

Dronina YE, Karpova TI, Sadretdinova OV, Didenko LV, Tartakovsky IS (2012) Features of the formation of legionella biofilms in artificial and natural water systems. J Microbiol Epidemiol Immunobiol 4:76–80. (In Russian)

El-Azizi M, Rao S, Kanchanapoom T, Khardon N (2005) In vitro activity of vancomycin, quinupristin/dalfopristin and linezolid against intact and disrupted biofilms of staphylococci. Ann Clin Microbiol Antimicrob 4:2

Enoch DA, Birkett CI, Ludlam HA (2007) Non-fermentative Gram-negative bacteria. Int J Antimicrob Agents 29:33–41

Flemming HC, Wingender J (2010) The biofilm matrix. Nat Rev Microbiol 8:623–633

Gabrilska RA, Rumbaugh KP (2015) Biofilm models of polymicrobial infection. Future Microbiol 10:1997–2015

Gales AC, Jones RN, Forward KR, Linares J, Sader HS, Verhoef J (2001) Emerging importance of multidrug-resistant *Acinetobacter* species and *Stenotrophomonas maltophilia* as pathogens in seriously ill patients: geographic patterns, epidemiological features, and trends in the SENTRY Antimicrobial Surveillance Program (1997–1999). Clin Infect Dis 32:104–113

Goloshva EV (2005) Changes in colonization resistance of intestines with dysbacteriosis caused by broad-spectrum antibiotics. Dissertation of the candidate of biological sciences, Rostov-on-Don, 180 p. (In Russian)

Goloshva EV (2017) Effect of biotic and abiotic factors on the biofilm formation of bacteria epidemiology and infectious diseases. Curr Issues 2:50–61. (In Russian)

Gordina EM, Horovits ES, Lemkina LM, Pospelova SV (2016) Effect of lysozyme on biofilm formation of coagulase-negative staphylococci isolated from healthy people. Probl Med Mycol 18(2):56–57. (In Russian)

Gostev VV, Sidorenko SV (2014) Bacterial biofilms and infections. J Infect 2(3):4–15. (In Russian)

Gruzina VD (2003) Communicative signals of bacteria. Antibiot Chemother 48(10):32–39. (In Russian)

Horovits ES, Gordina EM, Pospelova SV, Aliyev LO, Shchukin VP (2016) Effect of ciprofloxacin on 24-hour biofilms of *Staphylococcus aureus*. Probl Med Mycol 18(2):58–59. (In Russian)

Ilyina TS, Romanova YM, Gunzburg AL (2004) Biofilms as a way of the existence of bacteria in the environment and the host organism: the phenomenon, genetic control and systems for regulating their development. Genetics 40(11):1445–1456. (In Russian)

Khrenov PA, Chestnova TV (2013) Overview of methods for combating microbial biofilms in inflammatory diseases. Bull New Med Technol 1:72–79. (In Russian)

Lisovskaya SA, Khaldeeva EV, Glushko NI (2016) Features of biofilm formation by clinical strains of *Candida albicans*. Probl Med Mycol 18(2):88. (In Russian)

Litvinenko ZN (2015) The influence of organic substances on the formation of biofilms in aquatic systems. Dissertation of the candidate of biological sciences, Khabarovsk, 143 p. (In Russian)

Macchi A, Ardito F, Marchese A et al (2006) Efficacy of N-acetylcysteine in combination with thiamphenicol in sequential (intramuscular/aerosol) therapy of upper respiratory tract infections even if sustained by bacterial biofilms. J Chemother 18:507–513

Manukhov IV (2011) Lux-operon structure and regulation mechanisms of the Quorum sensing type in marine bacteria. Dissertation of the doctor of biological sciences, Moscow. (In Russian)

Martyanov SV, Zhurina MV, El-Registan GI, Plakunov VK (2015) The activating effect of azithromycin on the formation of bacterial biofilms and the fight against this phenomenon. Microbiology 84(1):27–27. (In Russian)

Maurice NM, Bedi B, Sadikot RT (2018) *Pseudomonas aeruginosa* biofilms: host response and clinical implications in lung infections. Am J Respir Cell Mol Biol 58(4):428–439

Mayansky AN, Chebotar IV, Evteeva NI, Rudneva EI (2011) Interspecific interaction of bacteria and the formation of a mixed (polymicrobial) biofilm. J Microbiol 1:93–101. (In Russian)

Nikolaev YA, Plakunov VK (2007) Is biofilm a "city of microbes" or an analogue of a multicellular organism? Microbiology 76(2):149–163. (In Russian)

Osipov GA, Rodionov GG (2013) Human microecology in health and disease according to microbial marker mass spectrometry. Biomed Sociopsychol Probl Safety Emerg Situations 2:43–53. (In Russian)

Patrusheva EV (2000) Microecological changes at an experimental dysbacteriosis and a role of bactericidal systems of cells of a host organism. Dissertation of a candidate of biological sciences, Volgograd, 130 p. (In Russian)

Plakunov VK, Strelkova EA, Zhurina MV (2010) Persistence and adaptive mutagenesis in biofilms. Microbiology 79(4):447–458. (In Russian)

Pronina EA, Shvidenko IG, Shub GM (2010) The formation of bacterial biofilms under the influence of electromagnetic radiation. Basic Res 10:40–45. (In Russian)

Radzig MA (2013) Interaction of bacterial cells with silver and gold compounds: effect on growth, biofilm formation, mechanisms of action, nanoparticle biogenesis. Dissertation of the candidate of biological sciences, Moscow, 200 p. (In Russian)

Rattanaumpawan P, Ussavasodhi P, Kiratisin P, Aswapokee N (2013) Epidemiology of bacteremia caused by uncommon non-fermentative Gram-negative bacteria. BMC Infect Dis 13(1):167

Riise GC, Qvarfordt I, Larsson S, Eliasson V, Andersson BA (2000) Inhibitory effect of N-acetylcysteine on adherence of Streptococcus pneumoniae and Haemophilus influenzae to human oropharyngeal epithelial cells in vitro. Respiration 67(5):552–558

Risman BV, Rybalchenko OV, Bondarenko VM, Ryzhankova AV (2011) Suppression of bacterial biofilms in case of purulent-necrotic complications of diabetic foot syndrome by ultrasound cavitation. J Microbiol Epidemiol Immunobiol 4:14–19. (In Russian)

Romanova YM, Ginzburg AL (2011) Bacterial biofilms as a natural form of the existence of bacteria in the environment and the host organism. J Microbiol Epidemiol Immunobiol 3:99–109. (In Russian)

Rybalchenko OV, Bondarenko VM (2013) The formation of biofilms symbiotic representatives of the intestinal microbiota as a form of existence of bacteria. Bull St Petersburg Univ Ser 11 Med 1:189–186. (In Russian)

Rybalchenko OV, Stepanova OM, Astafyev AM, Kudryavtsev AA, Orlova OG, Kapustina VV (2015) Influence of a pulse-periodic corona discharge on the viability of Escherichia coli M17 cells in biofilms. J Microbiol Epidemiol Immunobiol 6:17–23. (In Russian)

Shaginyan IA, Chernukha MY (2005) Non-fermenting Gram-negative bacteria in the etiology of nosocomial infections: clinical, microbiological and epidemiological features. Clin Microbiol Antimicrob Chemotherapy 7(3):271–285. (In Russian)

Shenderov BA (2001) Medical microbial ecology and functional nutrition. T. 3: probiotics and functional nutrition: biofilm. Skin and mucous membranes. Immunol Mech, 287 p. (In Russian)

Shlepotina NM, Plotkin LL, Belov VV (2014) Microbiological and clinical significance of biofilm infections (literature review). Ural Med J 4:106–112. (In Russian)

Smirnova TA, Didenko LV, Azizbekyan RR, Romanova YM (2010) Structural and functional characteristics of bacterial biofilms. Microbiology 79(4):435–446. (In Russian)

Sreeremya S (2017) A review on microbial biofilm. Int J Adv Res Dev 2:7–10

Sutherland IW (2001) Biofilm exopolysaccharides: a strong and sticky framework. Microbiology 147(1):3–9

Vorobev AA, Nesvizhsky YV, Zudenkov AE et al (2001) Comparative study of the parietal and luminal microflora of the colon in an experiment on mice. J Microbiol Epidemiol Immunobiol 1:62–67. (In Russian)

Zhurlov OS, Perunova NB, Ivanova EV, Egorova OS (2012) Effect of human platelet antimicrobial peptides on biofilm formation of Staphylococcus aureus. J Microbiol Epidemiol Immunobiol 4:66–70. (In Russian)

Zubkov MN (2003) Non-fermenting bacteria: classification, general characteristics, role in human pathology. Identification Pseudomonas spp. and similar microorganisms. Infect Antimicrob Ther 1:170–177. (In Russian)

Zueva LP, Aslanov BI, Akimkin VG (2014) A modern view on the role of bacteriophages in the evolution of hospital strains and the prevention of infections associated with the provision of medical care. J Microbiol Epidemiol Immunobiol 3:100–107. (In Russian)

Chapter 12
Anti-quorum Sensing Systems and Biofilm Formation

Sarangam Majumdar

Contents

Abstract Bacterial communication process can be broadly classified into two categories such as chemical communication process (quorum sensing mechanism) and electrical communication system (controlled by potassium ion channels). Quorum sensing is a well-known density-dependent optimal survival strategy, which is mediated by chemical signalling molecules (autoinducers). Collective bacterial behaviour regulates the bacterial lifestyle on surface (i.e. biofilms). Bacterial cell-to-cell communication process and biofilms formation cause several infectious diseases. In this chapter, we mainly focus on anti-quorum sensing mechanism and biofilm formation (including nanofabrication) that form the point of view of experimental approaches as well as mathematical models. In the end, we point out some significant and fundamental experimental observations on anti-quorum sensing and biofilm formation.

Keywords Bacteria · Biofilms · Anti-quorum sensing · Nanofabrication

S. Majumdar (✉)
Dipartimento di Ingegneria Scienze Informatiche e Matematica, Università degli Studi di L' Aquila, L' Aquila, Italy

© Springer Nature Switzerland AG 2020
R. Prasad et al. (eds.), *Nanostructures for Antimicrobial and Antibiofilm Applications*, Nanotechnology in the Life Sciences,
https://doi.org/10.1007/978-3-030-40337-9_12

12.1 Introduction

Bacteria can communicate with their surrounding bacteria by using chemical signalling communication systems. This chemical signalling mechanism is formally known as quorum sensing (Fuqua et al. 1996; Gray et al. 1994). Microbiologists intensively studied this critical biochemical phenomenon to understand the information processing system of different bacteria and their collective behaviour in the last few decades. Bacterial communication system is controlled by autoinducers (chemical signalling molecules). Bacteria prepare their optimal survival strategies to survive in a different environment by using different quorum sensing circuits (Miller and Bassler 2001; Williams et al. 2007; Shapiro 1998). Quorum sensing bacteria eject autoinducers in the environment and the surrounding bacteria receive autoinducers. In this fashion, the concentration of the autoinducers increases as a function of cell number density (Shapiro 2007; Majumdar and Mondal 2016; Majumdar and Pal 2016, 2017a, b; Majumdar et al. 2017). When the concentration reaches as minimal threshold, a collective bacterial behaviour is initiated, which triggers cascade of signalling events and regulate an array of biochemical process such as biofilm formation, swarming, virulence, bioluminescence, symbiosis, competence, antibiotic production, sporulation, conjugation and gene expression (Majumdar and Pal 2018; Majumdar and Roy 2018a, b).

Bacterial biofilms are considered as a collective bacterial living form, where bacterial cells are embedded in an extracellular polymeric substance (EPS) that are adherent to each other and a surface (Vert et al. 2012). Bacterial biofilms have different emergent properties such as localized gradient, sorption, enzyme retention, tolerance and resistance, competition and cooperation (Flemming et al. 2016). The mechanical stability of the biofilm is provided by EPS. EPS are lipids, nucleic acid, proteins and polysaccharides (Flemming and Wingender 2010). Pathogenic bacteria are harmful for human. This bacteria community living culture (i.e. biofilms) is a cause of different infectious diseases such as urinary tract, synthetic vascular grafts, gastrointestinal tract, dental implant, cardiac implant and many more (Majumdar and Roy 2018a, b).

12.2 Anti-quorum Sensing Model

We are discussing anti-quorum sensing models, which describe anti-quorum sensing treatment in biofilms and batch cultures. This model is based on LasI/R system of *P. aeruginosa* and applicable for LuxI/R homolog systems. The bacterial population can be assumed as two subpopulations such as up-regulated cells and down-regulated cells (Ward 2008; Anguige et al. 2004, 2005, 2006).

- N_d represents the down-regulated cell density. The cells contain empty *lux*-box. *P. aeruginosa* produce autoinducers and EPS matrix at a low rate. A nonvirulent activity is observed at that situation.

- N_u represents the up-regulated cell density. Cells have complex (LasR-autoinducer) bound *lux*-box. The autoinducers and EPS are produced at enhanced rate. Virulent activity is observed.
- Total bacterial population density is $N_T = N_d + N_u$.
- Down-regulated cell divides into two down-regulated cells.
- Up-regulated cell divides into one down-regulated and one up-regulated cell.
- We assume that LasR (with concentration R) is generated at rate R_0 and binds with autoinducer A (reversable reaction) and form complex P (LasR-autoinducer). We get,
- $$\frac{dR}{dt} = R_0 - k_{ra}AR + k_pP - \lambda_RR \quad \text{and} \quad \text{LasR-autoinducer} \quad \text{complex} \quad \text{equation}$$

$$\frac{dP}{dt} = k_{ra}AR - k_pP - \lambda_pP. \text{ Output of LasI (up-regulated cells) occurs at}$$

constant rate and decays as follows $\frac{dL}{dt} = L_0 - \lambda_LL$ (see Fig. 12.1).
- Autoinducers are generated with a background level k_d and decay with constant λ. The rate of change of autoinducer (down-regulated cells) $=k_d - k_{ra}AR + k_pP - \lambda A$.
- The rate of change of autoinducers (up-regulated cells) $=k_aL + k_d - k_{ra}AR + k_pP - \lambda A$, where k_aL shows a massive increase (up-regulated cells) of autoinducers production. $-\lambda A$ describes the rate of change of autoinducers (external media).
- If $\frac{dR}{dt} = \frac{dP}{dt} = \frac{dL}{dt} = 0$ (equilibrium condition) then $L = L_\infty$,

$$P = \frac{P_\infty}{R_\infty}RA, \quad R = \frac{R_\infty}{1 + \mu_RA}, \quad \text{where} \quad L_\infty = \frac{L_0}{\lambda_L}, \quad R_\infty = \frac{R_0}{\lambda_R}, \quad \mu_R = \frac{\lambda_pP_\infty}{\lambda_RR_\infty}$$

and $P_\infty = \frac{R_\infty k_{ra}}{(k_p + \lambda_p)}$.
- With the substitution $R = R_\infty$, $\mu_RA \approx 0$ and $P = P_\infty A$, we have the rate of change of autoinducers (down-regulated cells) $=k_d - \sigma A - \lambda A$ and the rate of change of autoinducers (up-regulated cells) $=k_u + k_d - \sigma A - \lambda A$, where $\sigma = \lambda_pP_\infty$ and $k_u = k_aL_\infty$.
- We assume that up-regulation rate of bacterial cells is proportional to $P_\infty A$ (complex concentration). So, up-regulation rate is αA, where $\alpha = \alpha_aP_\infty$ and α_a is proportionality constant. The down-regulation rate of cells is $\beta = \lambda_p$.
- Let Q_1, Q_2, Q_3 be the concentration of the anti-LuxR (homolog) agent, anti-autoinducer agent and anti-LuxI (homolog) agent respectively which follows.

$$\text{LasR} + Q_1 \xrightarrow{k_1} (1 - v_1)Q_1 + \text{by product}$$

$$\text{Autoinducer} + Q_2 \xrightarrow{\mu_2} (1 - v_2)Q_2 + \text{by product}$$

$$\text{Autoinducer} + Q_3 \xrightarrow{\mu_3} (1 - v_3) Q_3 + \text{by product}$$

- where v_1, v_2, v_3 represent the average amount of Q_1, Q_2, Q_3 lost by the reaction respectively.
- So, we find LasR concentration (at equilibrium) is $R = \dfrac{R_\infty}{(1 + \gamma_1 Q_1)}$, where $\gamma_1 = \dfrac{k_1}{k_R}$. Moreover, LasR-autoinducer binding rate and up-regulation rate is reduced by the factor $(1 + \gamma_1 Q_1)$.
- We find $-\mu_2 Q_2 A$ as an additional term in equation of rate of change of autoinducers (up-regulated and down-regulated cells).
- LasI equilibrium level reduces to $L_\infty/(1 + \gamma_3 Q_3)$, where $\gamma_3 = \dfrac{k_3}{\lambda_L}$. The new autoinducer output rate term is $k_u/(1 + \gamma_3 Q_3)$.

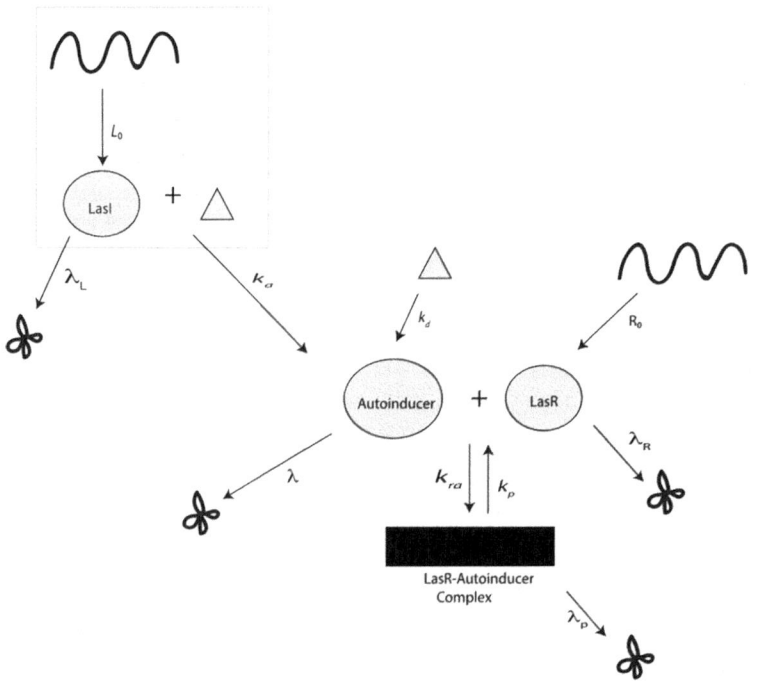

Fig. 12.1 Schematic visualization of *P. aeruginosa* quorum sensing process (LasI/LasR system), which is used for the mathematical modelling approach. The rectangular box represents a reaction for up-regulated cells only and wavy line represents the transcription of protein

12.2.1   Mathematical Model of Anti-quorum Sensing Treatment (in Batch Culture)

Now, we assume that the parameters of mathematical model are continuous in space and time. In this modelling approach, we neglect the stochastic effects. The following set of equations are based on the above assumptions (Ward 2008; Anguige et al. 2004, 2005, 2006):

$$\frac{dN_d}{dt} = rN_T - \frac{\alpha A}{1+\gamma_1 Q_1}N_d + \beta N_u \tag{12.1}$$

$$\frac{dN_u}{dt} = \frac{\alpha A}{1+\gamma_1 Q_1}N_d - \beta N_u \tag{12.2}$$

$$\frac{dA}{dt} = \frac{k_u}{1+\gamma_3 Q_3}N_u + k_d N_T - \frac{\sigma A}{1+\gamma_1 Q_1}N_T - \lambda A - \mu_2 Q_2 A \tag{12.3}$$

$$\frac{dQ_1}{dt} = \phi_1 - \frac{\mu_1 Q_1}{1+\gamma_1 Q_1}N_T - \lambda_1 Q_1 \tag{12.4}$$

$$\frac{dQ_2}{dt} = \phi_2 - \mu_2 v_2 A Q_2 - \lambda_2 Q_2 \tag{12.5}$$

$$\frac{dQ_3}{dt} = \phi_3 - \frac{\mu_3 Q_3}{1+\gamma_3 Q_3}N_u - \lambda_3 Q_3 \tag{12.6}$$

Drug can be introduced at the beginning or being drip-fed at a rate ϕ_i (for $i = 1$, 2, 3). The parameters $\mu_1 = v_1 k_1$ and $\mu_3 = v_3 k_3$ represent the drug loss rates.

12.2.2   Mathematical Model of Anti-quorum Sensing Treatment (in Biofilms)

Now, we consider bacterial cells distribution as a function of space z and time t. z is a perpendicular distance from the bacteria biofilm base with $z = H(t)$. M represents a volume fraction, which is occupied by death cells. The rest of the space is captured by EPS (E) and water (W). Thus $N_T + M + E + W = 1$. The pore space is increasing at the time of EPS production. So we get $W = W_0 + \theta E$ where θ and W_0 are constant. Finally, we have $N_T + M + (1 + \theta)E = 1 - W_0$. Furthermore, we assume that quorum sensing process regulates the nutrient concentration (c) and EPS production. The following set of equations give a detail mathematical framework of anti-quorum sensing treatment in biofilms (see details in Table 12.1) (Anguige et al. 2006; Ward 2008).

Table 12.1 Model parameters and its description

Description	Parameter
Oxygen consumption constant	ρ
Sets minimal death rate	τ
Birth rate oxygen concentration (Half max.)	c_1
Death rate oxygen concentration (Half max.)	c_2
Dissolved oxygen concentration	c_{ext}
Maximum birth rate	B_1
Maximum death rate	B_2
Surface autoinducer mass transfer rate	Q_a
Diffusion rate of autoinducer	D_a
Diffusion rate of species	D_i
Diffusion rate of oxygen	D_c
Background EPS production rate	E_0
Ma. EPS production rate by up-regulated cells	k_E
EPS decay rate	λ_E
Maximum density of cells in biofilms	ω
EPS generated pore space constant	θ
Initial biofilm depth	H_0
Void fraction at maximum bacterial packing	W_0
Decay rate of quorum sensing inhibitor	λ_i
Drip rate of quorum sensing inhibitor	φ_i
Mean Q_2 loss in reaction with autoinducer	v_1
1/conc. When quorum sensing inhibitor is 50% effective	γ_1, γ_3
Drug loss rate (due to quorum sensing inhibitor action)	μ_2
Drug loss rate (due to quorum sensing inhibitor action)	μ_1, μ_3
Autoinducer loss rate by LasR/Autoinducer binding	σ
Autoinducer decay rate	λ
Autoinducer production rate by down-regulated cells	k_d
Autoinducer production rate by up-regulated cells	k_u
Down-regulation rate	β
Maximal up-regulation rate	α
Cell birth rate	r

$$\frac{\partial N_T}{\partial t} + \frac{\partial v N_T}{\partial z} = N_T \left(F_b(c) - F_d(c) \right) \tag{12.7}$$

$$\frac{\partial M}{\partial t} + \frac{\partial v M}{\partial z} = N_T F_d(c) \tag{12.8}$$

$$\frac{\partial N_u}{\partial t} + \frac{\partial v N_u}{\partial z} = \frac{\alpha A}{1 + \gamma_1 Q_1} N_d - \beta N_u \tag{12.9}$$

$$\frac{\partial E}{\partial t} + \frac{\partial v E}{\partial z} = \left(E_0 N_T + k_E N_u \right) F_b \left(c \right) - \lambda_E E \tag{12.10}$$

$$0 = D_a \frac{\partial^2 A}{\partial z^2} + \frac{k_u^*}{1 + \gamma_3 Q_3} N_u + k_d^* N_T - \frac{\sigma^* A}{1 + \gamma_1 Q_1} N_T - \lambda A - \mu_2 Q_2 A \tag{12.11}$$

$$0 = D_1 \frac{\partial^2 Q_1}{\partial z^2} - \frac{\mu_1^* Q_1}{1 + \gamma_1 Q_1} N_T - \lambda_1 Q_1 \tag{12.12}$$

$$0 = D_2 \frac{\partial}{\partial z} \left(W \frac{\partial Q_2}{\partial z} \right) - \mu_2 v_2 A W Q_2 - \lambda_2 W Q_2 \tag{12.13}$$

$$0 = D_3 \frac{\partial^2 Q_3}{\partial z^2} - \frac{\mu_3^* Q_3}{1 + \gamma_3 Q_3} N_u - \lambda_3 Q_3 \tag{12.14}$$

$$0 = D_c \frac{\partial^2 c}{\partial z^2} - \rho N_T F_b \left(c \right) \tag{12.15}$$

$$\frac{\partial v}{\partial z} = \frac{1}{1 - W_0} \left(N_T F_b \left(c \right) + \left(1 + \theta \right) \left(E_0 N_T + k_E N_u \right) F_b \left(c \right) - \lambda_E E \right) \tag{12.16}$$

$$\frac{dH}{dt} = v \left(H, t \right) \tag{12.17}$$

One can stimulate the mathematical model with Michaelis-Menten kinetics

$$F_b \left(c \right) = B_1 \frac{c}{c_1 + c} \quad F_d \left(c \right) = B_2 \left(1 - \tau \frac{c}{c_2 + C} \right)$$ where $F_d(c)$ and $F_b(c)$ represent bacterial death and birth rate respectively (Ward 2008).

12.2.3 Model Predictions

- Up-regulation occurs after the initial period. We observe rapid up-regulation after a certain time (around 3 h) and 12–13% up-regulated cells at any time. Up regulation bacteria are dependent on the growth phase (in batch culture) (Ward 2008).

- Bacterial colony virulence can be measured by

$$N_u^{frac} = 1 - \frac{\sigma(\beta+r)}{\alpha k_u} \left(\text{for exponential growth phase} \right)$$

$$N_u^{frac} = 1 - \frac{\beta(\sigma K + \lambda)}{\alpha k_u K} \left(\text{for stationary phase} \right) \left(K \text{ is the population size} \right) \left(\text{Ward 2008} \right)$$

- Anti-LasI agent is the effective treatment than others. Anti-LasR treatment is the most effective QSI (Ward 2008).
- Bacterial biofilm is slowed down after the initial acceleration of growth. Up-regulated cell fraction $U(t) = \dfrac{\int_0^H N_u(z,t)dz}{\int_0^H N_T(z,t)dz}$. Living cells are located near the surface (Ward 2008).
- We find a shift in biofilm growth rates with scale μM. Anti-LasR and anti-LasI agents are similar (Ward 2008).
- QSI is required for suppressing quorum sensing for biofilms and batch culture (Ward 2008).

12.3   Nanofabrication

Bacteriology and nanotechnology are the rapidly growing research field. It has been evidenced that bacteria experience spatial structure in different scales. Microfluidic devise and nanofabrication are useful for those scales. We uncover several long-standing questions using nanofabrication, which includes bacterial growth, development, density-dependent behaviour any many more. Bacteria can also grow in a nanofabricated chamber. Dynamics of a bacterial community can be explored by nanofabrication and microfluids (i.e. synthetic ecosystems). Moreover, we can study the completion and cooperation in bacteria communities and shed a new light into the dark matter of biology (Hol and Dekker 2014).

12.4   Significant and Fundamental Experimental Observations

- We can find the anti-quorum sensing compounds in six different plants such as *Conocarpus erectus* L., *Quercus virginiana* Mill., *Callistemon viminalis* G. Don, *Bucida burceras* L., *Chamaecyce hypericifolia* (L.) Millsp. and *Tetrazygia bicolor* (Mill.) Cogn. (Adonizio et al. 2006).

- Biofilm formation is regulated by quorum sensing, which is a fundamental cause of urinary tract infection. Curcumin (anti-quorum sensing agent) from *Curcuma longa* inhibit *E. coli* and *P. aeruginosa* biofilm formation (Packiavathy et al. 2014).
- Anti-quorum sensing activity is exhibited by malabaricone C (Chong et al. 2011).
- Quorum sensing inhibitors (QSI) play a crucial role in the biofilm formation. This quorum sensing inhibitors (QSI) are important anti-biofilm agents (Brackman and Coenye 2015).
- Quorum sensing blocker is an important strategy to switch off virulence factor (Finch et al. 1998).
- Essential oils are potential inhibitor of quorum sensing process and prevent biofilm formation (Kerekes et al. 2013).
- *Kigelia africana* extracts have anti-quorum sensing potential (Chenia 2013).
- Diterpene phytol has anti-quorum sensing activity, which reduces *P. aeruginosa* biofilm formation (Pejin et al. 2015).
- *P. aeruginosa* virulence activity can be blocked by small molecules in MvfR communication process (Starkey et al. 2014).
- Punicalagin has anti-quorum sensing potential (Li et al. 2014).
- Parthenolide is a potential anti-biofilm and anti-quorum agent (Kalia et al. 2018).
- A time-sharing behaviour (nutrient competition) is observed between biofilms (Liu et al. 2017).
- *B. cereus* is a quorum sensing and opportunistic human pathogen bacteria. A set of synthetic peptides are discovered, which are potential anti-virulence agents. We can find out several anti-virulence agents using single and multiple amino acid replacements method (Yehuda et al. 2019).

References

Adonizio AL, Downum K, Bennett BC, Mathee K (2006) Anti-quorum sensing activity of medicinal plants in southern Florida. J Ethnopharmacol 105(3):427–435

Anguige K, King JR, Ward JP, Williams P (2004) Mathematical modelling of therapies targeted at bacterial quorum sensing. Math Biosci 192(1):39–83

Anguige K, King JR, Ward JP (2005) Modelling antibiotic-and anti-quorum sensing treatment of a spatially-structured Pseudomonas aeruginosa population. J Math Biol 51(5):557–594

Anguige K, King JR, Ward JP (2006) A multi-phase mathematical model of quorum sensing in a maturing Pseudomonas aeruginosa biofilm. Math Biosci 203(2):240–276

Brackman G, Coenye T (2015) Quorum sensing inhibitors as anti-biofilm agents. Curr Pharm Des 21(1):5–11

Chong YM, Yin WF, Ho CY, Mustafa MR, Hadi AHA, Awang K, Narrima P, Koh CL, Appleton DR, Chan KG (2011) Malabaricone C from Myristica cinnamomea exhibits anti-quorum sensing activity. J Nat Prod 74(10):2261–2264

Chenia H (2013) Anti-quorum sensing potential of crude Kigelia africana fruit extracts. Sensors 13(3):2802–2817

Finch RG, Pritchard DI, Bycroft BW, Williams P, Stewart GS (1998) Quorum sensing: a novel target for anti-infective therapy. J Antimicrob Chemother 42(5):569–571

Flemming HC, Wingender J, Szewzyk U, Steinberg P, Rice SA, Kjelleberg S (2016) Biofilms: an emergent form of bacterial life. Nat Rev Microbiol 14(9):563

Flemming HC, Wingender J (2010) The biofilm matrix. Nat Rev Microbiol 8(9):623

Fuqua C, Winans SC, Greenberg EP (1996) Census and consensus in bacterial ecosystems: the LuxR LuxI family of quorum-sensing transcriptional regulators. Annu Rev Microbiol 50:727–751

Gray KM, Passador L, Iglewski BH, Greenberg EP (1994) Interchangeability and specificity of components from the quorum-sensing regulatory systems of Vibrio fischeri and Pseudomonas aeruginosa. J Bacteriol 176:3076–3080

Hol FJ, Dekker C (2014) Zooming in to see the bigger picture: microfluidic and nanofabrication tools to study bacteria. Science 346(6208):1251821

Kalia M, Yadav VK, Singh PK, Sharma D, Narvi SS, Agarwal V (2018) Exploring the impact of parthenolide as anti-quorum sensing and anti-biofilm agent against Pseudomonas aeruginosa. Life Sci 199:96–103

Kerekes EB, Deák É, Takó M, Tserennadmid R, Petkovits T, Vágvölgyi C, Krisch J (2013) Anti-biofilm forming and anti-quorum sensing activity of selected essential oils and their main components on food-related micro-organisms. J Appl Microbiol 115(4):933–942

Li G, Yan C, Xu Y, Feng Y, Wu Q, Lv X et al (2014) Punicalagin inhibits Salmonella virulence factors and has anti-quorum-sensing potential. Appl Environ Microbiol 80(19):6204–6211

Liu J, Martinez-Corral R, Prindle A, Dong-yeon DL, Larkin J, Gabalda-Sagarra M, Garcia-Ojalvo J, Süel GM (2017) Coupling between distant biofilms and emergence of nutrient time-sharing. Science 356(6338):638–642

Majumdar S, Mondal S (2016) Conversation game: talking bacteria. J Cell Commun Signal 10(4):331–335

Majumdar S, Pal S (2016) Quorum sensing: a quantum perspective. J Cell Commun Signal 10(3):173–175

Majumdar S, Roy S, Llinas R (2017) Bacterial conversations and pattern formation. bioRxiv. https://doi.org/10.1101/098053

Majumdar S, Pal S (2017a) Cross-species communication in bacterial world. J Cell Commun Signal 11(2):187–190

Majumdar S, Pal S (2017b) Bacterial intelligence: imitation games, time-sharing and long-range quantum coherence. J Cell Commun Signal 11(3):281–284

Majumdar S, Pal S (2018) Information transmission in microbial and fungal communication: from classical to quantum. J Cell Commun Signal 12(2):491–502

Majumdar S, Roy S (2018a) Relevance of quantum mechanics in bacterial communication. NeuroQuantology 16(3):1–6

Majumdar S, Roy S (2018b) Mathematical model of quorum sensing and biofilm. In: Bramhachari PV (ed) Implication of quorum sensing system in biofilm formation and virulence. Springer, Singapore, pp 351–368

Miller MB, Bassler BL (2001) Quorum sensing in bacteria. Annu Rev Microbiol 55(1):165–199

Packiavathy IASV, Priya S, Pandian SK, Ravi AV (2014) Inhibition of biofilm development of uropathogens by curcumin—an anti-quorum sensing agent from Curcuma longa. Food Chem 148:453–460

Pejin B, Ciric A, Glamoclija J, Nikolic M, Sokovic M (2015) In vitro anti-quorum sensing activity of phytol. Nat Prod Res 29(4):374–377

Shapiro JA (1998) Thinking about bacterial populations as multicellular organisms. Annu Rev Microbiol 52:81–104

Shapiro JA (2007) Bacteria are small but not stupid: cognition, natural genetic engineering and socio-bacteriology. Stud Hist Philos Biol Biomed Sci 38(4):807–819

Starkey M, Lepine F, Maura D, Bandyopadhaya A, Lesic B, He J, Kitao T, Righi V, Milot S, Tzika A, Rahme L (2014) Identification of anti-virulence compounds that disrupt quorum-sensing regulated acute and persistent pathogenicity. PLoS Pathog 10(8):e1004321

Vert M, Doi Y, Hellwich KH, Hess M, Hodge P, Kubisa P, Rinaudo M, Schué F (2012) Terminology for biorelated polymers and applications (IUPAC recommendations 2012). Pure Appl Chem 84(2):377–410

Ward J (2008) Mathematical modeling of quorum-sensing control in biofilms. In: Balaban N (ed) Control of biofilm infections by signal manipulation. Springer, Berlin, pp 79–108

Williams P, Winzer K, Chan WC, Camara M (2007) Look who's talking: communication and quorum sensing in the bacterial world. Philos Trans R Soc Lond B Biol Sci 362(1483):1119–1134

Yehuda A, Slamti L, Malach E, Lereclus D, Hayouka Z (2019) Elucidating the hot spot residues of quorum sensing peptidic autoinducer PapR by multiple amino acids replacements. Front Microbiol 10:1246

Chapter 13
Nanostructures for Antimicrobial and Antibiofilm Photodynamic Therapy

V. T. Anju, Busi Siddhardha, and Madhu Dyavaiah

Contents

Abstract Unprecedented explosion of research in the field of nanotechnology has gained importance in the treatment, prevention, and eradication of antibiotic-resistant bacterial strains. The emerging multidrug-resistant bacteria (MDRB) pose a major threat to the modern health-care system. The MDRB strains cannot be treated with conventional antibiotics due to their rapid mutations and resistance. Antibiotic-resistant bacteria that produce biofilm are responsible for approximately 700,000 deaths each year. One of the biggest problems faced by research society is

V. T. Anju · M. Dyavaiah (✉)
Department of Biochemistry and Molecular Biology, School of Life Sciences,
Pondicherry University, Pondicherry, India

B. Siddhardha
Department of Microbiology, School of Life Sciences, Pondicherry University,
Pondicherry, India

© Springer Nature Switzerland AG 2020
R. Prasad et al. (eds.), *Nanostructures for Antimicrobial and Antibiofilm Applications*, Nanotechnology in the Life Sciences,
https://doi.org/10.1007/978-3-030-40337-9_13

to find alternatives to combat the increasing number of resistant variants. Photodynamic therapy (PDT) was established recently and remains a successful treatment modality for infectious diseases caused by microbial strains and biofilms. Light-mediated inactivation through photodynamic therapy provides new dimensions to eradicate antibiotic-resistant microbes. Antimicrobial photodynamic therapy (aPDT) has gained interest in nanotechnology where the effectiveness of photosensitizers (PS) can be enhanced by the use of nanoparticles (NPs). In the last two decades, different techniques have been raised for aPDT in combination with nanoparticles. Nanoparticles are used in aPDT either as photosensitizing agents or as PS delivery agents. Nanoparticles used in aPDT improve the dispersion and selective delivery of PS to the target cells. Over last decades, various nanoparticles are utilized in aPDT as nanocarriers. Polymeric nanovehicles, nanomicelles, and liposome are used to encapsulate PS molecules. The inorganic metallic nanoparticles are extensively studied for the photoinactivation of resistant microorganisms and their biofilms. The four types of combinations between nanoparticles and PS are categorized as nanoparticles embedded with PS, nanoparticles with PS bound to the surface, nanoparticles as the PS, and PS alongside nanoparticles. Nanoparticles have enhanced the activity of aPDT by encapsulating the PS in nanoparticles or binding the PS on the surface of nanoparticles covalently. The photoactive nanoparticles were successful as antimicrobial agents and more effective against antibiotic-resistant microbial strains and their biofilms.

Keywords Multidrug-resistant bacteria · Biofilms · Nanoparticles · Antimicrobial photodynamic therapy · Photosensitizer

13.1 Introduction

13.1.1 The Emergence of Multidrug-Resistant Bacteria

Antibiotic resistance in microorganisms and associated infections are serious and growing challenge in health-care system. Antibiotic-resistant bacteria were observed in the patients with weak health condition, regular and repeated exposure to antibiotics, and frequent hospitalization. For instance, methicillin-resistant *Staphylococcus aureus* (MRSA) is prevalent in hospitalized patients and even exhibits resistance to alternative antimicrobial agents especially vancomycin. The resistance to vancomycin is reported in several bacteria including hospital-acquired pathogenic bacteria *Enterococcus* sp. Lack of effective therapies for multidrug-resistant bacteria urges the immediate attention in antimicrobial drug discovery process (McAdam et al. 2012).

Even though antibiotics have saved billions of lives, current antimicrobial resistance (AMR) patterns among the bacterial pathogens become a reason for thousands of deaths worldwide. Scientists concluded that optional use of antimicrobial agents, window of treatment period, and choice of antibiotics and dosage of antibiotics are the important factors that can reduce the consequences of AMR (Holmstrup and Klausen 2018). Antibiotic resistance is observed in both hospital settings and primary health-care centers as leading threat to human health (National Collaborating Centre for Infectious Diseases 2010).

The diseases associated with AMR strains in health-care system include gonorrhea, tuberculosis, typhoid fever, and group B *Streptococcus* infections (Llor and Bjerrum 2014). AMR is known as silent pandemic with profound effect on treatment procedures. Microorganisms are adapted to the available antimicrobial agents through evolution which resulted in the difficulties in disease treatment. Still now, there is no "single silver bullet" to reduce the AMR strains to avoid innumerable future victims of antibiotic-resistant diseases. According to the specific Sustainable Development Goals (SDGs) agenda, it is the peak time to find a permanent solution to this evolving threat caused by antibiotic-resistant bacteria (Jasovský et al. 2016).

13.1.2 Emerging Infectious Diseases and Antibiotic-Resistant Biofilms

Microorganisms thrive in diverse habitats as they are having the ability to adhere to substrates and form communities. These microorganisms living in populations can able to cooperate, compete, and interact to carry out complex processes. Microbial biofilms are represented by single species or multispecies encased in a self-produced exopolysaccharide matrix (EPS) (Bjarnsholt 2013). Biofilms also contribute to the emerging problem of antibiotic resistance in microorganisms. Horizontal gene transfer in bacteria mediates the evolution and genetic diversity among microbial communities. The highly protected structure of biofilms formed by EPS prevents the contact of antimicrobial agents to bacteria by selectively restricting the diffusion (Chadha 2012).

The possible reasons for the antibiotic resistance in biofilms include restricted penetration of drugs, decreased growth rate of cells, and resistance gene expressions. According to the global report made by National Institutes of Health, biofilms are accountable for more than 60% of microbial infections which are difficult to eradicate. There are common bacterial infections caused by biofilms, and frequently relapsing are urinary tract infections caused by *E. coli*, dental plaques, gingivitis, middle ear infections caused by *Haemophilus influenza*, catheter-related infections caused by *Staphylococcus aureus*, and other Gram-positive organism. Complications associated with antibiotic-resistant and biofilm-forming bacteria are life-threatening which causes grave morbidity and mortality (Lewis 2001). Other life-threatening infections by antibiotic-resistant bacteria are diseases in cystic fibrosis patients caused by *P. aeruginosa*, indwelling-device-associated infections by *S. aureus*, and endocarditis by *S. aureus*. The growing incidence of nosocomial infections is the major burden on health-care system which are contributed mainly by antimicrobial-resistant pathogens.

A group of nosocomial pathogens termed as "ESKAPE" with antibiotic resistance are the major reason for increased treatment costs and high morbidity and mortality. The ESKAPE pathogens cover both the Gram-negative and Gram-positive bacteria such as *Enterococcus faecium, S. aureus, Klebsiella pneumoniae, Acinetobacter baumannii, Pseudomonas aeruginosa,* and *Enterobacter* species (Santajit and Indrawattana 2016). ESKAPE pathogens causing nosocomial infections exhibit different profiles of antibiotic resistance, infections, and disease transmission. The array of resistance mechanism utilized by pathogens includes enzymatic inactivation, drug target modification, change in the integrity of cell membrane, expression of efflux pump mechanism, and protection by biofilms. In the near future, antimicrobial resistance profiles and biofilms of these pathogens seem to be a major challenge to global health-care systems. This alters the scientific community regarding the scarcity of potential antimicrobial agents in the pipeline and the need for developing new therapeutic methods or novel antimicrobial agents against infections caused by pathogens (Wilson 2014; Li and Nikaido 2009; Wright 2005; Aziz et al. 2014, 2015, 2016).

13.1.3 Survival Mechanisms of Bacterial Biofilms

Biofilms are population or group of microorganisms having complex architecture made of proteins, polysaccharide, and DNA which are self-protected by EPS matrix from adverse conditions. Biofilms that establish on medical devices, living tissues, and water bodies are difficult to eradicate. They are recalcitrant to antibiotics, phagocytosis, and other disinfectants. The structure of biofilms is supported by several interstitial voids such as macro- and microvoids that allow the diffusion of drugs, but the cells modulate the architecture of biofilm communities in response to the external and internal stimuli which pump out the antimicrobial agents. The cells in close proximity exchange their chromosomal plasmids and exhibit bacterial communication pathways called quorum sensing (Singh et al. 2017).

The stable and compact nature of biofilms, reduced rates of cell growth, protection by matrix polymers, and respiration of biofilm bacteria prevent the attack and destruction of biofilms by natural and artificial chemical agents. The enhanced resistance is a profound trait associated with biofilms. Biofilms show 10–1000-fold resistance to several antimicrobial agents than the planktonic bacteria. For instance, *S. aureus* cells are susceptible to chlorine, an oxidizing biocide, but demand 600-fold increased concentration of antimicrobial agents to kill biofilms of *S. aureus* (Jing et al. 2014; Costerton et al. 1995; Donlan 2002). The factors contributing to antibiotic resistance include poor penetration of antibiotic, reduced growth of cells, and unique biofilm structure. The antimicrobial agents are unable to penetrate through all areas of biofilms and especially unable to reach the deeper layers of biofilm. The capacity of antibiotic to penetrate the biofilm layers is purely dependent on the concentration and choice of antibiotic used (Vrany et al. 1997). Nondividing cells of biofilms escape from the detrimental effects of antibiotics targeted against their growth specific factors due to the reduced growth in biofilms. Scientists have reported that the unusual resistance of biofilms is contributed by the above mentioned factor (Davies 2003). Other important factor governing the antimicrobial resistance is the heterogeneity of cells within the biofilm. Signaling factors, waste products, and nutrient gradients can determine the formation of heterogeneity within the biofilm. Studies reveal that the location of cells in the biofilm community exhibits varying response to antimicrobial agents. The slow growth rate of some cells may lead to the initiation of general stress response within the biofilm. The stress response faced by biofilm cells protects them from extreme environmental conditions including cold shock, heat shock, chemical agents, and changes in the pH (Mah and O'Toole 2001). Persister and phenotypic variants are responsible for high levels of resistance in biofilms. Most of biofilm cells are sensitive to antibiotics similar to planktonic cells, whereas the persister cells are responsible for the survival of cells. Persister cells survive from the action of antimicrobial agents, while phenotypic variants survive even in elevated levels of antibiotics (Drenkard and Ausubel 2002). Overproduction and expression of multidrug efflux pumps in prokaryotic and eukaryotic cells are the reason for the extrusion of drugs, toxic metabolites, and other compounds from the cell which contributes resistance to the antibiotics (Drenkard 2003). The most important feature of biofilms and planktonic cells toward antimicrobial agents is their high resistance profiles. The infections caused by microbial biofilms are problematic and hard to eradicate. There are several factors contributing to the resistance, and the antibiotic resistance in biofilms is multifactorial (Fig. 13.1).

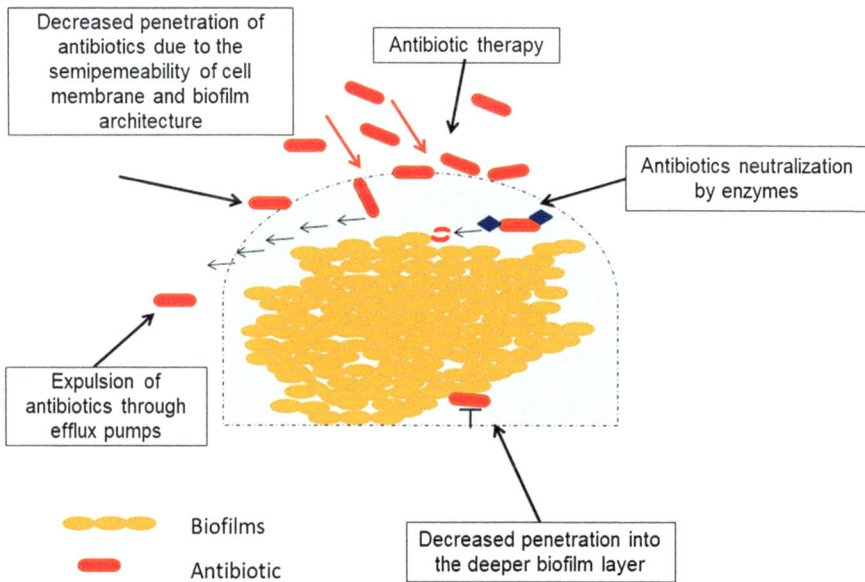

Fig. 13.1 Biofilm-mediated antibiotic resistance and associated factors

13.1.4 Antimicrobial and Antibiofilm Strategies

Recently health-care system has reached "post antibiotic era" where many of the antimicrobial agents are no longer effective and useful. This is because of the failure in the discovery of antibiotics; so far no new classes of antibiotics were designed from the past 45 years for infections caused by Gram-negative bacteria. According to the reports, only 37 antimicrobial drugs were successful in phase II or phase III clinical trials of 2016. Therefore, the needs to develop alternative therapeutic methods are urged in research to treat infections caused by drug-resistant microbes. Moreover along with antibiotic resistance, antibiotics may show accidental consequences such as stimulating hyper-inflammatory effects or off-target effects by disturbing beneficial microbiota. The disturbances to the normal microbiota can be detrimental to our health resulting in disease. The critical aspect taken into consideration during the design of next-generation antimicrobials is to preserve the normal microbiota (Fernandes and Martens 2017). Next-generation antimicrobials for infectious diseases include nucleic acid-based approaches for selectively targeting microorganisms. CRISPR-Cas system has been developed to selectively target antibiotic-resistant genes in complex microbial consortia and pure cultures. Other than nucleic acid-based systems, peptide-based antimicrobials are also under research with enhanced specificity. Peptide molecules are easily modified to achieve specific functions especially their biophysical properties such as net charge, amphiphilicity, and hydrophobicity to enhance the antimicrobial action. Many synthetic variants are designed based on their structure-activity relationship with observed broad-spectrum antimicrobial action. A synthetic peptide-based antibiotic, e.g., lysins, showed antimicrobial activity against Gram-positive and Gram-negative bacterial pathogens (de la Fuente-Nunez et al. 2017).

The next-generation antimicrobial therapies based on peptides and nucleic acid systems are still in infancy due to the limitations in their clinical applications. These therapeutic methods can be upgraded for better antimicrobial activities by protecting them from host degradation by proteases and nucleases. New drug carrier agents are required for the efficient and enhanced delivery to the target sites. A paradigm shift in the design of new and effective antimicrobials includes selection of effective antibiofilm molecules, better use of existing antimicrobials, development of next-generation antibiotics against biofilms, and regulatory issues for the establishment of new diagnostic assays (Ceri et al. 2010). Novel strategies are required to control the biofilm formation by considering the health-related problems due to biofilms, which include prevention of cell adhesion to the substrates, reduction in the EPS production, and disruption of quorum-sensing phenomenon in bacteria (Majik and Parvatkar 2013). Novel antibiofilm strategies include the use of natural products (plant extracts, honey, essential oil, cumin oil, cinnamon oil), bacteriophage therapy, quorum quenching molecules, new surfaces for prevention of biofilm (new coatings and paint, new surface materials, and surface modifications), nanotechnology (micro- and nanoemulsion) and controlling biofilms with enzymes (deoxyribonuclease 1, lysostaphin, α-amylase, lyase, and lactonase), photodynamic therapy, biosurfactants, bacteriocin, bioelectric approach, and ultrasonic treatment (Sadekuzzaman et al. 2015). The design and development of potential antibiofilm strategies have gained much interest in the field of medicine. Still, biomedical researchers are in urge for development of alternative methods to inhibit biofilm-associated infections.

13.1.5   The Emergence of Antimicrobial Photodynamic Therapy

In the array of promising novel therapeutic methods, light-based therapy known as photodynamic therapy (PDT) is introduced. PDT was discovered by Oskar Raab and Hermann von Tappeiner 100 years ago. Now PDT is known as alternative treatment modality for localized infections other than tumor cells. Antimicrobial photodynamic inactivation (aPDI) is a potential antimicrobial method that works on several factors of multidrug resistance patterns such as virulence traits, biofilms, spores, and efflux pumps (St. Denis et al. 2011). aPDI involves the integration of harmless light of appropriate wavelength with light-sensitive dye or photosensitizer in presence of oxygen present in/around the cells. The photoactivated photosensitizer is excited to an energetic state where it undergoes collision with oxygen and results in the generation of cytotoxic reactive oxygen species. There are many advantages of aPDI over conventional therapies which make it superior. aPDI never leads to the development of resistance in microorganisms, and the effective action is more rapid compared to other methods. This method is effective for the photoinactivation of both native bacterial strains and multidrug-resistant strains (Dai et al. 2010). The photophysical processes behind aPDI are classified into type I and type II mechanisms. Both type I and II can occur together based on the type of photosensitizer (PS) and the microenvironment of PS used. PS excited by light is entered to a long-lived triplet state where it interacts with molecular oxygen in type I (energy transfer) or type II (electron transfer) mechanism. Free radicals such as superoxide anion and hydroxyl radicals are produced by type I, and singlet oxygen is produced by type II process (Huang et al. 2012a).

Several life-threatening pathogens of humans that are resistant to the available antibiotics can be treated by aPDI. There are several categories of PS available for the effective photoinactivation of pathogens. New methylene blue and dimethyl-methylene blue from phenothiazinium family of PS were reported for phototoxicity against MRSA, fungi, and some Gram-negative bacteria, revealing their broad spectrum of action (Sperandio et al. 2013). The photodamage caused by PS against microorganisms includes morphological and functional changes (Fig. 13.2). It can be loss of enzyme activity, protein oxidation, lipid peroxidation, DNA damage, formation of protein-protein cross-links, inhibition of metabolic process, change in the mesosome structure, damage to the cell membrane, cytoplasmic leakage, and inactivation of membrane transport system.

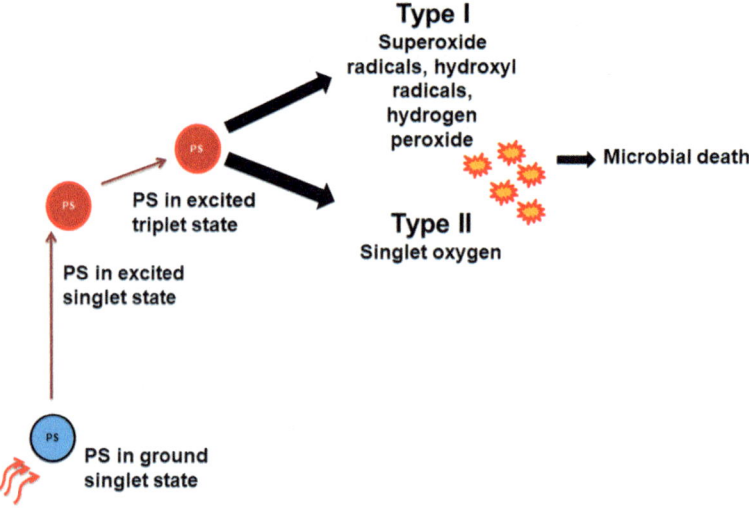

Fig. 13.2 Different mechanisms involved in antimicrobial photodynamic activity of photosensitizers and the generation of reactive oxygen species through different photophysical processes

13.2 Antimicrobial Photodynamic Therapy and Nanotechnology

Ideal characteristics of PS for optimal photoinactivation include purity of the PS at a stable composition; solubility in aqueous solutions; high quantum yield of ROS for maximum potential; safe, nontoxic nature; and selectivity toward target cells instead of host cells. Still, many photosensitizers such as phthalocyanines, chlorins, and porphyrins that are highly lipophilic tend to aggregate in physiological solutions that results in poor aPDI (Jori et al. 2006). aPDI requires various tools for the optimal delivery of PS molecules to the target cells. The field of nanotechnology enhanced the antimicrobial photodynamic therapy and eliminated the limitations related to PS as the nanoparticles can be employed as PS carrier to the microbial cells. Several suitable PS nanocarriers and conjugates were studied with maximum antimicrobial action. Liposomes, micelles, and nanoparticles are studied widely as carriers with improved PS delivery (Kashef et al. 2017) (Table 13.1).

The unique features of nanoparticles include their nanoscale size, high surface-to-volume ratio, and tunable morphology which allows the application of nanotechnology in aPDI. The physical and chemical properties of nanoparticles (NPs) enable them to cross the biological barriers and gain the access to target site. Manipulation of NPs in their shape, chemical properties, and size facilitates the molecular interaction of the drug or PS by increasing the potential in therapy. Also they are engineered to deliver the therapeutic agents precisely to the target site. Application of nanotechnology gained importance in aPDI as the effectiveness of therapy increases with NPs. The bacterial inactivation kinetics of PS is significantly improved by the

Table 13.1 Different types of nanoparticles employed for antimicrobial and antibiofilm photodynamic therapy

Nanoparticles	Photosensitizer/ nanocarrier	Antimicrobial/ antibiofilm activity	Test microorganisms
Silver/gold NPs-Zn phthalocyanine-polylysine conjugates	Nanocarrier	Antimicrobial	*S. aureus* (Nombona et al. 2012)
PEGylated zinc oxide NPs-doxorubicin	Nanocarrier	Antimicrobial	Gram-positive bacteria (Lucky et al. 2015)
Silver-iron bimetallic NPs with indium phthalocyanines	Nanocarrier	Antimicrobial	*E. coli* (Magadla et al. 2019)
Multiwalled carbon nanotubes-rose bengal	Nanocarrier	Antimicrobial and antibiofilm	*E.coli* (Vt et al. 2018)
Mesoporous silica NPs-rose bengal	Nanocarrier	Antifungal and antibiofilm	*C. albicans* (Parasuraman Paramanantham et al. 2018)
Gold nanorods-toluidine blue	Nanocarrier	Antimicrobial	MRSA (Kuo et al. 2016)
Cationic fullerenes	Photosensitizer	Antimicrobial	*S. aureus, E. coli,* and *C. albicans* (Kashef et al. 2017)
Gold NPs with tiopronin and toluidine blue	Nanocarriers	Antimicrobial	*S. aureus* (Pissuwan et al. 2010)
Titanium oxide suspension	Photosensitizer	Antimicrobial	*Acinetobacter* sp. (Yin et al. 2015)

use of nanoparticles (St. Denis et al. 2011). The application of nanotechnology in aPDI includes delivery of PS by encapsulating PS in NPs, increased quantum yield of ROS by covalent attachment of PS to NPs, and NPs acting as photosensitizers (fullerenes, titanium oxide, etc.) (Fig. 13.3).

13.2.1 Fullerenes

Fullerenes are allotropic form of carbon with a closed cage shape, and the number of carbon molecules can be of 60, 70, 72, 76, 84, and up to 100. Fullerenes gained much attention in aPDI due to their tunability to impart higher ROS generation for effective photodestruction. Scientists have explored fullerenes for the delivery of PS by attaching to the hydrophilic or amphiphilic groups added. Fullerenes are superior to the conventional PS in many cases. They are more photostable and less photoinactivated compared to the tetrapyrroles. Fullerenes generate ROS through both type I and II pathways. Light-harvesting antenna can be attached to the C60 fullerenes to enhance the quantum yield of ROS. Moreover, self-assembly of fullerenes into vesicles enhances the tissue-targeting property and also allows multivalent drug deliver to cells (Huang et al. 2012b). A novel *N,N*-dimethyl-2-(4'-*N,N,N*-trimethyl-

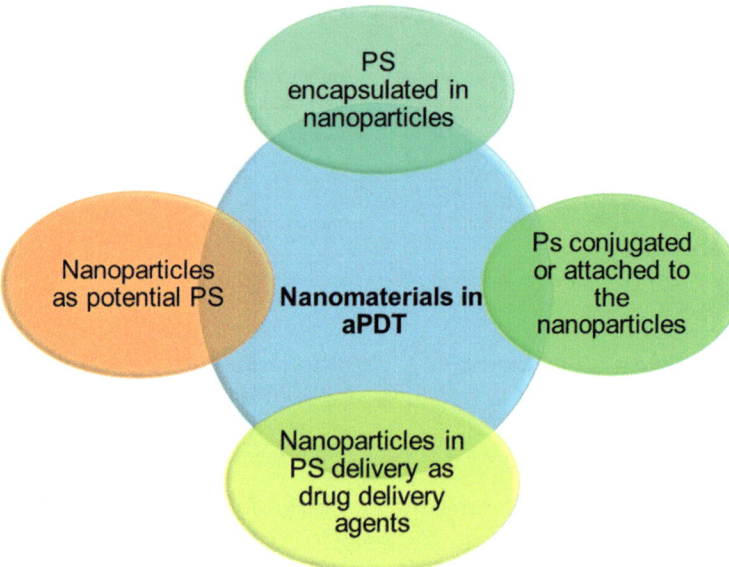

Fig. 13.3 Potential applications of nanostructures in antimicrobial and antibiofilm photodynamic therapy

aminophenyl)fulleropyrrolidinium iodide (DTC(60)(2+)) synthesized effectively and photoinactivated *E. coli* with a 3.5 log decrease in the cells when treated with 1 μM of fullerene derivative for 30 min exposure (Mroz et al. 2007). Buckminster fullerene (C60) photoinactivated the enveloped virus, Semliki Forest virus or vesicular stomatitis virus, when irradiated with visible light for 5 h and resulted in the reduction of virion units to 7 logs (Käsermann and Kempf 1998). Fullerenes have been tested for their in vivo photoinactivation capacity in mice model infected with bacteria. An excision mice wound infection model inoculated with *Proteus mirabilis* and *Pseudomonas aeruginosa* was photoinactivated using tris-cationic BF6 fullerenes and irradiated with a white light of 400–700 nm. Mice infected with *P. mirabilis* showed an 82% survival by fullerene-mediated aPDI, and mice infected with *P. aeruginosa* showed survival of 60% when tobramycin is combined with aPDI for synergistic therapeutic effect (Sharma et al. 2011).

Zhao et al. (2019) described aPDI of plant pathogenic fungi, *Sclerotinia sclerotiorum* and *Fusarium graminearum*, with self-n-doped fullerene ammonium halides. Studies have verified that improved antifungal activity was not dependent on ROS production but because of the reducing property of halide anions (Zhao et al. 2019). Photoactivated C60 fullerenes reduced the infectious titer of iridovirus by 4.5 lg ID 50/ml units by a light exposure for 1 h (Rud et al. 2012). Scientists have reported the importance of charged moieties on fullerenes and their photoinactivation ability. In the study, a series of cationic fullerenes were used as PS against *S. aureus*, *E. coli*, and *C. albicans*. Tetra cationic derivative of fullerene was found to be the most active against *S. aureus*, whereas *C. albicans* showed resistance to aPDI, and intermediate activity was noticed against *E. coli* (Mizuno et al. 2011). Antimicrobial photodynamic activity of fullerenes and their hydroxylated derivatives, fullerenols, showed highest activity against *Malassezia furfur*, *C. albicans*, *Staphylococcus epidermidis*, and *Propionibacterium acnes*. The antimicrobial activity of fullerenes against fungi was higher than the bacteria due to their greater interaction with the fungal cell wall components (Aoshima et al. 2009).

13.2.2   Carbon Nanotubes and Graphene

Carbon nanotubes (CNT) are carbon allotropes with cylinder structure made of graphene sheets. CNTs are categorized as single-walled and multiwalled carbon nanotubes, where single-walled are made by wrapping a single layer of graphene sheet onto a seamless cylinder and multiwalled with multiple layers of graphene (Wang et al. 2009). Graphene oxide (GO) is made of sp2 and sp3 hybridized carbon atoms and gained much interest in aPDI due to the well-dispersed stability in solution, large surface area, and excellent biocompatibility (Wang et al. 2011). The unique optical properties of CNT and graphene oxide (GO) allow them to absorb light in the range of 700–1100 nm. The NIR irradiation makes them suitable for all biological applications, targeted treatment, and deeper penetration into tissues. These nanomaterials exhibit photoinactivation properties and antibacterial activities (Weissleder 2001). Graphene nanowalls with stainless steel as substrates damaged cell membrane of *S. aureus* within 1 h of interaction with bacteria (Akhavan and Ghaderi 2010). GO/titanium oxide films exhibited phototoxicity against *E. coli* after 4 h of solar light exposure (Akhavan and Ghaderi 2009). The ability of carbon nanotubes and GO to inactivate bacteria after few minutes of interaction is well studied. Magnetic reduced GO functionalized with glutaraldehyde effectively killed *E. coli* and *S. aureus* under NIR laser irradiation (Wu et al. 2013). It reduced 99% of Gram-negative and Gram-positive bacterial cells upon 10 min of exposure. Graphene quantum dots synthesized with two photon excitation, two photon stability, strong two photon luminescence, and two photon absorption in NIR region showed increased phototoxicity against *E. coli* and MRSA (Kuo et al. 2016). The toxicity mechanisms of CNTs include disruption of metabolic pathways, oxidative stress, damage to the cell membrane, generation of ROS, and oxidative stress.

Bacteria were damaged when contact with CNTs due to the interaction with the cell membrane. Single-walled CNTs were more toxic in photoinactivation of bacteria than multiwalled due to their size (Kang et al. 2008). Reports are available on the antibacterial and antiviral applications of CNT-drug conjugates. Multiwalled carbon nanotubes conjugated with protoporphyrin IX (PPIX) were inactivated 80% of *S. aureus* colonies within 1 h of visible light exposure. The photodamage was due to the combined effect, i.e., mechanical strength of CNT and biocidal nature of porphyrins (Spagnul et al. 2015). Banerjee et al. (2012) investigated antiviral photodynamic inactivation of PPIX-conjugated multiwalled carbon nanotubes against influenza virus. It was found that visible-light-mediated photoinactivation reduced the ability of virus particles to infect mammalian cells without any development of resistance (Banerjee et al. 2012).

13.2.3 Metal NPs in aPDT

Among metals, gold (Au) and silver (Ag) are involved in the plasmonic enhancement of aPDT. These metal NPs (AuNPs and AgNPs) potentiate the optical field near to the surface due to the electron conductance by local electric field and finally increase the photoactivity of PS attached to it (Huang et al. 2012b). Methylene blue-conjugated AuNPs reduced the viable count of *S. aureus* to three percentages on photoinactivation with diode laser of 660 nm. Gold NPs were found to be effective dye carrier (Tawfik et al. 2015). Methylene blue-conjugated AuNPs photoinactivated nosocomially acquired *C. albicans* biofilms through type I toxicity (Khan et al. 2012). AuNPs with PPIX was employed previously for the photoinactivation of *S. aureus* using xenon light. Protein damage that occurred after aPDT was one of the bactericidal effect noticed (Fathi et al. 2013). Gold nanorods functionalized with a series of boron dipyrromethene (BODIPY) dyes were proposed as good PS for antimicrobial photodynamic inactivation which results in high quantum yield of singlet oxygen and low fluorescent yields (Kubheka et al. 2016).

Silver is a known antimicrobial agent and has good affinity toward sulfur and phosphorus atoms in biomolecules. AuNPs can cause damage to the cell membrane affecting the cell viability, involved in inhibition of enzyme required for DNA replication, and cause cell death. The antimicrobial activity of AuNPs depends on the size, shape, surface area, and dose of particles employed (Dakal et al. 2016; Durán et al. 2016; Matsumura et al. 2003). Misba et al. (2016) studied antibacterial and antibiofilm efficacy of novel AuNPs-toluidine blue conjugates against *S. mutans*. The study revealed that the viability of cells was reduced by 4 log 10 reductions. The conjugates exhibited bactericidal effect with different mechanisms of action such as antibiofilm activity, cell membrane damage, and leakage of cellular contents (Misba et al. 2016).

13.2.3.1 Titanium Oxide and Zinc Oxide

Zinc oxide (ZnO) and titanium oxide (TiO$_2$) are effective photocatalysts of aPDI which produces ROS upon light exposure by transferring electrons. The most abundantly produced ROS are hydroxyl radicals. These metal NPs are able to excite with UV light and visible light (Rehman et al. 2009). TiO$_2$ was found to be effective photocatalyst than ZnO for antimicrobial activity in disinfecting the wastewater (Foster et al. 2011). The antimicrobial activities of ZnO and TiO$_2$ have been reported against broad spectrum of organisms along with the combined effects of metal oxide NPs under UV irradiation. Reported photoactivated ZnO and TiO$_2$ nanoparticles significantly inhibited pathogens prevalent in infected wounds such as *S. aureus* and *S. epidermidis* (Lipovsky et al. 2011). An orthodontic adhesive containing cationic curcumin conjugated to ZnONPs was successfully used to control biofilms formed by cariogenic bacteria *Lactobacillus acidophilus*, *Streptococcus sobrinus*, and *Streptococcus mutans* and reduced their metabolic activity to 100% (Pourhajibagher et al. 2019).

ZnONPs showed antimicrobial potential against nosocomial pathogens especially imipenem- and colistin-resistant *A. baumannii* and *K. pneumonia*. The photodestruction activity of ZnONPs was due to membrane disruption, ROS generation, protein dysfunction, imbalanced metal homeostasis, and DNA damage (Yang et al. 2018; Bhuyan et al. 2015). In another study, combined action of photosensitizer and titanium oxide at very low concentration of nanoparticles was employed for effective antimicrobial activity. TiO-methylene blue conjugate showed good antimicrobial activity against *E. coli*, *S. aureus*, and *C. albicans* when irradiated with light at two wavelengths (405 and 625 nm) (Tuchina and Tuchin 2010). TiO$_2$ nanoparticles photoactivated by UVA light produced reactive oxygen species and exhibited antimicrobial reduction up to 6 logs against broad-spectrum microorganisms such as fungi, Gram-positive bacteria, and Gram-negative bacteria (Huang et al. 2016). Nitrogen doped onto TiO$_2$NPs used in dental resin formulations showed antibacterial activity against *E. coli* after 1 h of exposure with blue light (Zane et al. 2016).

13.3 Nanocarriers in aPDT

A nanocarrier in antimicrobial photodynamic therapy offers wide applications. Nanocarriers used in aPDT help in the controlled and sustained release of PS, targeted PS delivery, and less toxicity to host cells. Both hydrophilic and hydrophobic drugs can be encapsulated into NPs that results in the successful penetration of drugs. Several PS tend to form aggregates in aqueous solutions and form inactive photosensitizers. Some PSs is unable to cross the biofilms and penetrate into the bacterial cell. The advantages of nanocarriers in aPDT include the enhanced inactivation kinetics of PS, improved quantum ROS yield, and use as photosensitizers. Several metal-based NPs and carbon-based NPs are successful nanocarriers of PSs to bacteria. Carboxyl-functionalized carbon nanotubes were used as carriers of methylene blue for the effective photodestruction of *E. coli* and *S. aureus* (Parasuraman et al. 2019). There are other inorganic NPs such as upconversion NPs, magnetic NPs, and calcium phosphate NPs that act as good drug carriers for antimicrobial photodynamic inactivation. Nanocarriers used in aPDT include polymeric NPs, nanomicelles, liposomes, and nanoemulsions for improved delivery of drugs and reduced minimum inhibitory concentration of drugs.

13.3.1 Liposomes

There are several nanocarriers used for the effective and improved delivery of drugs. Liposomes are having micro- or nanostructured vesicles formed by lipid bilayer around the aqueous core with size ranging from 50 to 1000 nm (Derycke 2004). Liposomes are widely used nanocarriers as they are able to carry hydrophilic drugs in aqueous core and hydrophobic drugs in fatty acid chains in the lipid bilayers (Torchilin 2005). Liposomes in conjugation with different PS with targeted delivery are successfully applied against many bacteria in vitro. The enhanced PS delivery to the target bacteria is mediated by combining liposomes with cationic phospholipids such as dimethyldioctadecylammonium bromide or sterilamine. Liposomes imparted with positive charges help in the stronger attachment to the negatively charged bacterial surface. This helps in the selective delivery of drugs to the bacteria over host cells (Paulo et al. 2011).

In a study, liposomes loaded with hematoporphyrin eradicated Gram-positive pathogens such as MSSA and MRSA (Tsai et al. 2009). Modified liposomes carrying novel antimicrobial peptide and potent PS temoporfin were used for the effective delivery of drug to bacteria. The targeted delivery of temoporfin loaded in liposomes eradicated MRSA and reduced *P. aeruginosa* by 3.3 log 10 when irradiated with 625 nm laser (Yang et al. 2011). Methylene blue- and toluidine blue-encapsulated liposomes exhibited 3 log reductions in the growth of *S. aureus* and *E. coli* compared to free liposomes when excited under blue light (Nakonechny et al. 2010).

13.3.2 Polymeric Nanoparticles

Polymeric nanoparticles are well-known drug delivery systems in aPDI. The polymeric nanoparticles are synthesized using different substrates including silica, proteins (collagen, human seroalbumin), poly lactic-co-glycolic acid (PLGA), and polysaccharides (alginate, dextran, and chitosan) (Prasad et al. 2017). Among all the polymeric NPs, PLGA exhibited more photodestruction on bacteria (Pagonis et al. 2010). These NPs are capable of concentrating more PS on bacterial cell wall resulting in the lethal damage to the bacteria. The polymeric NPs are synthesized in such a way that PS binds firmly to the membrane. The different chemical functionalization of polymeric NPs helps in the controlled release of drugs and targeted delivery (Hah et al. 2011). Methylene blue-loaded PLGA NPs when irradiated with a red light of 660 nm inhibited dental plaque pathogen of humans. Methylene blue PLGA NPs eliminated the dental plaque microorganisms in the biofilm stage by 25% more than that of biofilms treated with MB (de Freitas et al. 2016). In another study, PLGA nanoparticles encapsulated with methylene blue showed phototoxicity against planktonic cells and biofilms of dental plaque microorganisms isolated from chronic periodontitis patients (Klepac-Ceraj et al. 2011). Curcumin-encapsulated polymeric NPs (PLGA and dextran sulfate) with enhanced water solubility were employed for antimicrobial photodynamic inactivation of *C. albicans* and for oral candidiasis (Sakima et al. 2018).

13.3.3 Nanoemulsions

Nanoemulsions (NEs) are made of two immiscible liquids where one liquid is continuous phase and other as noncontinuous droplets. The diameter of these droplets is smaller than 300 nm. Like liposomes, nanoemulsions are also used for improving the properties of hydrophobic PS. These emulsions are employed particularly for topical drug delivery systems. NEs are known as skin permeation inducers. NEs with skin permeation ability are applied in aPDT for superficial skin infections because they can penetrate the stratum corneum to reach the deeper side of the epidermis. The inability of some hydrophilic drugs or PS to reach lower epidermis layer can be incorporated into nanoemulsion (Maisch et al. 2009). NEs exhibit broad-spectrum antimicrobial action with less generation of resistance and are suitable for surface contamination and wound treatment (Hwang et al. 2013). Ribeiro et al. (2015) described photodynamic antibiofilm therapy of MRSA and MSSA using chloroaluminum phthalocyanine encapsulated in NEs on irradiation with LED source (Ribeiro et al. 2015). The cell metabolism of MSSA and MRSA was reduced by 80% and 73%, respectively, with significant reduction in the CFU/ml of bacteria (Ribeiro et al. 2015). Scientists have reported that cationic NEs with zinc 2,9,16,23-tetrakis(phenylthio)-29H,31H-phthalocyanine reduced the biofilms formed by *C. albicans* and other fungi such as *T. mucoides* and *K. ohmeri*. The biofilms of *C. albicans* were reduced by 0.45 log 10 and other fungi reduced to 0.85 and 0.84 log 10 (Junqueira et al. 2012).

13.3.4 Nanomicelles

Nanomicelles are more appropriate for the delivery of PS than liposomes as they improve the solubility of PS and prevent the aggregation of PS in aqueous solutions. Preparation and scale up of micelles are much easier than that of liposomes and are widely used drug carriers (Tsai et al. 2009). Aluminum chloride phthalocyanine formulated in block copolymers of micelles proved to be potential drug delivery system and antimicrobial photodynamic agent against *S. aureus* and *C. albicans* (Vilsinski et al. 2015). Hypericin-encapsulated pluronic P123 micelles were reported as promising PS for the treatment of chronic infections caused by Gram-positive bacteria. The biofilms of *E. faecalis* and *S. aureus* were reduced to 2.86 and 2.30 CFU log reductions (Vilsinski et al. 2015). Chitosan micelles encapsulated with thymol were synthesized for the effective aPDT against biofilms. Light controllable micelles are formulated for controlled release of thymol which generated ROS with strong bactericidal effect on biofilms of *Listeria monocytogenes* and *S. aureus* (Wang et al. 2019).

13.4 Conclusions

In this chapter, we have summarized the recent advancement and developments occurred in the photoinactivation of microorganisms using nanostructures. Nanostructures alone or in conjugation with conventional photosensitizers are photoactivated under visible light and employed as a promising and alternative therapeutic strategies for the eradication and elimination of antimicrobial resistance bacteria and biofilms. Several nanostructures are applied as dye or drug carriers to the target bacteria for antimicrobial and antibiofilm therapy. The disadvantages of traditional photosensitizers have forced the integration of nanotechnology in aPDT. The solubility and stability of PS are improved through the use of dye delivery agent. These dye delivery agents enhance the antimicrobial photoinactivation of PS by increasing the solubility of dyes and improving the photoinactivation kinetics and quantum yield of ROS.

References

Akhavan O, Ghaderi E (2009) Photocatalytic reduction of graphene oxide nanosheets on TiO_2 thin film for photoinactivation of bacteria in solar light irradiation. J Phys Chem C 113:20214–20220. https://doi.org/10.1021/jp906325q

Akhavan O, Ghaderi E (2010) Toxicity of graphene and graphene oxide nanowalls against bacteria. ACS Nano 4:5731–5736. https://doi.org/10.1021/nn101390x

Aoshima H, Kokubo K, Shirakawa S et al (2009) Antimicrobial activity of fullerenes and their hydroxylated derivatives. Biocontrol Sci 14:69–72. https://doi.org/10.4265/bio.14.69

Aziz N, Fatma T, Varma A, Prasad R (2014) Biogenic synthesis of silver nanoparticles using Scenedesmus abundans and evaluation of their antibacterial activity. J Nanopart 2014:689419. https://doi.org/10.1155/2014/689419

Aziz N, Faraz M, Pandey R, Sakir M, Fatma T, Varma A, Barman I, Prasad R (2015) Facile algae-derived route to biogenic silver nanoparticles: Synthesis, antibacterial and photocatalytic properties. Langmuir 31:11605–11612. https://doi.org/10.1021/acs.langmuir.5b03081

Aziz N, Pandey R, Barman I, Prasad R (2016) Leveraging the attributes of Mucor hiemalis-derived silver nanoparticles for a synergistic broad-spectrum antimicrobial platform. Front Microbiol 7:1984. https://doi.org/10.3389/fmicb.2016.01984

Banerjee I, Douaisi MP, Mondal D, Kane RS (2012) Light-activated nanotube–porphyrin conjugates as effective antiviral agents. Nanotechnology 23:105101. https://doi.org/10.1088/0957-4484/23/10/105101

Bhuyan T, Mishra K, Khanuja M, Prasad R, Varma A (2015) Biosynthesis of zinc oxide nanoparticles from *Azadirachta indica* for antibacterial and photocatalytic applications. Mater Sci Semicond Process 32:55–61

Bjarnsholt T (2013) The role of bacterial biofilms in chronic infections. APMIS 121:1–58. https://doi.org/10.1111/apm.12099

Ceri H, Olson ME, Turner RJ (2010) Needed, new paradigms in antibiotic development. Expert Opin Pharmacother 11:1233–1237. https://doi.org/10.1517/14656561003724747

Chadha T (2012) Antibiotic resistant genes in natural environment. Agrotechnology 01:1–3. https://doi.org/10.4172/2168-9881.1000104

Costerton JW, Lewandowski Z, Caldwell DE et al (1995) Microbial biofilms. Annu Rev Microbiol 49:711–745. https://doi.org/10.1146/annurev.mi.49.100195.003431

Dai T, Tegos GP, Zhiyentayev T et al (2010) Photodynamic therapy for methicillin-resistant *Staphylococcus aureus* infection in a mouse skin abrasion model. Lasers Surg Med 42:38–44. https://doi.org/10.1002/lsm.20887

Dakal TC, Kumar A, Majumdar RS, Yadav V (2016) Mechanistic basis of antimicrobial actions of silver nanoparticles. Front Microbiol 7:1–17. https://doi.org/10.3389/fmicb.2016.01831

Davies D (2003) Understanding biofilm resistance to antibacterial agents. Nat Rev Drug Discov 2:114–122. https://doi.org/10.1038/nrd1008

De Freitas LM, Calixto GMF, Chorilli M et al (2016) Polymeric nanoparticle-based photodynamic therapy for chronic periodontitis *in vivo*. Int J Mol Sci 17:769. https://doi.org/10.3390/ijms17050769

De la Fuente-Nunez C, Torres MD, Mojica FJ, Lu TK (2017) Next-generation precision antimicrobials: towards personalized treatment of infectious diseases. Curr Opin Microbiol 37:95–102. https://doi.org/10.1016/j.mib.2017.05.014

Derycke A (2004) Liposomes for photodynamic therapy. Adv Drug Deliv Rev 56:17–30. https://doi.org/10.1016/j.addr.2003.07.014

Donlan RM (2002) Biofilms: microbial life on surfaces. Emerg Infect Dis 8:881–890. https://doi.org/10.3201/eid0809.020063

Drenkard E (2003) Antimicrobial resistance of *Pseudomonas aeruginosa* biofilms. Microbes Infect 5:1213–1219. https://doi.org/10.1016/j.micinf.2003.08.009

Drenkard E, Ausubel FM (2002) *Pseudomonas* biofilm formation and antibiotic resistance are linked to phenotypic variation. Nature 416:740–743. https://doi.org/10.1038/416740a

Durán N, Durán M, de Jesus MB et al (2016) Silver nanoparticles: a new view on mechanistic aspects on antimicrobial activity. Nanomed Nanotechnol Biol Med 12:789–799. https://doi.org/10.1016/j.nano.2015.11.016

Fathi A, Rasekaran R et al (2013) Influence of hydrogen peroxide or gold nanoparticles in protoporphyrin IX mediated antimicrobial photodynamic therapy on *Staphylococcus aureus*. Afr J Microbiol Res 7:4617–4624. https://doi.org/10.5897/2013.5885

Fernandes P, Martens E (2017) Antibiotics in late clinical development. Biochem Pharmacol 133:152–163. https://doi.org/10.1016/j.bcp.2016.09.025

Foster HA, Ditta IB, Varghese S, Steele A (2011) Photocatalytic disinfection using titanium dioxide: spectrum and mechanism of antimicrobial activity. Appl Microbiol Biotechnol 90:1847–1868. https://doi.org/10.1007/s00253-011-3213-7

Hah HJ, Kim G, Lee Y-EK et al (2011) Methylene blue-conjugated hydrogel nanoparticles and tumor-cell targeted photodynamic therapy. Macromol Biosci 11:90–99. https://doi.org/10.1002/mabi.201000231

Holmstrup P, Klausen B (2018) The growing problem of antimicrobial resistance. Oral Dis 24:291–295. https://doi.org/10.1111/odi.12610

Huang L, Xuan Y, Koide Y et al (2012a) Type I and type II mechanisms of antimicrobial photodynamic therapy: an *in vitro* study on Gram-negative and Gram-positive bacteria. Lasers Surg Med 44:490–499. https://doi.org/10.1002/lsm.22045

Huang Y-Y, Sharma SK, Dai T et al (2012b) Can nanotechnology potentiate photodynamic therapy? Nanotechnol Rev 1:111–146. https://doi.org/10.1515/ntrev-2011-0005

Huang Y-Y, Choi H, Kushida Y et al (2016) Broad-spectrum antimicrobial effects of photocatalysis using titanium dioxide nanoparticles are strongly potentiated by addition of potassium iodide. Antimicrob Agents Chemother 60:5445–5453. https://doi.org/10.1128/AAC.00980-16

Hwang YY, Ramalingam K, Bienek DR et al (2013) Antimicrobial activity of nanoemulsion in combination with cetylpyridinium chloride in multidrug-resistant *Acinetobacter baumannii*. Antimicrob Agents Chemother 57:3568–3575. https://doi.org/10.1128/AAC.02109-12

Jasovský D, Littmann J, Zorzet A, Cars O (2016) Antimicrobial resistance—a threat to the world's sustainable development. Ups J Med Sci 121:159–164. https://doi.org/10.1080/03009734.201 6.1195900

Jing H, Mezgebe B, Aly Hassan A et al (2014) Experimental and modeling studies of sorption of ceria nanoparticle on microbial biofilms. Bioresour Technol 161:109–117. https://doi.org/10.1016/j.biortech.2014.03.015

Jori G, Fabris C, Soncin M et al (2006) Photodynamic therapy in the treatment of microbial infections: basic principles and perspective applications. Lasers Surg Med 38:468–481. https://doi.org/10.1002/lsm.20361

Junqueira JC, Jorge AOC, Barbosa JO et al (2012) Photodynamic inactivation of biofilms formed by *Candida* spp., *Trichosporon mucoides*, and *Kodamaea ohmeri* by cationic nanoemulsion of zinc 2,9,16,23-tetrakis(phenylthio)-29H, 31H-phthalocyanine (ZnPc). Lasers Med Sci 27:1205–1212. https://doi.org/10.1007/s10103-012-1050-2

Kang S, Herzberg M, Rodrigues DF, Elimelech M (2008) Antibacterial effects of carbon nanotubes: size does matter. Langmuir 24:6409–6413. https://doi.org/10.1021/la800951v

Käsermann F, Kempf C (1998) Buckminsterfullerene and photodynamic inactivation of viruses. Rev Med Virol 8:143–151. https://doi.org/10.1002/(SICI)1099-1654 (199807/09)8:3<143::AID-RMV214>3.0.CO;2-B

Kashef N, Huang Y-Y, Hamblin MR (2017) Advances in antimicrobial photodynamic inactivation at the nanoscale. Nanophotonics 6:853–879. https://doi.org/10.1515/nanoph-2016-0189

Khan S, Khan AU, Azam A, Alam F (2012) Gold nanoparticles enhance methylene blue-induced photodynamic therapy: a novel therapeutic approach to inhibit *Candida albicans* biofilm. Int J Nanomedicine 2012:3245–3257. https://doi.org/10.2147/IJN.S31219

Klepac-Ceraj V, Patel N, Song X et al (2011) Photodynamic effects of methylene blue-loaded polymeric nanoparticles on dental plaque bacteria. Lasers Surg Med 43:600–606. https://doi.org/10.1002/lsm.21069

Kubheka G, Uddin I, Amuhaya E et al (2016) Synthesis and photophysicochemical properties of BODIPY dye functionalized gold nanorods for use in antimicrobial photodynamic therapy. J Porphyr Phthalocyanines 20:1016–1024. https://doi.org/10.1142/S108842461650070X

Kuo W-S, Chang C-Y, Chen H-H et al (2016) Two-photon photoexcited photodynamic therapy and contrast agent with antimicrobial graphene quantum dots. ACS Appl Mater Interfaces 8:30467–30474. https://doi.org/10.1021/acsami.6b12014

Lewis K (2001) Riddle of biofilm resistance. Antimicrob Agents Chemother 45:999–1007. https://doi.org/10.1128/AAC.45.4.999-1007.2001

Li X-Z, Nikaido H (2009) Efflux-mediated drug resistance in bacteria. Drugs 69:1555–1623. https://doi.org/10.2165/11317030-000000000-00000

Lipovsky A, Gedanken A, Nitzan Y, Lubart R (2011) Enhanced inactivation of bacteria by metal-oxide nanoparticles combined with visible light irradiation. Lasers Surg Med 43:236–240. https://doi.org/10.1002/lsm.21033

Llor C, Bjerrum L (2014) Antimicrobial resistance: risk associated with antibiotic overuse and initiatives to reduce the problem. Ther Adv Drug Saf 5:229–241. https://doi.org/10.1177/2042098614554919

Lucky SS, Soo KC, Zhang Y (2015) Nanoparticles in photodynamic therapy. Chem Rev 115:1990–2042. https://doi.org/10.1021/cr5004198

Magadla A, Oluwole DO, Managa M, Nyokong T (2019) Physicochemical and antimicrobial photodynamic chemotherapy (against *E. Coli*) by indium phthalocyanines in the presence of silver-iron bimetallic nanoparticles. Polyhedron 40:2710–2721. https://doi.org/10.1016/j.poly.2019.01.032

Mah T-FC, O'Toole GA (2001) Mechanisms of biofilm resistance to antimicrobial agents. Trends Microbiol 9:34–39. https://doi.org/10.1016/S0966-842X(00)01913-2

Maisch T, Santarelli F, Schreml S et al (2009) Fluorescence induction of protoporphyrin IX by a new 5-aminolevulinic acid nanoemulsion used for photodynamic therapy in a full-thickness ex vivo skin model. Exp Dermatol 19:e302–e305. https://doi.org/10.1111/j.1600-0625.2009.01001.x

Majik M, Parvatkar P (2013) Next generation biofilm inhibitors for *Pseudomonas aeruginosa*: synthesis and rational design approaches. Curr Top Med Chem 14:81–109. https://doi.org/10.2174/15680266136661311131152257

Matsumura Y, Yoshikata K, Kunisaki S, Tsuchido T (2003) Mode of bactericidal action of silver zeolite and its comparison with that of silver nitrate. Appl Environ Microbiol 69:4278–4281. https://doi.org/10.1128/AEM.69.7.4278-4281.2003

McAdam AJ, Hooper DC, DeMaria A et al (2012) Antibiotic resistance: how serious is the problem, and what can be done? Clin Chem 58:1182–1186. https://doi.org/10.1373/clinchem.2011.181636

Misba L, Kulshrestha S, Khan AU (2016) Antibiofilm action of a toluidine blue O-silver nanoparticle conjugate on *Streptococcus mutans*: a mechanism of type I photodynamic therapy. Biofouling 32:313–328. https://doi.org/10.1080/08927014.2016.1141899

Mizuno K, Zhiyentayev T, Huangv L et al (2011) Antimicrobial photodynamic therapy with functionalized fullerenes: quantitative structure-activity relationships. J Nanomed Nanotechnol 02:1–9. https://doi.org/10.4172/2157-7439.1000109

Mroz P, Tegos GP, Gali H et al (2007) Photodynamic therapy with fullerenes. Photochem Photobiol Sci 6:1139–1149. https://doi.org/10.1039/b711141j

Nakonechny F, Firer MA, Nitzan Y, Nisnevitch M (2010) Intracellular antimicrobial photodynamic therapy: a novel technique for efficient eradication of pathogenic bacteria. Photochem Photobiol 86:1350–1355. https://doi.org/10.1111/j.1751-1097.2010.00804.x

National Collaborating Centre for Infectious Diseases (2010) Proceedings of Community-Acquired Antimicrobial Resistance Consultation Notes, Winnipeg, MB, Canada, 10–11 February 2010 Available at: http://www.nccid.ca/files/caAMR_ConsultationNotes_final.pdf

Nombona N, Antunes E, Chidawanyika W et al (2012) Synthesis, photophysics and photochemistry of phthalocyanine-ε-polylysine conjugates in the presence of metal nanoparticles against *Staphylococcus aureus*. J Photochem Photobiol A Chem 233:24–33. https://doi.org/10.1016/j.jphotochem.2012.02.012

Pagonis TC, Chen J, Fontana CR et al (2010) Nanoparticle-based endodontic antimicrobial photodynamic therapy. J Endod 36:322–328. https://doi.org/10.1016/j.joen.2009.10.011

Paramanantham P, Antony AP, Sruthil Lal SB et al (2018) Antimicrobial photodynamic inactivation of fungal biofilm using amino functionalized mesoporous silica-rose bengal nanoconjugate against *Candida albicans*. Sci Afr e00007:1. https://doi.org/10.1016/j.sciaf.2018.e00007

Parasuraman P, Anju VT, Sruthil Lal S et al (2019) Synthesis and antimicrobial photodynamic effect of methylene blue conjugated carbon nanotubes on *E. coli* and *S. aureus*. Photochem Photobiol Sci 1(8):563–576. https://doi.org/10.1039/C8PP00369F

Paulo J, Longo F, Muehlmann LA, De Azevedo RB (2011) Nanostructured carriers for photodynamic therapy applications in microbiology, pp 189–196

Pissuwan D, Cortie CH, Valenzuela SM, Cortie MB (2010) Functionalised gold nanoparticles for controlling pathogenic bacteria. Trends Biotechnol 28:207–213. https://doi.org/10.1016/j.tibtech.2009.12.004

Pourhajibagher M, Salehi Vaziri A, Takzaree N, Ghorbanzadeh R (2019) Physico-mechanical and antimicrobial properties of an orthodontic adhesive containing cationic curcumin doped zinc oxide nanoparticles subjected to photodynamic therapy. Photodiagn Photodyn Ther 25:239–246. https://doi.org/10.1016/j.pdpdt.2019.01.002

Prasad R, Pandey R, Varma A, Barman I (2017) Polymer based nanoparticles for drug delivery systems and cancer therapeutics. In: Natural Polymers for Drug Delivery (eds. Kharkwal H and Janaswamy S), CAB International, UK. 53–70

Rehman S, Ullah R, Butt AM, Gohar ND (2009) Strategies of making TiO$_2$ and ZnO visible light active. J Hazard Mater 170:560–569. https://doi.org/10.1016/j.jhazmat.2009.05.064

Ribeiro APD, Andrade MC, Bagnato VS et al (2015) Antimicrobial photodynamic therapy against pathogenic bacterial suspensions and biofilms using chloro-aluminum phthalocyanine encapsulated in nanoemulsions. Lasers Med Sci 30:549–559. https://doi.org/10.1007/s10103-013-1354-x

Rud Y, Buchatskyy L, Prylutskyy Y et al (2012) Using C 60 fullerenes for photodynamic inactivation of mosquito iridescent viruses. J Enzyme Inhib Med Chem 27:614–617. https://doi.org/10.3109/14756366.2011.601303

Sadekuzzaman M, Yang S, Mizan MFR, Ha SD (2015) Current and recent advanced strategies for combating biofilms. Compr Rev Food Sci Food Saf 14:491–509. https://doi.org/10.1111/1541-4337.12144

Sakima V, Barbugli P, Cerri P et al (2018) Antimicrobial photodynamic therapy mediated by curcumin-loaded polymeric nanoparticles in a murine model of oral candidiasis. Molecules 23:2075. https://doi.org/10.3390/molecules23082075

Santajit S, Indrawattana N (2016) Mechanisms of antimicrobial resistance in ESKAPE pathogens. Biomed Res Int 2016:1–8. https://doi.org/10.1155/2016/2475067

Sharma SK, Chiang LY, Hamblin MR (2011) Photodynamic therapy with fullerenes *in vivo*: reality or a dream? Nanomedicine 6:1813–1825. https://doi.org/10.2217/nnm.11.144

Singh S, Singh SK, Chowdhury I, Singh R (2017) Understanding the mechanism of bacterial biofilms resistance to antimicrobial agents. Open Microbiol J 11:53–62. https://doi.org/10.2174/1874285801711010053

Spagnul C, Turner LC, Boyle RW (2015) Immobilized photosensitizers for antimicrobial applications. J Photochem Photobiol B Biol 150:11–30. https://doi.org/10.1016/j.jphotobiol.2015.04.021

Sperandio F, Huang Y-Y, Hamblin M (2013) Antimicrobial photodynamic therapy to kill Gram-negative bacteria. Recent Pat Antiinfect Drug Discov 8:108–120. https://doi.org/10.2174/1574891X113089990012

St. Denis TG, Dai T, Izikson L et al (2011) All you need is light. Virulence 2:509–520. https://doi.org/10.4161/viru.2.6.17889

Tawfik AA, Alsharnoubi J, Morsy M (2015) Photodynamic antibacterial enhanced effect of methylene blue-gold nanoparticles conjugate on *Staphylococcal aureus* isolated from impetigo lesions *in vitro* study. Photodiagn Photodyn Ther 12:215–220. https://doi.org/10.1016/j.pdpdt.2015.03.003

Torchilin V (2005) Fluorescence microscopy to follow the targeting of liposomes and micelles to cells and their intracellular fate. Adv Drug Deliv Rev 57:95–109. https://doi.org/10.1016/j.addr.2004.06.002

Tsai T, Yang Y-T, Wang T-H et al (2009) Improved photodynamic inactivation of Gram-positive bacteria using hematoporphyrin encapsulated in liposomes and micelles. Lasers Surg Med 41:316–322. https://doi.org/10.1002/lsm.20754

Tuchina ES, Tuchin VV (2010) TiO_2 nanoparticle enhanced photodynamic inhibition of pathogens. Laser Phys Lett 7:607–612. https://doi.org/10.1002/lapl.201010030

Vilsinski BH, Gerola AP, Enumo JA et al (2015) Formulation of aluminum chloride phthalocyanine in pluronic TM P-123 and F-127 block copolymer micelles: photophysical properties and photodynamic inactivation of microorganisms. Photochem Photobiol 91:518–525. https://doi.org/10.1111/php.12421

Vrany JD, Stewart PS, Suci PA (1997) Comparison of recalcitrance to ciprofloxacin and levofloxacin exhibited by *Pseudomonas aeruginosa* biofilms displaying rapid-transport characteristics. Antimicrob Agents Chemother 41:1352–1358

Vt A, Paramanantham P, Sb SL et al (2018) Antimicrobial photodynamic activity of rose bengal conjugated multi walled carbon nanotubes against planktonic cells and biofilm of *Escherichia coli*. Photodiagn Photodyn Ther 24:300–310. https://doi.org/10.1016/j.pdpdt.2018.10.013

Wang X, Li Q, Xie J et al (2009) Fabrication of ultralong and electrically uniform single-walled carbon nanotubes on clean substrates. Nano Lett 9:3137–3141. https://doi.org/10.1021/nl901260b

Wang Y, Li Z, Wang J et al (2011) Graphene and graphene oxide: biofunctionalization and applications in biotechnology. Trends Biotechnol 29:205–212. https://doi.org/10.1016/j.tibtech.2011.01.008

Wang Z, Bai H, Lu C et al (2019) Light controllable chitosan micelles with ROS generation and essential oil release for the treatment of bacterial biofilm. Carbohydr Polym 205:533–539. https://doi.org/10.1016/j.carbpol.2018.10.095

Weissleder R (2001) A clearer vision for in vivo imaging. Nat Biotechnol 19:316–317. https://doi.org/10.1038/86684

Wilson DN (2014) Ribosome-targeting antibiotics and mechanisms of bacterial resistance. Nat Rev Microbiol 12:35–48. https://doi.org/10.1038/nrmicro3155

Wright G (2005) Bacterial resistance to antibiotics: enzymatic degradation and modification. Adv Drug Deliv Rev 57:1451–1470. https://doi.org/10.1016/j.addr.2005.04.002

Wu M-C, Deokar AR, Liao J-H et al (2013) Graphene-based photothermal agent for rapid and effective killing of bacteria. ACS Nano 7:1281–1290. https://doi.org/10.1021/nn304782d

Yang K, Gitter B, Rüger R et al (2011) Antimicrobial peptide-modified liposomes for bacteria targeted delivery of temoporfin in photodynamic antimicrobial chemotherapy. Photochem Photobiol Sci 10:1593. https://doi.org/10.1039/c1pp05100h

Yang M, Chang K, Chen L et al (2018) Blue light irradiation triggers the antimicrobial potential of ZnO nanoparticles on drug-resistant *Acinetobacter baumannii*. J Photochem Photobiol B Biol 180:235–242. https://doi.org/10.1016/j.jphotobiol.2018.02.003

Yin R, Agrawal T, Khan U et al (2015) Antimicrobial photodynamic inactivation in nanomedicine: small light strides against bad bugs. Nanomedicine 10:2379–2404. https://doi.org/10.2217/nnm.15.67

Zane A, Zuo R, Villamena F et al (2016) Biocompatibility and antibacterial activity of nitrogen-doped titanium dioxide nanoparticles for use in dental resin formulations. Int J Nanomedicine 11:6459–6470. https://doi.org/10.2147/IJN.S117584

Zhao J, Lin Z, Fang S et al (2019) Photoexcitation of self-n-doped fullerene ammonium halides: the role of halide ion and a possible synergistic dual-redox cycle mechanism within their aggregate. J Photochem Photobiol A Chem 373:131–138. https://doi.org/10.1016/j.jphotochem.2019.01.008

Chapter 14
Recent Advancements in the Design and Synthesis of Antibacterial and Biofilm Nanoplatforms

Parasuraman Paramanantham and Busi Siddhardha

Contents

Abstract Bacterial infections are a major threat for the global population. Realizing the interaction between the infectious bacteria and antibacterial agents is fundamental to understand the novel therapeutic approaches that combat the bacterial infection leading to morbidity and mortality. At present situation, contemporary healthcare sectors and clinical microbiology deal with the current issue, discovering novel therapeutic strategies to overcome bacterial infections associated with rapidly accelerating multidrug-resistant bacteria. Currently, a remarkable advancement has been occurred in nanobiotechnology toward the formulation of several novel nanoparticles (NPs) that are actively participated as potential antibacterial agents. The control over morphologies and several significant features including size,

P. Paramanantham · B. Siddhardha (✉)
Department of Microbiology, School of Life Sciences, Pondicherry University, Puducherry, India

© Springer Nature Switzerland AG 2020
R. Prasad et al. (eds.), *Nanostructures for Antimicrobial and Antibiofilm Applications*, Nanotechnology in the Life Sciences,
https://doi.org/10.1007/978-3-030-40337-9_14

shape, and nature of the particles made it an extensive area of research in NP synthesis. This feasibility in the technology for the development of NPs facilitates utilization of NPs in the wide range of applications including biomedicine, biosensor, and catalyst with cost-effective manner. The NP-based antibacterial agents are out grouped from the fundamental antibacterial agents such as antibiotic and anti-infective. Nanoantibacterial agents are able to address the regular complication in mechanism of antibiotic resistance including multidrug efflux pumps, permeability regulation, degradation of antibacterial agents, and untargeted site binding affinity mutations. This highly spotted nanoantimicrobial agents are developed by means of different methodologies for both research and commercial utilization. These methodologies are widely classified into physical, chemical, and biological, which are gaining significant advancements in the recent years. The present chapter critically discusses the recent development in the nanobiotechnology, availability of various methods for the synthesis of NPs, field-specific formulation of NPs, wide range of application of NPs in the healthcare sectors specifically antibacterial and biofilm, and future perspective in the field of nanobiotechnology.

Keywords Nanoparticles · Nanobiotechnology · Multidrug-resistant bacteria · Physical, chemical, and biological methods

14.1 Introduction

Antibiotic resistance is a global challenge faced by healthcare system. Scientists have labeled the emerging crisis of antibiotic resistance as dreadful. The global threat experienced by the multidrug resistance strains is mainly due to the abuse of antimicrobial agents and lack of design and development of new antimicrobial agents. The infections associated with antibiotic resistance created a considerable health and financial problem over healthcare system (Ventola 2015a). The Infectious Diseases Society of America emphasized about a faction of antibiotic-resistant microorganism such as *Enterococcus faecium, Staphylococcus aureus, Klebsiella pneumoniae, Acinetobacter baumannii, Pseudomonas aeruginosa,* and *Enterobacter* spp., which are named as "ESKAPE pathogens." ESKAPE pathogens are able to escape from the activity of antibiotics and represents new patterns in pathogenesis, resistance, and transmission of diseases (Pendleton et al. 2013).

The world has now reached the middle of post-antibiotic era where two million patients have infections contributed by drug-resistant bacteria only and 23,000 people die per year. The antimicrobial-resistant strain has negative impact on morbidity and mortality. The situation of antimicrobial resistance (AMR) may cause ten million deaths by 2050, which cost up to $100 trillion (Luepke et al. 2017). The main causes of death in post-antibiotic era were due to the tuberculosis, pneumonia, and gastrointestinal infections, which accounts for 30% of all deaths. The infections caused by Gram-negative antibiotic-resistant bacteria are even difficult to treat with the conventional antimicrobials. In most healthcare settings, the early and proper identification of causative microbial agents and their antibiotic susceptibility patterns are lacking behind where antimicrobial agents are prescribed and used liberally. The resistant patterns dramatically increased when coupled with the poor disease control strategies, and this will results in the dissemination of resistant bacteria to other patients and environment (Akova 2016). The antibiotic resistance genes responsible for antibiotic resistance and associated infections are acquired by pathogenic bacteria through horizontal gene transfer (HGT). Horizontal gene transfer causes antibiotic resistance to disseminate from commercial and environmental species to pathogenic bacteria (Frieri et al. 2017). The antimicrobial resistance (AMR) is a silent pandemic, which is never expected to be cured on treatment with antimicrobials. AMR strains adapted to the adverse conditions through the evolution by escaping from the attack of antibiotics. So far, there is no single silver bullet to avoid innumerable victims of drug-resistant infections (Jasovský et al. 2016).

According to the reports, 80% of the pathogens associated with antibiotic resistance and persistent infections form biofilms (Algburi et al. 2017). Biofilms are formed by microorganisms where structured, coordinated, and functional thin layers of cell develop. The formation of biofilms is known as a survival strategy by microbes and for the establishment of infections. Host defense mechanism and antimicrobial activity are subsequently diminished against the bacteria surviving in the biofilm structure. The biofilm architecture is supported by components produced by the microorganism living inside it, and approximately 90% of the biofilm structure is made of exopolysaccharide (EPS) matrix, proteins, and DNA. The advantages of self-produced protective matrix include stable configuration to the cells, to act as a frame for the enzymes, antibiotics, and cells for attachment and mediate surface adhesion to the substrates (Vianna Santos 2014). The resistance of biofilms is related to their difficulty to penetrate the cell wall composed of EPS. This enables the unrestricted dissemination of the resistance genes among biofilm microorganism. Further, microorganisms present in the biofilms are in their stationary phase, which hinders the action of beta lactam drugs.

In addition, there is an urgent need to find and apply new antimicrobial agents or new therapeutic methods to reduce the microbial activity and combat associated infections (Algburi et al. 2017; Aziz et al. 2016). The development of resistant strains is occurring globally, which endangers the value of antibiotics. In order to escape from the antibiotic resistance crisis, enactment of necessary steps are required such as improving the tracking, diagnosis, and prescribing practices, optimizing therapeutic regimes, preventing the transmission of infections, and adopting antibiotic stewardship programs (Ventola 2015b). In this context, nanomaterials emerged as new approach for managing the microbial and biofilm infections. Several nanomaterials emerged as antimicrobial agents and drug carriers due to their unique physicochemical characteristics such as their sizes, shapes, and high surface-to-volume ratios. The ability of nanoparticles to interrupt the quorum-sensing networks present in the biofilms has been utilized as a strategy in antimicrobial therapy. Furthermore, nanoparticles have been employed as efficient carriers of lipophilic drugs and conventional antimicrobials with improved penetration to the microorganisms and biofilms. The field of nanotechnology has gained much attention for the design and synthesis of nonconventional antimicrobials with the use of nanomaterials and nano-sized carriers. The nanomaterials are designed to combat bacterial diseases by reducing the cell viability, inhibiting the bacterial communication pathways known as quorum sensing, and eliminating resistant biofilms (Singh et al. 2017).

14.2 Emergence of Nanobiotechnology

The field of nanobiotechnology offers a wide variety of applications that range from industrial applications to biomedical applications. One of the main applications of nanoparticles is in biology and biomedical research. The advantages of use of nanoparticles in biology are the ability to engineer the nanoparticles to gain unique composition and functionalities (Wang and Wang 2014). There are several nanomaterials with different sizes, shapes, functions, and compositions with wide range of applications. These nanoparticles can be fabricated with various molecules using nanoprecipitation and lithography techniques. From ancient times, metals like copper (Cu), zinc (Zn), gold (Au), silver (Ag), and titanium (TiO_2) are well studied for therapeutic purposes due to their broad spectrum of antimicrobial activities. Recent advancements in the field of nanotechnology have recognized these metals as potential inhibitory agents for the growth of pathogenic microorganisms. In order to combat the infections by microorganisms, nanoparticles exert multiple functionalities such as inhibition/disruption of biofilms and/or enhanced intracellular accumulation of antimicrobial agents. Various nanoparticles are studied and characterized for their antimicrobial activities. Antimicrobial activity of these metal and metal oxide nanoparticles are purely dependent on their physical and chemical features. The large surface-to-volume ratio of nanoparticles ensures a broad range of reactions with the bioorganics, which is available on the surface of bacteria. The surface area will be larger with smaller nanoparticles. The increased area of nanoparticle enhances the contact with bacteria, which can augment the chemical and biological activities. The use of nanomaterials for biological application has reduced the concentration required for improved antimicrobial properties to a hundred fold (Muzammil et al. 2018).

14.3 Antibacterial and Antibiofilm Mechanism of Nanomaterials

The nanoparticles in antibacterial therapy are important as it is complementary to antibiotics. These nanoparticles are having multimodes of action. The mechanism and efficacy of nanomaterials (NMs) depends of the cell wall structure, metabolic

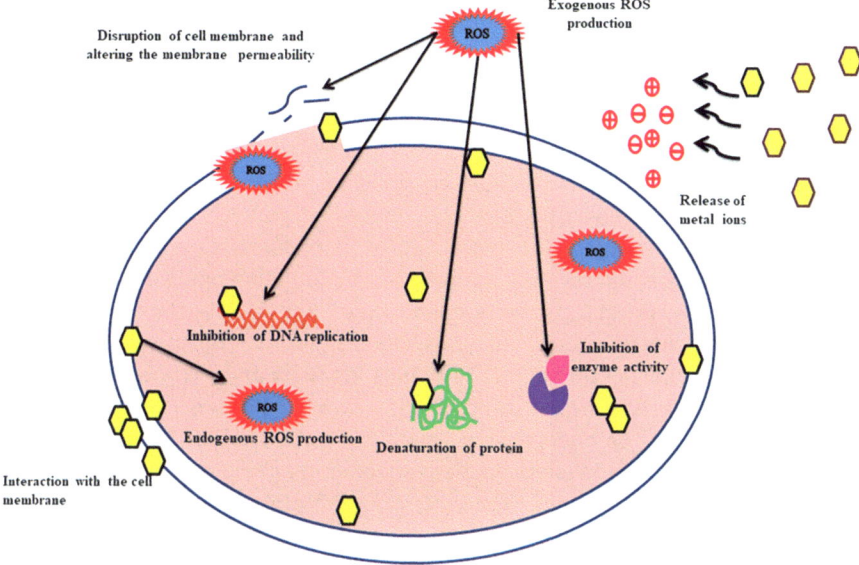

Fig. 14.1 Understanding the antimicrobial mechanisms of nanoparticles

pathways, and physiological state of bacteria and components of bacteria, which upon disruption could be lethal to the bacteria (Fig. 14.1). Other factors contributing the susceptibility of bacteria toward nanoparticles include the pH, aeration, temperature, and physicochemical properties of NMs. The major lethal pathways of NMs on bacteria are disruption of membrane integrity and cytotoxic death by production of reactive oxygen species (ROS) (Beyth et al. 2015; Aziz et al. 2014, 2015, 2019). Metal nanoparticles in contact with the bacteria can change the metabolic activity of bacteria. The ability of NMs to enter the biofilms and eliminate the bacteria represents an enormous advantage for curing the diseases. The nanomaterials in contact with bacteria form electrostatic attractions, vander Waals forces, hydrophobic interaction, and receptor-ligand interactions. A nanomaterial once gains the access to the bacteria, crosses the cell membrane, and interferes with the metabolic activity. This will influence and alter the shape and function of cell membrane. Further, these NMs on interaction with the bacterial biomolecules such as DNA, protein, ribosomes, enzymes, and lysosomes lead to the changes in cell membrane integrity and permeability, heterogeneous alterations, enzyme inhibition, protein deactivation, altering gene expression, electrolyte balance disorders, and oxidative stress. All these mechanism can be widely categorized as oxidative stress, non oxidative mechanism, and metal ion release (Wang et al. 2017). Metal oxide nanoparticles such as silver and zinc exhibit antibacterial mechanism through free metal ion toxicity, which arises from the release of metal ions from the surface of metal nanoparticles (Dizaj et al. 2014).

14.4 The Antibiofilm Mechanisms of Materials

The antibacterial and antibiofilm mechanism of nanomaterials depends on size, shape, surface area, charge, and competence of nanoparticles. Metal- and metal oxide-based nanoparticles are widely studied for their antimicrobial and antibiofilm properties. These nanoparticles exhibit microbicidal properties by the production of ROS, which disrupts physical structures, metabolic pathways, and synthesis of DNA, ultimately leading to cell death (Fig. 14.2). Nanoparticles are fabricated in different methods for different antimicrobial and antibiofilm applications. The fabrication techniques are categorized as top-down or bottom-up methods (Gold et al. 2018). The top-down approach is initiated with the bulk material to nanoscale products through attrition or ball milling. This method results in the production of nanocomposites with broad size distribution, increased impurities, and nonuniform particle geometries. The bottom-up approach exploits varied methods to synthesis nanoparticles from raw materials and environment for the production of particles with consistent shape, size, and geometry. Such fabrication techniques include sol-gel nanofabrication, atomic layer deposition, and colloidal methods (Biswas et al. 2012).

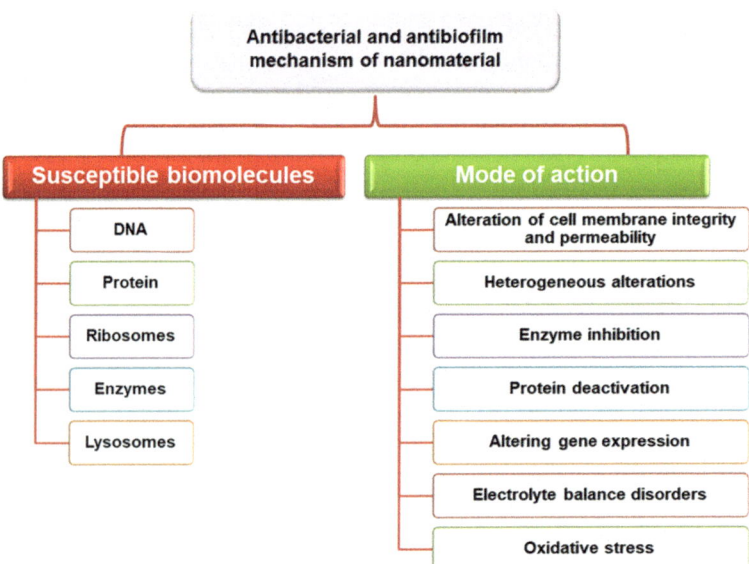

Fig. 14.2 Mode of actions and susceptible biomolecule in nanoparticles mediate antimicrobial and antibiofilm activities

14.5   Physical Properties of Nanoparticles Favoring the Antimicrobial Mechanisms

The antibacterial and antibiofilm mechanism of nanoparticles are not purely dependent on the material chemistry but also on their physical properties including the shape, size, solubility, agglomeration, and surface charge of nanoparticles.

14.5.1   Size

The nanoparticles with smaller size are able to penetrate the bacterial cell walls exhibiting higher antibacterial effect. Nanoparticles with smaller than 30 nm are effective antibacterial agents due to their higher surface-to-volume ratio. In a study, spherical nanoparticle with a diameter of 5 nm showed higher reactivity to the cell membrane of *E. coli* (Morones et al. 2005). The antimicrobial mechanism of nanoparticles is size dependent, which varies greatly from that of particles with larger diameter and also from their bulk counter parts. The nanoparticles with potential antibacterial mechanism and desired particle size are synthesized by controlling the synthesis parameters such as solvent polarity, concentration of substrates, density, and stabilizing agents (Jadhav et al. 2011). Platinum nanoparticles with 1–3 nm size showed bactericidal effect against *P. aeruginosa*, whereas those with 4–21 nm showed bacterial compatible properties (Gopal et al. 2013). Silver nanoparticles with an average size of 13.15 nm diameter proved to have better antibacterial and antibiofilm properties against imipenem-resistant *P. aeruginosa* than the bulk silver (Ali et al. 2018). Zinc oxide nanoparticles presented reduced viability and toxicity against *E. coli* and *S. aureus* when exposed to decreasing size of nanoparticles (Jiang et al. 2009). Authors examined the antibacterial activity of silver nanoparticles synthesized with different shapes and sizes. Better antibacterial activity was demonstrated by spherical-shaped smallest silver nanoparticles of 15–50 nm diameters against *P. aeruginosa* and *E. coli* (Raza et al. 2016). Nanosilver with 100 nm size did not show any bactericidal effect on methicillin-resistant *S. aureus* (MRSA), whereas 10 nm particles effectively inhibited MRSA (Ayala-Núñez et al. 2009). It was confirmed in another study that silver nanoparticles synthesized with two different sizes (10 and 100 nm) displayed different antimicrobial effect against *Methylobacterium* spp. The more antibacterial activity of 10 nm-sized nanoparticles was correlated with the larger surface area of silver nanoparticles. Particles with more surface area can possibly release more ions, which act as a vital factor in the antibacterial mechanism (Jeong et al. 2014).

14.5.2 Shape

Morphology or shapes of nanoparticles are correlated with the bactericidal activities exerted by nanoparticles. Raza et al. (2016) studied the varying antiseptic effect of spherical and triangular nanoparticles against *P. aeruginosa* and *E. coli* (Raza et al. 2016). The antibacterial and antibiofilm activity of nanoparticle changes according to their nature and shape. In a study, silver nanoparticles found to be a strong bactericidal agents when in truncated triangular shapes than in spherical or rod shape (Pal et al. 2007). Yttrium nanoparticles (Y_2O_3) synthesized in prismatic shape showed greater antibacterial activity against *P. desmolyticum* and *S. aureus*. This is probably due to the interaction of prismatic Y_2O_3 with the bacterial cell membrane, which leads to the breakage of membrane (Prasannakumar et al. 2015).

14.5.3 Zeta Potential

Increased bactericidal activity exerted by nanoparticles can also be contributed by their electronic effects or zeta potential. This can be well explained by the strong electrostatic interactions formed between the nanoparticles and negatively charged bacterial membrane. This attracts the bacteria to adhere or attach to the nanoparticles and finally permits the entry of these materials into the bacteria. The strong zeta potentials developed between bacterial membrane and nanoparticles cause membrane penetration, disruption of membrane, bacterial flocculation, and finally reduction of cell viability (Wang et al. 2017). Mostly, nanoparticles imparted with positive charge are involved in the development of strong electrostatic attraction with the bacterial cell membrane. In addition to the positively charged nanoparticles, negatively charged particles also involved in the antibacterial activity due to their molecular crowding, eventually leading to the interaction between nanoparticle and bacterial membrane (Arakha et al. 2015). Silica nanoparticles grafted with charged cobaltocenium was conjugated with penicillin for higher antibacterial activity against *P. aeruginosa*, *K. pneumonia*, *E. coli*, and *P. vulgaris*. Nanoconjugates with cationic charges and antibiotic exhibited synergistic activity leading to the enhanced lysis of bacteria when compared to the individual polymers (Pageni et al. 2018). Surface functionalization or modification imparts desired charges or electronic states to nanoparticles. Generally, polycationic chitosan can attach to the surface of bacteria, which leads to the decreased osmotic stability of cell and finally leakage of intracellular components. Authors discussed enhanced antimicrobial effect of chitosan silver nanoparticles than the free chitosan and silver. Chitosan was able to attract the bacteria due to their positive charge, and smaller-sized silver nanoparticles induced pores on cell membrane and hence together caused fragmentation of bacteria (Banerjee et al. 2010). Magnesium oxide synthesized through aerogel procedure showed enhanced bactericidal effect against *Bacillus megaterium*, spores of *Bacillus subtilis*, and *E. coli* due to the positive charge they possess (Koper et al. 2002).

14.5.4 Agglomeration and Doping Modification

The extensive use of nanoparticles in clinical trials is limited by their agglomeration effects. Agglomeration of nanoparticles has reduced their wide application in biology and medicine. Agglomerated nanoparticles showed lesser antibacterial effect than free particles (He et al. 2014). Solubility and agglomeration of nanoparticles play important role in the cytotoxicity and genotoxicity. Agglomeration of nanoparticles can be prevented by doping modification of nanoparticles. Solubility of nanoparticle influences the agglomeration effects of particles, and this controls the interaction of nanoparticle with the bacterial system. Agglomerated nanoparticles limit the ability to enter the cell and production of cytotoxic species. This is attributed by the general decrease in the surface area exposed to the bacterial system. In contrast, when nanoparticles are not agglomerated, they are able to penetrate the bacteria and produce reactive oxygen species, which are lethal to bacteria. The interactions of nanoparticles in their dispersed medium, pH, and surface charge also contribute to their agglomeration (He et al. 2014; Ahamed et al. 2008). The antibacterial activity of zinc oxide nanoparticles is altered when doped with fluorine. Zinc nanoparticles doped with fluorine produced more ROS and exhibited greater antibacterial effect than undoped nanoparticles (He et al. 2014).

14.6 Synthesis of Nanoparticles

There are numerous methods employed for the synthesis of different types of nanoparticles (Fig. 14.3). In general, all the methods are categorized into three different, physical, chemical, and biological methods. Physical methods apply usually high temperatures for the synthesis. The demerits of physical synthesis include the time required for synthesis, space, and detrimental environmental effects. The advantages of this method are the absence of solvent contamination and uniformity in the distribution of nanoparticles when compared to the chemical method of synthesis. The concentration and size of the nanoparticles vary with the temperature employed for the synthesis. Nanoparticles in solution are synthesized using laser ablation method of bulk metals (Abou El-Nour et al. 2010).

Chemical synthesis is widely used for the synthesis of metal nanoparticles. Chemical synthesis method includes three main components, namely, the metal precursor, reducing agent, and stabilizing/capping agent. In the chemical reduction method, reduction is attained by employing different organic and inorganic reducing agents. Chemical methods include the nucleation of metal ions and then formation of metallic colloidal nanoparticles by agglomeration into oligomeric clusters. Chemical synthesis of nanoparticles yields more nanoparticles than physical methods. Other type of chemical synthesis includes electrochemical synthesis, UV initiated photo reduction, photoinduced reduction, and microemulsion method (Zewde et al. 2016).

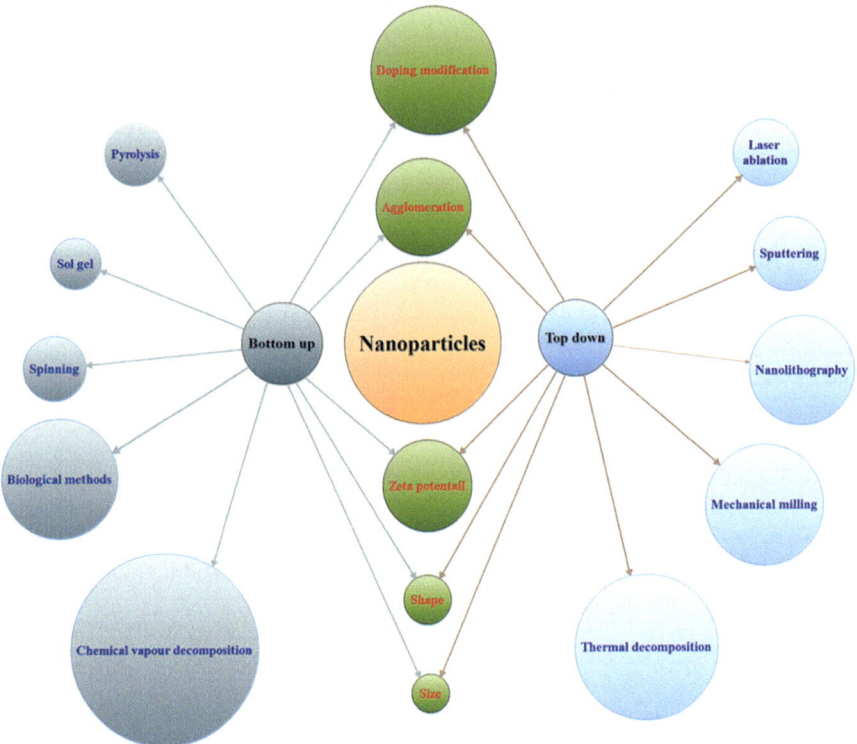

Fig. 14.3 Design and synthesis of nanoparticles through different synthesis routes

Biological method of nanoparticle synthesis also known as green synthesis is the most feasible method over physical and chemical method. Biological synthesis couples the reducing ability of microbial cells, biological molecules, and enzymes (Prasad et al. 2016). Green synthesis of nanotechnology integrates biological entities such as plants, bacteria, and fungi due to their capability to reduce metals to nanoparticles (Prasad et al. 2018a, b). Biological synthesis involves environmental friendly process and is advantageous over physical and chemical methods. Another merit of biological synthesis part is the production of the highly stable nanoparticle compared to other methods (Fernando et al. 2018). Plants are rich source for phytochemicals and enzymes. These phytochemicals are able to reduce various metal ions in to nanoscale materials. Greener synthesis offers a cost-effective renewable source for the large-scale synthesis of nanoparticles (Ahmed et al. 2016; Prasad 2014). Phytochemicals present in green tea leaf and black tea leaf such as polyphenols, flavonoids, and phenolic compounds were studied for the reduction of silver ions to silver nanoparticles and for their stabilization (Nishibuchi et al. 2012). Some of the bacteria like *Lactobacillus* sp., *Klebsiella pneumonia*, *Enterobacter cloacae*, *E. coli*, *Acinetobacter* sp., *P. aeruginosa*, and *S. aureus* are known for the synthesis of silver nanoparticles (Fernando et al. 2018). Fungi like *Fusarium* sp. produce nitrate reductase enzyme and some capping agents, which reduce silver ions and are responsible for the long-term stability of nanoparticles (Ingle et al. 2008; Prasad et al. 2016).

Altogether, physical, biological, and chemical synthesis routes are coming under either of two broad methods known as the bottom-up and top-down methods. Bottom-up methods include different synthesis routes such as biological synthesis, pyrolysis, spinning, sol-gel, and chemical vapor deposition methods. Laser ablation, sputtering, thermal decomposition, mechanical milling, and nanolithography are the some common top-down or destructive methods of nanoparticle synthesis (Anu Mary Ealia and Saravanakumar 2017).

14.7 Different Methods of Synthesis of Nanoparticles

14.7.1 *Physical*

Physical synthesis of nanoparticles involves the synthesis in the absence of a solvent, which allows the uniform size distribution of nanoparticles compared to the chemical methods. Some of the important physical routes are laser ablation, evaporation-condensation, thermal decomposition, arc discharge method, and ultrasonic spray pyrolysis (De Matteis et al. 2018). Among all physical routes, laser ablation is unique and important method, which synthesizes pure and metallic nanoparticles without the use of chemical reagents. Even though hazardous chemicals are not utilized in physical methods, limitations of large energy consumption and longer times to get thermal stability for nanoparticles make physical routes as unfeasible in synthesis (Zhang et al. 2016).

14.7.2 *Chemical*

Chemical reduction, microemulsion techniques, sonochemical method, sol-gel, and microwave-assisted synthesis are the common chemical synthesis methods. Chemical methods open up the possibility of obtaining monodispersed nanoparticles having tunable size with low cost and rapidity of steps in synthesis. However, the toxic elements and hazardous chemicals employed create the difficulty in purification of particles. Even though physical and chemical routes are proven to be appropriate for the synthesis of large-scale nanoparticles, recent developments made efforts to create eco-friendly approach through the greener synthesis of nanoparticles (De Matteis et al. 2018).

14.7.3 Biological

The emergence of eco-friendly methods in the area of material synthesis has acquired much attention and expanded more into biological applications. Diverse groups of inorganic nanoparticles with distinct chemical composition, shape, size, and morphology have been designed using different microorganisms and plants for their wide variety of applications in antimicrobial therapy (Li et al. 2011). Nanoparticles produced through biogenic process are always superior in characteristics than those by chemical routes. Despite the fact that chemical routes are always able to produce nanoparticles in large scale in short time, they are costly methods and harmful to the environment and human health. The biogenic enzymatic process for nanoparticles is supported by several microorganisms that thrive in ambient conditions of varying pH, pressure, and temperature. These nanoparticles produced through biological methods are having greater specific surface area and higher catalytic reactivity (Bhattacharya and Mukherjee 2008). The nanoparticle synthesis through microorganisms can be classified as intracellular and extracellular according to the cellular site where particles are synthesized. Usually, microorganisms grab the target metal ions from their surroundings and then convert them into respective nanoparticles through the action of enzymes produced by their metabolic activities (Li et al. 2011; Prasad et al. 2016).

14.8   Antimicrobial Applications of Nanoparticles

Metal and metal oxide nanoparticles are widely explored and studied for their great potentials as antimicrobials and antibiofilm agents. The antimicrobial effect exhibited by these nanoparticles is measured as the function of their surface area in contact with the microbial cells. The small size and high surface-to-volume ratio of nanoparticles augment the interaction between them and cell membrane of bacteria to perform wide range of antimicrobial applications (Table 14.1). The antimicrobial applications of nanoparticles range from water treatment plants to synthetic textiles, to biomedicine, to surgical devices, to food processing and packaging when nanoparticles are embedded on to the surfaces (Khan et al. 2017).

Silver nanoparticles are known as strong antibacterial and antifungal agents. It has shown strong cytotoxicity against wide variety of microorganism. Rai et al. (2009) studied the antibacterial activity of silver nanoparticles against *E. coli*, *P. aeruginosa*, and *S. aureus*. The antibacterial activity of silver nanoparticles was size dependent and showed cytotoxicity in the range of 1–10 nm size, which attach to the surface of bacterial cell. This interrupts the common cell membrane function like respiration and changes the membrane permeability. In a study, silver nanoparticles showed antifungal activity against two plant disease causing fungi, *Magnaporthe grisea* and *Bipolaris sorokiniana*. In vitro experiments revealed that silver nanoparticles inhibited the colony formation of these pathogenic fungi (Jo

Table 14.1 Antimicrobial mechanism and applications of nanoparticles synthesized through different methods

Nanoparticles	Antimicrobial mechanisms	Method of synthesis	Physiochemical characteristics of nanoparticles	Test microorganism	References
Silver	Inhibition of DNA replication, expression of ribosomal proteins, and bacterial electron transport chain	Green synthesis using leaves and fruits of *Calotropis procera*	90–100 nm	*Vibrio cholerae* and enterotoxic *E. coli* (ETEC)	Reidl et al. (2014)
	Interaction of silver ions with the bacterial cell wall, alter the membrane integrity by decaying lipopolysaccharide molecules, and form pits	Modified Tollens' method in conjunction with phytochemicals	Spherical shape, 45 nm	*E. coli and S. aureus*	Grinham et al. (2019)
Zinc	Inhibition of adenylyl cyclase activity and reduction in the levels of this second messenger	Green synthesis using leaves and fruits of *Calotropis procera*	90–100 nm	*Vibrio cholerae* and enterotoxic *E. coli* (ETEC)	Reidl et al. (2014)
	Damage to the cell wall due to the localized interaction of ZnO with the membrane, improved membrane permeability, uptake of NPs due to loss of proton motive force and uptake of toxic released zinc ions	Microwave decomposition, simple wet chemical route, hydrothermal synthesis, solvothermal method, deposition process, and simple precipitation method	2 μm, 8–10 nm, 12 nm, 45 nm, and 80 nm	*S. enterica serovar* Enteritidis, *P. aeruginosa, S. aureus, E. coli, C. jejuni,* and *E. coli O157:H7*	Seeni et al. (2015)
Gold	Alteration of cell membrane potential and reduced adenosine triphosphate (ATP) synthase activities, which results in the reduced metabolic activities. Interruption of tRNA binding to the ribosomal and disintegrating its biological function	Chemical synthesis using sodium borohydride as a reducing agent	6–40 nm	Enteric human pathogens; *E. coli, S. aureus, B. subtilis,* and *K. pneumonia*	Shamaila et al. (2016)
	Size-dependent mechanism	Green synthesis using leaves of *Pergularia daemia*	Spherical shape, 10 nm	*S. aurous, P. aeruginosa,* and *E. coli*	Rajendran (2017)
Titanium oxide	Interaction with the cell membrane, reactive oxygen species generation, and photooxidation of intracellular components causing cell death	Sol-gel method	Anatase type, 13 nm	*E. coli, P. aeruginosa, K. pneumoniae,* and *S. aureus*	Desai and Kowshik (2009)
	Antibacterial, antibiofilm mechanisms, and photooxidation		Less than 50 nm	Methicillin-resistant *S. aureus*	Jesline et al. (2015)

(continued)

Table 14.1 (continued)

Nanoparticles	Antimicrobial mechanisms	Method of synthesis	Physiochemical characteristics of nanoparticles	Test microorganism	References
Copper oxide	DNA degradation, reduction of bacterial respiration, alteration in the conformation, and electron transference of cytochrome reductases	Synthesis using soybean extract	Quasi-spherical shape, 5.67 nm	E. faecalis, E. coli, and S. aureus	Morales-Sánchez et al. (2017)
	Release of copper ions, penetration in to the cell membrane, and disruption of cell membrane. Disruption of biochemical pathway by chelating with the enzymes and DNA damage	Synthesis using sodium borohydride in the presence of ascorbic acid as antioxidant	Spherical shape, 5.3 nm	E. coli, S. aureus, and C. albicans	Bogdanović et al. (2014)

et al. 2009). Silver nanoparticles are known for their antiviral activities and exhibited antiviral activity against human immunodeficiency virus type 1 in their noncytotoxic concentrations (Ahmed et al. 2016).

Nanoparticles synthesized through green approach using aqueous extract of *Acorus calamus* rhizome inhibited the growth of phytopathogens affecting the yield of vegetable crops. In the above study, *P. aeruginosa* was more susceptible to iron oxide nanoparticles than manganese oxide nanoparticles. Iron oxide nanoparticles showed complete growth inhibition of *Aspergillus flavus* (Arasu et al. 2019). Magnesium oxide nanoparticles synthesized in an average size of 68.02 nm using the brown algae *Sargassum wightii* revealed broad spectrum of antibacterial and antifungal activities against human pathogens such as *S. aureus, E. coli, P. aeruginosa, S. typhimurium, S. marcescens, P. mirabilis, A. niger, N. oryzae, F. solani,* and *Chaetomium* sp. (Pugazhendhi et al. 2019).

Novel nanoconjugates were prepared using chitosan-polyvinyl chloride polymers comprising silver nanoparticles and exhibited excellent antibacterial activity against *S. aureus, E. coli, P. aeruginosa, S. typhimurium,* and *L. monocytogenes* after 30, 60, and 120 min of exposure (Gaballah et al. 2019). Biomaterial immobilized with subtilisin conjugated gold core and silver shell nanoparticles exhibited excellent antibiofilm potential against *S. aureus* and *E. coli* than the free silver-gold nanoparticles (Prabhawathi et al. 2019). Another biomaterial immobilized with silver and titanium dioxide nanoparticles was found to be excellent antibacterial agents for long-term applications. Biomaterial was prepared by incorporating the silver and titanium nanoparticles into cellulose acetate nanofibers. The nanocomposite nanofibers showed antibacterial activity against *E. coli* and *S. aureus* for 36 h, and the bacterial growth was inhibited up to 72 h (Jatoi et al. 2019). Metal fluoride nanoparticles were used as antibacterial and antibiofilm agents due to their low solubility and provided prolonged protection from pathogens. Yttrium fluoride nanoparticle-coated catheters were able to inhibit the bacterial colonization and finally biofilm formation by two common nosocomial pathogens such as *E. coli* and *S. aureus* compared to the uncoated catheters (Lellouche et al. 2012). Multifunctional nanoparticles were prepared containing amphiphilic silane and photosensitizers like coumarin 6 and chlorine 6 for investigating antimicrobial photodynamic therapy against periodontitis pathogens. Multifunctional nanoparticles reduced the biofilms of *Porphyromonas gingivalis, Fusobacterium nucleatum,* and *Streptococcus sanguinis* than the control groups. The CFU was reduced by 4–5 orders of magnitude after the photodynamic therapy of pathogens (Sun et al. 2019). Hybrid nanomaterials synthesized using two different antibacterial agents like graphene oxide and biogenic silver nanoparticles were used as anti-adhesion agent, which prevented the biofilm formation to 100% after 1 h of exposure to the nanomaterials (de Faria et al. 2014). Khalid et al. (2019) has synthesized silver and iron oxide nanoparticles coated with rhamnolipids having antibacterial, anti-adhesive, and antibiofilm properties. Biosurfactant coated nanoparticles showed synergistic antibacterial and anti-adhesive properties against biofilms of *P. aeruginosa* and *S. aureus* due to the production of reactive oxygen species by rhamnolipids (Khalid et al. 2019).

The antibacterial and antifungal efficacy of nitric oxide was enhanced by employing silica nanoparticles releasing nitric oxide. The nitric oxide-releasing silica nanoparticles inhibited the growth of biofilms formed by *E. coli, P. aeruginosa, S. aureus, S. epidermidis,* and *C. albicans* by 99% (Hetrick et al. 2009). Antimicrobial peptide conjugated with metallic nanoparticles gained much interest in the field of medicine, due to their increased solubility of peptide molecules and potential enhanced antimicrobial activity especially toward broad range of drug-resistant bacteria. Vancomycin-conjugated gold nanoparticles showed enhanced antimicrobial activity against vancomycin-resistant *S. aureus* (VRSA) (Rajchakit and Sarojini 2017). There are several groups of nanoparticles and their nanoconjugates, which are in high demand to replace the inefficient antibiotics for eliminating antibiotic-resistant microorganisms and their biofilms.

14.9   Future Prospects and Summary

Nanotechnology has become one of the promising alternatives applied in broad areas of science. The term nanobiotechnology attained much attention in the scientific filed due to their wide spread applications biomedical field. The potential use of nanoparticles for the treatment of microbial infections caused by drug-resistant microbial cells and their biofilms is accelerating in the modern research. More recent progression in the development of nanoparticles has enhanced the applications of nanosized products in healthcare products such as drug delivery scaffolds, burn dressings, water purification systems, and antimicrobial agents. Since the commencement of nanoparticle synthesis over the centuries ago, researchers were in the pursuit of exploiting novel methods of nanoparticle production in an optimized manner for various applications. The physicochemical characteristics of nanoparticles are having much significance on their function, and extensive studies are required to explore and fully understand the synthesis of these particles. Different kinds of physical, chemical, and biological methods are available to engineer, modulate, and fabricate ideal size, shape, and other morphology of nanoparticles for antimicrobial applications. The application of green chemistry in nanoparticle synthesis has revolutionized the field of nanobiotechnology. The green synthesis is a cost-effective and eco-friendly method for the large-scale synthesis of nanoparticles. Various methods of nanoparticle synthesis and their applications as antimicrobial and antibiofilm agents are explained in this chapter. The strong inhibitory and microbicidal activities of nanoparticles toward bacteria, fungi, and viruses and their mechanism of action was discussed with along with their antibiofilm mechanisms. Furthermore, there are many biogenic methods and other routes still not explored for the production of nanoscale products. Novel synthesis routes need to be figured out in the future with the goal of scalable production and fabrication of nanoparticles for their successful application in antimicrobial therapy.

References

Abou El-Nour KMM, Eftaiha A, Al-Warthan A, Ammar RAA (2010) Synthesis and applications of silver nanoparticles. Arab J Chem 3:135–140. https://doi.org/10.1016/j.arabjc.2010.04.008

Ahamed M, Karns M, Goodson M et al (2008) DNA damage response to different surface chemistry of silver nanoparticles in mammalian cells. Toxicol Appl Pharmacol 233:404–410. https://doi.org/10.1016/j.taap.2008.09.015

Ahmed S, Ahmad M, Swami BL, Ikram S (2016) A review on plants extract mediated synthesis of silver nanoparticles for antimicrobial applications: a green expertise. J Adv Res 7:17–28. https://doi.org/10.1016/j.jare.2015.02.007

Akova M (2016) Epidemiology of antimicrobial resistance in bloodstream infections. Virulence 7:252–266. https://doi.org/10.1080/21505594.2016.1159366

Algburi A, Comito N, Kashtanov D et al (2017) Control of biofilm formation: antibiotics and beyond. Appl Environ Microbiol 83:1–16. https://doi.org/10.1128/AEM.02508-16

Ali SG, Ansari MA, Khan HM et al (2018) Antibacterial and antibiofilm potential of green synthesized silver nanoparticles against imipenem resistant clinical isolates of P. aeruginosa. Bionanoscience 8:544–553. https://doi.org/10.1007/s12668-018-0505-8

Anu Mary Ealia S, Saravanakumar MP (2017) A review on the classification, characterisation, synthesis of nanoparticles and their application. IOP Conf Ser Mater Sci Eng 032019:263. https://doi.org/10.1088/1757-899X/263/3/032019

Arakha M, Pal S, Samantarrai D et al (2015) Antimicrobial activity of iron oxide nanoparticle upon modulation of nanoparticle-bacteria interface. Sci Rep 5:14813. https://doi.org/10.1038/srep14813

Arasu MV, Arokiyaraj S, Viayaraghavan P et al (2019) One step green synthesis of larvicidal, and azo dye degrading antibacterial nanoparticles by response surface methodology. J Photochem Photobiol B Biol 190:154–162. https://doi.org/10.1016/j.jphotobiol.2018.11.020

Ayala-Núñez NV, Lara Villegas HH, del Carmen Ixtepan Turrent L, Rodríguez Padilla C (2009) Silver nanoparticles toxicity and bactericidal effect against methicillin-resistant *Staphylococcus aureus*: nanoscale does matter. NanoBiotechnology 5:2–9. https://doi.org/10.1007/s12030-009-9029-1

Aziz N, Faraz M, Pandey R, Sakir M, Fatma T, Varma A, Barman I, Prasad R (2015) Facile algae-derived route to biogenic silver nanoparticles: synthesis, antibacterial and photocatalytic properties. Langmuir 31:11605–11612. https://doi.org/10.1021/acs.langmuir.5b03081

Aziz N, Fatma T, Varma A, Prasad R (2014) Biogenic synthesis of silver nanoparticles using Scenedesmus abundans and evaluation of their antibacterial activity. J Nanopart 2014:689419. https://doi.org/10.1155/2014/689419

Aziz N, Pandey R, Barman I, Prasad R (2016) Leveraging the attributes of Mucor hiemalis-derived silver nanoparticles for a synergistic broad-spectrum antimicrobial platform. Front Microbiol 7:1984. https://doi.org/10.3389/fmicb.2016.01984

Aziz N, Faraz M, Sherwani MA, Fatma T, Prasad R (2019) Illuminating the anticancerous efficacy of a new fungal chassis for silver nanoparticle synthesis. Front Chem 7:65. https://doi.org/10.3389/fchem.2019.00065

Banerjee M, Mallick S, Paul A et al (2010) Heightened reactive oxygen species generation in the antimicrobial activity of a three component iodinated chitosan−silver nanoparticle composite. Langmuir 26:5901–5908. https://doi.org/10.1021/la9038528

Beyth N, Houri-Haddad Y, Domb A et al (2015) Alternative antimicrobial approach: nano-antimicrobial materials. Evid Based Complement Alternat Med 2015:1–16. https://doi.org/10.1155/2015/246012

Bhattacharya R, Mukherjee P (2008) Biological properties of "naked" metal nanoparticles. Adv Drug Deliv Rev 60:1289–1306. https://doi.org/10.1016/j.addr.2008.03.013

Biswas A, Bayer IS, Biris AS et al (2012) Advances in top–down and bottom–up surface nanofabrication: techniques, applications & future prospects. Adv Colloid Interf Sci 170:2–27. https://doi.org/10.1016/j.cis.2011.11.001

Bogdanović U, Lazić V, Vodnik V et al (2014) Copper nanoparticles with high antimicrobial activity. Mater Lett 128:75–78. https://doi.org/10.1016/j.matlet.2014.04.106

De Matteis V, Cascione M, Toma C, Leporatti S (2018) Silver nanoparticles: synthetic routes, in vitro toxicity and theranostic applications for cancer disease. Nanomaterials 8:319. https://doi.org/10.3390/nano8050319

Desai VS, Kowshik M (2009) Antimicrobial activity of titanium dioxide nanoparticles synthesized by sol-gel technique. Res J Microbiol 4:97–103. https://doi.org/10.3923/jm.2009.97.103

Dizaj SM, Lotfipour F, Barzegar-Jalali M et al (2014) Antimicrobial activity of the metals and metal oxide nanoparticles. Mater Sci Eng C 44:278–284. https://doi.org/10.1016/j.msec.2014.08.031

Fernando S, Gunasekara T, Holton J (2018) Antimicrobial nanoparticles: applications and mechanisms of action. Sri Lankan J Infect Dis 8(2). https://doi.org/10.4038/sljid.v8i1.8167

Frieri M, Kumar K, Boutin A (2017) Antibiotic resistance. J Infect Public Health 10:369–378. https://doi.org/10.1016/j.jiph.2016.08.007

Fonseca de Faria A, Martinez DST, Meira SMM, Mazarin de Moraes AC, Brandelli A, Filho AGS, Alves OL (2014) Anti-adhesion and antibacterial activity of silver nanoparticles supported on graphene oxide sheets. Colloids and Surfaces B: Biointerfaces 113:115–124

Gaballah ST, El-Nazer HA, Abdel-Monem RA et al (2019) Synthesis of novel chitosan-PVC conjugates encompassing Ag nanoparticles as antibacterial polymers for biomedical applications. Int J Biol Macromol 121:707–717. https://doi.org/10.1016/j.ijbiomac.2018.10.085

Gold K, Slay B, Knackstedt M, Gaharwar AK (2018) Antimicrobial activity of metal and metal-oxide based nanoparticles. Adv Ther 1:1700033. https://doi.org/10.1002/adtp.201700033

Gopal J, Hasan N, Manikandan M, Wu H-F (2013) Bacterial toxicity/compatibility of platinum nanospheres, nanocuboids and nanoflowers. Sci Rep 3:1260. https://doi.org/10.1038/srep01260

Grinham C, AbuDalo MA, Al-Shurafat AW et al (2019) Synthesis of silver nanoparticles using a modified Tollens' method in conjunction with phytochemicals and assessment of their antimicrobial activity. PeerJ e6413:7. https://doi.org/10.7717/peerj.6413

He W, Kim H-K, Wamer WG et al (2014) Photogenerated charge carriers and reactive oxygen species in ZnO/Au hybrid nanostructures with enhanced photocatalytic and antibacterial activity. J Am Chem Soc 136:750–757. https://doi.org/10.1021/ja410800y

Hetrick EM, Shin JH, Paul HS, Schoenfisch MH (2009) Anti-biofilm efficacy of nitric oxide-releasing silica nanoparticles. Biomaterials 30 (14):2782–2789

Ingle A, Gade A, Pierrat S et al (2008) Mycosynthesis of silver nanoparticles using the fungus *Fusarium acuminatum* and its activity against some human pathogenic bacteria. Curr Nanosci 4:141–144. https://doi.org/10.2174/157341308784340804

Jadhav S, Gaikwad S, Nimse M, Rajbhoj A (2011) Copper oxide nanoparticles: synthesis, characterization and their antibacterial activity. J Clust Sci 22:121–129. https://doi.org/10.1007/s10876-011-0349-7

Jasovský D, Littmann J, Zorzet A, Cars O (2016) Antimicrobial resistance—a threat to the world's sustainable development. Ups J Med Sci 121:159–164. https://doi.org/10.1080/03009734.2016.1195900

Jatoi AW, Kim IS, Ni Q-Q (2019) Cellulose acetate nanofibers embedded with AgNPs anchored TiO$_2$ nanoparticles for long term excellent antibacterial applications. Carbohydr Polym 207:640–649. https://doi.org/10.1016/j.carbpol.2018.12.029

Jeong Y, Lim DW, Choi J (2014) Assessment of size-dependent antimicrobial and cytotoxic properties of silver nanoparticles. Adv Mater Sci Eng. https://doi.org/10.1155/2014/763807

Jesline A, John NP, Narayanan PM et al (2015) Antimicrobial activity of zinc and titanium dioxide nanoparticles against biofilm-producing methicillin-resistant *Staphylococcus aureus*. Appl Nanosci 5:157–162. https://doi.org/10.1007/s13204-014-0301-x

Jiang W, Mashayekhi H, Xing B (2009) Bacterial toxicity comparison between nano- and micro-scaled oxide particles. Environ Pollut 157:1619–1625. https://doi.org/10.1016/j.envpol.2008.12.025

Jo Y-K, Kim BH, Jung G (2009) Antifungal activity of silver ions and nanoparticles on phytopathogenic fungi. Plant Dis 93:1037–1043. https://doi.org/10.1094/PDIS-93-10-1037

Khalid HF, Tehseen B, Sarwar Y, Hussain SZ, Khan WS, Raza ZA, Bajwa SZ, Kanaras AG, Hussain I, Rehman A (2019) Biosurfactant coated silver and iron oxide nanoparticles with enhanced anti-biofilm and antiadhesive properties. Journal of Hazardous Materials 364:441–448

Khan I, Saeed K, Khan I (2017) Nanoparticles: properties, applications and toxicities. Arab J Chem. https://doi.org/10.1016/j.arabjc.2017.05.011

Koper OB, Klabunde JS, Marchin GL et al (2002) Nanoscale powders and formulations with biocidal activity toward spores and vegetative cells of *Bacillus* species, viruses, and toxins. Curr Microbiol 44:49–55. https://doi.org/10.1007/s00284-001-0073-x

Lellouche J, Friedman A, Gedanken A, Banin E (2012) Antibacterial and antibiofilm properties of yttrium fluoride nanoparticles. Int J Nanomedicine 7:5611–5624. https://doi.org/10.2147/IJN.S37075

Li X, Xu H, Chen Z-S, Chen G (2011) Biosynthesis of nanoparticles by microorganisms and their applications. J Nanomater 2011:1–16. https://doi.org/10.1155/2011/270974

Luepke KH, Suda KJ, Boucher H et al (2017) Past, present, and future of antibacterial economics: increasing bacterial resistance, limited antibiotic pipeline, and societal implications. Pharmacother J Hum Pharmacol Drug Ther 37:71–84. https://doi.org/10.1002/phar.1868

Morales-Sánchez E, Martínez-Castañón G-A, Compeán Jasso ME et al (2017) Antimicrobial properties of copper nanoparticles and amino acid chelated copper nanoparticles produced by using a soya extract. Bioinorg Chem Appl 2017:1–6. https://doi.org/10.1155/2017/1064918

Morones JR, Elechiguerra JL, Camacho A et al (2005) The bactericidal effect of silver nanoparticles. Nanotechnology 16:2346–2353. https://doi.org/10.1088/0957-4484/16/10/059

Muzammil S, Hayat S, Fakhar-E-Alam M et al (2018) Nanoantibiotics: future nanotechnologies to combat antibiotic resistance. Front Biosci (Elite Ed) 10:352–374

Nishibuchi M, Chieng NM, Loo YY (2012) Synthesis of silver nanoparticles by using tea leaf extract from *Camellia Sinensis*. Int J Nanomedicine 41:4263. https://doi.org/10.2147/IJN.S33344

Pageni P, Yang P, Chen YP et al (2018) Charged metallopolymer-grafted silica nanoparticles for antimicrobial applications. Biomacromolecules 19:417–425. https://doi.org/10.1021/acs.biomac.7b01510

Pal S, Tak YK, Song JM (2007) Does the antibacterial activity of silver nanoparticles depend on the shape of the nanoparticle? A study of the gram-negative bacterium *Escherichia coli*. Appl Environ Microbiol 73:1712–1720. https://doi.org/10.1128/AEM.02218-06

Pendleton JN, Gorman SP, Gilmore BF (2013) Clinical relevance of the ESKAPE pathogens. Expert Rev Anti-Infect Ther 11:297–308. https://doi.org/10.1586/eri.13.12

Prabhawathi V, Sivakumar PM, Boobalan T et al (2019) Design of antimicrobial polycaprolactam nanocomposite by immobilizing subtilisin conjugated Au/Ag core-shell nanoparticles for biomedical applications. Mater Sci Eng C 94:656–665. https://doi.org/10.1016/j.msec.2018.10.020

Prasad R (2014) Synthesis of silver nanoparticles in photosynthetic plants. J Nanoparticles. Article ID 963961. https://doi.org/10.1155/2014/963961

Prasad R, Pandey R, Barman I (2016) Engineering tailored nanoparticles with microbes: quo vadis. WIREs Nanomed Nanobiotechnol 8:316–330. https://doi.org/10.1002/wnan.1363

Prasad R, Jha A, Prasad K (2018a) Exploring the realms of nature for nanosynthesis. Springer, New York. ISBN 978-3-319-99570-0. https://www.springer.com/978-3-319-99570-0

Prasad R, Kumar V, Kumar M, Wang S (2018b) Fungal nanobionics: principles and applications. Springer, Singapore. ISBN 978-981-10-8666-3. https://www.springer.com/gb/book/9789811086656

Prasannakumar JB, Vidya YS, Anantharaju KS et al (2015) Bio-mediated route for the synthesis of shape tunable Y_2O_3: Tb^{3+} nanoparticles: photoluminescence and antibacterial properties. Spectrochim Acta Part A Mol Biomol Spectrosc 151:131–140. https://doi.org/10.1016/j.saa.2015.06.081

Pugazhendhi A, Prabhu R, Muruganantham K et al (2019) Anticancer, antimicrobial and photocatalytic activities of green synthesized magnesium oxide nanoparticles (MgONPs) using aqueous extract of *Sargassum wightii*. J Photochem Photobiol B Biol 190:86–97. https://doi. org/10.1016/j.jphotobiol.2018.11.014

Rai M, Yadav A, Gade A (2009) Silver nanoparticles as a new generation of antimicrobials. Biotechnol Adv 27:76–83

Rajchakit U, Sarojini V (2017) Recent developments in antimicrobial-peptide-conjugated gold nanoparticles. Bioconjug Chem 28:2673–2686. https://doi.org/10.1021/acs. bioconjchem.7b00368

Rajendran A (2017) Antibacterial properties and mechanism of gold nanoparticles obtained from *Pergularia Daemia* leaf extract. J Nanomed Res 6:1–5. https://doi.org/10.15406/ jnmr.2017.06.00146

Raza M, Kanwal Z, Rauf A et al (2016) Size- and shape-dependent antibacterial studies of silver nanoparticles synthesized by wet chemical routes. Nano 6(74). https://doi.org/10.3390/ nano6040074

Reidl J, Leitner DR, Goessler W et al (2014) Antibacterial activity of silver and zinc nanoparticles against *Vibrio cholerae* and enterotoxic *Escherichia coli*. Int J Med Microbiol 305:85–95. https://doi.org/10.1016/j.ijmm.2014.11.005

Seeni A, Kaus NHM, Sirelkhatim A et al (2015) Review on zinc oxide nanoparticles: antibacterial activity and toxicity mechanism. Nano-Micro Lett 7:219–242. https://doi.org/10.1007/ s40820-015-0040-x

Shamaila S, Zafar N, Riaz S et al (2016) Gold nanoparticles: an efficient antimicrobial agent against enteric bacterial human pathogen. Nano 6(71). https://doi.org/10.3390/nano6040071

Singh BN, Prateeksha, Upreti DK et al (2017) Bactericidal, quorum quenching and anti-biofilm nanofactories: a new niche for nanotechnologists. Crit Rev Biotechnol 37:525–540. https://doi. org/10.1080/07388551.2016.1199010

Sun X, Wang L, Lynch CD et al (2019) Nanoparticles having amphiphilic silane containing chlorin e6 with strong anti-biofilm activity against periodontitis-related pathogens. J Dent 81:70–84. https://doi.org/10.1016/j.jdent.2018.12.011

Ventola CL (2015a) The antibiotic resistance crisis: part 1: causes and threats. P T 40:277–283

Ventola CL (2015b) The antibiotic resistance crisis: part 2: management strategies and new agents. P T 40:344–352. https://doi.org/10.1289/ehp.01109785

Vianna Santos RC (2014) Antibiofilm applications of nanotechnology. Fungal Genom Biol 04:1– 3. https://doi.org/10.4172/2165-8056.1000e117

Wang EC, Wang AZ (2014) Nanoparticles and their applications in cell and molecular biology. Integr Biol 6:9–26. https://doi.org/10.1039/c3ib40165k

Wang L, Hu C, Shao L (2017) The antimicrobial activity of nanoparticles: present situation and prospects for the future. Int J Nanomedicine 12:1227–1249. https://doi.org/10.2147/IJN. S121956

Zewde B, Ambaye A, Iii JS, Raghavan D (2016) A review of stabilized silver nanoparticles– synthesis, biological properties, characterization, and potential areas of applications. JSM Nanotechnol Nanomed 4:1043–1057

Zhang X-F, Liu Z-G, Shen W, Gurunathan S (2016) Silver nanoparticles: synthesis, characterization, properties, applications, and therapeutic approaches. Int J Mol Sci 17:1534. https://doi. org/10.3390/ijms17091534

Chapter 15
Nanoemulsions for Antimicrobial and Anti-biofilm Applications

Pattnaik Subhaswaraj and Busi Siddhardha

Contents

Abstract Over the last few decades, though there is a significant upliftment in the technological interfaces and advancement in the biotechnological and biomedical sectors to combat microbial infections, the increasing prevalence of multidrug resistance (MDR) to conventional antimicrobials remains a serious public health issue. The inherent ability of pathogenic bacteria to form highly resistant biofilm matrix further aggravates the situation emphatically. In this context, the current drug discovery programmes focus on the development of novel bioactive compounds to combat bacterial virulence and biofilm formation. Though the natural bioactive compounds exhibit characteristic influence on biofilm development, their solubility, bioavailability and biocompatibility limit their widespread application. In this regard, nanotechnological intervention has revolutionized the current drug discovery

P. Subhaswaraj · B. Siddhardha (✉)
Department of Microbiology, School of Life Sciences, Pondicherry University, Pondicherry, India

© Springer Nature Switzerland AG 2020
R. Prasad et al. (eds.), *Nanostructures for Antimicrobial and Antibiofilm Applications*, Nanotechnology in the Life Sciences, https://doi.org/10.1007/978-3-030-40337-9_15

347

programmes to combat microbial infections and associated health consequences. Among the different nanomaterials used, nanoemulsions hold unique advantages to be used as novel antimicrobials and anti-biofilm agents owing to their unique physicochemical properties, improved stability and enhanced biocompatibility. Nanoemulsions are nano-sized emulsions, engineered and designed to characteristically improve the targeted and localized delivery of drug moieties or bioactive compounds in the physiological conditions with prolonged therapeutic efficacy without any adverse effect and related toxicity. Owing to the physical stability and improved bioavailability of nanoemulsions, it can be exploited for efficient and localized delivery of antimicrobial drugs and anti-biofilm agents to target sites for combating microbial infections, biofilm formation and associated MDR phenomenon. The antimicrobial nanoemulsions have also exhibited widespread application in the pharmaceutical and food processing industries and agricultural sectors suggesting new avenues and dimensions for the development of novel antimicrobials in the post-antibiotic era.

Keywords Nanoemulsion · Nanotechnology · Antimicrobials · Biofilm · Drug delivery · Multidrug resistance (MDR)

15.1   Introduction

In the last few decades, there is a marked increase in the emergence of several chronic microbial infections and related infectious diseases despite the evidence-based development in the biotechnological and biomedical sectors (Morens and Fauci 2013). As per recent trends, a broad spectrum of infectious diseases, disorders and microbial infections such as cancer, diabetes, cardiovascular diseases, cystic fibrosis, tuberculosis, AIDS and several other life-threatening complications have established themselves as serious public health concern. The epidemiological profile of these diseases and disorders is generally associated with high rate of mortality and morbidity in human beings (Dye 2014; Michael et al. 2014). The emergence of infectious diseases and microbial infections are characteristically driven by environmental and socio-economic factors thereby increasing the potential risks of global health issues as well as affecting the socio-economic status of developing countries (Jones et al. 2008). In recent times, developed countries also are facing not only health risks but also economic burden in an exponential manner, though a marked development in technological advancement has been achieved over the last few decades.

15.1.1 Golden Age of Antibiotics and Antibiotic Therapy

The discovery of antibiotics to treat microbial infections and associated infectious diseases in the twentieth century is arguably the most important event in the history of human civilization that radically transformed the healthcare and biomedical sectors (Zhu et al. 2014). The widespread therapeutic efficacy of antibiotics in controlling the microbial infections and several other diseases has proven their role as "wonder drugs" or "magic bullets" (Zaman et al. 2017). In majority of developing countries, the introduction of antibiotics has greatly reduced the potential risk of mortality and morbidity. Meanwhile in developed countries, the introduction of antibiotics has significantly improved the life expectancy as well as decreased the rate of mortality resulting from chronic microbial infections (Ventola 2015). In addition, the discovery of antibiotics strictly improves the conventional biomedical practices by strictly minimizing the risks associated with complicated operations and invasive recitations (Ashkenazi 2013). The discovery of antibiotics has resulted in a paradigm shift in drug discovery and developmental research in the treatment of infectious diseases and microbial infections (Aminov 2010; Aziz et al. 2016).

15.1.2 Emergence of Antibiotic Resistance

The world has witnessed the emergence of antibiotics in the last few decades, which has revolutionized the understanding of the treatment of several diseases and microbial infections. However, the irrational and indiscriminate use of antibiotics with poor infection control practices worldwide resulted in the emergence of resistance to the antibiotics (Frieri et al. 2017). According to the recent reports published by WHO's Global Antimicrobial Resistance Surveillance System (GLASS) and US Centers for Disease Control and Prevention (CDC), antimicrobial resistance has become a serious global health crisis posing serious threat to human health resulting in prolonged chronic infections and high rate of mortality. In the last few years, a number of pathogenic bacteria, particularly ESKAPE group of pathogens (*Enterococcus faecium*, *Staphylococcus aureus*, *Klebsiella pneumoniae*, *Acinetobacter baumannii*, *Pseudomonas aeruginosa* and *Enterobacter* sp.), are accounted for a majority of nosocomial infections by acquiring resistance to wide spectrum of available antibiotics leading to multidrug resistance (MDR) and extensive drug resistance (XDR) phenomenon (Roca et al. 2015; Santajit and Indrawattana 2016). In particular, *A. baumannii* and *P. aeruginosa* are considered as the most clinically important pathogens owing to their ability to exhibit resistance against fourth-generation antibiotics including carbapenem and β-lactam group of antibiotics as evidenced from the priority pathogen's list of WHO published in 2017 (Lupo et al. 2018). The development of antibiotic resistance in developing countries including India has risen considerably in the last decade owing to poor public health system and lack of awareness in using antibiotics, though the global consumption of antibiotics has increased dramatically in the developing countries (Laxminarayan and Chaudhury 2016). The resistance shown by pathogenic microorganisms could be attributed to the development of specific features such as decreased membrane permeability and increased efflux machinery (Blair et al. 2015; Lin et al. 2015). Besides, the rapid emergence of antibiotic resistance could be due to the increasing gap between the enhanced resistance to available antibiotics and limited availability and development of new drugs (Ferri et al. 2017).

In majority of microbial infections, the quorum sensing (QS) regulatory network controls the expression of an array of virulence phenotypes and pathogenic determinants during the development of severe chronic infections. The QS regulatory network has an inherent property to regulate the production of typical biomarkers which are responsible for the development of extracellular matrix called biofilms (Hall and Mah 2017). The biofilms are surface-attached heterogenous assemblages of microorganisms typically encased in an extracellular matrix, mainly composed of carbohydrates and proteins. The biofilms account for the sessile life style of pathogenic microorganisms and typically function as a cooperative consortium for stress management and antibiotic resistance (Qi et al. 2016; Banin et al. 2017). The microbial biofilms not only improved the survival of the encased microbial cells from environmental stress conditions but also conferred increased resistance to antimicrobial agents by minimizing the susceptibility of microorganisms to antimicrobial agents. According to estimation, biofilm-forming microorganisms have the inherent ability to exhibit 10–10,000 times more resistance to antimicrobial agents as compared to their free-living planktonic counterparts (Balcazar et al. 2015). The key features that promote biofilm-specific antibiotic resistance are the inability of antibiotics to penetrate biofilm matrix, altered metabolism, slow growth rate, presence of persister cells, antibiotic tolerance determinants and QS-regulated biomarker phenotypes (Olsen 2015; Flemming et al. 2016). As the microbial biofilms are generally associated with enhanced antimicrobial resistance, decreased susceptibility to antimicrobial therapies and altered host immune responses; targeting microbial biofilm and QS-regulated biofilm markers could be a promising alternative in downregulating the chronic microbial infections and increasing the therapeutic efficacy of conventional antibiotics (Wu et al. 2015).

15.1.3 Development of Alternative Therapeutics

Over the last few decades, microbial infections remain the most apocalyptic public health issues leading to high rate of mortality and morbidity in developing as well as developed countries. Besides, the increasing gap between the emergence of antibiotic resistance and development of new class of antibiotics has further aggravated the health risks to human beings as well as animals (Rangel-Vega et al. 2015). In this context, it is imperative to quest for alternative approaches to develop anti-infective agents specifically targeting microbial communication, biofilm dynamics and antibiotic resistance and complementing the therapeutic efficacy of the available antibiotics. The need for effective therapeutic strategies beyond the conventional antimicrobial therapies in the post-antibiotic era should be centered around the design and development of efficient, cost-effective, eco-friendly therapeutic strategies targeting bacterial infections and related MDR and XDR phenomenon instead of targeting microbial growth process. In the last few years, the pursuit of development of novel and effective non-conventional therapies has provided a platform for intensive and innovative research to discover effective drug candidates targeting microbial infections and drug resistance (Rather et al. 2017; Yuan et al. 2017).

In the development of alternative therapeutics to conventional antibiotics, several non-antibiotic approaches have been regarded as possible therapeutic strategies. These non-conventional strategies include the development of antibiotic adjuvants, antimicrobial peptides, efflux pump inhibitors, anti-infectives and anti-biofilm agents (Fig. 15.1) (Gill et al. 2015). The advent of combinatorial drug therapy termed as "antibiotic adjuvants" gained considerable interest in recent times owing to their efficacy in ameliorating the drug resistance. The development of antibiotic adjuvants relies upon two basic features, namely combating the bacterial resistance profile and enhancing the antimicrobial efficacy of the co-administered antibiotics along with the adjuvant. The efflux pump inhibitors (EPIs) and β-lactamase

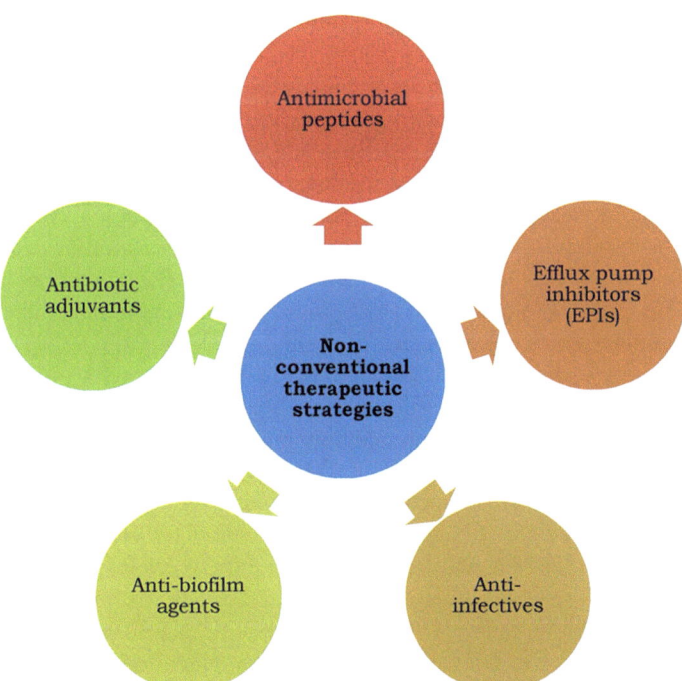

Fig. 15.1 Schematic illustration of the available alternative/non-conventional therapeutic strategies to treat microbial infections and to fight against antibiotic resistance phenomenon

inhibitors are the commonly used adjuvants to be exploited along with classical antibiotics to control antibiotic resistance phenomenon (Gonzalez-Bello 2017).

The concept of anti-infectives as a therapeutic alternative to classical antibiotic therapies is generally associated with targeting microbial virulence and biofilm dynamics instead of affecting the microbial growth, thereby minimizing the selective pressure associated with antibiotics. The decreased selective pressure is generally resulting in the alteration of pathogenic microorganisms in developing tolerance and resistance against therapeutic antibiotics. The advent of anti-infectives also exhibited promising characteristics to be used in combination with conventional antibiotics, to increase the efficacy of classical antibiotics by providing synergistic effect (Czaplewski et al. 2016). Apart from new drug development strategies, the notable development in the field of anti-infectives or antivirulence or anti-pathogenic drugs seeks considerable interest in the last few decades specifically targeting bacterial pathogenicity without affecting the microbial growth thereby minimizing the selective pressure responsible for antibiotic resistance (Courvalin 2016). It is evident that QS regulatory network in pathogenic bacteria coordinates an array of social behaviours such as expression of virulence phenotypes, production of pathogenic determinants, secretion of cytotoxic elements and antibiotic tolerance determinants. The multifaceted dimension of QS network gained considerable attention by providing a suitable target for the development of anti-infective strategies thereby providing ample avenues for the development of alternative therapeutic strategies complementing the exhausted supply of effective antibiotics (Antunes et al. 2010; Schuster et al. 2013). In the last few years, synthetic as well as natural compounds were exploited for their ability to downregulate the QS regulatory network in MDR and XDR pathogens. In particular, plant-derived phytochemicals and microbial metabolites were also exploited for their efficacy in attenuating the QS-mediated bacterial virulence, biofilm formation and drug resistance phenomenon (Kalia 2013).

The ability of pathogenic microorganisms to implement the highly complex QS network for the development of highly persistent microbial biofilms significantly affects the conventional antimicrobial therapies. The recalcitrant nature of microbial biofilms could be attributed to their efficacy in activating the antibiotic tolerance determinant genes, different stress response genes and expression of different biofilm markers such as exopolysaccharides, rhamnolipids, eDNA and alginates. These processes coordinate the efficacy of pathogenic microorganisms in fighting against environmental stress and antimicrobial therapy (Jolivet-Gougeon and Bonnaure-Mallet 2014; Kumar et al. 2017). In this context, targeting the QS-controlled biofilm formation could be a promising therapeutic approach. As per the recent trends, a number of quorum sensing inhibitors (QSIs) were used to target the formation, development and maturation of biofilm dynamics by modulating the production of biofilm markers such as exopolysaccharides, rhamnolipids, alginates, eDNA and adhesion proteins (Wilkins et al. 2014). In majority of the cases, naturally occurring bioactive compounds, in particular plant-derived phytochemicals followed by microbial metabolites, showed promising response in combating microbial biofilms by disrupting the membrane dynamics and increasing the susceptibility of encased microorganisms to the antimicrobial drugs (Taylor 2013).

15.1.4   Limitations Associated with Alternative Therapeutics

From a historical perspective, several non-conventional therapeutic approaches targeting microbial infections, biofilm formation and antibiotic resistance have been developed with differential specificity and efficacy. For example, recent trends in the development of QSIs and anti-biofilm agents have provided new dimensions to combat microbial pathogenicity, QS-controlled biofilm integrity and drug resistance phenomenon. Similarly, the development of potent efflux pump inhibitors also complemented the therapeutic efficacy of antibiotics by specifically inhibiting the functional role of efflux pumps thereby altering the susceptibility of pathogenic microorganisms to the classical antibiotics (Chatterjee et al. 2016; Hauser et al. 2016). The use of non-conventional antimicrobial therapy, in particular the development of anti-infective therapy, is considered to be promising and effective in combating microbial infections-associated health issues. However, the use of natural drug moieties, in particular bioactive phytochemicals and microbial metabolites, as anti-infective agents is frequently associated with certain limitations such as relative toxicity, bioavailability and biocompatibility issues. These limitations hinder the widespread application of novel anti-infective agents. In this context, a highly effective drug delivery platform is required to enhance the efficacy of these non-conventional drugs against microbial infections and associated drug resistance.

15.2   Nanotechnology in Drug Discovery and Development

Though, in the quest for development of novel, alternative and effective antimicrobial strategies natural plant-based compounds and bioactive microbial metabolites play crucial role, the emergence of nanotechnology as a promising strategy seems to be quite influential in shaping the future of drug discovery programmes (Khezerlou et al. 2018). In the last few years, the concept of nanotechnology has gained considerable attention with significant momentum for widespread application. The advent of nanotechnological platforms in the field of biomedicines has characteristically revolutionized the biomedical and pharmaceutical sectors with widespread application in the form of smart and early diagnosis of infectious diseases, targeted delivery of the therapeutic drug candidates and protection of medical devices from biofouling agents (Zhu et al. 2014; Zaidi et al. 2017; Prasad et al. 2017). The unique physicochemical properties, thermal stability, optical characteristics and widespread biological activities of the developed nanomaterials in the last few years have considerably revolutionized the current therapeutic options available in the pharmaceutical and biomedical sectors. The advent of nanotechnology has provided a multifaceted platform for the development of promising biosensing, bioimaging and therapeutic agents thereby providing research opportunities for early diagnosis of diseases and thus enhancing the therapeutic efficacy of the conventional drug moieties (Zhu et al. 2014). The concept of green nanotechnology also provides a new paradigm in developing eco-friendly, non-toxic and efficient nanomaterials as an alternative and effective therapeutics against microbial infections, drug resistance-associated health risks and other life-threatening diseases (Das et al. 2017; Prasad R et al. 2018). The advent of nanobiotechnology has several advantages such as controlled release profile, enhanced drug solubility, improved bioavailability, biocompatibility, increased therapeutic index, prolonged therapeutic effect and increased drug uptake efficiency. The sheer advantages of nanobiotechnology has provided new dimensions to employ them as efficient drug carriers for targeted and localized delivery of therapeutic drugs at the target sites without systemic toxicity (Zaidi et al. 2017).

The introduction of nanotechnology into the field of biology not only exhibited widespread antimicrobial and anti-biofilm properties but also critically complemented the classical antibiotics as well as bioactive drugs. The inherent bioactive potential of nanomaterials could be attributed to their ability to traverse the cellular barriers by disrupting the bacterial cell membrane, enhance the drug uptake by the bacterial cells, modulate the efflux kinetics and targeted delivery of drug candidates at the target sites (Abed and Couvreur 2014; Pelgrift and Friedman 2013). The unique physicochemical and optical properties of nanomaterials could also be exploited as effective contrasting agents in the process of biomedical diagnosis. This concept of bioimaging using nanomaterials as a contrasting agent has been used regularly in cancer therapeutics. Hence, it could also be implemented in the diagnosis of microbial infections and their effective clearance from the body of living organisms (Kumar and Das 2017). The nanotechnological interference opens up new possibilities and widespread avenues for the development of novel therapeutic drug candidates to control QS-regulated virulence profile, production of pathogenic determinants, biofilm formation and development and drug resistance profile (Nguyen and Goycoolea 2017). One of the important attributes of using nanoplatforms to fight against MDR and XDR biofilms is the stimuli-triggered mechanism, which enables the activation of encapsulated drugs at the target sites with high specificity without affecting the host tissues as well as other beneficial microorganisms (Koo et al. 2017).

The advent of nanomedicines has revolutionary implications in cancer therapy, diabetes, neurodegenerative disorders and microbial infections. Nanotherapeutic approaches could be designed and developed for the regulation of biofilm-forming MDR and XDR bacteria and associated drug resistance phenomenon (Prasad M et al. 2018). In the recent years, the advent of nanomedicines has been considered beyond cancer therapeutics particularly in combating both intracellular and extracellular microbial infections and associated health hazards (Lepeltier et al. 2015). As per the current trends, metal and metal oxide nanoparticles particularly silver nanoparticles (AgNPs), gold nanoparticles (AuNPs) and zinc oxide nanoparticles (ZnO NPs) synthesized by green chemistry approach have gained considerable attention as therapeutic arsenal against microbial chronic infections by specifically targeting the QS-regulated virulence and biofilm formation (Manju et al. 2016; Singh et al. 2016; Ishwarya et al. 2018). Besides, the biocompatibility, enhanced stability of polymeric nanoparticles could also be used to encapsulate bioactive compounds to downregulate QS-controlled virulence and biofilm dynamics (Pattnaik et al. 2018). The recent development of nanoliposomes also gained considerable attention in the battle against microbial biofilms and related chronic infections (Khan et al. 2017).

In the battle against microbial infections and antibiotic resistance, no doubt nanotechnological intervention has revolutionary impact on drug delivery and therapeutics; all the nanoformulations and nanomaterials have certain limitations. These limitations need to be overcome by developing novel and thermodynamically stable drug delivery systems. The concept of nanoemulsion is a recently developed strategy to enhance the therapeutic efficacy of bioactive drug candidates. Nanoemulsions are basically thermodynamically stable, colloidal particulate systems employing specific emulsifying agents whose function is to form a stable isotropic system by mediating the amalgamation of two immiscible liquids into a single phase (Jaiswal et al. 2015). Nanoemulsions are typically spherical and the size of the spheres range from 20 to 200 nm. The nanoemulsions act as promising drug carriers with widespread application in drug delivery, cancer therapy, vaccination and enzyme replacement therapy (Jaiswal et al. 2015; Singh et al. 2017; Prasad et al. 2017).

15.3 Nanoemulsion: A New Warrior in Drug Discovery

In the pursuit of developing novel drug delivery systems, colloidal drug delivery systems are considered to be promising and influential, particularly in the food and pharmaceutical industries. Colloidal drug delivery systems are basically designed to encapsulate functional lipophilic components with low solubility issues so that they can be dispersed within aqueous media. Colloidal drug delivery systems are basically of three categories, namely microemulsions, nanoemulsions and emulsions based on the differences in their thermodynamic stability and particle dimensions (Ostertag et al. 2012). The advent of nanoemulsions as novel and promising drug delivery system utilizes an emulsifying agent particularly pharmaceutical surfactants which are generally regarded as safe (GRAS). The surfactant type and concentration critically determine the stability and drug delivery efficacy of nanoemulsions. The major advantages of nanoemulsions are improved physical stability, greater surface area, increased drug loading, improved drug solubility, enhanced bioavailability, controlled release profile, protection from early degradation and most importantly the use of nanoemulsion is non-toxic and non-irritant in nature (Fig. 15.2) (Chime et al. 2014; Jaiswal et al. 2015). Colloidal nanoemulsions have been found to be influential in numerous applications in the field of biomedicines, drug delivery, food sectors and pharmaceutical and cosmetics industries (Karthik et al. 2018). The advent of nanoemulsions seems to be a promising tool for improving the antimicrobial potential of plant-based essential oils as compared to conventional emulsion systems. The characteristic features such as higher active surface area/volume ratio, small droplet size and improved stability are responsible for enhancing the delivery of bioactive compounds bypassing the biological membranes

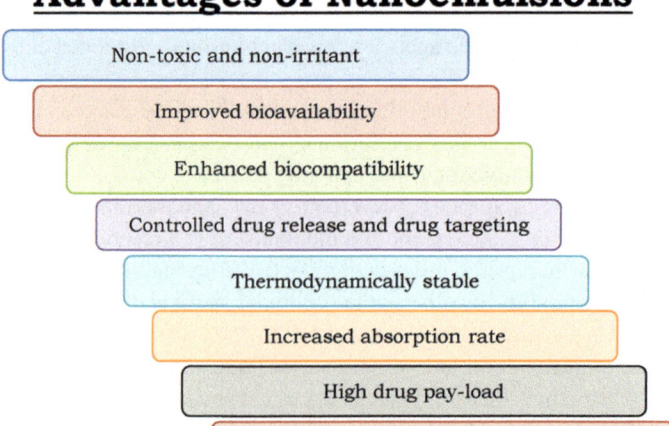

Advantages of Nanoemulsions

Fig. 15.2 Schematic overview of the widespread advantages of nanoemulsions in biomedical and pharmaceutical applications

by modulating the membrane dynamics. The use of nanoemulsions as carriers of antimicrobial essential oils would allow the reduction in the concentration to be used in order to achieve equivalent microbial inactivation levels to those of conventional emulsions or bulk oils which lead to overcome the low threshold values of essential oil incorporation for consumer acceptance (Artiga-Artigas et al. 2018a).

15.3.1   Types of Nanoemulsions

The nanoemulsions or microemulsions are basically composed of an oil, an emulsifier or preferably a surfactant with or without cosurfactant and water. The choice of the oil basically depends upon the solubility of the drug of choice in the oil, the toxicity of the oil and the proposed route of administration of the emulsion system. In a majority of cases, the essential oils that are considered as GRAS (generally regarded as safe for human consumption) and plant-derived edible oils that have inherent antimicrobial activity are the obvious choices for nanoemulsion formulations (Franklyne et al. 2016). One of the important parameters for the development of nanoemulsions is the use of an emulsifying agent. Basically, food-grade surfactants including polysorbates, sugar esters and lecithins and/or biocompatible biopolymers such as natural gums, vegetable or animal proteins, pectins, modified starches and chitosan are frequently employed as emulsifying agents. Besides, these emulsifying agents also greatly provided some added features into the nanoemulsions such as specific interfacial behaviour (electrostatic forces, steric repulsion and rheology), loading capability, improved stability as well as response to environmental stresses (Donsi and Ferrari 2016; Artiga-Artigas et al. 2018a). The design and development of nanoemulsions are basically categorized into three categories: (1) oil in water (O/W) nanoemulsion where oil is dispersed in the continuous aqueous phase, (2) water in oil (W/O) nanoemulsion where water is dispersed in continuous oil phase and (3) bi-continuous nanoemulsions (Jaiswal et al. 2015). The main components of nanoemulsions are oil, emulsifying agents, and aqueous phases.

15.3.2 Synthesis and Characteristics of Nanoemulsions

The preparation of desirable and characteristic nanoemulsions for a variety of applications involves two basic methods: (1) high-energy emulsification method and (2) low-energy emulsification method. The high-energy emulsification method basically involves the use of mechanically intense disruptive forces to break up the oil phase into tiny droplets. This method generally includes several procedures such as high-energy stirring, ultrasonic emulsification, high-pressure homogenization, microfluidization and membrane emulsification to achieve the intense disruptive forces. Ultrasonic emulsification is one of the most commonly used high-energy emulsification method utilizing ultrasound energy to tune the droplet size of the nanoemulsion system thereby improving the antimicrobial and other bioactivities (Rostami et al. 2018). Besides, the low-energy emulsification method involves the spontaneous emulsification technique to titrate a mixture of oil and surfactant into aqueous system. The low-energy emulsification method includes phase inversion temperature (PIT), emulsion inversion point (EIP) and spontaneous emulsification and is comparatively advantageous for commercial applications due to simple and easy-to-use strategies (Fig. 15.3) (Ghosh et al. 2014). Emulsion phase inversion (EPI) method is a commonly used low-energy emulsification approach, which utilizes the chemical energy released from the emulsification process as a consequence of change in the spontaneous curvature of surfactant molecules from negative to positive (obtaining oil-in-water nanoemulsions, O/W) or from positive to negative (obtaining water-in-oil nanoemulsions, W/O). The EPI method also advocates about the use of bicontinuous microemulsions where the mean curvature of the surfactant molecules is zero to achieve minimum droplet sizes with improved biological potential (Zhang et al. 2017). In addition, both high-energy and low-energy emulsification methods could also be combined to prepare reverse nanoemulsion in a highly viscous system (Jaiswal et al. 2015; Ryu et al. 2018a).

In the development of desirable nanoemulsions for improved bioactivity, it is important to maintain the balance between the lipophilic and hydrophilic contents in the naoemulsions i.e. hydrophilic–lipophilic balance (HLB). Besides, the optimal balance of the use of specific emulsifiers, in particular surfactant at an optimal concentration seems to be influential in enhancing biological activity (Krishnamoorthy et al. 2018; Lu et al. 2018). The stability and biological activity of nanoemulsions containing selected bioactive compounds or essential oils are greatly affected by the nature as well as concentration of the surfactant used during

Preparation Methods of Nanoemulsions

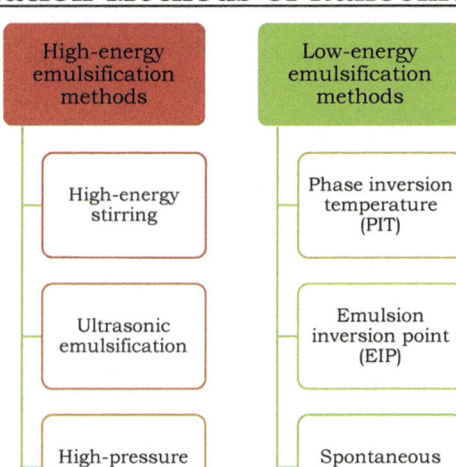

Fig. 15.3 Schematic illustration of different modes of preparation of nanoemulsions for widespread biomedical and pharmaceutical applications

the formulation of nanoemulsions (Artiga-Artigas et al. 2018b). One of the unique characteristics of the use of nanomaterials in biomedical applications is their selective modification for improved efficacy. In the development of nanoemulsions for antimicrobial and anti-biofilm properties, the stability of the nanoemulsions containing bioactive compounds plays a pivotal role in enhancing the biological activities of nanoemulsions. The stability of nanoemulsions could be improved by the addition of some texture modifiers such as biopolymers including sodium alginates to the aqueous phase. The addition of alginates greatly improved the nanoemulsions stability by characteristically reducing the Brownian movement of the droplets. These texture modifiers also interact with the surfactant chains causing steric and/or electrostatic repulsions between droplet interfaces droplets coalescence or gravitational separation (Artiga-Artigas et al. 2017). The advent of nanoemulsions as promising drug delivery systems for the encapsulation of bioactive compounds and targeted delivery of the bioactive compounds have several advantages such as (1) maintenance of stability of lipophilic bioactive compounds, (2) protection of encapsulated bioactive compounds against early enzymatic degradation, (3) sustained release profile by tuning the dimension and composition of droplets and (4) enhancement of cellular uptake and bioavailability for improved therapeutic efficacy (Sessa et al. 2014).

15.4 Applications of Nanoemulsions

The kinetic stability, biocompatibility and non-toxic features of nanoemulsions were exploited for widespread application in biomedical, pharmaceutical, food processing and cosmetics sectors (Fig. 15.4). Besides, the ability of nanoemulsions to increase the therapeutic index of the encapsulated drugs by minimizing the solubility issues could be taken into consideration for various biomedical applications. The advent of nanoemulsions as prolific drug delivery systems seems to be promising owing to their slow and sustained release profile, enhanced drug uptake, improved therapeutic index of the encapsulated drug moieties, enhanced bioavailability and enhanced physicochemical stability during the therapeutic process (Gadkari et al. 2017). Polyphenols are the structurally diverse group of phytochemicals with widespread application as antioxidant, antimicrobial, anti QS, anti-biofilm, anti-diabetic and anti-cancer agents. However, their relative instability and low solubility and bioavailability greatly limit their potential applications. In this context, it is imperative to improve the stability and bioavailability of polyphenols for potential biomedical applications. The introduction of nanoemulsions for the encapsulation of polyphenols greatly enhanced the bioavailability of polyphenols, thereby improving the bioactivity of polyphenols on a long-term basis (Lu et al. 2016). No doubt plant-derived essential oils have tremendous biological potential, but their solubility issues hinder their applicability on a long-term basis. In this context, the efficacy of

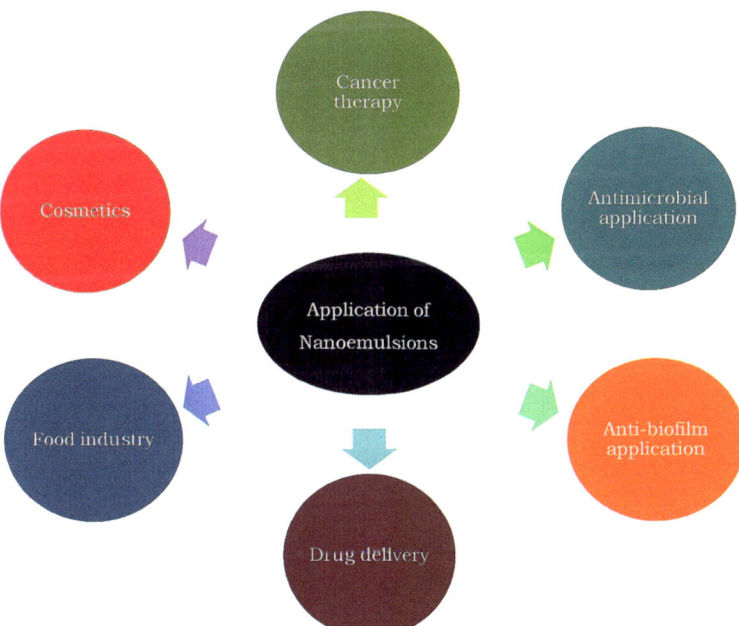

Fig. 15.4 An overview of the widespread applications of nanoemulsions

Rosmarinus officinalis essential oil encapsulated into nanoemulsion system in targeting dengue vector, *Aedes aegypti*, was evaluated. The O/W nanoemulsion containing the essential oil exhibited promising larvicidal activities suggesting their efficacy in vector control programmes (Duarte et al. 2015). The ability of nanoemulsions in scavenging free radicals also contributed their role as promising antioxidants which complemented the endogenous antioxidant machinery for clearance of highly reactive free radicals from the body (Lou et al. 2017).

15.4.1 Antibacterial Potential of Nanoemulsions

From ancient times, plant-derived essential oils are reported for their extensive antimicrobial, anticancer, anti-biofilm and antioxidant properties in vitro. However, the low solubility and bioavailability issues associated with the essential oils limit their widespread application. In this context, it is imperative to encapsulate the essential oils into a suitable delivery system that not only enhances the solubility and dispersibility but also characteristically improves the antimicrobial properties by promoting contact with bacteria (Majeed et al. 2016). In the development of novel antimicrobial and anti-biofilm strategies as an alternative to conventional antibiotics, nanotechnological intervention in particular the widespread application of nanoemulsions could be used to target microbial infections. The use of cationic and non-cationic lipidic nanoemulsions to target *Escherichia coli* infections has given a new dimension to the use of nanoemulsions as therapeutic agents in the clinical setup to combat microbial infections (Singh et al. 2015a, b). The application of nanoemulsions in the field of biomedicines provided aided protection to the thermolabile and light-sensitive bioactive components from degradation thereby increasing the bioavailability and stability of the encapsulated drug moieties for improved bioactivity. The technological limitations such as hydrophobic nature, less stability, volatility and less bioavailabiity associated with bioactive essential oil could be improved by utilizing the O/W nanoemulsions which not only counteract the limitations associated with native essential oils but also greatly improved the therapeutic efficacy of these essential oils (Nirmal et al. 2018).

The volatile nature of plant-derived essential oils naturally hinders its widespread biomedical applications. In this context, the development of kinetically stable and desirable nanoemulsions could be a promising aspect in counteracting the associated limitations and subsequently improving the antimicrobial efficacy (Guerra-Rosas et al. 2017). Eugenol is an important constituent of plant-derived essential oils and is reported for widespread biomedical applications. However, its low solubility limits its extended applications, and O/W nanoemulsions could be of promising implications as nanocarriers for the delivery of eugenol by improving the solubility and increased therapeutic efficacy. The eugenol-loaded O/W nanoemulsions were reported for their promising antibacterial applications against biofilm-forming *S. aureus* by modulating the bacterial membrane integrity (Ghosh et al. 2014). Besides eugenol, capsaicin is also an important bioactive lipophilic compound with biomedical applications, i.e., in the treatment of cardiovascular and respiratory diseases. The development of nanoemulsions containing capsaicin has greatly improved the antimicrobial properties of capsaicin against *E. coli* and *S. aureus* (Akbas et al. 2018). The O/W nanoemulsion prepared from *Zingiber officinale* Roscoe leaves oil showed characteristic antibacterial properties against the biofilm-forming pathogen *Streptococcus mutans* (Mostafa 2018). Recently, O/W nanoemulsions containing *Eugenia brejoensis* essential oil were developed, which showed significant antibacterial activity against *P. fluorescens* by greatly minimizing the technological limitations associated with essential oil thereby suggesting the applicability of non-toxic nanoemulsions in food processing and packaging industries (Mendes et al. 2018). The widespread potential of nanoemulsions in the food processing industries could be attributed to their promising antibacterial activities against several food-borne pathogens including *E. coli* (Salvia-Trujillo et al. 2015). Besides, nanoemulsions of anise oil and cinnamon bark oil showed promising antibacterial potential against food-borne pathogens, *Listeria monocystogenes*, *S. enterica* and *E. coli* with improved physicochemical stability suggesting their long-term effect (Topuz et al. 2016; Ma et al. 2016).

The *Thymus daenensis* essential oil was formulated into water-dispersible nanoemulsions and antibacterial activity was evaluated as compared to native essential oil. The nanoemulsions containing essential oil showed superior antibacterial activity against *E.coli* by disrupting the bacterial cell membrane integrity thereby improving the drug uptake (Moghimi et al. 2016a). Recently, the effect of nanoemulsions droplet size and the composition of oil and emulsifier or surfactant on antibacterial activity of nanoemulsions were advocated. The nanoemulsions containing *Cleome viscosa* essential oil showed promising antibacterial properties against drug resistant MRSA, *Streptococcus pyogenes*, *K. pneumoniae*, *E. coli* and *P. aeruginosa* with a droplet size of 7 nm and the use of oil and surfactant (1:3). This result suggested that the antimicrobial properties of nanoemulsions are indirectly proportional to the size of the droplet (Krishnamoorthy et al. 2018). In the recent years, various nanoemulsion formulations were designed and developed to combat microbial infections and found to be influential in biomedical, pharmaceutical and food technology-based sectors (Table 15.1).

In recent years, the concept of minimally processed fruits and vegetables (MPFV) emerged as a growing sector in food industries owing to their improved nutritional value. However, their shorter shelf life and unnoticed microbiological safety remains a great concern for human health in the twenty-first century. Though, a variety of plant-based essential oils are used for their antimicrobial properties, limitations such as low water solubility, bioavailability and volatility hinder their efficacy when used in MPFV. In this context, the highly stable and biocompatible nature of nanoemulsions is of great use in foods, which could be instrumental in their applications in combating the microflora associated with MPFV (Prakash et al. 2018).

Table 15.1 List of different nanoemulsion formulations, emulsifiers used for fabrication and their widespread antimicrobial activity

Sl no.	Nanoemulsion type	Loaded drugs	Method of preparation	Emulsifiers	Mean droplet size (nm)	Target microorganism	References
1.	O/W	Essential oils (EOs)	Microfluidization	Tween 80, sodium alginate (1:1)	190	*Escherichia coli*	Acevedo-Fani et al. (2015)
2.	O/W	Lemongrass Essential oils	Microfluidization	Tween 80	11	*E. coli*	Guerra-Rosas et al. (2017)
3.	O/W	Citral essential oils	Ultrasonic emulsification	Ethylene glycol	<100	*S. aureus, E. coli, P. aeruginosa, S. typhimurium, L. monocytogenes*	Lu et al. (2018)
4.	O/W	Cetylpyridinium chloride	Microfluidization	Triton X-100	170	*A. baumannii*	Hwang et al. (2013)
5.	O/W	Sage essential oils	High-intensity sonication	Tween 80, Span 80	222	*E. coli, S. typhi*	Moghimi et al. (2016b)
6.	O/W	Oregano oil	Ultrasonic emulsification	Tween 80	148	*L. monocytogenes, S. typhi*	Bhargava et al. (2015)
7.	O/W	Oregano essential oil	Low-energy emulsification	Pluronic F127	39.54	*Propionibacterium acnes, S. epidermidis*	Taleb et al. (2018)
8.	O/W	Sage essential oil	Low-energy emulsification	Tween 80, Span 80	89.45	*B. cereus, K. pneumoniae, E. faecalis*	Gharenaghadeh et al. (2017)
9.	O/W	Carvacrol nanoemulsion	High-pressure homogenization	Quillaja saponin	150	*S. enterica*	Ryu et al. (2018b)
10.	O/W	Thymol essential oil	Homogenization	Tween 80, Lauric arginate	100	*Zygosacchharomyces bailii*	Chang et al. (2015)

15.4.2 Antifungal Potential of Nanoemulsion

From the early evidence, the introduction of nano-based drug delivery systems significantly improved the therapeutic efficacy of the drugs by improving the intra-cellular penetration in biological tissues, sustained release profile thereby prolonging the drug effect, decreasing tissue irritability and enhancing the bioavailability, stability and specificity (Danielli et al. 2013). In recent years, the advent of nano-emulsions gained considerable attention to be used for prolific biological activities. The synthesis of nanoemulsions containing bioactive volatile oils of *Stenachaentum megapotamicum* exhibited promising antifungal properties against *Epidermophyton floccosum* and *Trichophyton rubrum* with improved minimum fungicidal concentration (MFC) as compared to native volatile oils and their constituents (Danielli et al. 2013). The encapsulation of the antibiotic amphotericin B into suitable nanoemulsion systems greatly enhanced the antifungal potential of amphotericin B against *Aspergillus fumigatus* and *Candida albicans* (Hussain et al. 2017).

15.4.3 Anti-quorum Sensing Activity of Nanoemulsions

A majority of bacterial pathogens showing typical drug resistance phenomenon typically constitute a highly complex, cell-to-cell communication cascade termed as QS. The QS regulatory network plays pivotal role in inducing bacterial virulence, production of pathogenic determinants, toxic byproducts, cytotoxic elements, biofilm formation and development and most importantly in showing resistance against conventional antimicrobial agents. In the recent years an array of naturally derived compounds particularly plant phytochemicals and microbial metabolites were reported for their in vitro anti QS activity. However, a majority of such natural compounds are found to be associated with bioavailability, solubility and biocompatibility issues. In this context, to maintain the anti-infective potential of such compounds, nano-based platforms are being developed. The spice oil nanoemulsions were synthesized by implementing the ultrasonic emulsification technique and evaluated for their ability to downregulate the QS-controlled production of different virulence phenotypes and biofilm formation in *Chromobacterium violaceum* CV026 (Venkadesaperumal et al. 2016). This result was further corroborated when *Cuminum cyminum*- and *Piper nigrum*- based nanoemulsions inhibited the QS-regulated violacein production in *C. violaceum* CV026. Besides, the developed nanoemulsions also significantly downregulated the QS-regulated virulence phenotypes in *E. coli* and *S. enterica* (Amrutha et al. 2017).

15.4.4 Anti-biofilm Potential of Nanoemulsions

According to the WHO's priority pathogen list, *A. baumannii* is considered to be the most important pathogenic bacteria with the ability to show resistance against more or less to every other available antibiotic. As per the WHO's guidelines, *A. baumannii* has presented an uphill task in front of the scientific community to develop alternative and effective strategies to counteract the *A. baumannii* infection and biofilm formation. Recently, nanotechnological intervention has provided a new dimension to combat microbial infections efficiently as compared to conventional antimicrobial therapy. Recently, *T. daenensis* essential oil containing nanoemulsions were formulated as an arsenal against multidrug-resistant biofilm-forming *A. baumannii*. The developed nanoemulsions exhibited prolific anti-biofilm potential against the recalcitrant biofilm dynamics of *A. baumannii* and thus provided novel avenues in targeting biofilm-related microbial infections as promising therapeutic strategies (Moghimi et al. 2018).

Earlier, *Citrus medica* L. var. sarcodactylis essential oil containing nanoemulsions were shown to exhibit promising anti-biofilm activities against biofilm-forming ESKAPE group of pathogens, *S. aureus* and *E. coli* (Lou et al. 2017). The pathogenicity profile of methicillin-resistant *S. aureus*(MRSA) is quite widespread resulting in severe burn wound, organ dysfunction, hospital-related infections, and most importantly MDR to available antibiotics. In this context, the advent of nanotechnology has provided a new dimension to treat MRSA infections and associated drug resistance. The chlorohexidine acetate nanoemulsions were designed to target biofilm dynamics in MRSA by disrupting the cell wall integrity and membrane dynamics (Song et al. 2016). Earlier to this report, chlorohexidine acetate nanoemulsions were designed to improve the antibacterial and anti-biofilm activity against *S. mutans* by characteristically improving the solubility and stability in the aqueous phase (Li et al. 2015). The introduction of nanoemulsions as promising nanocarriers for potential antimicrobial and anti-biofilm actions has gained considerable attention as compared to liposomal nanocarriers. This concept was eventually validated when soyaethyl morpholinium ethosulfate (SME), a cationic amphiphile immobilized into nanoemulsions and liposomes. The SME-immobilized O/W nanoemulsions exhibited characteristic disruption in the biofilm dynamics of MRSA and *S. epidermidis* as compared to SME-immobilized liposomes. This result suggested nanoemulsions as potential and efficient nanocarriers in controlling microbial infections, biofilm formation and drug resistance phenomenon (Lin et al. 2017).

P. aeruginosa belongs to the ESKAPE group of pathogens causing severe chronic infections and other hospital-acquired infections in the immunocompromised individuals. *P. aeruginosa* utilizes the highly complex network of QS to regulate the expression of pathogenic determinants and biofilm formation. The intensity and severity of *P. aeruginosa* infection and drug resistance profile is relatively increasing at an alarming rate and it is imperative to develop alternative strategies to combat *P. aeruginosa* infection. In 2017, nanoemulsions containing *Eucalyptus globules* oil exhibited promising antimicrobial as well as anti-biofilm potential against biofilm-forming *P. aeruginosa* suggesting their efficacy in controlling the microbial infections and associated drug resistance (Quatrin et al. 2017). The use of naturally occurring antimicrobial compounds has gained considerable attention as alternative therapeutic strategies. However, their bioavailability and solubility issues limit their applications. In this context, an effective carrier system is required to deliver the drugs at the target sites by neutralizing the limitations associated with native bioactive compounds. Recently the solubility issues associated with bioactive trans-cinnamic acid was neutralized by formulating desirable nanoemulsions which characteristically improved the efficacy of trans-cinnamic acid. The O/W nanoemulsions containing trans-cinnamic acid showed tremendous anti-biofilm potential against *P. aeruginosa*, *S. aureus* and food-borne pathogen *Salmonella typhimurium* (Letsididi et al. 2018). Recently, the nanoemulsions containing the essential oil of *Cymbopogon flexuosus* were also reported for their biofilm eradication properties against biofilm-forming *P. aeruginosa* and *S. aureus* by characteristically improving the stability of the essential oil (Gundel et al. 2018).

Earlier, the anti-biofilm properties of nanoemulsions formulated with *Zataria multiflora* Boiss essential oil were evaluated against *S. typhimurium* and *Listeria monocystogenes*. The enhanced anti-biofilm efficacy of the nanoemulsions as compared to native essential oil from *Z. multiflora* suggested the efficacy of nanoemulsions as promising nanocarriers in combating bifilm dynamics of pathogenic bacteria (Shahabi et al. 2017). Recently, patchouli essential oil from *Pogostemon cablin*, *P. heyneanus* and *P. plectranthoides* was employed to develop desirable nanoemulsions for antimicrobial and anti-biofilm properties. The nanoemulsions containing the essential oils from *Pogostemon* spp. significantly eradicated the biofilm dynamics of *S. aureus*, *S. mutans* and *Candida albicans* suggesting their widespread application in combating drug resistant microbial pathogens (Adhavan et al. 2017).

15.5 Current Trends and Future Avenues

In the last few years, the advent of synthetic chemistry and combinatorial chemistry applications in the field of nanobiotechnology has given new avenues for the development of novel nanocomposites with enhanced antimicrobial and anti-biofilm efficacy (Ng et al. 2014). It is evident from extensive research orientation that nanoemulsions are known for their antimicrobial properties by improving therapeutic index, enhancing the bioavailability and biocompatibility. Similar trend is also observed in the case of polymeric nanoparticles where the polymers provide a promising platform for the encapsulation of drug candidates and their targeted delivery without systemic toxicity. In this context, designing a potent nanocomposite comprising of nanoemulsions containing plant-derived essential oils specifically coated with modified biocompatible chitosan biopolymer for improved antimicrobial properties (Severino et al. 2015).

The modification of bioactive compounds encapsulated nanoemulsions by coating the biocompatible and biodegradable natural biopolymer chitosan greatly improved the stability profile of the nanoemulsions leading to improved antimicrobial and anti-biofilm properties. The improved stability profile of nanoemulsions using chitosan biopolymer proved to be of promising importance in food and beverages industries (Li et al. 2016). Among the different nanomaterials being used for localized delivery of antibiotics, drug candidates and bioactive compounds, the exploitation of nanoemulsions seems to be interesting owing to their suitability for food applications. Though a number of reports were available depicting the promising aspect of nanoemulsions in combating microbial infections, biofilm dynamics and drug resistance phenomenon, systematic research on the biomedical and pharmaceutical applications of nanoemulsions remain obscured and need to be taken consideration from laboratory to clinical setup (Donsi et al. 2012). The concept of nanoemulsions containing essential oils and/or bioactive compounds used in combination with conventional antimicrobial agents greatly improved the antimicrobial properties by means of synergistic activity against a wide range of pathogenic microorganisms with improved stability (Zhang et al. 2014). In the twenty-first century, it is important to maintain food quality and consumer safety by sheer avoidance of food spoilage by food-borne pathogens. In the recent years, food industries have shown considerable interest in developing novel antimicrobial nanoemulsions using plant-derived essential oils targeting food spoilage by pathogenic microorganisms. The typical antimicrobial properties of these nanoemulsions could be promising in maintaining food quality and improving their shelf life (Sugumar et al. 2016).

15.6 Conclusion

The advent of nanotechnology and its intervention into the field of biology is regarded as a promising discovery in the history of science. The field of nanotechnology finds its trends and applications in distinct disciplines such as physics, chemistry, biology, engineering pharmaceuticals and biomedical sectors. Among the different nanomaterials used, the advent of nanoemulsions has gained considerable attention in recent times in the food processing and packaging industries, pharmaceutical sectors and most importantly in combating microbial infections, biofilm formation and drug resistance. Though, the nanoemulsions showed promising potential in attenuating microbial infections and antimicrobial drug resistance phenomenon, it is imperative to shift the promising applications from laboratory conditions to clinical setup for improved therapeutics for human healthcare.

References

Abed N, Couvreur P (2014) Nanocarriers for antibiotics: a promising solution to treat intracellular bacterial infections. Int J Antimicrob Agent 43:485–496. https://doi.org/10.1016/j.ijantimicag.2014.02.009

Acevedo-Fani A, Salvia-Trujillo L, Rojas-Grau MA, Martin-Belloso O (2015) Edible films from essential-oil-loaded nanoemulsions: physiochemical characterization and antimicrobial properties. Food Hydrocolloid 47:168–177. https://doi.org/10.1016/j.foodhyd.2015.01.032

Adhavan P, Kaur G, Princy A, Murugan R (2017) Essential oil nanoemulsions of wild patchouli attenuate multi-drug resistant Gram-positive, Gram-negative and *Candida albicans*. Ind Crop Prod 100:106–116. https://doi.org/10.1016/j.indcrop.2017.02.015

Akbas E, Soyler B, Oztop MH (2018) Formation of capsaicin loaded nanoemulsions with high pressure homogenization and ultrasonication. LWT Food Sci Technol 96:266–273. https://doi.org/10.1016/j.lwt.2018.05.043

Aminov RI (2010) A brief history of the antibiotic era: lessons learned and challenges for the future. Front Microbiol 1:134. https://doi.org/10.3389/fmicb.2010.00134

Amrutha B, Sundar K, Shetty PH (2017) Spice oil nanoemulsions: potential natural inhibitors against pathogenic *E. coli* and *Salmonella* spp. from fresh fruits and vegetables. LWT Food Sci Technol 79:152–159. https://doi.org/10.1016/j.lwt.2017.01.031

Antunes LCM, Ferreira RBR, Buckner MMC, Finlay BB (2010) Quorum sensing in bacterial virulence. Microbiology 156:2271–2282. https://doi.org/10.1099/mic.0.038794-0

Artiga-Artigas M, Acevedo-Fani A, Martin-Belloso O (2017) Effect of sodium alginate incorporation procedure on the physicochemical properties of nanoemulsions. Food Hydrocolloid 70:191–200. https://doi.org/10.1016/j.foodhyd.2017.04.006

Artiga-Artigas M, Guerra-Rosas MI, Morales-Castro J, Salvia-Trujillo L (2018a) Influence of essential oils and pectin on nanoemulsion formulation: a ternary phase experimental approach. Food Hydrocolloid 81:209–219. https://doi.org/10.1016/j.foodhyd.2018.03.001

Artiga-Artigas M, Lanjari-Perez Y, Martin-Belloso O (2018b) Curcumin-loaded nanoemulsions stability as affected by the nature and concentration of surfactant. Food Chem 266:466–474. https://doi.org/10.1016/j.foodchem.2018.06.043

Ashkenazi S (2013) Beginning and possibly the end of the antibiotic era. J Paediatr Child Health 49:E179–E182. https://doi.org/10.1111/jpc.12032

Aziz N, Pandey R, Barman I, Prasad R (2016) Leveraging the attributes of Mucor hiemalis-derived silver nanoparticles for a synergistic broad-spectrum antimicrobial platform. Front Microbiol 7:1984. https://doi.org/10.3389/fmicb.2016.01984

Balcazar JL, Subirats J, Borrego CM (2015) The role of biofilms as environmental reservoirs of antibiotic resistance. Front Microbiol 6:1216. https://doi.org/10.3389/fmicb.2015.01216

Banin E, Hughes D, Kuipers OP (2017) Editorial: bacterial pathogens, antibiotics and antibiotic resistance. FEMS Microbiol Rev 41(3):450–452. https://doi.org/10.1093/femsre/fux016

Bhargava K, Conti DS, da Rocha SRP, Zhang Y (2015) Application of oregano oil nanoemulsion to the control of food borne bacteria on fresh lettuce. Food Microbiol 47:69–73. https://doi.org/10.1016/j.fm.2014.11.007

Blair JMA, Webber MA, Baylay AJ, Ogbolu DO, Piddock LJV (2015) Molecular mechanisms of antibiotic resistance. Nat Rev Microbiol 13:42–51. https://doi.org/10.1038/nrmicro3380

Chang Y, McLandsborough L, McClements DJ (2015) Fabrication, stability and efficacy of dual-component antimicrobial nanoemulsions: essential oil (thyme oil) and cationic surfactant (lauric arginate). Food Chem 172:298–304. https://doi.org/10.1016/j.foodchem.2014.09.081

Chatterjee M, Anju CP, Biswas L, Anil Kumar V, Gopi Mohan C, Biswas R (2016) Antibiotic resistance in *Pseudomonas aeruginosa* and alternative therapeutic options. Int J Med Microbiol 306:48–58. https://doi.org/10.1016/j.ijmm.2015.11.004

Chime SA, Kenechukwu FC, Attama AA (2014) Nanoemulsions—advances in formulation, characterization and applications in drug delivery. In: Sezer AD (ed) Application of nanotechnology in drug delivery. InTech, Istanbul, pp 77–126. https://doi.org/10.5772/15371

Courvalin P (2016) Why is antibiotic resistance a deadly emerging disease? Clin Microbiol Infect 22:405–407. https://doi.org/10.1016/j.cmi.2016.01.012

Czaplewski L, Bax R, Clokie M, Dawson M, Fairhead H, Fischetti VA, Foster S, Gilmore BF, Hancock REW, Harper D, Henderson IR, Hilpert K, Jones BV, Kadioglu A, Knowles D, Olafsdottir S, Payne D, Projan S, Shaunak S, Silverman J, Thomas CM, Trust TJ, Warn P, Rex JH (2016) Alternatives to antibiotics—a pipeline portfolio review. Lancet Infect Dis 16(2):239–251. https://doi.org/10.1016/S1473-3099(15)00466-1

Danielli LJ, dos Reis M, Bianchini M, Camargo GS, Bordignon SAL, Gurreiro IK, Fuentefria A, Apel MA (2013) Antidermatophytic activity of volatile oil and nanoemulsion of *Stenachaenium megapotamicum* (Spreng.) Baker. Ind Crop Prod 50:23–28. https://doi.org/10.1016/j. indcrop.2013.07.027

Das B, Dash SK, Mandal D, Ghosh T, Chattopadhyay S, Tripathy S, Das S, Dey SK, Das D, Roy S (2017) Green synthesized silver nanoparticles destroy multidrug resistant bacteria via reactive oxygen species mediated membrane damage. Arab J Chem 10:862–876. https://doi. org/10.1016/j.arabjc.2015.08.008

Donsi F, Ferrari G (2016) Essential oil nanoemulsions as antimicrobial agents in food. J Biotechnol 233:106–120. https://doi.org/10.1016/j.jbiotec.2011.07.001

Donsi F, Annunziata M, Vincensi M, Ferrari G (2012) Design of nanoemulsion-based delivery systems of natural antimicrobials: effect of the emulsifier. J Biotechnol 159:342–350. https:// doi.org/10.1016/j.jbiotec.2016.07.005

Duarte JL, Amado JRR, Oliveira AEMFM, Cruz RAS, Ferreira M, Souto RNP, Falcao DQ, Carvalho JCT, Fernandes CP (2015) Evaluation of larvicidal activity of a nanoemulsion of *Rosmarinus officinalis* essential oil. Rev Bras Farmacogn 25:189–192. https://doi.org/10.1016/j. bjp.2015.02.010

Dye C (2014) After 2015: infectious diseases in a new era of health and development. Phil Trans R Soc B 369:20130426. https://doi.org/10.1098/rstb.2013.0426

Ferri M, Ranucci E, Romagnoli P, Giaccone V (2017) Antimicrobial resistance: a global emerging threat to public health systems. Crit Rev Food Sci Nutr 57(13):2857–2876. https://doi.org/10. 1080/10408398.2015.1077192

Flemming HC, Wingender J, Szewzyk U, Steinberg P, Rice SA, Kjelleberg S (2016) Biofilms: an emergent form of bacterial life. Nat Rev Microbiol 14:563–575. https://doi.org/10.1038/ nrmicro.2016.94

Franklyne JS, Mukherjee A, Chandrasekaran N (2016) Essential oil micro- and nanoemulsions: promising roles in antimicrobial therapy targeting human pathogens. Lett Appl Microbiol 63:322–334. https://doi.org/10.1111/lam.12631

Frieri M, Kumar K, Boutin A (2017) Antibiotic resistance. J Infect Public Health 10(4):369–378. https://doi.org/10.1016/j.jiph.2016.08.007

Gadkari PV, Shashidhar MG, Balaraman M (2017) Delivery of green tea catechins through oil-in-water (O/W) nanoemulsion and assessment of storage stability. J Food Eng 199:65–76. https:// doi.org/10.1016/j.jfoodeng.2016.12.009

Gharenaghadeh S, Karimi N, Forghani S, Nourazarian M, Gharehnaghadeh S, Jabbari V, Khiabani MS, Kafil HS (2017) Application of Salvia multicaulis essential oil-containing nanoemulsion against food-borne pathogens. Food Biosci 19:128–133. https://doi.org/10.1016/j. fbio.2017.07.003

Ghosh V, Mukherjee A, Chandrasekaran N (2014) Eugenol-loaded antimicrobial nanoemulsion preserves fruit juice against, microbial spoilage. Colloid Surf B Biointerface 114:392–397. https://doi.org/10.1016/j.colsurfb.2013.10.034

Gill EE, Franco OL, Hancock REW (2015) Antibiotic adjuvants: diverse strategies for controlling drug-resistant pathogens. Chem Biol Drug Des 85:56–78. https://doi.org/10.1111/cbdd.12478

Gonzalez-Bello C (2017) Antibiotic adjuvants—a strategy to unlock bacterial resistance to antibiotics. Bioorg Med Chem Lett 27:4221–4228. https://doi.org/10.1016/j.bmcl.2017.08.027

Guerra-Rosas MI, Morales-Castro J, Cubero-Marquez MA, Salvia-Trujillo L, Martin-Belloso O (2017) Antimicrobial activity of nanoemulsions containing essential oils and high methoxyl pectin during long-term storage. Food Cont 77:131–138. https://doi.org/10.1016/j. foodcont.2017.02.008

Gundel SS, de Souza ME, Quatrin PM, Klein B, Wagner R, Gundel A, Vaucher RA, Santos RCV, Ourique AF (2018) Nanoemulsions containing *Cymbopogon flexuosus* essential oil: development, characterization, stability study and evaluation of antimicrobial and antibiofilm activities. Microb Pathog 118:268–276. https://doi.org/10.1016/j.micpath.2018.03.043

Hall CW, Mah TF (2017) Molecular mechanisms of biofilm-based antibiotic resistance and tolerance in pathogenic bacteria. FEMS Microbiol Rev 41(3):276–301. https://doi.org/10.1093/femsre/fux010

Hauser AR, Mecsas J, Moir DT (2016) Beyond antibiotics: new therapeutic approaches for bacterial infections. Clin Infect Dis 63(1):89–95. https://doi.org/10.1093/cid/ciw200

Hussain A, Singh S, Webster TJ, Ahmad FJ (2017) New perspectives in the topical delivery of optimized amphotericin B loaded nanoemulsions using excipients with innate anti-fungal activities: a mechanistic and histopathological investigation. Nanomed Nanotechnol Biol Med 13:1117–1126. https://doi.org/10.1016/j.nano.2016.12.002

Hwang YY, Ramalingam K, Blenek DR, Lee V, You T, Alvarez R (2013) Antimicrobial activity of nanoemulsion in combination with cetylpyridinium chloride in multidrug-resistant *Acinetobacter baumannii*. Antimicrob Agent Chemother 57:3568–3575. https://doi.org/10.1128/AAC.02109-12

Ishwarya R, Vaseeharan B, Kalyani S, Banumathi B, Govindarajan M, Alharbi NS, Kadaikunnan S, Al-anbr MN, Khaled JM, Benelli G (2018) Facile green synthesis of zinc oxide nanoparticles using *Ulva lactuca* seaweed extract and evaluation of their photocatalytic, antibiofilm and insecticidal activity. J Photochem Photobiol Biol 178:249–258. https://doi.org/10.1016/j.jphotobiol.2017.11.006

Jaiswal M, Dudhe R, Sharma PK (2015) Nanoemulsion: an advanced mode of drug delivery system. 3 Biotech 5(2):123–127. https://doi.org/10.1007/s13205-014-0214-0

Jolivet-Gougeon A, Bonnaure-Mallet M (2014) Biofilm as a mechanism of bacterial resistance. Drug Dis Today Technol 11:49–56. https://doi.org/10.1016/j.ddtec.2014.02.003

Jones KE, Patel NG, Levy MA, Storeygard A, Balk D, Gittleman JL, Daszak P (2008) Global trends in emerging infectious diseases. Nature 451:990–993. https://doi.org/10.1038/nature06536

Kalia VC (2013) Quorum sensing inhibitors: an overview. Biotechnol Adv 31:224–245. https://doi.org/10.1016/j.biotechadv.2012.10.004

Karthik P, Padma IS, Anandharamakrishnan C (2018) Droplet coalescence as a potential marker for physicochemical fate of nanoemulsions during *in-vitro* small intestine digestion. Colloid Surf A Physicochem Eng Aspect 553:278–287. https://doi.org/10.1016/j.colsurfa.2018.05.066

Khan SN, Khan S, Iqbal J, Khan R, Khan AU (2017) Enhanced killing and antibiofilm activity of encapsulated cinnamaldehyde against *Candida albicans*. Front Microbiol 8:1641. https://doi.org/10.3389/fmicb.2017.01641

Khezerlou A, Alizadeh-Sani M, Azizi-Lalabadi M, Ehsani A (2018) Nanoparticles and their antimicrobial properties against pathogens including bacteria, fungi, parasites and viruses. Microb Pathog 123:505–526. https://doi.org/10.1016/j.micpath.2018.08.008

Koo H, Allan RN, Howlin RP, Stoodley P, Hall-Stoodley L (2017) Targeting microbial biofilms: current and prospective therapeutic strategies. Nat Rev Microbiol 15:740–755. https://doi.org/10.1038/nrmicro.2017.99

Krishnamoorthy R, Athinarayanan J, Periasamy VS, Adisa AR, Al-Shuniaber MA, Gassem MA, Alshatwi AA (2018) Antimicrobial activity of nanoemulsion on drug-resistant bacterial pathogens. Microb Pathog 120:85–96. https://doi.org/10.1016/j.micpath.2018.04.035

Kumar MS, Das AP (2017) Emerging nanotechnology based strategies for diagnosis and therapeutics of urinary tract infections: a review. Adv Colloid Interf Sci 249:53–65. https://doi.org/10.1016/j.cis.2017.06.010

Kumar A, Alam A, Rani M, Ehtesham NZ, Hasnain SE (2017) Biofilms: survival and defence strategy for pathogens. Int J Med Microbiol 307:481–489. https://doi.org/10.1016/j.ijmm.2017.09.016

Laxminarayan R, Chaudhury RR (2016) Antibiotic resistance in India: drivers and opportunities for action. PLoS Med 13(3):e1001974. https://doi.org/10.1371/journal.pmed.1001974

Lepeltier E, Nuhn L, Lehr CM, Zentel R (2015) Not just for tumor targeting: unmet medical needs and opportunities for. Nanomedicine 10(20):3147–3166. https://doi.org/10.2217/nnm.15.132

Letsididi KS, Lou Z, Letsididi R, Mohammed K, Maguy BL (2018) Antimicrobial and antibiofilm effects of trans-cinnamic acid nanoemulsion and its potential application on lettuce. LWT Food Sci Technol 94:25–32. https://doi.org/10.1016/j.lwt.2018.04.018

Li YF, Sun HW, Gao R, Liu KY, Zhang HQ, Fu QH, Qing SL, Guo G, Zou QM (2015) Inhibited biofilm formation and improved antibacterial activity of a novel nanoemulsion against cariogenic *Streptococcus mutans in vitro* and *in vivo*. Int J Nanomedicine 10:447–462. https://doi.org/10.2147/IJN.S72920

Li J, Hwang IC, Chen X, Park HJ (2016) Effects of chitosan coating on curcumin loaded nanoemulsion: study on stability and *in vitro* digestibility. Food Hydrocolloid 60:138–147. https://doi.org/10.1016/j.foodhyd.2016.03.016

Lin J, Nishino K, Roberts MC, Tolmasky M, Aminov RI, Zhang L (2015) Mechanisms of antibiotic resistance. Front Microbiol 6:00034. https://doi.org/10.3389/fmicb.2015.00034

Lin MH, Hung CF, Aljuffali IA, Sung CT, Huang CT, Fang JY (2017) Cationic amphiphile in phospholipid bilayer or oil–water interface of nanocarriers affects planktonic and biofilm bacteria killing. Nanomed Nanotechnol Biol Med 13:353–361. https://doi.org/10.1016/j.nano.2016.08.011

Lou Z, Chen J, Yu F, Wang H, Kou X, Ma C, Zhu S (2017) The antioxidant, antibacterial, antibiofilm activity of essential oil from *Citrus medica* L. var. sarcodactylis and its nanoemulsion. LWT Food Sci Technol 80:371–377. https://doi.org/10.1016/j.lwt.2017.02.037

Lu W, Kelly AL, Miao S (2016) Emulsion-based encapsulation and delivery systems for polyphenols. Trend Food Sci Technol 47:1–9. https://doi.org/10.1016/j.tifs.2015.10.015

Lu WC, Huang DW, Wang CCR, Yeh CH, Tsai JC, Huang YT, Li PH (2018) Preparation, characterization, and antimicrobial activity of nanoemulsions incorporating citral essential oil. J Food Drug Anal 26:82–89. https://doi.org/10.1016/j.jfda.2016.12.018

Lupo A, Haenni M, Madec JY (2018) Antimicrobial resistance in *Acinetobacter* spp. and *Pseudomonas* spp. Microbiol Spectr 6(3). https://doi.org/10.1128/microbiolspec.ARBA-0007-2017

Ma Q, Davidson PM, Zhong Q (2016) Antimicrobial properties of microemulsions formulated with essential oils, soybean oil, and Tween 80. Int J Food Microbiol 226:20–25. https://doi.org/10.1016/j.ijfoodmicro.2016.03.011

Majeed H, Liu F, Hategekimana J, Sharif HR, Qi J, Ali B, Bian YY, Ma J, Yokoyama W, Zhong F (2016) Bactericidal action mechanism of negatively charged food grade clove oil nanoemulsions. Food Chem 197:75–83. https://doi.org/10.1016/j.foodchem.2015.10.015

Manju S, Malaikozhundan B, Vijayakumar S, Shanthi S, Jaishabanu A, Ekambaram P, Vaseeharan B (2016) Antibacterial, antibiofilm and cytotoxic effects of *Nigella sativa* essential oil coated gold nanoparticles. Microb Pathog 91:129–135. https://doi.org/10.1016/j.micpath.2015.11.021

Mendes JF, Martins HHA, Otoni CG, Santana NA, Silva RCS, Da Silva AG, Silva MV, Correia MTS, Machado G, Pinheiro ACM, Piccoli RH, Oliveira JE (2018) Chemical composition and antibacterial activity of *Eugenia brejoensis* essential oil nanoemulsions against *Pseudomonas fluorescens*. LWT Food Sci Technol 93:659–664. https://doi.org/10.1016/j.lwt.2018.04.015

Michael CA, Dominey-Howes D, Labbate M (2014) The antimicrobial resistance crisis: causes, consequences, and management. Front Public Health 2:145. https://doi.org/10.3389/fpubh.2014.00145

Moghimi R, Ghaderi L, Rafati H, Aliahmadi A, McClements DJ (2016a) Superior antibacterial activity of nanoemulsion of *Thymus daenensis* essential oil against *E coli*. Food Chem 194:410–415. https://doi.org/10.1016/j.foodchem.2015.07.139

Moghimi R, Aliahmadi A, McClements DJ, Rafati H (2016b) Investigations of the effectiveness of nanoemulsions from sage oil as antibacterial agents on some food borne pathogens. LWT Food Sci Technol 71:69–76. https://doi.org/10.1016/j.lwt.2016.03.018

Moghimi R, Aliahmadi A, Rafati H, Abtahi HR, Amini S, Feizabadi MM (2018) Antibacterial and anti-biofilm activity of nanoemulsion of *Thymus daenensis* oil against multi-drug resistant *Acinetobacter baumannii*. J Mol Liquid 265:765–770. https://doi.org/10.1016/j.molliq.2018.07.023

Morens DM, Fauci AS (2013) Emerging infectious diseases: threat to human health and global stability. PLoS Pathog 9(7):e1003467. https://doi.org/10.1371/journal.ppat.1003467

Mostafa NM (2018) Antibacterial activity of ginger (*Zingiber officinale*) leaves essential oil nano-emulsion against the cariogenic *Streptococcus mutans*. J Appl Pharmaceut Sci 8(09):034–041. https://doi.org/10.7324/JAPS.2018.8906

Ng VWL, Chan JMW, Sardon H, Ono RJ, Garcia JM, Yang YY, Hedrick JL (2014) Antimicrobial hydrogels: a new weapon in the arsenal against multidrug-resistant infections. Adv Drug Del Rev 78:46–62. https://doi.org/10.1016/j.addr.2014.10.028

Nguyen HT, Goycoolea FM (2017) Chitosan/cyclodextrin/TPP nanoparticles loaded with querce-tin as novel bacterial quorum sensing inhibitors. Molecules 22:1975. https://doi.org/10.3390/molecules22111975

Nirmal NP, Mereddy R, Li L, Sultanbawa Y (2018) Formulation, characterisation and antibacte-rial activity of lemon myrtle and anise myrtle essential oil in water nanoemulsion. Food Chem 254:1–7. https://doi.org/10.1016/j.foodchem.2018.01.173

Olsen I (2015) Biofilm-specific antibiotic tolerance and resistance. Eur J Clin Microbiol Infect Dis 34:877–886. https://doi.org/10.1007/s10096-015-2323-z

Ostertag F, Weiss J, McClements DJ (2012) Low-energy formation of edible nanoemulsions: fac-tors influencing droplet size produced by emulsion phase inversion. J Colloid Interface Sci 388:95–102. https://doi.org/10.1016/j.jcis.2012.07.089

Pattnaik S, Barik S, Muralitharan G, Busi S (2018) Ferulic acid encapsulated chitosan tripoly-phosphate nanoparticles attenuate quorum sensing regulated virulence and biofilm forma-tion in *Pseudomonas aeruginosa* PAO1. IET Nanobiotechnol 12(8):1056–1061. https://doi.org/10.1049/iet-nbt.2018.5114

Pelgrift RY, Friedman AJ (2013) Nanotechnology as a therapeutic tool to combat microbial resis-tance. Adv Drug Del Rev 65:1803–1815. https://doi.org/10.1016/j.addr.2013.07.011

Prakash A, Baskaran R, Paramasivam N, Vadivel V (2018) Essential oil based nanoemulsions to improve the microbial quality of minimally processed fruits and vegetables: a review. Food Res Int 111:509–523. https://doi.org/10.1016/j.foodres.2018.05.066

Prasad R, Pandey R, Varma A, Barman I (2017) Polymer based nanoparticles for drug delivery systems and cancer therapeutics. In: Kharkwal H, Janaswamy S (eds) Natural polymers for drug delivery. CAB International, Wallingford, pp 53–70

Prasad R, Jha A, Prasad K (2018) Exploring the realms of nature for nanosynthesis. Springer, New York. ISBN 978-3-319-99570-0. https://www.springer.com/978-3-319-99570-0

Prasad M, Lambe UP, Brar B, Shah I, Manimegalai J, Ranjan K, Rao R, Kumar S, Mahant S, Khurana SK, Iqbal HMN, Dhama K, Misri J, Prasad G (2018) Nanotherapeutics: an insight into healthcare and multi-dimensional applications in medical sector of the modern world. Biomed Pharmacother 97:1521–1537. https://doi.org/10.1016/j.biopha.2017.11.026

Qi L, Li H, Zhang C, Liang B, Li J, Wang L, Du X, Liu X, Qiu S, Song H (2016) Relationship between antibiotic resistance, biofilm formation, and biofilm-specific resistance in *Acinetobacter bau-mannii*. Front Microbiol 7:483. https://doi.org/10.3389/fmicb.2016.00483

Quatrin PM, Verdi CM, de Souza ME, de Godoi SN, Klein B, Gundel A, Wagner R, Vaucher RA, Ourique AF, Santos RCV (2017) Antimicrobial and antibiofilm activities of nanoemulsions containing *Eucalyptus globulus* oil against *Pseudomonas aeruginosa* and *Candida* spp. Microb Pathog 112:230–242. https://doi.org/10.1016/j.micpath.2017.09.062

Rangel-Vega A, Bernstein LR, Mandujano-Tinoco EA, Garcia-Contreras SJ, Garcia-Contreras R (2015) Drug repurposing as an alternative for the treatment of recalcitrant bacterial infections. Front Microbiol 6:282. https://doi.org/10.3389/fmicb.2015.00282

Rather IA, Kim BC, Bajpai VK, Park YH (2017) Self-medication and antibiotic resistance: crisis, current challenges, and prevention. Saudi J Biol Sci 24:808–812. https://doi.org/10.1016/j.sjbs.2017.01.004

Roca I, Akova M, Bacquero F, Carlet J, Cavaleri M, Coenen S, Cohen J, Findlay D, Gyssens I, Heure OE, Kahlmeter G, Kruse H, Laxminarayan R, Liebana E, Lopez-Cerero L, MacGowan A, Martins M, Rodriguez-Bano J, Rolain JM, Segovia C, Sigauque B, Tacconelli E, Wellington E, Vila J (2015) The global threat of antimicrobial resistance: science for intervention. New Microbe New Infect 6:22–29. https://doi.org/10.1016/j.nmni.2015.02.007

Rostami H, Nikoo AM, Rajabzadeh G, Niknia N, Salehi S (2018) Development of cumin essential oil nanoemulsions and its emulsion filled hydrogels. Food Biosci 26:126–132. https://doi.org/10.1016/j.fbio.2018.10.010

Ryu V, McClements DJ, Corradini MG, McLandsborough L (2018a) Effect of ripening inhibitor type on formation, stability, and antimicrobial activity of thyme oil nanoemulsion. Food Chem 245:104–111. https://doi.org/10.1016/j.foodchem.2017.10.084

Ryu V, McClements DJ, Corradini MG, Yang JS, McLandsborough L (2018b) Natural antimicrobial delivery systems: formulation, antimicrobial activity, and mechanism of action of quillaja saponin-stabilized carvacrol nanoemulsions. Food Hydrocolloid 82:442–450. https://doi.org/10.1016/j.foodhyd.2018.04.017

Salvia-Trujillo L, Rojas-Grau A, Soliva-Fortuny R, Martin-Belloso O (2015) Physicochemical characterization and antimicrobial activity of foodgrade emulsions and nanoemulsions incorporating essential oils. Food Hydrocolloid 43:547–556. https://doi.org/10.1016/j.foodhyd.2014.07.012

Santajit S, Indrawattana N (2016) Mechanisms of antimicrobial resistance in ESKAPE pathogens. Biomed Res Int 2016:2475067. https://doi.org/10.1155/2016/2475067

Schuster M, Sexton DJ, Diggle SP, Greenberg EP (2013) Acyl-homoserine lactone quorum sensing: from evolution to application. Annu Rev Microbiol 67:43–63. https://doi.org/10.1146/annurev-micro-092412-155635

Sessa M, Balestrieri ML, Ferrari G, Servillo L, Castaldo D, D'Onofrio N, Donsi F, Tsao R (2014) Bioavailability of encapsulated resveratrol into nanoemulsion-based delivery systems. Food Chem 147:42–50. https://doi.org/10.1016/j.foodchem.2013.09.088

Severino R, Ferrari G, Vu KD, Donsi F, Salmieri S, Lacroix M (2015) Antimicrobial effects of modified chitosan based coating containing nanoemulsion of essential oils, modified atmosphere packaging and gamma irradiation against *Escherichia coli* O157:H7 and *Salmonella typhimurium* on green beans. Food Cont 50:215–222. https://doi.org/10.1016/j.foodcont.2014.08.029

Shahabi N, Tajik H, Moradi M, Forough M, Ezati P (2017) Physical, antimicrobial and antibiofilm properties of *Zataria multiflora* Boiss essential oil nanoemulsion. Int J Food Sci Technol 52:1645–1652. https://doi.org/10.1111/ijfs.13438

Singh N, Verma SM, Singh SK, Verma PRP, Ahsan MN (2015a) Antibacterial activity of cationised and non-cationised placebo lipidic nanoemulsion using transmission electron microscopy. J Exp Nanosci 10(4):299–309. https://doi.org/10.1080/17458080.2013.830199

Singh N, Verma SM, Singh SK, Verma PRP (2015b) Antibacterial action of lipidic nanoemulsions using atomic force microscopy and scanning electron microscopy on *Escherichia coli*. J Exp Nanosci 10(5):381–391. https://doi.org/10.1080/17458080.2013.838702

Singh BR, Singh BN, Singh A, Khan W, Naqvi AH, Singh HB (2016) Mycofabricated biosilver nanoparticles interrupt *Pseudomonas aeruginosa* quorum sensing systems. Sci Rep 5:13719. https://doi.org/10.1038/srep13719

Singh Y, Meher JG, Raval K, Khan FA, Chaurasia M, Jain NK, Chourasia MK (2017) Nanoemulsion: concepts, development and applications in drug delivery. J Contr Release 252:28–49. https://doi.org/10.1016/j.jconrel.2017.03.008

Song Z, Sun H, Yang Y, Jing H, Yang L, Tong Y, Wei C, Wang Z, Zou Q, Zeng H (2016) Enhanced efficacy and anti-biofilm activity of novel nanoemulsions against skin burn wound multi-drug resistant MRSA infections. Nanomed Nanotechnol Biol Med 12:1543–1555. https://doi. org/10.1016/j.nano.2016.01.015

Sugumar S, Singh S, Mukherjee A, Chandrasekaran N (2016) Nanoemulsion of orange oil with non ionic surfactant produced emulsion using ultrasonication technique: evaluating against food spoilage yeast. Appl Nanosci 6:113–120. https://doi.org/10.1007/s13204-015-0412-z

Taleb MH, Abdeltawab NF, Shamma RN, Abdelgayed SS, Mohamed SS, Farag MA, Ramadan MA (2018) *Origanum vulgare* L. essential oil as a potential anti-acne topical nanoemulsion—*in vitro* and *in vivo* study. Molecules 23:2164. https://doi.org/10.3390/molecules23092164

Taylor PW (2013) Alternative natural sources for a new generation of antibacterial agents. Int J Antimicrob Agent 42:195–201. https://doi.org/10.1016/j.ijantimicag.2013.05.004

Topuz OK, Ozvural EB, Zhao Q, Huang Q, Chikindas M, Golucku M (2016) Physical and antimicrobial properties of anise oil loaded nanoemulsions on the survival of foodborne pathogens. Food Chem 203:117–123. https://doi.org/10.1016/j.foodchem.2016.02.051

Venkadesaperumal G, Rucha S, Sundar K, Shetty PH (2016) Anti-quorum sensing activity of spice oil nanoemulsions against food borne pathogens. LWT Food Sci Technol 66:225–231. https://doi.org/10.1016/j.lwt.2015.10.044

Ventola CL (2015) The antibiotic resistance crisis. Pharm Therapeut 40(4):277–283

Wilkins M, Hall-Stoodley L, Allan RN, Faust SN (2014) New approaches to the treatment of biofilm-related infections. J Infect 69:S47–S52. https://doi.org/10.1016/j.jinf.2014.07.014

Wu H, Moser C, Wang HZ, Hoiby N, Song ZJ (2015) Strategies for combating bacterial biofilm infections. Int J Oral Sci 7:1–7. https://doi.org/10.1038/ijos.2014.65

Yuan YG, Peng QL, Gurunathan S (2017) Effect of silver nanoparticles on multiple drug-resistant strains of *Staphylococcus aureus* and *Pseudomonas aeruginosa* from mastitis-infected goats: an alternative approach for antimicrobial therapy. Int J Mol Sci 18:569. https://doi.org/10.3390/ijms18030569

Zaidi S, Misba L, Khan AU (2017) Nano-therapeutics: a revolution in infection control in post antibiotic era. Nanomed Nanotechnol Biol Med 13(7):2281–2301. https://doi.org/10.1016/j.nano.2017.06.015

Zaman SB, Hussain MA, Nye R, Mehta V, Mamun KT, Hossain N (2017) A review on antibiotic resistance: alarm bells are ringing. Cureus 9(6):e1403. https://doi.org/10.7759/cureus.1403

Zhang Z, Vriesekoop F, Yuan Q, Liang H (2014) Effects of nisin on the antimicrobial activity of D-limonene and its nanoemulsion. Food Chem 150:307–312. https://doi.org/10.1016/j.foodchem.2013.10.160

Zhang S, Zhang M, Fang Z, Liu Y (2017) Preparation and characterization of blended cloves/cinnamon essential oil nanoemulsions. LWT Food Sci Technol 75:316–322. https://doi.org/10.1016/j.lwt.2016.08.046

Zhu X, Radovic-Moreno AF, Wu J, Langer R, Shi J (2014) Nanomedicine in the management of microbial infection—overview and perspectives. Nano Today 9:478–498. https://doi.org/10.1016/j.nantod.2014.06.003

Chapter 16
Mesoporous Silica Nanomaterials as Antibacterial and Antibiofilm Agents

Pethakamsetty Lakshmi and Sudhakar Pola

Contents

P. Lakshmi
Department of Microbiology, AUCST, Andhra University,
Visakhapatnam, Andhra Pradesh, India

S. Pola (✉)
Department of Biotechnology, AUCST, Andhra University,
Visakhapatnam, Andhra Pradesh, India
e-mail: sudhakar@andhrauniversity.edu.in

© Springer Nature Switzerland AG 2020
R. Prasad et al. (eds.), *Nanostructures for Antimicrobial and Antibiofilm Applications*, Nanotechnology in the Life Sciences,
https://doi.org/10.1007/978-3-030-40337-9_16

Abstract Antimicrobial agents are vital to fight infectious diseases which are pooling up day by day. The treatment of microbial infections is increasingly getting convoluted by the ability of microorganisms to develop resistance towards a wide range of antimicrobial agents. Resistance is most often an evolutionary process taking place either through lateral gene transfer or during antibiotic therapy, thereby contributing to the emergence of diseases that were under good control for many years.

Further, drug resistance enforces high-dose administration of antibiotics leading to adverse side effects and intolerable toxicity. This has prompted the search for alternative strategies to treat microbial infections either by controlling their growth or by preventing the formation of bacterial biofilms. Recently tremendous developments in the field of nanotechnology have been recorded with nanoscale materials emerging as novel antimicrobial agents.

Nanotechnology is an interdisciplinary area of science with promising interests across the globe steering into nanoindustrial revolution with innumerable applications. The enormous diversity of the nanoparticles that exhibit new and enhanced size-dependent properties compared to their bulk material are being exploited as antimicrobials for treating infectious diseases. Numerous nanodevices like carbon nanotubes, quantum dots, and polymeric micelles have been reported as potential antibacterial candidates. In the present scenario, mesoporous silica nanoparticles (MSNs) are emerging for their widespread applications as antibacterial and antibiofilm agents. MSNs are constituted of an amorphous silica matrix with ordered porous molecular sieves characterized by periodic arrangements of uniformly sized mesopores (diameter between 2 and 50 nm). MSNs with uniform and tailorable pore dimensions with high surface areas are currently being employed in a number of applications such as wastewater remediation, indoor air cleaning, bio-catalysis, drug delivery, CO_2 capture, bioanalytical sample preparation, pervaporation membrane improvement, etc. MSNs with their unique properties like chemical stability, surface functionality, and biocompatibility are used in quorum quenching as well as prospective antibacterial agents. The present book chapter deals with MSNs and their applications as possible antibacterial and antibiofilm agents.

Keywords Mesoporous silica nanoparticles · MSNs · Antibacterial · Antibiofilm · Biocompatibility

16.1 Introduction

Antimicrobial agents are vital to fight infectious diseases which are pooling up day by day. The treatment of microbial contaminations is progressively getting convoluted by the ability of microorganisms to develop resistance towards a wide range of antimicrobial agents (Aziz et al. 2016). Further drug resistance enforces enhanced-dose administration of antibiotics rendering contrary side effects. It leads to the search for novel approaches to treat microbial infections either by controlling their growth or by preventing the formation of bacterial biofilms. Recently tremendous developments in the field of nanotechnology have been recorded employing nano-sized materials as newer emerging antibacterial agents (Aziz et al. 2016).

Nanoscience is an area of science with full of promising interests across the globe steering into nanoindustrial revolution with innumerable applications in catalysis, cosmetics, diagnostics, and targeted drug delivery systems (Juan et al. 2015; Prasad et al. 2016, 2017). The enormous diversity of the nanoparticles with their novel and improved size-dependent possessions compared to their bulk material is being exploited for treating infectious diseases. Numerous nanodevices like carbon nanotubes, quantum dots, and polymeric micelles have been reported as potential antibacterial candidates (Albanese et al. 2012; Lakshmi et al. 2017). However, recent trends involved molecular manufacturing of nanomaterials studied under molecular nanotechnology.

Molecular nanotechnology involves theoretical manipulation of single molecules to produce the desired structure or an atom in a finely controlled way, using the principles of mechanosynthesis operating on a nanoscale. Particle size, porosity, and surface properties of nanomaterials can be certainly monitored to match the physicochemical characteristics of guest components with intended applications (Bayir et al. 2018). Further by conjugating the functional groups, stimuli-sensitive molecules and targeting molecules to both the inner and outer surfaces of the silica pores lead to the improvement of disparity and loading and subsequent release of transporters to targeted places (Chan et al. 2016).

Nanoporous materials can be made up of an amorphous or crystalline framework of cage type or cylindrical structures with void spaces. According to IUPAC, based on pore size, nanoporous materials are of three types, namely, macroporous (pore sizes between 50 and 1000 nm), microporous (0.2–5 nm), and mesoporous (pore size ranging between 2 and 50 nm). Porous polymeric beads that allow easy access to the internal pores at relative ease are the macroporous materials. Carbons and amorphous glasses, zeolites, and metal-organic frameworks (MOFs) with high thermal stability and catalytic activity are examples of microporous materials which are employed in cracking processes and also be served as ion exchange media, gas separation, and drying agents. MOFs are currently considered as the fast-growing classes of microporous solids. Comparatively mesoporous materials with an intermediate pore size such as porous inorganic solids with the controllable large internal surface area are currently exploited at an atomic, molecular, and nanometer scales leading to alternate disease treatment strategies (Chen et al. 1993).

Mesoporous silica nanoparticles (MSNs) with their well-known applications as antibacterial and antibiofilm agents are constituted of a porous amorphous silica matrix with uniformly sized mesopores arranged periodically in the form of a molecular sieve (diameter between 2 and 50 nm). MSNs with their unique properties like chemical stability, surface functionality, and biocompatibility are used in quorum quenching as well as prospective antibacterial agents (An et al. 2016). The present book chapter deals with MSNs and their applications as possible antibacterial and antibiofilm agents.

16.2   Mesoporous Silica Nanoparticles (MSNs)

Mesoporous silica is a very popular inorganic nanomaterial made up of two most copious (silicon and oxygen) elements in the environment, existing as silicon dioxide (SiO_2). Silica molecule exists in a complex of interconnected silicon atoms in a tetrahedral arrangement linked covalently with four oxygen atoms. Based on extensive physicochemical, ecotoxicological safety, and epidemiology data, it is evident that there were no ecological or health hazards allied with these materials. Further US FDA has regarded silica as a material that is "generally recognized as safe" and has been approved by the EU for their usage in cosmetics and food additives (Bobo et al. 2016).

The discovery of silica nanoparticles dates back to late 1970s. Silica nanoparticles with 4.6–30 nm pores are arranged in an hexagonal array termed SBA (Santa Barbara Amorphous material) produced by the California University, Santa Barbara (Sakai-Kato et al. 2011). Later MSNs were synthesized independently by Mobil Corporation laboratory, Japan, in 1997 under the trade name MCM (Mobil crystalline materials) or Mobil composition of matter (WU et al. 2016). Table 16.1 shows some of the morphologies of mesoporous silica (MS) and their associated materials (Bagwe et al. 2006).

MSU-n (Michigan State University silica), KIT-1 (Korean Institute of technology silica), and FSM-16 (folded sheet-derived mesoporous silica), HMM-33 (Hiroshima Mesoporous Material-33), TUD-1 (Technical Delft University), and COK-12 (Centrum voor Oppervlaktechemie en Katalyse/Centre for Research Chemistry and Catalysis) are synthesized newly with varied sizes and pore symmetry (Fruijtier-Pölloth 2012). Tozuka et al. (2005) have demonstrated the usage of quaternary ammonium surfactants with layered polysilicate kanemite as a template for FSM-16, which is used in pharmaceutical applications besides as an adsorbent and catalyst.

Currently MSNs with a wide range of pore geometries (hexagonal, cubic, cylindrical) and particle morphologies (discs, spheres, rods) have been synthesized and exploited in the field of medical sciences (Fig. 16.1). MSNs have a honeycomb-like structure with narrow pore size distributions and high surface areas (>500 m^2/g).

Table 16.1 List of mesoporous silica nanoparticles

S. no	MSN type	Pore symmetry	Pore volume (cm^3/g)	Pore size (nm)
1	SBA-11	3D cubic	0.68	2.1–3.6
2	SBA-12	3D hexagonal	0.83	3.1
3	SBA-15	2D hexagonal	1.17	6–0
4	SBA-16	Cubic	0.91	5–15
5	MCM-41	2D hexagonal	>1.0	1.5–8
6	MCM-48	3D cubic	>1.0	2–5
7	MCM-50	Lamellar	>1.0	2–5
8	KIT-5	Cubic	0.45	9.3
9	COK-12	Hexagonal	0.45	5.8

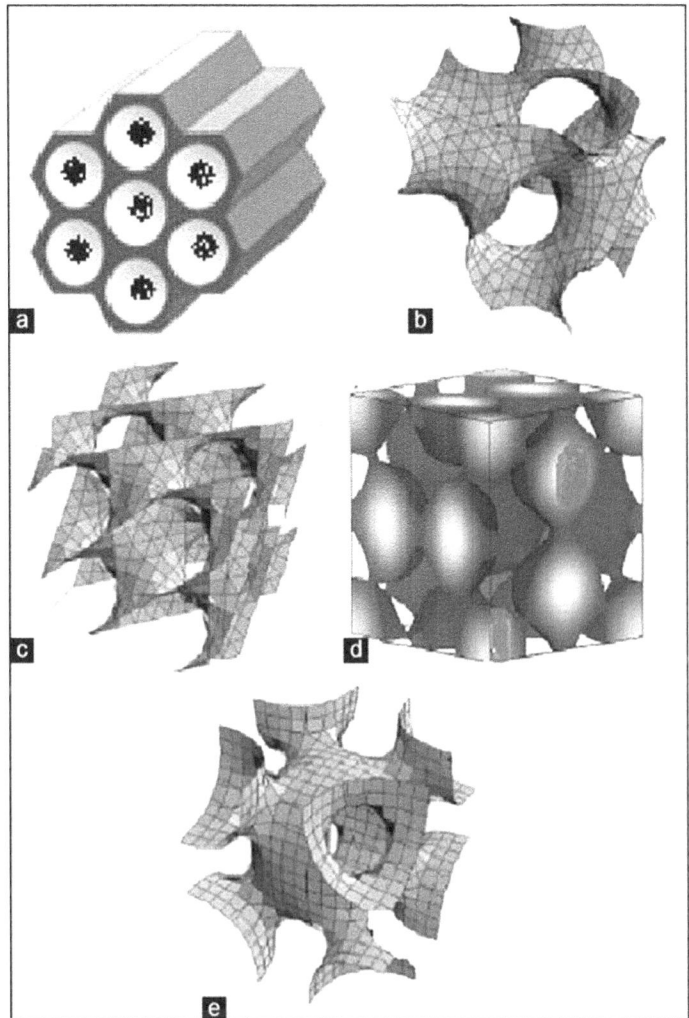

Fig. 16.1 Shapes of MSNs. (**a**) Hexagonal 2D, (**b**) cubic bicontinuous, (**c**) bicontinuous cubic, (**d**) cage type, (**e**) cage type, respectively

The strong Si-O bonds render them stable against external mechanical stress and degradation thereby making them unique and significant in the field of biotechnology. Mesoporous materials have many advantages such as:

- Tunable pore diameter
- The unique, customizable mesoporous structure
- Low cytotoxicity
- Better binding ability with organic ligands
- Enhanced surface properties to bond with therapeutic molecules

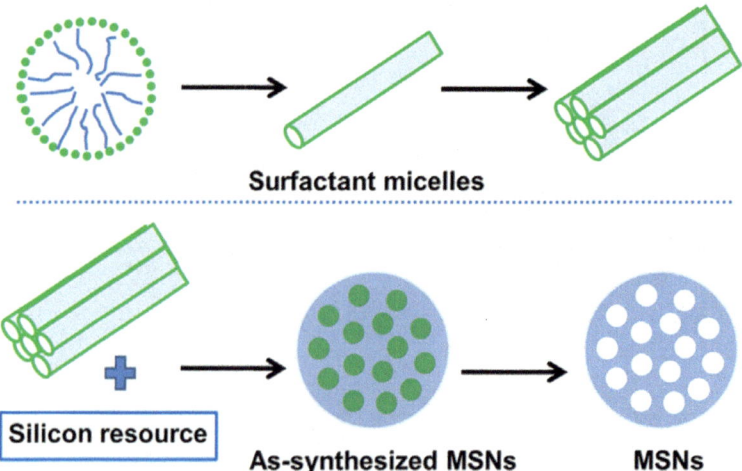

Fig. 16.2 Schematic representation of MSNs synthesis

- Uniform adsorption and subsequent drug delivery
- Biocompatibility with large pore volume to surface area
- Ecofriendly and regarded as safe

Owing to the above advantages, MSNs were exploited over a wide array of applications in industrial, therapeutic, food, and cosmetic industry. Today, MSNs have been used as adsorbents, drug delivery vehicles, biosensor, bioimaging and biosignal probes, and other critical diagnostic applications (Hom et al. 2010; Grumezescu et al. 2013).

16.3   Synthesis of MSNs

The synthesis of silica nanoparticles can be carried out by a wide range of approaches which can be by physical techniques (e.g., sputtering, sonochemical, and microwave-assisted), mechanical methods (e.g., ball milling and attrition), and chemical routes (precipitation, micelles, solvothermal, and vapor phase synthesis). The mesoporous particle synthesis can be done by a spray drying method or simple sol-gel method with slight modifications in their procedures (Ashraf et al. 2015).

Currently, MSNs are synthesized (Fig. 16.2) with a template made of micellar rods reacting with tetraethyl orthosilicate (TEOS), which results in nanosized spheres or rods consisting of a regular arrangement of pores. Presently TEOS is replaced by a better precursor **MPTMS** (3-mercaptopropyl trimethoxysilane) which ensures uniform sphere formation and also reduces the chance of aggregation. Further, the rate of aggregation can be completely reduced either by capping or plating of the MSNs with gold nanoparticles (Paul et al. 2017; Cicily 2017).

Table 16.2 List of chemical constituents used in the synthesis of MSNs

Substrate	Function
N-dodecanonyl-β-alanine	Surfactant with an amino acid residue
CTAB	Increase water solubility of hydrophobic ligand
PEG	Improve biocompatibility and functional characteristics of silica matrix
Tween 80	Surfactant
PVA	Settle down gel in THEOS-containing solution
PEO	Induce hydration PEO/sol ratio regulates size
Sodium hydroxide	Catalyst
Hydrogen fluoride	Catalyst
Hydrogen chloride	Catalyst
Nonionic triblock copolymer	Structure-directing agent
Trihydroxysilylpropylmethyl phosphate	Prevent-interplace aggregation
Ammonium nitrate	Surfactant removal
Methanol	Solvent in TMOS
Ethanol	Solvent in TEOS

CTAB N-cetyl trimethyl ammonium bromide, *PEG* Polyethylene glycol, *PVA* Polyvinyl alcohol, *PEO* Polyethylene oxide, *TMOS* Tetramethoxysilane, *TEOS* Tetraethyl orthosilicate, *MSN* Mesoporous silica nanoparticles, *THEOS* Tetrakis (2-hydroxyethyl) orthosilicate, *SBA* Santa barbara amorphous
Pharmaceutics 2018, 10, 118; https://doi.org/10.3390/pharmaceutics10030118

The following are the various chemical constituents employed in the formation of mesoporous silica nanoparticles (Table 16.2).

The mechanism of MSN synthesis is a complex multistep protocol that involves extreme conditions like high temperature, pH, and use of highly toxic precursors. Till date, three simultaneous technologies for the synthesis of MSNs, such as the invention of "Stober" synthesis for monodisperse nanoparticles (in 1968), as well as the other two methods including MCM-41 (in 1992) and SBA-15 (in 1998) have been equally exploited. These three protocols collectively received more attention, confirming their widespread usage in the synthesis of MSNs employed in biomedical research for drug delivery and toxicity studies. The versatile applications of MSNs can be attributed to their unique mesoporous ordered structures with large pore volumes and high internal surface areas (Moller et al. 2007).

16.4 Characterization of MSNs

Characterization of synthesized MSNs was carried out by microscopic (Fig. 16.3) and spectral analysis. MSNs were studied through transmission electron microscopy (TEM) with bright field imaging operating at 200 kV. TEM is useful to determine the pore characteristics as well as the shape and dimension of the particles. Micrograph Digital TM software is used to measure pore size and the particle

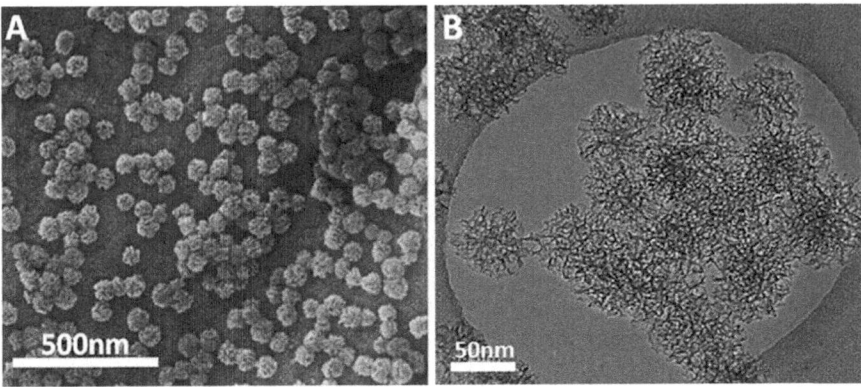

Fig. 16.3 (**a**) MSN SEM image and (**b**) transmission electron microscopic image of MSNs (https://doi.org/10.1155/2014/176015)

characteristics, whereas channel diameters were studied by line plot display (Chan et al. 2016). Scanning electron microscopy reveals the topography of the MSNs where the particle morphology and pore directions were filmed. Recent studies through SEM analysis showed bacterial cell elongation to nearly twice to tenfold increase in the presence of MSN particles supporting the hypothesis of cell division impairment by mesoporous silica nanoparticles in bacterial population possibly by interacting with FtsZ (Jorge et al. 2016). Crystallographic symmetry of MSNs by XRD analysis is not very clear and ambiguous due to similar short-range peaks which might overlap and appear at low angles (Huang et al. 2013). Further the MSNs were also analyzed by FTIR spectroscopy using KBr pellets which are used to identify organic and inorganic materials and their bonding patterns by measuring the absorption peaks of infrared radiations (Cicily 2017).

16.5 MSNs as Antibacterial Agents

Antibacterial agents are chemical constituents which can selectively kill or control the growth of bacterial population, without affecting the surrounding tissue. In general, the agents employed to retard the growth of the bacterial strains are referred to as bacteriostatic, while those which kill the organism are called bactericidal drugs. Antibacterial agents are principal constituents essential to combat infectious diseases. Currently numerous synthetic and semisynthetic chemical substances like β-lactams, aminoglycosides, tetracyclines, sulfa drugs, etc. are employed to treat a wide variety of bacterial diseases. However despite the incidence of a wide series of antimicrobials to fight infections, there is still a tremendous need for the potential novel antibiotics, since most of the antibiotics rendered are ineffective and will have to be used in higher doses due to the emergence of resistance in bacteria.

Resistance in bacteria is inheritable and is acquired either through vertical or lateral gene transfer and which might be chromosomal or extra chromosomal (plasmids). Today the emergence of superbugs such as MRSA (methicillin-resistant *Staphylococcus aureus*), MDR-TB (multidrug-resistant TB), of late VRSA (vancomycin-resistant *Staphylococcus aureus)*, etc. in the medical world is due to the resistance conferred by the bacteria through use and abuse of antibiotic therapy (Worthington et al. 2012). It has been recently noted that half a million of new incidences of MDR TB are recorded annually (Webster and Seil 2012). Further development of new degradative enzymes such as β-lactamases (NDM-1) in certain bacterial strains has led to the failure of an entire set of β-lactam antibiotics which constitute a major share of antibacterial agents (Xia et al. 2009; Ventola et al. 2015). Therefore drug resistance not only enforces the usage of high doses of synergistic drugs but also ends up with adverse side effects.

Drug resistance has impelled the search for development of substitute strategies to treat microbial infections. Among others, nanoscale materials have been synthesized as novel antimicrobial agents with numerous classes of nanosized carriers for treating infectious diseases (Allahverdiyev et al. 2011). Nanomaterials offer enhanced properties to traditional organic antibacterial agents and can control the resistance property of bacterial superbugs. Nanomaterials exert antimicrobial activity by accumulating on the cell membranes, affecting their permeability and transport mechanisms thereby leading to cell leakage and eventual cell death (Aziz et al. 2014, 2015, 2016, 2019). Further nanoparticles in the presence of oxygen trigger the generation of ROS (reactive oxygen species) such as OH^-, O_2^-, and H_2O_2, disrupting the normal metabolic functions of microbes leading to cell death (Hudson et al. 2008). Figure 16.4 depicts the toxic effects of nanoparticles on bacterial structures such as capsule (polysaccharide), cell wall (peptidoglycans), cell membranes, ribosomes (protein synthesis), nucleic acid synthesis (DNA damage), etc. (Kar 2016).

Two major studies, namely, Sahoo et al. (2015) and Tarn et al. (2013), have suggested the exploitation of silica nanoparticles as potential antimicrobials with selective toxicity against bacteria. Later silica-coated silicon nanotubes and silver-doped silica nanoparticles with their biocompatibility and chemical and high thermal stability have proved to display exceptional antimicrobial activity against bacterial populations. The ability of MSNs to encapsulate with inorganic materials (silver, gold, palladium, or iron oxide) creating nanocrystals with a yolk/shell architecture gives them additional functionality in binary system against resistant bacterial strains (Cicily 2017). Trewyn et al. (2007) have synthesized MSNs coated with bactericidal cationic surfactants to treat bacterial diseases, while Juan et al. (2015) used metal nanoparticle coatings on MSNs for microbial infections. The silver nanoparticles encapsulated with MSNs (Ag@MESs) in a core shell structure is used as the best source of antimicrobial silver ions where hydrophobic silver nanocrystals are dispersed uniformly without aggregation by the mesoporous silica shell lattice structures (Sen Karaman et al. 2016).

Mesoporous silica nanostructures are considered as one of the forerunners among nanoparticles to be exploited as antibacterial agents which substitute the broad

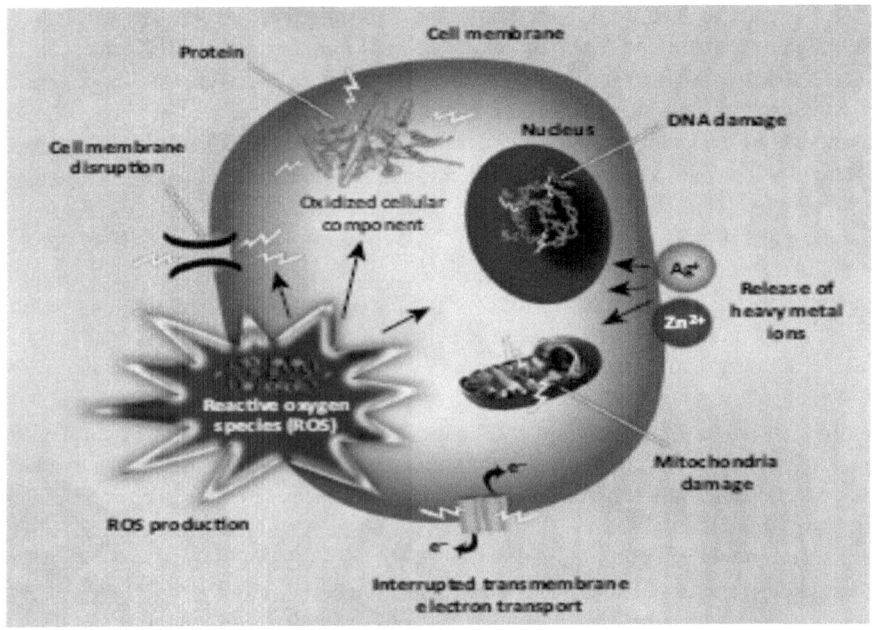

Fig. 16.4 Toxicity mechanisms of antibacterial nanoparticles (NPs)

usage and amassing of metal nanoparticles in the environment. Literature on MSNs in a wide variety of sizes, shapes, and versatile applications showed explosive usage in biomedical research. Also, their characteristic high surface area ($\geq$1000 m^2/g) to volume ratio and adaptable surface functionalization with controlled release of incorporated agents makes them efficient lead molecules to combat antibacterial resistance and to subsequent eradication of biofilm formations (Cheng et al. 2016). Engaging mesoporous silica nanostructures as antibacterial agents is eco-friendly and environmentally safe, since they are easily biodegraded into undisruptive products in the presence of water. Recent studies on mesoporous silica particles are mainly on the drug loading and drug delivery mechanisms owing to their biocompatibility and low cytotoxicity. MSNs with polymer coatings of proteins, DNA, RNA, antibiotics, and other biomolecules are a great choice as carrier vehicles in therapeutic nanomedicine (Wang et al. 2010).

16.5.1 Silver Ion (Ag⁺) and Chitosan-Doped Mesoporous Silica Nanoparticles

Synthesis of multifunctional MSNs was carried out by dispersing MSNs doped with silver ions using chitosan. Chitosan is used for surface modifications, to prevent the leakage of silver ions, and to increase the dispersibility of MSNs. The antimicrobial effectiveness was found to show a profound effect on the tested pathogens twice when compared to the normal Ag NPs (Sen Karaman et al. 2016).

16.5.2   Surface Modifications of MSNs to Enhance Antimicrobial Activity

It has been identified that nude MSNs showed mild activity or no activity against microbial strains as such their surface functionalization has been improved by grafting methods to develop ordered secondary structures with new exciting antibacterial and antifungal properties. Polypeptide polymer-grafted mesoporous silica nanoparticles displayed excellent antimicrobial activity against clinical pathogens. Poly-L-lysine which was covalently attached to the surfaces of MSNs using polyvinyl benzyl tributyl phosphonium chloride is responsible for effective destruction of the peptidoglycan in the cell wall resulting in bacterial cell lysis.

Wang et al. (2015) were successful to synthesize MSNs (2.5 nm) with high surface area and maximum loading ability for antibiotics like doxorubicin or gefitinib. Jorge et al. (2016) have developed the surfactant template method to optimize the parameters controlling pore size, particle dimension, and surface modifications. XRD analysis along with FTIR spectra reveals the presence of the characteristic functional groups adsorbed to the surface of MSNs (Paul et al. 2017).

Recently essential oils (EO) encapsulated into MSN matrices were intensively exploited in antimicrobial applications and considered as ideal substances to stabilize volatile compounds and to guarantee their systematic release (Zhao et al. 2017). Encapsulation of EOs into the MSNs enhances their half-life, prolonging their circulation time followed by controlled delivery. Sousa et al. (2014) have synthesized silica mesoporous nanostructures loaded with EOs using multiple emulsion processes which are effective against different clinical bacterial pathogens. However there was no systematic evidence addressing how surface modifications could control the antibacterial activities of MSNs, but nonetheless previous studies favored surface modifications of MSNs for enhanced antimicrobial activity (Ispas et al. 2009).

16.5.3   Mode of Action of MSNs

The mesoporous silica nanoparticles with controllable structural parameters besides huge surface area and high porosity are the ideal antibacterial agents. The mode of action of MSNs on bacteria is by damaging the cell membrane integrity through hydrogen bonding between bacterial lipopolysaccharides and surface hydroxyl groups of MSNs and also by adsorption of membrane lipid molecules onto the MSNs surface. The exact mechanism was attributed to the electrostatic interaction between the cationic head groups of the hollow MSNs with the phosphate groups of the microbial cell walls, thereby leading to the outflow of electrolytes causing bacterial lysis (Sharmila et al. 2016). Besides membrane damage, membrane gelation and fluidization with MSN attachment are some of the possible destruction mechanisms. Conversely MSN concentration gets lowered rendering them less effective in the presence of bacterial debris, causing precipitation of MSNs, which can be prevented by constant shaking in a shaking incubator.

16.5.4   Biocompatibility of Mesoporous Silica Nanoparticles

MSN biocompatibility as drugs or drug vehicles is mainly based on their cellular uptake and cytotoxic properties, which can be studied using fluorescence and confocal microscopy. Karin Möller and Thomas Bein (2017) have demonstrated that the biocompatibility of MSNs depends on the particle shape, size, surface chemistry, and the presence of functional ligands. Saladino (2016) synthesized a series of antibiotic-loaded MSNs with specific toxicity to kill bacterial populations. Recently MSNs were surface coated by vancomycin (MSNsⲤVan) for selective detection and killing of clinically pathogenic bacteria (Chun Xu et al. 2018).

16.6   Antibacterial Tests

Advances in molecular biology combined with biochemical, serological, staining, and microscopic techniques have led to the successful identification and culturing of microorganisms. Bacteria may be exposed to nanometer-sized particles of sediment in their natural environment without adverse effects. However, the synthetic nanoparticles interact with bacteria acting as antimicrobials. To understand the impact of nanomaterials on the physiology and metabolism of the microorganisms, in vivo measurements of bacterial communities can be made where they are susceptible to nanomaterial exposure. For example, the normal flora of the skin may be exposed to large quantities of nanomaterials that are incorporated into topical preparations including sunscreen and cosmetics (Kar 2016). However, measuring whole communities of bacteria is problematic and cumbersome. Further, most environmental bacteria are not easily cultured in the laboratory, and culture-independent techniques, including DNA sequence-based identification, are semiquantitative. As such accurate in vivo measurements are difficult to achieve. The great diversity of bacterial communities, both spatially and temporally, might further make data misrepresentative in small-scale studies. The alternate approach is to study nanoparticle interactions with a well-characterized model system that is easily manipulated in the laboratory and has an international standard that can be made consistent between research groups. Different antibacterial tests can be carried out under in vitro conditions such as the following.

16.6.1   Agar Diffusion Method

Agar cup plate method or Kirby-Bauer disk diffusion method is the most common preliminary test to study the antibacterial activity under in vitro conditions. The freshly revived bacterial cultures (18 h old) were inoculated in a nutrient medium and were transferred into Petri plates. Petri plates thus prepared were incubated at 30 °C for 16–18 h and examined for antibacterial activity by measuring the zones of inhibition (Kavanagh 1992).

16.6.2   Determination of MIC by Dilution Broth Method

Minimum inhibitory concentration or MIC was determined either by macrodilution or microdilution broth method using McFarland nephelometer standards. Multifunctional microplate reader is used to determine the bacterial viability (Tecan Infinite M200) at 600 nm OD (Obeidat et al. 2012).

16.6.3   Bacterial Testing of the Growth Curve

Bacterial growth rate was determined by incubating the cultures in a shaking incubator at 200 rpm at 37 °C and later the percentage of growth inhibition was calculated by % inhibition = (OD of untreated – OD of MSN treated) OD of untreated ×100 (Balouiri et al. 2016).

16.7   MSNs as Antibiofilm Agents

The major drawback of antimicrobial agents is their failure to fight against resistant microbial strains that can produce biofilms. Biofilms are formed by a complex microbial community glued to a solid surface, emancipating an extracellular polymeric matrix (EPM) that covers and protects the bacterial cell community (Fig. 16.5). Of late it has been identified that many microbes frequently form biofilms around commonly used medical devices resulting in appalling diseases. Common clinical pathogens like *S. aureus, E. coli, P. aeruginosa,* etc. are found to form biofilms on catheters, medical shunts, prosthesis, breast implants, orthopedic devices, surgical equipment, etc. causing chronic sinusitis, burn wounds, urinary tract infections, biliary tract infections, prostatitis, and other trauma infections. Similarly *E. faecalis, Proteus mirabilis, K. pneumoniae, S. mutans* were also reported to form biofilms on medical devices leading to serious nosocomial infections (Percival et al. 2008).

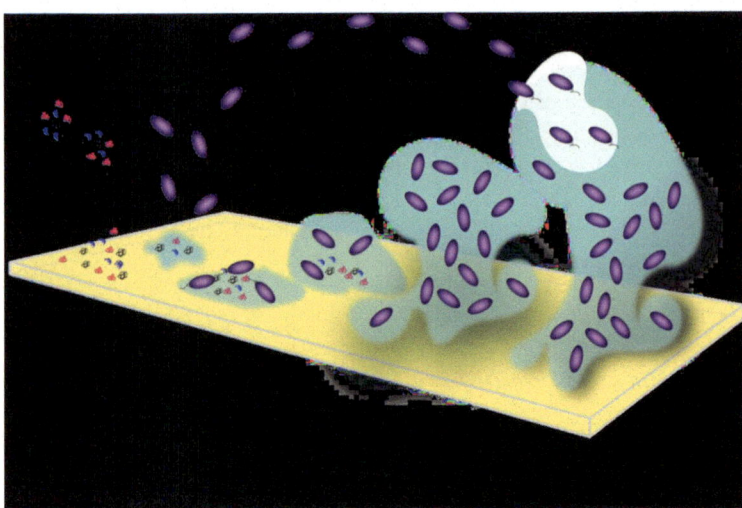

Fig. 16.5 Biofilm growth cycle

Previously control of bacterial contamination and biofilm formation was carried out either by physical (UV) or chemical (flushing, chlorination) disinfection. Nowadays, surface functionalization with broad-spectrum antimicrobial coatings is effective in killing or controlling the bacterial infections (Renwick et al. 2016). The need for novel drugs which can prevent bacterial colonization and biofilm formation without promoting resistance has led to the following developments:

- Micro-topographic surfaces: Nano-engineered materials with surface topography were developed to prevent bacterial adhesion and biofilm formation. Anti-adhesive coatings using hydrophilic polymers and their derivatives (hyaluronic acid, polyethylene glycol, heparin, etc.) gained much attention recently for the development of bacterial repellent and anti-adhesive surfaces (Reema et al. 2018).
- Antimicrobials with covalently immobilized surfaces such as cationic QACs (quaternary ammonium compounds) and phosphonium moieties were identified as contact-killing surfaces. However their antimicrobial activity diminishes in the presence of bacterial debris (Portin 2012).
- Biocide releasing antimicrobials: Metal nanoparticles (NPs) with surface leaching antibiotics have been designed for a specific delivery of the bactericidal agents into the targeted zones. But leaching materials short lifetime is the major limitation of these antibacterial/antibiofilm agents (Ventola 2015).

Despite of the above innovative strategies, there is still a need to develop antibiofilm agents considering drug resistance and combination therapies. Recent strategies include the use of surface modified medical devices with designs of antibiofilm coatings developed either by graft polymerization, layer-by-layer (LbL) assembly, self-assembled monolayers, or surface covalent modifications. Of all the above methods, LbL deposition of surface coatings with bactericidal and anti-adhesive properties without the need for chemical modifications and cross-linking agents has proved successful (Song et al. 2017).

16.7.1 MSNs for Controlling Bacterial Biofilms

Recently, mesoporous silica nanoparticles with their unique physicochemical characteristics such as easy functionalization, thermal stability, excellent biocompatibility, and low cytotoxicity compared to their solid/nonporous counterparts have gained much attention (Spataru et al. 2016). Single or mixed populations of Gram-positive bacteria (e.g., *S. aureus, Bacillus* spp., *Streptococcus* spp.) and Gram-negative bacteria (e.g., *E. coli, Serratia* sp., *Pseudomonas* spp.) usually result in biofilm formation leading to persistent microbial infections that are resistant to antibiotic therapy. Antibiotics which were earlier efficient against bacterial species may not be currently effective against biofilm embedded bacteria (Merezeanu et al. 2016). Nanomaterials served as a potential platform to solve the limitations of traditional therapies in preventing biofilm formations or in treating the preexisting biofilm infections. Various metals, metal oxides, hybrid polymer, and biopolymer silica nanomaterials have been recommended as next-generation antimicrobials with maximum activity against biofilm-resistant bacterial populations (Zhang et al. 2012).

Biofilm eradication needs competent penetration and accumulation of the nanoparticles into the biofilm complex. The interactions between MSNs and biofilms are complex, and upon attachment, the silica nanoparticles (<10 nm) diffuse easily through pores in the biofilm structure affecting the membrane integrity. They also inactivate the surface proteins developing into spatiotemporal aggregation patterns among bacterial population resulting in cell lysis and eventual biofilm destruction. MSN deposition within the biofilms and their subsequent action depend on the heterogeneity of the charges and electrostatic interactions across the entire biofilm structure. The cationic quantum dots were able to pass the matrix barrier and accumulate inside bacteria, whereas hydrophilic groups affect mainly the EPM surrounding the cells (Lee et al. 2016). The penetration of MSNs into the bacterial biofilm matrix depends on the size of the EPM pores, high repulsive forces between oppositely charged NPs and biofilm matrix components, hydrophobicity of the surrounding environment, and existence of chemical gradients within the EPM (extracellular polymeric matrix). Further MSN surface capping with small ligands or polymers (polysaccharides, PEG, glycolipids) enhances stability and surface functionality (Ammer et al. 2016).

The ability of mesoporous nanomaterials to penetrate the EPM makes them efficient against resistant bacterial clones within the biofilm depths (Malone et al. 2017). MSN coatings on medical devices recently proved successful in reducing bacterial colonization and biofilm formation. Techniques such as UV irradiation, ultrasound sonochemistry, and LbL assembly are used to develop MSNs incorporated with functional materials or coatings (Li et al. 2012). Recent reports of antibacterial MSNs with cationic biopolymer loadings such as aminocellulose or thiolated chitosan have proved more successful than the biopolymer itself in destroying planktonic bacteria affecting the cell membrane integrity (Li and Wang 2013). Further essential oils (EOs) are definite candidates to decrease the selection of resistant bacterial species, but are rendered inefficient due to their high hydrophobicity and volatility. To protect and preserve the effects of these active substances (EOs), the microencapsulation technique has been developed by loading these bioactive volatile substances into mesoporous silica nanoparticles, thereby converting them into strong chemosterilant compounds (Zhang et al. 2016).

Amino-decorated SiO_2 NPs were synthesized which can easily penetrate and eradicate pathogenic biofilms (*P. aeruginosa* and *E. coli)* through regulated release of bactericidal components (Merezeanu et al. 2016). The antibiotic-encapsulated MSNs (vancomycin, kanamycin) were synthesized for effective biofilm degradation of *S. aureus* (Qi et al. 2013). Further synchronized application of matrix-degrading enzymes like lysosomes with mesoporous nanomaterials has also been proposed as an alternative strategy to facilitate the easy penetration of MSNs to eliminate biofilms (Gupta and Variyar 2016).

16.7.2 Mechanisms of MSNs Against Bacteria

The mechanisms of MSNs toxicity towards bacterial biofilms are vague and have to be understood completely. Primarily the nanosilica materials adhere to the membranes of bacteria in the biofilm cloud through electrostatic interactions and disrupt the integrity of the bacterial membrane network and the entire biofilm complex eventually. The oxidative stress induction generally triggers nanotoxicity through free radical formation. Once inside the cells, the metallic MSNs, either by themselves or by the released ions, will interact with proteins or DNA or RNA molecules, affecting the vital metabolic activities in the microbes. The silica nanomaterials with preferential binding sites to phosphorus- and sulfur-containing proteins and enzymes modify their activity by generating ROS (reactive oxygen species) inducing cell death ultimately. Furthermore, mesoporous silica nanoparticles with their high surface-area-to-volume ratio help to maximize the bioavailability of the loaded antimicrobials during their exposure to microbes (Balaure et al. 2017).

16.7.3 Lysosome-Coated MSNs

Lysosome-coated MSNs are currently employed as potential antibiofilm agents due to their competent antimicrobial activity. They exhibit minimum cytotoxicity and almost insignificant hemolytic side effects under both in vitro and in vivo conditions (Fig. 16.6). Lysozyme (Lys), a natural enzyme found abundantly in mammalian secretions (tears, saliva, etc.), displays remarkable antibacterial activity by destroying the 1,4- β-linkages between N-acetyl muramic acid and *N*-acetyl- D-

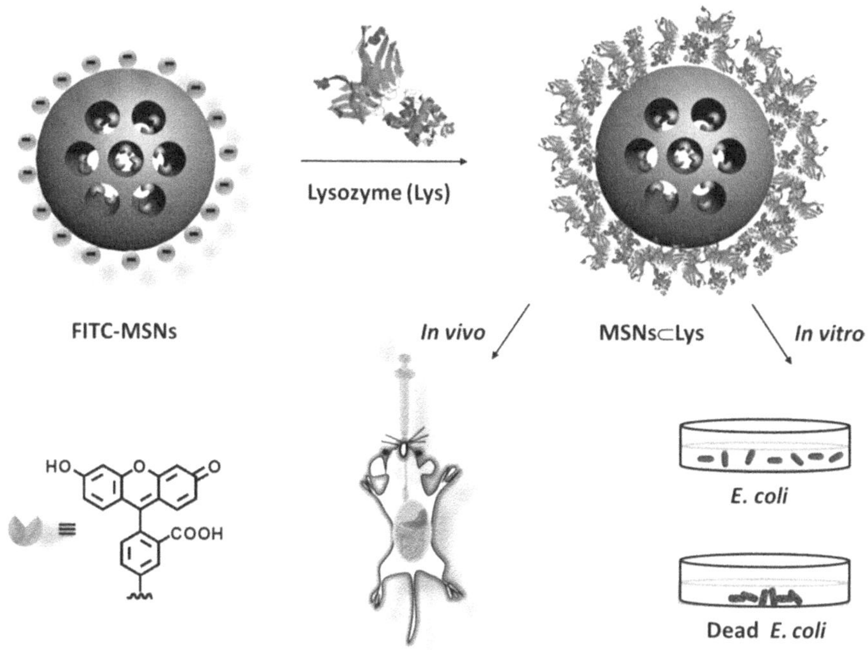

Fig. 16.6 Schematic representation of antibacterial activity of lysozyme-coated MSNs

glucosamine (NAM–NAG) residues of cell wall peptidoglycan (Song et al. 2017). However, due to their instability and weak binding affinity with peptidoglycan, this ubiquitous enzyme rendered useless for antibacterial defense.

MSNs with their wide range applications in biomedical sciences demonstrated stability and enhanced biological activity with an enzyme or protein conjugate surface immobilization. MSNs coated with lysosomes display enhanced stability of Lys with selective toxicity, thereby reducing the risk of development of resistance compared to conventional antibiotics. MSNs⊂Lys corona increases the membrane perturbation properties by enhancing the local concentration of Lys on the surface cell walls which is responsible for peptidoglycan hydrolysis.

16.8 Antibiofilm Tests

16.8.1 Initial Bacterial Adhesion

Initial bacterial adhesion test was performed using non-treated silicone and acylase-coated sheets and observed with bright field microscopy. To allow bacterial adhesion, the bacterial samples inoculated in TSB (tryptic soy broth) were incubated for 3 h at 37 °C (Malone et al. 2017).

16.8.2   Single Species Biofilm Inhibition

The biofilm inhibition activity or the total biomass reduction was evaluated using crystal violet (CV) assay. Quantification was carried out both in static and dynamic conditions, against the bacterial population. The total biomass was estimated by the amount of CV bound to each sample by measuring the absorbance at 595 nm (Gurunathan et al. 2014).

16.8.3   Mixed-Species Biofilm Inhibition

The biofilm inhibition potential for mixed species biofilms was studied using a combination of the two enzymes (acylase and amylase) into a hybrid multifunctional coating system, and the CV assay was used to measure the total biofilm mass (Gurunathan et al. 2014).

16.8.4   Bacterial Viability in the Biofilms

Single- and multiple-species biofilms were cultured with acylase on hybrid silicone samples using a mixture of two dyes, namely green fluorescent Syto 9 and red-fluorescent propidium iodide (1:1), and were subjected to analysis for bacterial viability with Live/Dead® BacLight™ kit (Coenye and Nelis 2010).

16.8.5   Biocompatibility Tests

The biocompatibility of the silicone samples coated with enzymes (acylase/amylase and both) was studied using human foreskin fibroblasts cell lines and cell viability was determined using alamarBlue® assay (Gurunathan et al. 2014).

16.8.6   MSN Effect on the Protein Leakage in Bacteria

Effect of MSNs on protein leakage was studied by suspending the bacterial cultures in phosphate-buffered saline (PBS) and incubated for 4 h at 37 °C in a shaking incubator (200 rpm). After incubation, the cultures were centrifuged and the supernatant was checked for the amount of protein leakage by Folin-Lowry method.

16.8.7 Membrane Fluidity Assay

Membrane fluidity assay is carried out using a fluorescent probe DPH (1,6-dipheny l-1,3,5-hexatriene). Bacterial cultures with MSNs were suspended in PSB and incubated for 90 min at 37 °C. After incubation, the cultures are centrifuged and resuspended in 5 μM DPH solution and incubated in dark for 1 h at 37 °C. The cells were washed thoroughly to eliminate excess DPH and were finally suspended in PBS. Bacterial sample without MSNs was used as a control, and sodium dodecyl sulfate (SDS) known for membrane damage is taken as a positive control. Fluorescence spectrophotometry was used to measure the fluorescence and the polarization index was later calculated (Sen Karaman et al. 2016).

SEM analysis reveals bacterial cell membrane damage and intracellular protein leakage reflects a significant degree of antimicrobial activity. According to Coenye and Nelis (2010), SEM observation has revealed the alteration of the majority of bacterial cells into elongated filamentous cells when treated with MSNs which was predicted to be one of the defensive mechanisms of microbes against antibiotics and harsh environmental conditions.

16.8.8 TCP Assay

Tissue culture plate (TCP) method was used to determine the biofilm activity based on the incorporation of the crystal violet by sessile cells through colorimetric measurements. Biofilm activity and the inhibition percentage can be calculated by [1 − (A595 of cells treated with MSNs/A595 of non-treated control cells)] × 100 (Kong et al. 2011), and the colony-forming unit was calculated by multiplying viable colonies with the dilution factor and expressed as CFU mL^{-1} (Chun Xu et al. 2018).

16.9 Conclusions

Recent developments in nanotechnology have led to the synthesis of efficient nanomaterials loaded with available antimicrobials with improved functionality. The nanomaterials exhibit unique properties, owing to their large surface area/volume and differ from those of their free molecules and bulk compositions. MSNs with their unique properties of size, shape, pore physiology, and surface chemistry are considered as excellent antibacterial and antibiofilm agents. The interaction of antimicrobials with the MSNs damages the bacterial cell membrane resulting in intracellular protein leakage. Enhanced antibacterial and antibiofilm effects were noted when MSNs were used synergistically with other prevalent antibiotics, enzymes, and other bioactive molecules. MSNs have proved to be efficient drug vehicles in delivering unstable, hydrophobic, volatile essential oils as potential antimicrobial compounds. They significantly improved the compounds' antimicrobial activity, thereby decreasing the opportunity for natural drug resistance to arise. Further the delivery platform could also be potentially extended to conventional biocides and other traditional antimicrobial agents with directed and controlled release of drugs to the target microbes.

Hence, this book chapter provides a novel approach into the antimicrobial effectiveness of MSNs, which holds promise for the advancement of future generation antibiotics with non-toxicity and supple design options. Currently antibiotic modifications at nanoscale can be considered as uncomplicated methods to progress the management of severe infections and are still practical alternatives to reduce the resource and time-consuming selection procedures for new drugs.

References

Albanese A, Tang PS, Chan WCW (2012) The effect of nanoparticle size, shape, and surface chemistry on biological systems. Annu Rev Biomed Eng 14:1–16

Allahverdiyev AM, Kon KV, Abamor ES, Bagirova M, Rafailovich M (2011) Coping with antibiotic resistance: combining nanoparticles with antibiotics and other antimicrobial agents. Expert Rev Anti-Infect Ther 9:1035–1052

Ammer MR, Zaman S, Khalid M, Bilal M, Erum S, Huang D, Che S (2016) Optimization of antibacterial activity of Eucalyptus tereticornis leaf extracts against Escherichia coli through response surface methodology. J Radiat Res Appl Sci 9(4):376–385

An N, Lin H, Yang C, Zhang T, Tong R, Chen Y, Qu F (2016) Gated magnetic mesoporous silica nanoparticles for intracellular enzyme-triggered drug delivery. Mater Sci Eng 69:292–300

Ashraf MA, Khan AM, Ahmad M, Sarfraz M (2015) Effectiveness of silica-based sol-gel microencapsulation method for odorants and flavors leading to sustainable environment. Front Chem 3:42

Aziz N, Fatma T, Varma A, Prasad R (2014) Biogenic synthesis of silver nanoparticles using Scenedesmus abundans and evaluation of their antibacterial activity. J Nanopart 2014:689419. https://doi.org/10.1155/2014/689419

Aziz N, Faraz M, Pandey R, Sakir M, Fatma T, Varma A, Barman I, Prasad R (2015) Facile algae-derived route to biogenic silver nanoparticles: synthesis, antibacterial and photocatalytic properties. Langmuir 31:11605–11612. https://doi.org/10.1021/acs.langmuir.5b03081

Aziz N, Pandey R, Barman I, Prasad R (2016) Leveraging the attributes of Mucor hiemalis-derived silver nanoparticles for a synergistic broad-spectrum antimicrobial platform. Front Microbiol 7:1984. https://doi.org/10.3389/fmicb.2016.01984

Aziz N, Faraz M, Sherwani MA, Fatma T, Prasad R (2019) Illuminating the anticancerous efficacy of a new fungal chassis for silver nanoparticle synthesis. Front Chem 7:65. https://doi.org/10.3389/fchem.2019.00065

Bagwe RP, Hilliard LR, Tan W (2006) Surface modification of silica nanoparticles to reduce aggregation and nonspecific binding. Langmuir 22:4357–4362

Balaure PC, Boarca B, Popescu RC, Savu D, Trusca R, Vasile BȘ, Grumezescu AM, Holban AM, Bolocan A, Andronescu E (2017) Bioactive mesoporous silica nanostructures with antimicrobial and anti-biofilm properties. Int J Pharm 531:35–46

Balouiri M, Sadiki M, Ibnsouda SK (2016) Methods for in vitro evaluating antimicrobial activity: a review. J Pharm Anal 6(2):71–79

Bayir S, Barras A, Boukherroub R, Szunerits S, Raehm L, Richeter S, Durand J (2018) Mesoporous silica nanoparticles for recent photodynamic therapy applications. Photochem Photobiol Sci 17(11):1651–1674. https://doi.org/10.1039/C8PP00143J

Bein KMT (2017) Talented mesoporous silica nanoparticles. Chem Mater 29(1):371–388

Bobo D, Robinson KJ, Islam J, Thurecht KJ, Corrie SR (2016) Nanoparticle-based medicines: a review of FDA-approved materials and clinical trials to date. Pharm Res 33:2373–2387

Chan AC, Cadena MB, Townley HE, Fricker MD, Thompson IP (2016) Effective delivery of volatile biocides employing mesoporous silicates for treating biofilms. J R Soc Interface 14:20160650

Chen CY, Burkett SL, Li HX, Davis ME (1993) Ordered mesoporous and macro porous inorganic film and membrane. Microporous Mater 2:27–34

Cheng G, Dai M, Ahmed S, Hao H, Wang X, Yuan Z (2016) Antimicrobial drugs in fighting against antimicrobial resistance. Front Microbiol 7(470):1–11

Cicily J (2017) Biomedical applications of mesoporous silica particles. University of Iowa, PhD thesis, p 2017

Coenye T, Nelis HJ (2010) In vitro and in vivo model systems to study microbial biofilm formation. J Microbiol Methods 83(2):89–105

Fruijtier-Pölloth C (2012) The toxicological mode of action and the safety of synthetic amorphous silica-a nanostructured material. Toxicology 294:61–79

Grumezescu AM, Andronescu E, Ficai A, Voicu G, Cocos O, Chifiriuc MC (2013) Eugenia caryophyllata essential oil- SiO_2 biohybrid structure for the potentiation of antibiotics' activity. Rom J Mater 43(2):160–166

Gupta S, Variyar PS (2016) Nanoencapsulation of essential oils for sustained release: application as therapeutics and antimicrobials. In: Mihai Grumezescu A (ed) Nanotechnology in the agri-food industry. Academic, Amsterdam, pp 641–672. (Encapsulations.)

Gurunathan S, Han JW, Deug-Nam KW, Kim J-H (2014) Enhanced antibacterial and anti-biofilm activities of silver nanoparticles against Gram-negative and Gram-positive bacteria. Nanoscale Res Lett 9:373

Hom C, Lu J, Liong M, Luo H, Li Z, Zink JI, Tamanoi F (2010) Mesoporous silica nanoparticles facilitate delivery of siRNA to shutdown signaling pathways in mammalian cells. Small 6:1185–1190

Huang X, Young NP, Townley HE (2013) Characterization and comparison of mesoporous silica particles for optimized drug delivery. Nanomater Nanotechnol 4:1. https://doi.org/10.5772/58290

Hudson SP, Padera RF, Langer R, Kohane DS (2008) The biocompatibility of mesoporous silicates. Biomaterials 29:4045–4055

Ispas C, Sokolov I, Andreescu S (2009) Enzyme functionalized mesoporous silica for bioanalytical applications. Anal Bioanal Chem 393:543–554

Jorge J, Verelst M, de Castro GR, Martines MAU (2016) Synthesis parameters for control of mesoporous silica nanoparticles (MSNs). Biointerface Res Appl Chem 6(5):1520–1524

Juan L. Paris, M. Victoria Cabañas, Miguel Manzano, María Vallet-Regí (2015) Polymer-Grafted mesoporous silica nanoparticles as ultraso/cund-responsive drug carriers ACS nano 9(11):11023–11033

Kar S (2016) Development of nano mullite based mesoporous silica biocer with incorporated bacteria for arsenic remediation. Ceram Silik 60(3):1–10

Kong B, Seog JH, Graham LM, Lee SB (2011) Experimental considerations on the cytotoxicity of nanoparticles. Nanomedicine 6(5):929–941

Kavanagh F (1992) Analytical microbiology-II. Academic, New York, pp 241–243

Lakshmi P, Sravani K, Swetha MK, Jhansi Rani S, Kollu P, Parine NR, Murali Krishna R, Pammi S (2017) *Diospyros assimilis* root extract assisted bio synthesized silver nanoparticles and their evaluation of antimicrobial activity. IETnanotechnology 12(2):133–137

Lee B-Y, Li Z, Clemens DL, Dillon BJ, Hwang AA, Zink JI (2016) Redox-triggered release of moxifloxacin from mesoporous silica nanoparticles functionalized with disulfide snap tops enhances efficacy against pneumonic tularemia in mice. Small 12(27):3690–3702

Li L, Wang H (2013) Enzyme-coated mesoporous silica nanoparticles as efficient antibacterial agents in vivo. Adv Healthc Mater 2(10):1351–1360

Li Z, Jonathan CB, Aleksandr Bosoy J, Stoddart F, Jeffrey IZ (2012) Mesoporous silica nanoparticles in biomedical applications. Chem Soc Rev 41:2590–2605

Malone M, Goeres DM, Gosbell I, Vickery K, Jensen S, Stoodley P (2017) Approaches to biofilm-associated infections: the need for standardized and relevant biofilm methods for clinical applications. Expert Rev Anti-Infect Ther 15(2):147–156

Merezeanu N, Gheorghe I, Popa M, Chifiriuc MC, Lazar V, Pantea O, Banu O, Bolocan A, Grigore R, Beresteanu VS (2016) Virulence and resistance features of Pseudomonas aeruginosa strains isolated from patients with cardiovascular diseases. Biointerface Res Appl Chem 6(2):1117–1121

Moller K, Kobler J, Bein T (2007) Colloidal suspensions of nanometer-sized mesoporous silica. Adv Funct Mater 17(4):605–612

Obeidat M, Shatnawi M, Al-alawi M, Al-Zubi E, Al-Dmoor H, Al-Qudah M, El-Qudah J, Otri I et al (2012) Antimicrobial activity of crude extracts of some plant leaves. Res J Microbiol 7:59–67

Paul Catalin Balaure, Bianca Boarca, Roxana Cristina Popescu, Diana Savu, Roxana Trusca, Bogdan Stefan Vasile, Alexandru Mihai Grumezescu, Alina Holban, Alexandra Bolocan, Ecaterina Andronescu (2017) Bioactive mesoporous silica nanostructures with anti-microbial and anti-biofilm properties. Int J Pharm 531:35–46

Percival SL, Bowler R, Woods EJ (2008) Assessing the effect of an antimicrobial wound dressing on biofilms. Wound Repair Regen 16:52–57

Portin L (2012) Layer by layer assembly of the polyelectrolyte on mesoporous silicon. In: Biosciences. University of Eastern Finland, Finland, pp 1–59

Prasad R, Pandey R, Barman I (2016) Engineering tailored nanoparticles with microbes: quo vadis. WIREs Nanomed Nanobiotechnol 8:316–330. https://doi.org/10.1002/wnan.1363

Prasad R, Pandey R, Varma A, Barman I (2017) Polymer based nanoparticles for drug delivery systems and cancer therapeutics. In: Kharkwal H, Janaswamy S (eds) Natural polymers for drug delivery. CAB International, UK, pp 53–70

Qi G, Li L, Yu F, Wang H (2013) Vancomycin-modified mesoporous silica nanoparticles for selective recognition and killing of pathogenic Gram-positive bacteria over macrophage-like cells. ACS Appl Mater Interfaces 5:10874–10881

Reema N, Usha YN, Ashok MR, Garg S (2018) Mesoporous silica nanoparticles: a comprehensive review on synthesis and recent advances. Pharmaceutics 10(3):118. https://doi.org/10.3390/pharmaceutics10030118

Renwick MJ, Brogan DM, Mossialos E (2016) A systematic review and critical assessment of incentive strategies for discovery and development of novel antibiotics. J Antibiot (Tokyo) 69(2):73–88

Sahoo B, Devi KS, Dutta S, Maiti TK, Pramanik P, Dhara D (2015) Biocompatible mesoporous silica-coated superparamagnetic manganese ferrite nanoparticles for targeted drug delivery and MR imaging applications. Nanomedicine 11(2):313–327

Sakai-Kato K, Hasegawa T, Takaoka A, Kato M, Toyo'oka T, Utsunomiya-Tate N (2011) Controlled structure and properties of silicate nanoparticle networks for incorporation of biosystem components. Nanotechnology 22:205–702

Saladino ML, Rubino S, Colomba P, Girasolo MA, Chillura Martino DF, Demirbag C, Caponetti E (2016) Pt(II) complex @mesoporous silica: preparation, characterization and study of release. Biointerface Res Appl Chem 6(6):1621–1626

Sen Karaman D, Sarwar S, Desai D, Björk E, Magnus Odén P, Chakrabarti J, Rosenholm M, Chakraborti S (2016) Shape engineering boosts antibacterial activity of chitosan-coated mesoporous silica nanoparticle doped with silver: a mechanistic investigation. J Mater Chem 4(19):3292–3304

Sharmila Devi S, Shanmuga Priya S, Sujitha MV (2016) Synthesis of mesoporous silica nanoparticles and drug loading for Gram-positive and Gram-negative bacteria. Int J Pharm Pharm Sci 8(5):196–201

Song Y, Li Y, Qien X, Liu Z (2017) Mesoporous silica nanoparticles for stimuli-responsive controlled drug delivery: advances, challenges, and outlook. Int J Nanomed 12:87–110

Sousa FL, Santos M, Rocha SM, Trindade T (2014) Encapsulation of essential oils in SiO2 micro-capsules and release behaviour of volatile compounds. J Microencapsul 31(7):627–635

Spataru CI, Ianchis R, Petcu C, Nistor CL, Purcar V, Trica B, Nitu SG, Somoghi R, Alexandrescu E, Oancea F, Donescu D (2016) Synthesis of non-toxic silica particles stabilized by molecular complex oleic-acid/sodium oleate. Int J Mol Sci 17(11):19–36

Tarn D, Ashley CE, Xue M, Carnes EC, Zink JI, Brinker CJ (2013) Mesoporous silica nanoparticle nanocarriers: biofunctionality and biocompatibility. Acc Chem Res 46:792–801

Tozuka Y, Wongmekiat A, Kimura K, Moribe K, Yamamura S, Yamamoto K (2005) Effect of pore size of FSM-16 on the entrapment of Flurbiprofen in mesoporous structures. Chem Pharm Bull 53:974–977

Trewyn GB, Giri S, Igor IS, Victor SYL (2007) Mesoporous silica nanoparticle based controlled release, drug delivery, and biosensor systems. Chem Commun 2007:3236–3245

Ventola CL (2015) The antibiotic resistance crisis: part 1: causes and threats. Pharm Ther 40(4):277–283

Wang F, Guo C, Yang L-R, Liu CZ (2010) Magnetic mesoporous silica nanoparticles: fabrication and their laccase immobilization performance. Bioresour Technol 101:8931–8935

Wang Y, Zhao Q, Han N, Bai L, Li J, Liu J, Che E, Hu L, Zhang Q, Jiang T, Wang S (2015) Mesoporous silica nanoparticles in drug delivery and biomedical applications. Nanomed Nanotechnol Biol Med Mol Med 11(2):313–327

Webster TJ, Seil I (2012) Antimicrobial applications of nanotechnology: methods and literature. Int J Nanomedicine 7:2767–2781

Worthington RJ, Richards JJ, Melander C (2012) Small molecule control of bacterial biofilms. Org Biomol Chem 10(37):7457

Wu W, Ye C, Xiao H, Sun X, Qu W, Li X, Chen M, Li J (2016) Hierarchical mesoporous silica nanoparticles for tailorable drug release. Int J Pharm 511(1):65–72

Xia TA, Kovochich M, Liong M, Meng H, Kabehie S, George S, Zink JI, Nel AE (2009) Polyethyleneimine coating enhances the cellular uptake of mesoporous silica nanoparticles and allows safe delivery of siRNA and DNA constructs. ACS Nano 3:3273–3286

Xu C, Hea Y, Lia Z, Yusilawatic AN, Ye Q (2018) Nanoengineered hollow mesoporous silica nanoparticles for the delivery of anteimicrobial protein into biofilm. J Mater Chem B 6:1899–1902

Zhang Y, Yuan Q, Chen T, Zhang X, Tan YCW (2012) DNA-capped mesoporous silica nanoparticles as an ion-responsive release system to determine the presence of mercury in aqueous solutions. Anal Chem 84(4):1956–1962

Zhang Y, Liu X, Wang Y, Jiang P, Quek SY (2016) Antibacterial activity and mechanism of cinnamon essential oil against Escherichia coli and Staphylococcus aureus. Food Control 59:282–289

Zhao Q, Han N, Bai L, Li J, Liu J, Che E, Hu L, Zhang Q, Jiang T, Wang S (2017) Mesoporous silica nanoparticles in drug delivery and biomedical applications. Nanomedicine 11(2):313–327

Chapter 17
Recent Trends in Antimicrobial or Biofilms with Advanced Specificity at Gene Level Treatment

Bojjibabu Chidipi, Samuel Ignatious Bolleddu, Achanta Jagadeesh, and Alalvala Mattareddy

Contents

Abstract Antimicrobials are lifesaving molecules helpful in eliminating different kinds of infections in the past 100 years and are still having prominent role in the medical practice. The broad-spectrum antibiotics are important even though they can cause a lot of side effects. The future of antimicrobials lies on the specific targeting without eliminating the beneficial flora and fauna of the body. The next-generation antibiotics and biofilms show both the specificity and the broad-spectrum antimicrobial activity without affecting the good bacteria of the gut. Nucleic acid- and peptide-based antibiotics are considered as the future antibiotics or biofilms with advanced specificity at gene level.

Keywords Disinfection · Photo catalysis · Fullerenes · Transduction · Reactive oxygen species

B. Chidipi · S. I. Bolleddu
Morsani College of Medicine, University of South Florida, Tampa, FL, USA

A. Jagadeesh
College of Pharmacy, Seoul National University, Seoul, South Korea

A. Mattareddy (✉)
Department of Zoology, School of Life and Health Sciences, Adikavi Nannaya University, Rajahmundry, Andhra Pradesh, India

© Springer Nature Switzerland AG 2020
R. Prasad et al. (eds.), *Nanostructures for Antimicrobial and Antibiofilm Applications*, Nanotechnology in the Life Sciences,
https://doi.org/10.1007/978-3-030-40337-9_17

17.1 Antimicrobial Nanotechnology

Nanoparticles play a new role in fields like cosmetics, medicine, engineering, and technology. Silver nanoparticle-based drug products have an antimicrobial property with promising outcomes in the new pharmaceutical manufacturing (Aziz et al. 2015, 2016). Some studies show that liver cancer patients are successfully treated with silver nanoparticles proving their anti-carcinogenic activity; they cause programmed cell death to carcinogenic liver cells (Aziz et al. 2019). Silver nanoparticles did not show any toxic outcomes in humans, unlike animal studies, and there is inadequate information about the toxic mechanisms of silver nanoparticles in animal studies (Ahmadian et al. 2018).

Recently many infections of *Staphylococcus aureus* were identified in patients using biofilms; this bacterial infection is proved to be cured by silver nanoparticles in the experimental studies. The functional genes of *Staphylococcus* bacteria respond to silver ions breaking the cell wall followed by the death of bacteria; only those genes which code for polysaccharides respond to silver ions but not the genes encoding a protein, DNA, and sugars. That is why the latest medical devices are coated with silver nanoparticles to avoid the *Staphylococcus* infection in treated patients. Biofilms coated with silver nanoparticles have different morphology and appearance compared to standard biofilms which can be clearly identified using the scanning electron microscope (Singh et al. 2018).

Food technology is also widely implementing the use of antimicrobial nanoparticles to prevent the spread of foodborne pathogens like salmonella, which previously caused many deaths due to infections in humans and cattle (Prasad et al. 2017a). Some specific salmonella species like enteritidis cannot be easily eliminated from the human body due to their resistance to antibiotics. Peroxidase enzymes could lyse these bacterial cell walls without the antibiotics by increasing the reactive oxygen levels to destroy the vacuoles in the bacteria, so recently nanoparticle-based enzymes were produced to carry out this process. Iron oxide nanoparticles carrying the peroxidase enzyme to the bacterial cell wall can eliminate the bacteria like *Salmonella enteritidis* from the mammalian body; these iron oxide nanoparticles will deliver peroxidase enzymes to the bacterial cell wall thereby lysing the vacuoles to release lysozymes which will destroy the cell organelles thereby causing cell apoptosis (Shi et al. 2018).

Surface modification (Fig. 17.1) is the most widely implemented technique to develop nanotechnology-based medical devices; titanium nanoparticles are coated on implants and biofilms for antimicrobial activity against *Staphylococcus* and *Pseudomonas* species. The degree of roughness on the implant surface is modified to create more surface area to carry the antimicrobial drugs. These implants are bacterial resistant and will not involve in the development of any infection or lesion on the body surface. Dental implants are administered to patients with mandibular or dental fractures or sinuses; these commonly produce infection or lesions due to the presence of *Pseudomonas aeruginosa*, *Staphylococcus aureus*, and other streptococcal species growing in the human saliva (Di Salle et al. 2018).

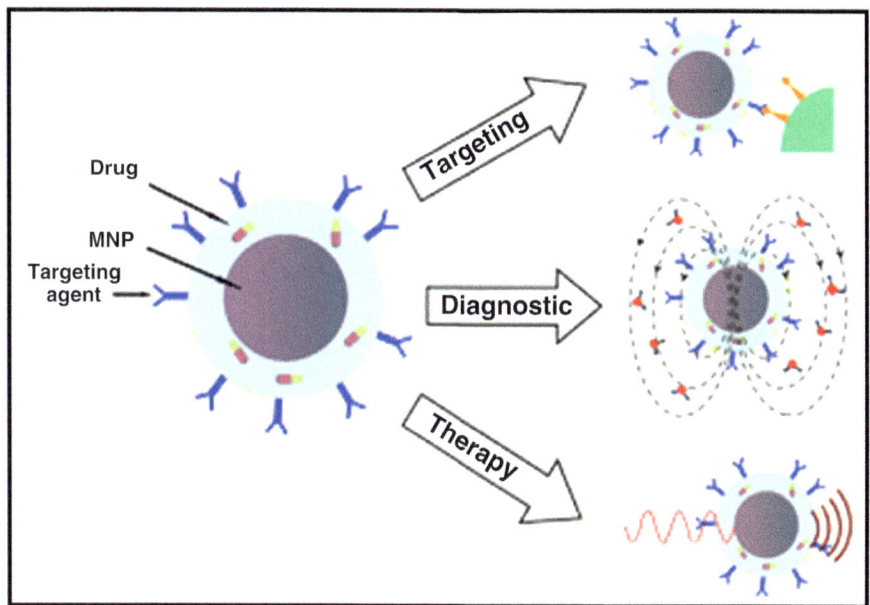

Fig. 17.1 Surface modification of nanoparticles used in antimicrobial therapy

17.2   Bacterial Resistance and Genome Sequencing

Deaths due to antibiotic resistant Gram-negative bacteria are proliferating, and treatment involves enormous medical costs. Novel therapeutics are focused on developing many antimicrobials to overcome these barriers; many investigational drugs are being developed and examined in humans during clinical practice to combat the antibacterial resistance (Avery and Nicolau 2018). The standard classes of multidrug-resistant Gram-negative bacteria are Enterobacteriaceae, Pseudomonadaceae, and Moraxellaceae family; primary pediatric infections are caused by Gram-negative bacteria like *Pseudomonas* and *Acinetobacter* species. The treatment considerations for antibiotic resistance population is different compared to the actual population (Hsu and Tamma 2014).

The novel antimicrobial therapy regimens are totally focused on combing the traditional medications with the novel antibiotics to safely reduce the toxic outcomes and resistance to the body. The traditional antibiotics have higher tolerance compared to novel antibiotics like beta-lactams, quinolones, cephalosporins, aminoglycosides, and beta-lactamase. For instance, combination therapy to overcome the Gram-negative bacterial resistance includes avibactam, cefotaxime, and phleomycin as infusion which are available to the public recently. According to some data published by the Infectious Diseases Society of America, there are about ten investigational antimicrobials against bacteria resistant to antibiotics that can be released into the market in 2020 (Bassetti et al. 2018).

Whole genome sequencing (Fig. 17.2) is a novel approach to combat the antibiotic resistance developed by bacteria for successful determination of the resistance expressing genes in the bacterial gene pool. This approach is a part of bioinformatics to identify the determinants of bacterial resistance and the relationship with the clones. The genome sequencing of *Clostridium difficile*, a known pathogen of several outbreaks of nosocomial infections facilitated to identify genes associated with pathogenesis. This became successful in identifying the genes encoding the bacterial resistance and the susceptibility of species to the various antibiotics like clindamycin, vancomycin, erythromycin, and tetracyclines. Point mutations at these homologous genes will prevent the antibacterial drug action; the multidrug resistance developed by these bacteria is mainly due to the expression of genes like *ermG*, *mefA*, and *msrD*. Many studies show how these Gram-negative bacteria share the above-mentioned genes to neighboring bacteria, especially in the gut to prevent the mechanism of antibiotics (Isidro et al. 2018).

This genetic approach will help to identify the mechanism of resistance, also the bacterial lineage in the spread of infection, which will help in targeting the bacterial genes directly without expressing the genes that promote multidrug resistance. Many opportunistic infection outbreaks in hospitalized patients are mainly caused by the *Pseudomonas aeruginosa* species, which is believed to possess spreading clones with advanced drug resistance mechanisms to drugs like carbapenem and other beta-lactamases. Whole genome sequencing helps to locate the multi-loci sequence and phylogeny analysis or evolutionary relationship of the bacterial strain

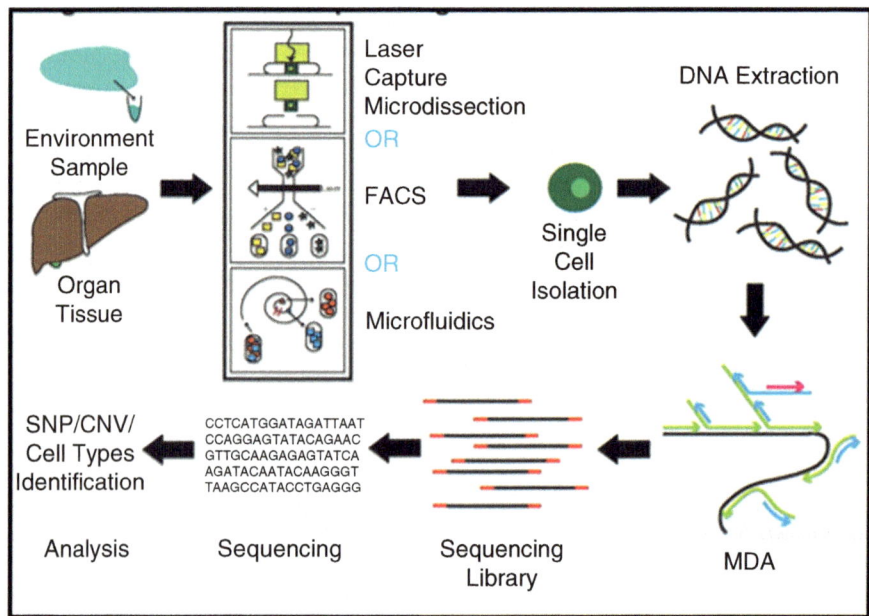

Fig. 17.2 Single cell bacterial genome sequencing

or species. The identified multidrug-resistant genes in *Pseudomonas* species are *GyrA*, *ParC*, *OprD*, *Arm*, and *Rmt* (Telling et al. 2018).

Global dissemination of hospital-acquired infections are prevalent due to an increased hospital stay of the patients for better treatment. Another set of opportunistic pathogens like *Enterococcus faecium* shows vancomycin resistance and reduced susceptibility to daptomycin treatment. The prevalence of this nosocomial infection is about 35.5% of the total hospital-acquired infections, and daptomycin is a commonly prescribed antibiotic in the USA for staphylococcal and enterococcal infections. The alterations in the genetic level to develop multidrug resistance are identified in these bacteria by gene sequencing and multi-locus typing. The data obtained by whole genome sequencing reveals mutations in *LiaR*, *LiaS*, and *Cls* genes; point mutations at any of these genes will express multidrug resistance. Comparative genome analysis of different potential harmful bacteria will help to identify the bacterial colonization and their susceptibility to antimicrobial drugs. The evolving changes in the process of developing antimicrobial resistance can be clearly depicted by this genetic analysis (Wang et al. 2018).

17.3   Biofilm Formation

Biofilms are communities of bacterial and fungal microbes that form protective matrix via adherening with the any surface. A biofilm are described as colonies of bacteria embedded in a thick, slimy barrier of sugars and proteins which can protect colonies from external factors. The surfaces of medical devices are vulnerable to the formation of biofilms, for example, contact lenses and orthopedic implants. They are significant contributors to diseases that are characterized by an underlying bacterial infection like osteomyelitis, periodontal disease, cystic fibrosis, and chronic acne, and biofilms are also found in wounds and are suspected of delaying healing in some.

Planktonic bacteria grow in microcolonies in 24 h of culturing. They can rapidly recover from their mechanical disruption and reform a mature biofilm of fewer than 1–2 days. A unique property of polymicrobial colonies is that they have many protective effects that different species of bacteria can provide to each other individually.

In their favorable conditions, antimicrobial resistant bacteria can be shielded with protective enzymes or antimicrobial binding proteins. As a result, it transfers virulent gene and proteins to the neighboring bacteria and facilitates the antimicrobial resistance. In many hospitalized patients who are unconscious or unable to take food orally and who are treated using the catheters in the veins especially in the arms, legs, groin, and chest may acquire additional blood borne infections by forming biofilms (Fig. 17.3). In the USA, more than four million catheters are inserted into patients to treat various cardiovascular ailments, in which nearly 10% of the treated population will show biofilm formation and infection susceptibility. The formation of biofilm inside the human body is a very complicated process and involves

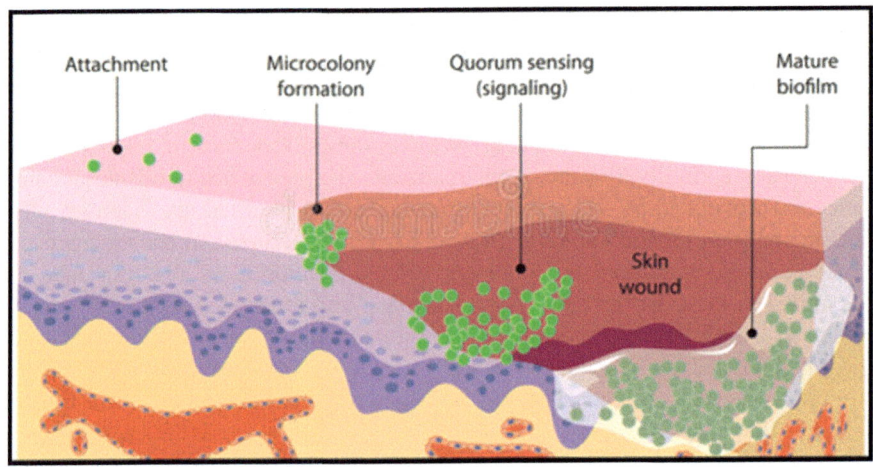

Fig. 17.3 Formation of biofilm after skin injury

different stages like immobilization and surface adherence along with intercellular interactions and microbial colony formation. The adherence of circulating microbes to the surface of the catheters is assisted by electrostatic interactions and forces between the catheter substrate and microbial cell wall leading to biofilm formation (Yousif et al. 2015).

These microbial colonies on the biofilm surface are different from the other microbes and have very complex physiology, and mechanism, and based on this, the biofilm-forming strains of bacteria have more virulence compared to the non-biofilm-forming strains. For example, infection due to *Candida albicans* is widespread worldwide with a prevalence level of 5–70% in the ICU-admitted patients receiving invasive antimicrobial or chemotherapy procedures for immunosuppressive disorders, and this increases the period of hospital stay (3–30 days) and medical expenses (40,000 USD) which proved to be a financial burden. The estimated mortality rate due to these infections is about 75% per annum globally. The pathogenicity of the candida biofilms is high and have greater virulence compared to the non-biofilm-forming strains; these biofilms are very resistant to antifungal agents and can lead to severe outcomes like bloodstream infection or candidemia if left unnoticed (Treviño-Rangel et al. 2018).

Sometimes antimicrobial biofilms will serve as a reservoir for the antibiotic resistance genes, thereby making it difficult for traditional antimicrobials to show their therapeutic effect at minimum sufficient concentrations. This is because of the extensive use of antibiotics and their accumulation in the environment as agricultural wastes as well as toxins in human body organs like kidney and liver; in addition, bacteria develop resistance by altering their genetic integrity suitable to face the antibiotic activity. These antibiotic resistance genes are created because of the extensive use of antibiotics, and the diseases caused by these pathogens are very hard to treat; nowadays, these communicable diseases are getting common throughout hospital and dining facilities. These unwanted antibiotic-resistant genes are in present in bacterial chromosomal DNA and plasmids. These genes are carried into the atmosphere by natural or manmade vectors and spread the infection along with water and air pollution (Guo et al. 2018).

The spread of infection causing antibiotic-resistant bacteria is mainly through food, water, and air pollution caused by the antibiotic resistance genes. *Salmonella typhi* is a typical example to mention about the functioning of these resistant genes, adaption to aerobic external environment from the anaerobic host environment maintaining the same pathogenicity until it is carried to another host body. This adaptivity, virulence, and biofilm forming capability are made possible by the small RNA molecules developed inside the host environment because of some mutations during the bacterial life cycle. These mutations are because of the availability of high concentrations of antibiotics in the blood in proportion to the bacterial population. These bacteria are classified into different groups based on the type of environment they choose to grow. For example, a group of bacteria called as *Salmonella* grows on chicken broth, sugar substances and dairy products. The upregulation of the biofilm-forming genes is noticed in the growth medium where there is an aerobic atmosphere. Biofilm formation by these bacteria is mainly seen on glass, steel, and plastic materials that come in contact, and maintenance of proper sanitary conditions is required to prevent possible infection by these organisms (Lamas et al. 2018).

17.4   New Trends in Biofilm Formation

The environment in which the pathogens and commensals interact play an essential role in the physiology and composition of the biofilms. The polymicrobial population embedded in the extracellular matrix (Fig. 17.4) forms the biofilm, which can generate many infections. Dental caries is a typical example for the biofilm formed on the teeth by the interaction between food and the microbial population in the oral cavity. These oral biofilms are the reservoirs for many harmful bacteria that nourish on the human food and develop their virulence accordingly; constant changing of diet or following of a fixed diet schedule combined with good oral hygiene will reduce the chances of oral cavities and infections.

The virulence potential of these microbiotas totally depends on the type of food they grow on like sugars, proteins, and fats. The microbial community is classified into the film-forming bacteria and infection-causing bacteria based on which they are involved in causing infection after forming the biofilms or vice versa. Many pharmaceutical oral products will only temporarily eliminate these populations but will not focus on disrupting their further chances to regrow forming new biofilms. New trends in treating such virulent bacteria should focus on understanding the physiology of matrix formation and developing an approach to prevent the formation of biofilms (Bowen et al. 2018).

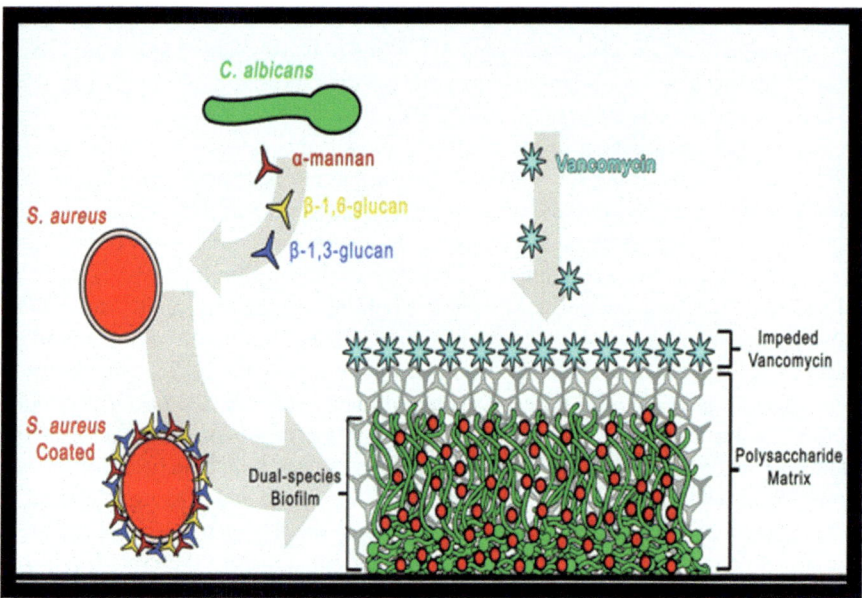

Fig. 17.4 Polysaccharide matrix during biofilm formation

17.5 Antimicrobial Drug Delivery

The small, colloidal polymeric nanoparticles are used as excellent vehicles to agents such as biomolecules and drugs. These nanoparticles, when tagged with imaging agents, offer various additional opportunities to exploit optical imaging in cancer diagnosis and guided hyperthermia therapy (Mahapatro and Singh 2011).

Biodegradable polymers PLGA and several different polymers, both synthetic and natural, have been utilized in formulating biodegradable nanoparticles (Prasad et al. 2017b). These polymers have been tested for toxicity and safety in extensive animal studies and are currently being used in humans for resorbable sutures, bone implants and screws, and contraceptive implants. These polymers are also used as graft materials for artificial organs and recently as supporting scaffolds in tissue engineering research (Panyam and Labhasetwar 2003). The commonly used material for surface modification is poly(ethylene glycol) (PEG), which is a hydrophilic, non-ionic polymer that has excellent biocompatibility. PEG attachment to the particle surface can reduce protein and enzyme adsorption on the surface and will reduce polymer degradation. The degree of protein adsorption can be reduced by modifying the density and molecular weight of polymer on the surface (Xie et al. 2017).

Drug loading into the NPs is achieved by two methods: first, by incorporating the drug at the time of NP production and, second, by adsorbing the drug after the formation of NPs by incubating them in the drug solution. In addition to the adsorption and incorporation, a new method of drug loading for the water-soluble drugs is chemical conjugation. The widely used surface-coating materials are polyethylene glycol, polyethylene oxide, poloxamers, and lauryl ethers (Soppimath et al. 2001). Synthesis and encapsulation of drugs in polymeric nanoparticles are extensively used for the nanoencapsulation of various useful bioactive molecules and medicinal drugs to develop nanomedicines. Cancer-related drugs paclitaxel, doxorubicin, cisplatin, triptorelin, dexamethasone, xanthine, etc. have been successfully encapsulated on PLGA nanoparticles (Kumari et al. 2010).

Surface functionalization of polystyrene NPs by adsorption of tunable PPE surfactants significantly reduces unspecific protein binding. PPE chemistry allows a straightforward adjustment of hydrophilicity in a homopolymer or the controlled synthesis of amphiphilic block copolymers. Structurally versatile poly(phosphoester) can be used for non-covalent surface modification of model nanocarriers providing an easier approach than covalent linkage (Muller et al. 2017). Three different methods like chemical coupling, surface activation, and coupling reaction were investigated for nanoparticle surface modification and co-incorporation of surface-modifying agents into nanoparticles. Since, the unmodified nanoparticles are anionic in nature, and mixing of these surface-modifying agents with the nanoparticle suspension could result in their ionic bonding with DMAB which optimally enhances arterial U-86 levels compared to other modifications and the unmodified nanoparticles (Song et al. 1998).

Nanoparticles have many applications in both microbiology and drug delivery using the nanoparticles (Table 17.1).Tumor uptake of PEG5000-TA-coated AuNPs could have been a result of the AuNPs' prolonged resident time in the blood and their ability to extravagate from tumor blood vessels. AuNPs coated with PEG5000-TA had the highest colloidal stability, as they did not form aggregates in PBS containing 10 mM DTT or 10% FBS as readily as AuNPs coated with PEG-SH. AuNPs in 20 nm also appeared to be excreted from the body, and 20 nm gold particles are coated with titanium terminated PEG5000 which potential drug delivery vehicles and diagnostic imaging agents (Zhang et al. 2009). SiO_2 NPs are essential materials, and the advantages of using it include low cost of production

Table 17.1 Applications of nanoparticles in microbiology and nanotechnology

Nanoparticle		Applications
Ag	⇒	Home appliances as antimicrobial agents
	⇒	Clothing for odor resistance
TiO_2	⇒	Paints and coating for antimicrobial properties
	⇒	Cosmetics as a UV absorber
Carbon	⇒	Consumer electronics
Nanotubes (CNT)	⇒	Sports equipment for light weight and durability
Fe_2O_3	⇒	Contrast agent for targeted tumor imaging
Fullerenes	⇒	Drug delivery
Fe	⇒	Environmental remediation
Au	⇒	Medical diagnostics

and having high-performance features. This is applied to manufacture materials such as chemical sensors, polishing material, food industry, and cosmetic materials (Telling et al. 2018). Nanobiopesticides like silica nanocomposites (NCs) have shown that nanosilica coated with gold, which is attached to breast carcinoma cells and exposed to the near-infrared light in vitro, show remarkable potential in killing the cancer cells (Esposito et al. 2004; Gref et al. 2000).

Polymeric NPs can be further endowed to target specific organs and tissues, overcoming certain biological barriers such as the blood-brain barrier. The lactic acid (LA) homopolymer (PLA) and its glycolic acid (GA) copolymer [poly(D,L-lactide-glycolide) (PLGA)] are among the most frequently used polymers for the carriers because of their biocompatibility, biodegradability, and drug release from the PLGA NPs which can be easily manipulated adjusting the ratio of LA to GA (Zhou et al. 2010). The linoleic acid–avidin conjugate yielded nanoparticles with enhanced ability to increase ligand density on anti-CD4-targeted nanoparticles formulated with the linoleic acid–avidin conjugate which resulted in a 5% increase in binding to CD4+ T cells. The novel avidin–linoleic acid conjugate facilitates enhanced ligand density on PLGA nanoparticles, resulting in functional enhancement of cellular targeting (Park et al. 2011).

Functionalized SiO_2 NPs with polyaniline as coupling agents were synthesized and the surface of SiO_2 NPs was treated with them. The surface of SiO_2 NPs modified by organic DAs to improve dispersion of them in nonpolar and weak polar organic solvents such as chloroform and acetone. FT-IR spectral measurements allowed us to conclude that the modified SiO_2 NPs have formed, and there is also an intermolecular interaction between the DAs and SiO_2 NPs (Mallakpour and Marefatpour 2015). To increase the therapeutic potential, targeted delivery system CAT has been developed involving direct modification or polyethylene glycol conjugation (Ahmadizadegan 2017; Saraf et al. 2013). For surface morphology, the mannose coupling is an attractive approach for targeting of CAT to macrophages, and it has a significant potential to detoxify ROS. Therefore this approach could be further explored for the treatment of liver disorders due to ROS (Ahmadizadegan 2017; Saraf et al. 2013).

17.6   Green Nanotechnology for Antimicrobial Delivery

The application and usage of antimicrobial drugs are needed to be treated safe, which is evident that usage of nanotechnology application can be safer in the delivery of the harmful drugs (Al Thaher et al. 2017). Nanoparticles have optimal bioavailability with the rapid onset of therapeutic activity (Uppal et al. 2018). The bioavailability confers the stability and non-reactiveness with the metabolism (Howick et al. 2018). The nano-pharmaceuticals have great potential in drug carrier which provide a platform to curb multidrug resistance, toxicity, target specificity, solubility, and low bioavailability (Uddin 2016). Some green chemistry polymers derived from natural polyphenols like tea can act as nanocarriers to deliver drugs at the gene level (Mahata et al. 2018).

The nanoparticles which are applied for antiseptic and disinfectant in chronic skin such as micellar nanoparticles (Jahromi et al. 2018). They are usually formulated as nanoemulsions as they have functional drug carrying capacity (<=20%) for a wide range of water solubility of antimicrobial drugs that have barriers with the systemic administration (Hörmann and Zimmer 2016). MNPs are evident that they are characterized to be the best vehicles with great potency of skin penetration with minimal adverse effects (Senapati et al. 2018). These are stable at room temperature for more than 5 years (Amodwala et al. 2017). These nanoparticles take support from skin framework and enhances the drug delivery in the stratum corneum with high therapeutic levels of drugs like acyclovir in the blood; when accompanied with nanocarriers, many antimicrobial drugs can be delivered to the bacteria at the gene level (Loh 2016) (Fig. 17.5).

The antibiotics are readily obtained in the form of multiphasic composition with supporting solvent, oil, and other surfactants (Prud'homme et al. 2017). The multiphasic composition is processed with ultra-sonicator and homogenizes with emulsifier and electric overload (Swager et al. 2018). The composition and stability in the antimicrobial nanoparticles can be analyzed by permeation technique or the HPLC (Moronshing and Subramaniam 2018). Polysporin is the best formulation with high stability and analyzed with various experiments (Nani et al. 2018).

The mode of drug delivery through MNPs are limited via transdermal delivery (Zou et al. 2016). Thus, MNPs are evident to have potential modes of delivery such as nasal, vaginal, rectal, and other parenteral routes (Carvalho et al. 2018). The other way of application of MNP technology is a nonconventional mode of delivery such as the behavior of nicotine transdermal patch and raloxifene estrogen patch which is termed as pseudo-patc or a patchless strip (Sharma and Chandra 2017).

Delivery of various pharmaceuticals such as antimicrobials at gene level are facilitated by supercritical fluids (SCF). They are highlighting the processing and the potentiality of the chemical industry for the sustainability and addressing the potential asstets for future (Fages et al. 2004). Pharmaceutical waste is measured by

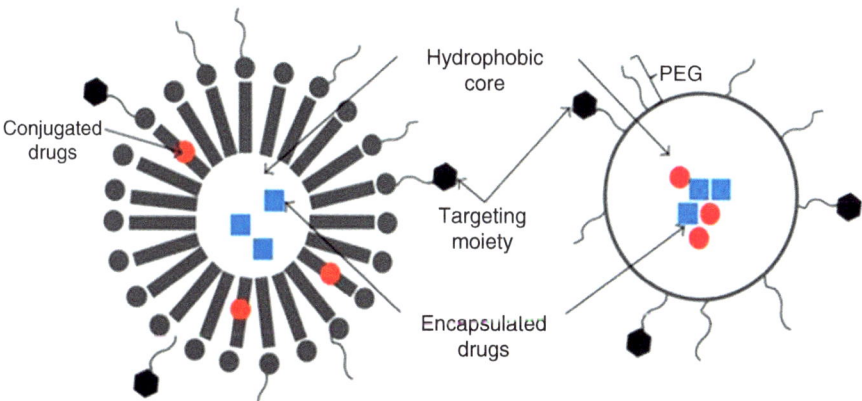

Fig. 17.5 Antimicrobial drug delivery using the micellar nanoparticles (Loh 2016)

E-factor (Kg waste per Kg product) with the help of chemical price-volume relation (Dias-Ferreira et al. 2016). As per some statistics in oil refiners, there is only 10% waste, but it is greater than 25% in pharmaceutical industries (Nickum 2018). The super critical fluids can protect environment by treating pharmaceutical waste.The purification of solid wastes with the supercritical fluids will avoid reliable incondensable waste products, thereby saving the environment (Lemasson et al. 2016). The treatment with supercritical fluids gets highlighted and various technologies are in good progress with ample research and publications (da Costa Lopes et al. 2016). The main objective of SCF is applicable in fields like micro-encapsulated particles and aerogels constituting silica particles (Bu et al. 2016). There are different techniques like impregnation and supercritical extraction through gaseous and supercritical fluids (Padrela et al. 2018).

The physical properties of SCF can be similar to any fluid which has critical pressure and temperature that can change physical state (Gorbaty and Bondarenko 2017). For example, CO_2 can behave as an SCF at a temperature of 31.1° and atmospheric pressure of 73 whereas water gains SCF properties at a temperature of 374.1° and pressure of 218.3pressure (Gandhi et al. 2017). The SCF effects on solubility and viscocity depend on temperature, and exceptions like porosity make SCF a unique from other solvents (Kumar et al. 2017). The physical properties are widely exploited in the SCF (Aliev and Abdulagatov 2017). The phase equilibrium of solid substance is more significant compared to gaseous solvents like air and it is measured by EOS (Eckert et al. 2017). The typical application of SCF extraction was processing of coffee using SCF (Hrnčič 2018).

Valencia is a plant product which have pharmaceutical and nutraceutical characters, which is also considered as SCF plant. For the extraction, aerogels are made with solvent exchange and pressurization. SCFs are widely used in material synthesis (Aliev and Abdulagatov 2017). There are different coating methods which are used globally with various coating designs (Song 2017). The various types of coating methods like basic dimensional analysis between water and Sc-CO_2 use chitosan as an encapsulant in chitosan-DMSO-CO_2 system encapsulated TiO_2 (Hao et al. 2017). Application of SCF are in impregnation of NSAIDs, polymeric matrix development and extraction of green biomass by counter current immiscible fluid extraction (Chen et al. 2019).

Super critical extraction has diverse applications in extraction of cannabinoids in medical use for controlled delivery using sorptive process and waste management and SCF in cosmetics by algae tissue engineering. Such applications are TiO_2 and Cao encapsulations in chitosan for self-heating applications (Freund et al. 2018). Pressure and temperature are the key determinants for SCF, and for rocket fuel, the warm water is not SC, so pressure and temperature make it SCF, and pre-polymer is used for encapsulation. Chinese SCF plant was cheaper compared to valencia big plants (Jovel et al. 2017). *Elsholtzia ciliata* is a bacteria whose extract acts as a supercritical fluid used in the antimicrobial delivery for infections like fever, colds, and gastric and renal disorders; this supercritical extract has both antimicrobial and antioxidant activity (Ma et al. 2018).

Bacterial colonization in body tissues create infections by forming biofilms; scaffolds that are synthetically made have porous properties that give them the ability to regenerate dead tissues by carrying antimicrobials like vancomycin that are processed using supercritical carbon dioxide foam (Garcia-Gonzalez 2018). Antimicrobial extracts of paraguariensis leaves and fruits are made using supercritical carbon dioxide, and compressed propane gas showed to reduce the minimum inhibitory concentrations of *Escherichia coli-* and *Staphylococcus aureus*-related infections (Fernandes et al. 2017). The supercritical oil extracts from citrus seeds and peels are proved to have antimicrobial properties when modified with supercritical carbon dioxide for antimicrobial by-products (Ndayishimiye et al. 2018). Herbal extracts have always proved to have antimicrobial properties as well as antioxidant activity; supercritical extracts obtained from these herbs are used in the preservation of food (Vieitez et al. 2018). The expression of the antimicrobial peptide genes will initiate the pathogenesis of periodontal diseases; the higher the expression of these genes, the more chronic the disease will be (Jourdain 2018).

The goal of drug delivery is to find a useful alternative for traditional oral drug regimens, which are worsening the viral infection to chronic disease due to lack of medication adherence and bioavailability in the patients. The multidrug resistance of bacteria is the main problem addressed by nanotechnology in drug delivery (Table 17.2). The controlled level of infection can be maintained for an extended period using the nano-formulations that can achieve targeted drug delivery to deeper tissues. The adolescent patients have increased the scope of complete recovery from the infection using nanomedicines compared to older patients. Patient medication adherence challenges and side effects can be minimized to a greater extent by adding these formulations to the drug regimen.

Nanoparticle formulation can be prepared using nanocarrier like gold or a polymer which is attached to the different combinational drug molecules on the surface of the nanoparticle. This nanoparticle formulation is administered in subcutaneous route to the HIV patient for prolonged bioavailability and extended drug release into the lymph tissues where the viral fragments are mainly located for future relapse of

Table 17.2 Drug resistance in treating biofilms

Gene	Function	Drug
P. falciparum chloroquine-resistance transporter (Pfcrt)	Reduces chloroquine levels in the digestive vacuole.	Chloroquine
P. falciparum multidrug resistance 1 (Pfmdr1)	Member of the ABC family of transporters. Associated with increased chloroquine resistance in the presence of Pfcrt	Chloroquine, Mefloquine
P. falciparum multidrug resistance protein (Pfmrp)	Transporter protein located in the parasite plasma membrane that regulates the movement of toxic metabolic products.	Quinoline
Target enzymes dhfr, dhps	4 mutations in dhfr and 2 mutations in dhps are associated with clinical resistance to antifolate combination therapies.	Sulfadoxine Pyrimethamine

infection cycle (Aziz et al. 2018). The formulation is made with cheap and reliable excipients and adjuvants to market it as a cost-effective medication for HIV (Cheng et al. 2018). Through this strategy, a new method of diagnosing and treating the virus in the lymph nodes can be possible even if the plasma viral load is zero (Bowen et al. 2018).

References

Ahmadian E et al (2018) Effect of silver nanoparticles in the induction of apoptosis on human hepatocellular carcinoma (HepG2) cell line. Mater Sci Eng C Mater Biol Appl 93:465–471

Ahmadizadegan H (2017) Surface modification of TiO2 nanoparticles with biodegradable nanocellolose and synthesis of novel polyimide/cellulose/TiO2 membrane. J Colloid Interface Sci 491:390–400

Al Thaher Y et al (2017) Nano-carrier based drug delivery systems for sustained antimicrobial agent release from orthopaedic cementous material. Adv Colloid Interface Sci 249:234–247

Aliev AM, Abdulagatov IM (2017) The study of microalgae Nannochloropsis salina fatty acid composition of the extracts using different techniques. SCF vs conventional extraction. J Mol Liquid 239:96–100

Amodwala S, Kumar P, Thakkar HP (2017) Statistically optimized fast dissolving microneedle transdermal patch of meloxicam: A patient friendly approach to manage arthritis. Eur J Pharm Sci 104:114–123

Avery LM, Nicolau DP (2018) Investigational drugs for the treatment of infections caused by multidrug-resistant Gram-negative bacteria. Expert Opin Investig Drugs 27(4):325–338

Aziz N et al. (2015) Facile algae-derived route to biogenic silver nanoparticles: Synthesis, antibacterial and photocatalytic properties. Langmuir 31:11605–11612. https://doi.org/10.1021/acs.langmuir.5b03081

Aziz N et al. (2016) Leveraging the attributes of Mucor hiemalis-derived silver nanoparticles for a synergistic broad-spectrum antimicrobial platform. Front Microbiol 7:1984. https://doi.org/10.3389/fmicb.2016.01984

Aziz M et al. (2018) Predictors of therapeutic outcome to nucleotide reverse transcriptase inhibitor in hepatitis B patients. Viral Immunol 31(9):632–638

Aziz N, Faraz M, Sherwani MA, Fatma T, Prasad R (2019) Illuminating the anticancerous efficacy of a new fungal chassis for silver nanoparticle synthesis. Front Chem 7:65. https://doi.org/10.3389/fchem.2019.00065

Bassetti M et al (2018) Antimicrobial resistance and treatment: an unmet clinical safety need. Expert Opin Drug Saf 17(7):669–680

Bowen WH et al (2018) Oral biofilms: pathogens, matrix, and Polymicrobial interactions in microenvironments. Trends Microbiol 26(3):229–242

Bu X et al (2016) Chiral analysis of poor UV absorbing pharmaceuticals by supercritical fluid chromatography-charged aerosol detection. J Supercrit Fluid 116:20–25

Carvalho AM et al (2018) Redox-responsive micellar nanoparticles from glycosaminoglycans for CD44 targeted drug delivery. Biomacromolecules 19(7):2991–2999

Chen A et al (2019) Fabrication of superrepellent microstructured polypropylene/graphene surfaces with enhanced wear resistance. J Mater Sci 54(5):3914–3926

Cheng YW et al (2018) The histone deacetylase inhibitor panobinostat exerts anticancer effects on esophageal squamous cell carcinoma cells by inducing cell cycle arrest. Cell Biochem Funct 36(8):398–407

da Costa Lopes AM et al (2016) Extraction and purification of phenolic compounds from lignocellulosic biomass assisted by ionic liquid, polymeric resins, and supercritical CO2. ACS Sustainable Chem Eng 4(6):3357–3367

Di Salle A et al (2018) Effects of various prophylactic procedures on titanium surfaces and biofilm formation. J Periodontal Implant Sci 48(6):373–382

Dias-Ferreira C et al (2016) Practices of pharmaceutical waste generation and discarding in households across Portugal. Waste Manag Res 34(10):1006–1013

Eckert CA, Brennecke JF, Ekart MP (2017) Molecular analysis of phase equilibria in supercritical fluids. In: Bruno TJ, Ely JF (eds) Supercritical fluid technology (1991). CRC, Boca Raton, FL, pp 163–192

Esposito K et al (2004) Effect of lifestyle changes on erectile dysfunction in obese men: a randomized controlled trial. JAMA 291(24):2978–2984

Fages J et al (2004) Particle generation for pharmaceutical applications using supercritical fluid technology. Powder Technnol 141(3):219–226

Fernandes CE et al (2017) Phytochemical profile, antioxidant and antimicrobial activity of extracts obtained from erva-mate (Ilex paraguariensis) fruit using compressed propane and supercritical CO2. J Food Sci Technol 54(1):98–104

Freund R et al (2018) Multifunctional efficiency: extending the concept of atom economy to functional nanomaterials. ACS Nano 12(3):2094–2105

Gandhi K, Arora S, Kumar A (2017) Industrial applications of supercritical fluid extraction: a review. Int J Chem Stud 5(3):336–340

Garcia-Gonzalez CA et al (2018) Antimicrobial properties and osteogenicity of vancomycin-loaded synthetic scaffolds obtained by supercritical foaming. ACS Appl Mater Interfaces 10(4):3349–3360

Gorbaty Y, Bondarenko GV (2017) Transition of liquid water to the supercritical state. J Mol Liquid 239:5–9

Gref R et al (2000) 'Stealth'corona-core nanoparticles surface modified by polyethylene glycol (PEG): influences of the corona (PEG chain length and surface density) and of the core composition on phagocytic uptake and plasma protein adsorption. Colloids Surf B: Biointerfaces 18(3–4):301–313

Guo X-p et al (2018) Biofilms as a sink for antibiotic resistance genes (ARGs) in the Yangtze Estuary. Water Res 129:277–286

Hao J et al (2017) Encapsulation of the flavonoid quercetin with chitosan-coated nano-liposomes. LWT - Food Sci Technol 85:37–44

Hörmann K, Zimmer AJJoCR (2016) Drug delivery and drug targeting with parenteral lipid nano-emulsions—a review. J Control Release 223:85–98

Howick K et al (2018) Sustained-release multiparticulates for oral delivery of a novel peptidic ghrelin agonist: formulation design and in vitro characterization. Int J Pharm 536(1):63–72

Hrnčič MK et al (2018) Application of supercritical and subcritical fluids in food processing. Food Qual Saf 2(2):59–67

Hsu AJ, Tamma PD (2014) Treatment of multidrug-resistant Gram-negative infections in children. Clin Infect Dis 58(10):1439–1448

Isidro J et al (2018) Genomic study of a Clostridium difficile multidrug resistant outbreak-related clone reveals novel determinants of resistance. Front Microbiol 9:2994

Jahromi MAM et al (2018) Nanomedicine and advanced technologies for burns: preventing infection and facilitating wound healing. Adv Drug Deliv Rev 123:33–64

Jourdain ML et al (2018) Antimicrobial peptide gene expression in periodontitis patients: a pilot study. J Clin Periodontol 45(5):524–537

Jovel DR, Walker ML, Yang V (2017) Research capability in combustion and propulsion at the Georgia Institute of Technology. In 53rd AIAA/SAE/ASEE joint propulsion conference. 2017

Kumar SJ et al (2017) Sustainable green solvents and techniques for lipid extraction from microalgae: a review. Algal Res 21:138–147

Kumari A, Yadav SK, Yadav SC (2010) Biodegradable polymeric nanoparticles based drug delivery systems. Colloids Surf B Biointerfaces 75(1):1–18

Lamas A et al (2018) Influence of milk, chicken residues and oxygen levels on biofilm formation on stainless steel, gene expression and small RNAs in Salmonella enterica. Food Control 90:1–9

Lemasson E, Bertin S, West C (2016) Use and practice of achiral and chiral supercritical fluid chromatography in pharmaceutical analysis and purification. J Sep Sci 39(1):212–233

Loh XJ et al (2016) Utilising inorganic nanocarriers for gene delivery. Biomater Sci 4(1):70–86

Ma J et al (2018) Composition, Antimicrobial and Antioxidant Activity of Supercritical Fluid Extract of Elsholtzia ciliata. J Essential Oil Bear Plant 21(2):556–562

Mahapatro A, Singh DK (2011) Biodegradable nanoparticles are excellent vehicle for site directed in-vivo delivery of drugs and vaccines. J Nanobiotechnology 9:55

Mahata D et al (2018) Self-assembled tea tannin graft copolymer as nanocarriers for antimicrobial drug delivery and wound healing activity. J Nanosci Nanotechnol 18(4):2361–2369

Mallakpour S, Marefatpour F (2015) An effective and environmentally friendly method for surface modification of amorphous silica nanoparticles by biodegradable diacids derived from different amino acids. Synth React Inorgan Metal-Organ and Nano-Metal Chem 45(3):376–380

Moronshing M, Subramaniam C (2018) Room temperature, multi-phasic detection of explosives and volatile organic compounds using thermodiffusion driven Soret colloids. ACS Sustainable Chem Eng 6:7

Muller J et al (2017) Coating nanoparticles with tunable surfactants facilitates control over the protein corona. Biomaterials 115:1–8

Nani M et al (2018) Evaluation and comparison of wound healing properties of an ointment (AlpaWash) containing Brazilian micronized propolis and *Peucedanum ostruthium* leaf extract in skin ulcer in rats. Int J Pharm Compd 22(2):154–163

Ndayishimiye J et al (2018) Antioxidant and antimicrobial activity of oils obtained from a mixture of citrus by-products using a modified supercritical carbon dioxide. J Ind Eng Chem 57:339–348

Nickum JE (2018) Et al. Pharmaceuticals in the environment 43(3):336–348

Padrela L et al (2018) Supercritical carbon dioxide-based technologies for the production of drug nanoparticles/nanocrystals–A comprehensive review. Adv Drug Deliv Rev 131:22–78

Panyam J, Labhasetwar V (2003) Biodegradable nanoparticles for drug and gene delivery to cells and tissue. Adv Drug Deliv Rev 55(3):329–347

Park J et al (2011) Enhancement of surface ligand display on PLGA nanoparticles with amphiphilic ligand conjugates. J Control Release 156(1):109–115

Prasad R, Kumar V, Kumar M (2017a) Nanotechnology: Food and Environmental Paradigm. Springer Nature Singapore Pte Ltd. ISBN 978-981-10-4678-0

Prasad R, Pandey R, Varma A, Barman I (2017b) Polymer based nanoparticles for drug delivery systems and cancer therapeutics. In: Natural Polymers for Drug Delivery (eds. Kharkwal H and Janaswamy S), CAB International, UK. 53–70

Prud'homme RK et al (2017) Polymer nanoparticles. Google Patents

Saraf A, Dasani A, Soni A (2013) Targeted delivery of catalase to macrophages by using surface modification of biodegradable nanoparticle. IJPSR 4(8):2221–2229

Senapati S et al (2018) Controlled drug delivery vehicles for cancer treatment and their performance. Signal Transduct Target Ther 3(1):7

Sharma M, Chandra SBJJCM (2017) The critical role of estrogen in menopausal osteoporosis menapoz Osteoporozunda Östrojenin Kritik Rolü. J Contemp Med 7(4):418–427

Shi S et al (2018) Iron oxide nanozyme suppresses intracellular Salmonella Enteritidis growth and alleviates infection in vivo. Theranostics 8(22):6149–6162

Singh N, Rajwade J, Paknikar KM (2018) Transcriptome analysis of silver nanoparticles treated Staphylococcus aureus reveals potential targets for biofilm inhibition. Colloids Surf B Biointerfaces 175:487–497

Song CX et al (1998) Arterial uptake of biodegradable nanoparticles for intravascular local drug delivery: results with an acute dog model. J Control Release 54(2):201–211

Song M et al (2017) Binary superlattice design by controlling DNA-mediated interactions. Langmuir 34(3):991–998

Soppimath KS et al (2001) Biodegradable polymeric nanoparticles as drug delivery devices. J Control Release 70(1–2):1–20

Swager TM et al (2018) Compositions and methods for arranging colloid phases. Google Patents

Telling K et al (2018) Multidrug resistant Pseudomonas aeruginosa in Estonian hospitals. BMC Infect Dis 18(1):513

Treviño-Rangel RdJ et al (2018) Association between Candida biofilm-forming bloodstream isolates and the clinical evolution in patients with candidemia: An observational nine-year single center study in Mexico. Rev Iberoam Micol 35(1):11–16

Uddin I et al (2016) Nanomaterials in the pharmaceuticals: Occurrence, behaviour and applications. Curr Pharm Des 22(11):1472–1484

Uppal S et al (2018) Nanoparticulate-based drug delivery systems for small molecule anti-diabetic drugs: an emerging paradigm for effective therapy. Acta Biomater 81:20–42

Vieitez I et al (2018) Antioxidant and antibacterial activity of different extracts from herbs obtained by maceration or supercritical technology. J Supercrit Fluid 133:58–64

Wang G et al (2018) Evolution and mutations predisposing to daptomycin resistance in vancomycin-resistant Enterococcus faecium ST736 strains. PLoS One 13(12):e0209785

Xie Z et al (2017) Immune cell-mediated biodegradable theranostic nanoparticles for melanoma targeting and drug delivery. Small 13(10)

Yousif A, Jamal MA, Raad I (2015) Biofilm-based central line-associated bloodstream infections. Adv Exp Med Biol 830:157–179

Zhang GD et al (2009) Influence of anchoring ligands and particle size on the colloidal stability and in vivo biodistribution of polyethylene glycol-coated gold nanoparticles in tumor-xenografted mice. Biomaterials 30(10):1928–1936

Zhou J et al (2010) Folic acid modified poly(lactide-co-glycolide) nanoparticles, layer-by-layer surface engineered for targeted delivery. Macromol Chem Phys 211(4):404–411

Zou Y et al (2016) Self-crosslinkable and intracellularly decrosslinkable biodegradable micellar nanoparticles: a robust, simple and multifunctional nanoplatform for high-efficiency targeted cancer chemotherapy. J Control Release 244:326–335

Chapter 18
Enterococcal Infections, Drug Resistance, and Application of Nanotechnology

Abhijit Banik, Suman Kumar Halder, Chandradipa Ghosh, and Keshab Chandra Mondal

Contents

A. Banik
Department of Microbiology, Vidyasagar University, Midnapore, West Bengal, India

Department of Human Physiology with Community Health, Vidyasagar University, Midnapore, West Bengal, India

S. K. Halder (✉) · K. C. Mondal (✉)
Department of Microbiology, Vidyasagar University, Midnapore, West Bengal, India
e-mail: sumanmic@mail.vidyasagar.ac.in

C. Ghosh
Department of Human Physiology with Community Health, Vidyasagar University, Midnapore, West Bengal, India
e-mail: ch_ghosh@mail.vidyasagar.ac.in

© Springer Nature Switzerland AG 2020
R. Prasad et al. (eds.), *Nanostructures for Antimicrobial and Antibiofilm Applications*, Nanotechnology in the Life Sciences,
https://doi.org/10.1007/978-3-030-40337-9_18

Abstract Enterococci are Gram-positive facultative anaerobes that have changed over epoch as highly modified represener of the gastrointestinal (GI) consortia of an extensive array of organisms like insects, birds, reptiles, mammals, and human. These commensal microorganisms have grossed resistance to all the antimicrobial drugs that currently exist. Multidrug-resistant (MDR) enterococci shows an extensive repertoire of mechanisms of drug resistance including drug target modification, overexpression of efflux pumps, inactivation of antibacterial agents, and cell membrane adaptive response that helps to persist in the body of the host and nosocomial atmosphere. MDR enterococci are renewed to persist in the GI environment and predisposing to invasive infections in those patients who are severely ill and immunocompromised. This chapter mainly focuses the resistance mechanisms of antimicrobial drugs and also role of certain new antimicrobial genes like *optr*A and *cfr* in enterococci. Moreover different strategies to control and therapeutic approaches for controlling MDR enterococci especially using nanotechnology are also highlighted.

Keywords Enterococci · Pathogenesis · Antibiotics · Multidrug resistance (MDR) · Endocarditis · Cephalosporin

18.1   Introduction

Enterococci are a primitive genus of microorganisms that are highly adapted to surviving in heterogeneous and harsh environmental conditions. In the ending of nineteenth century, a saprophytic and infectious cocci found in intestine was described as "*Enterococcus*" (Thiercelin 1899). MacCallum and Hastings also characterized an enterococcal organism, *Enterococcus faecalis*, from a fatal endocarditis case, and provide first comprehensive picture of its pathogenesis (MacCallum and Hastings 1899). Early report attested that enterococcal pathogens are basically commensal opportunist (Lebreton et al. 2014). With the development of genomic technologies, an array of enterococcal species has explored. Enterococci are the principal causes of healthcare-associated infections (HAIs) all around the globe. The last few decades have witnessed the development of multidrug-resistant (MDR) enterococci which extensively complicates this issue and also enhances the chance of treatment failure and sometimes leads to death. In the last decade, antibiotic-resistant enterococci have become familiar as the prime cause of nosocomial bacteremia, postsurgical wound infection, urinary tract infections, and device-associated infections (García-Solache and Rice 2019; Prabaker and Weinstein 2011; Emori and Gaynes 1993; McDonald et al. 1997).

In this section, we will explain the overall characteristics of the genus *Enterococcus* species, diseases induced by it, and the historical viewpoint behind the creation of MDR enterococci as pathogens and current knowledge of the molecular foundation of drug resistance in *Enterococcus*. Finally we addressed briefly the necessity to advance new drug targets, development of new approaches of nanobiotechnological methods against these dangerous and insubordinate organisms as well as difficulties and opportunities for the future.

18.1.1   Features of Enterococcus Genus

Enterococcus species are catalase negative Gram-positive bacteria, are natural inhabitants, and can be isolated easily from their habitats. They are also an important intestinal microfloral component of humans and animals (Van and Willems 2010). The basic physiological and morphological characteristics of all enterococcal strain include Gram-positive, ovoid/spherical cells organized in pairs/chains; among them a few strains exhibit pathogenic potential (Thiercelin 1899). Different salient feature of genus streptococci is represented in Fig. 18.1. They are obligatory fermentative chemoorganotrophs and non-spore-forming facultative anaerobes. They usually grow at an optimal temperature of 35 °C and can growth in the range of 10–45 °C. They normally have an optimal growth in medium having 6.5% sodium chloride (Facklam 1973). They are generally unable to produce catalase and all cytochromes. A few species are able to produce nominal catalase with weak effervescence. Usually they are homofermentative and produce only lactic acid as end product by fermenting glucose (Klein 2003).

18.1.2   Phylogenetic Diversity of Enterococcus Genus

In recent decades, knowledge regarding the biology, ecology, virulence, and genetics of the genus *Enterococcus* has sharply increased. The enterococci's taxonomy has changed considerably since the end of the twentieth century when the genus had only 20 species. As a consequence of improvements in differentiation techniques coupled with enhanced interest in enterococci, many fresh species have been well documented. Being ubiquitous, three *Enterococcal* species, namely, *E. durans*, *E. faecalis*, and *E. faecium*, were documented before 1950. *E. faecium and E. faecalis*

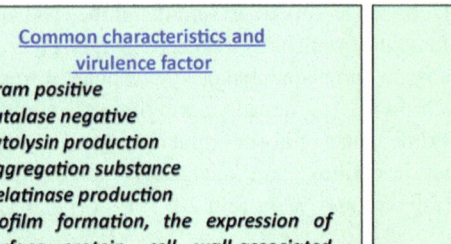

Common characteristics and virulence factor	Most Pathogenic strains
• **Gram positive** • **Catalase negative** • **Cytolysin production** • **Aggregation substance** • **Gelatinase production** • **Biofilm formation, the expression of surface protein, cell wall-associated enterococcal surface protein, phosphotransferase system, pathogenicity island**	• **E. faecalis** • **E. faecium** • **E. durans** • **E. avium** • **E. gallinarum** • **E. casseliflavus** • **E. hirae** • **E. mundtii** • **E. raffinosus**
Resistant to	Diseases
• **Beta-lactam antibiotics, Cephalosporin** • **Glycopeptide (Vancomycin), Aminoglycosides (Gentamycin, Sisomicin, Kanamycin, Netilmicin, Tobramycin), Oxazolidinones, Linezolid** • **Streptogramins, Macrolides, Lincosamides, Daptomycin** • **Tetracyclines, Glycylcyclines, Tigecycline** • **Quinolones, Fluoroquiolone** • **Rifampicin, Trimethoprim, Sulfamethoxazole**	• **Urinary tract infections (UTI)** • **Intra-abdominal infection with pelvic and soft tissue** • **Bacteremia** • **Endocarditis** • **Meningitis, hematogenous osteomyelitis, septic arthritis, and pneumonia**

Fig. 18.1 Different salient features of Enterococci

account for most of the enterococcal diseases (Malani et al. 2002). Other species like *E. avium, E. casseliflavus, E. hirae, E. gallinarum, E. mundtii*, and *E. raffinosus* have also been isolated from human infection (Devriese et al. 1994; Hammerum 2012; Murray 1990; Lebreton et al. 2014). In the era of 1992–2012, about 30 species of *Enterococcus* were documented, and only four of them were associated with human infection and pathogenesis (*E. sanguinicola, E. pallens, E. gilvus*, and *E. canintestini*). Till date, there are 52 species available that belong to *Enterococcus* genus.

18.1.3 Enterococci-Associated Infections

Over the last couple of eras, enterococci emerged as significant pathogens (Arias and Murray 2012). The variety of diseases caused by streptococci becomes devastating which is attributed to their tendency to become increased antibiotic resistance. Although other microorganisms are often isolated from the source site with enterococci, it is often not well understood and remains a paradox that the enterococci are directly associated with the manifestations of the diseases or whether they are avirulent and opportunistic one and suppose to play an insignificant role in the manifestation of diseases (Higuita and Huycke 2014). Among the several types of enterococcal infection, endocarditis and bacteremia are the leading life-threatening disease.

18.1.3.1   Urinary Tract Infections (UTI)

Urinary tract is the most susceptible area of enterococci infection. Lower urinary tract portions, especially cystitis, prostatitis, and epididymitis, are the frequent sites of UTI caused by enterococci in older man. Young women are also affected by uncomplicated cystitis, infected by enterococci. Occurrences of bacteremia in upper UTI are most often reported in older men. Enterococci-induced UTIs are more likely to be acquired in hospitals or in long-term settings, making them more resistant to antibiotics. Moreover, ICU setting also contributes to 15% of healthcare-associated UTI. Among the ICU patients, enterococci resistant to vancomycin have become the major urinary tract pathogens associated with healthcare (Hidron et al. 2008).

18.1.3.2   Intra-abdominal Infection with Pelvic and Soft Tissue

Intra-abdominal infection with pelvic and soft tissue is also the site of enterococcal infections. Enterococci are isolated from these samples often associated with other microbial flora and infrequently cause mono-microbial infection at the above sites. Bacteremia caused by enterococci is mainly associated with abscesses and wounds in the intra-abdominal and pelvic regions (Graninger and Ragette 1992; Maki and Agger 1988). Though most of the physician routinely follows antibiotic regimens to treat such type of infections, drainage of abscesses and debridement of wounds are also essential adjuncts to antibiotic therapy. Moreover, conjunction of liver cirrhosis or patients receiving chronic peritoneal dialysis most often suffered from an infection called peritonitis. Peritonitis mainly occurs in the abdominal lining. Moreover, abdominal or pelvic mixed aerobic-anaerobic infections should be considered separately. Though, in this type of cases, enterococci show monomicrobial infection, *Escherichia coli*, coagulase-negative *Staphylococci,* and *Staphylococcus aureus* are also responsible for bacterial peritonitis and dialysis-associated peritonitis. Over and above, enterococci are also often isolated in cultures from decubitus and foot ulcers. However, their roles in causing such site-specific infections are not clearly understood till date.

18.1.3.3 Bacteremia

Enterococci are presently one of the major causes of bacteremia associated with healthcare. Over the last couples of years, bacteremia is usually associated with gastrointestinal tract, although sources of bacteremia also reported from biliary and intra-abdominal regions, indwelling central lines, or infections in soft tissues. Polymicrobial bacteremia is associated mainly with enterococci, though there are also other microorganisms partially involved in the occurrence of such type of infection. Enterococci-associated bacteremia causes metastatic abscesses. The rate of overall mortality in enterococci associated bacteremia is varied (Maki and Agger 1988; Patterson et al. 1995; Higuita and Huycke 2014). Several scientific reviews regarding bloodstream infections clearly reported that enterococci is the only Gram-positive bacteria associated with high risks of death. Moreover, higher mortality rate was reported in the case of *E. faecium*-associated bacteremia than *E. faecalis* (Noskin et al. 1995a, b; Higuita and Huycke 2014). The chances of occurrence of enterococci-associated bacteremia are higher in the case of elderly people with multiple underlying diseases like malignancy, diabetes mellitus, cardiovascular diseases, transplantation, and postsurgery infection.

18.1.3.4 Endocarditis

Among different types of infection caused by enterococci, endocarditis is the most fatal enterococcal infections. The alimentary or urinogenital tract is the primary bacteremia source which leads to endocarditis. In reality, left-sided participation is much more prevalent than right-sided participation. Prosthetic valve enterococcal endocarditis has been increasingly marked. This is mainly associated with increasing application of prostheses in aged persons who have higher risks for bacteremia caused by enterococci (Anderson et al. 2004; Rice et al. 1991). Enterococcal endocarditis is more common in men compared to women (McDonald et al. 2005). Several retrospective analysis reported that between 15 and 39% of enterococcal endocarditis are healthcare associated (Anderson et al. 2004; McDonald et al. 2005). Endocarditis associated with enterococci is a subacute infection followed by cardiac failure, rather than an embolic effect (McDonald et al. 2005). Though death rates are low (9–15%) in enterococcal endocarditis in comparison to other infective endocarditis (McDonald et al. 2005; Rice et al. 1991; Wilson et al. 1984; Higuita and Huycke 2014), selection of effective therapy against the multidrug-resistant enterococci is definitely a challenging task.

18.1.3.5 Uncommon Infections

Meningitis, septic arthritis, hematogenous osteomyelitis, and pneumonia are the less common or rarely seen infections caused by enterococci. Pneumonia is quite rare even in the presence of ventilators, and it is reported in significantly weakened or in immunocompromised patients who have received antibiotic drugs of a broad spectrum. Antibiotic-resistant enterococci (VRE) are likely to be responsible for such types of infection than antibiotic-susceptible enterococcal isolates.

18.2 Expansion of Antibiotic Resistance

With the innovation of antimicrobial drugs discovery and understanding the microbiological foundation of diseases, infection became remediable with remarkable recovery. Clinicians, however, quickly understood that certain microbes appear to be less effective in responding to specific antimicrobials and that's why generations of antibiotics had come. It was also documented that penicillin and aminoglycoside are less effective against many enterococcal species, and conjugation of aminoglycosides with penicillin was prescribed which showed synergistic response that improved enterococcal endocarditis cure rates from 40 to 88% (Robbins and Tompsett 1951). Thus the particular combination of a cell-wall-active agent (i.e., penicillin/ampicillin) along with an aminoglycoside will be the solution for the treatment of deep-rooted *Enterococcus*-associated diseases, and this combination remains effective (Baddour et al. 2005).

Unknowingly the seeds of the resistance of enterococci against an array of drug were already being sown and propagated. By the help of comparative genomics, it was documented that the modern MDR *Enterococcus faecium* is a part of a genetic class that seems to have divergent root of ancestry from animal-adapted *E. faecium* strains in clinical practice approximately 75 years ago, corresponding to the introduction of antibiotics (Lebreton et al. 2013). This was achieved by various means which include an upsurge in horizontal gene transfer, metabolic bypass, and hypermutability in the enterococcal strains. The acquisition of genes for vancomycin resistance is one of the utmost examples of this adaptability. Vancomycin-resistance enterococci (VRE) was first time documented in 1988, and within two decades, more than 80% of *E. faecium* acquired the said property in the USA (Arias and Murray 2012). Of particular concern, *E. faecium* is also increasingly reported to cause nosocomial infection, which now occurs as often frequently as *E. faecalis* (Hidron et al. 2008). Recently, enterococci have also reported to share the vancomycin-resistant gene clusters with potential pathogens (such as methicillin-resistant *Staphylococcus aureus*) through horizontal gene transfer, which is matter of a great health risk (Chang et al. 2003; Ray et al. 2003). Enterococci have adapted rapidly despite the abundance of anti-Gram-positive antimicrobials, and the emergence of resistance against these agents has been theorized. This becomes a clinical challenge to treat enterococcal MDR infections. The following sections give a picture of the mechanisms and prevalence of antimicrobial resistance in enterococci which is summarized in Table 18.1.

Table 18.1 Antibiotics and the resistance mechanism of enterococci against them

Class/name of antibiotics	Basic mode of action	Specific gene(s)/operon responsible for resistance	Possible mechanism of antibiotic resistance
Ampicillin, Penicillin, Mezlocillin, Piperacillin	Inhibit the synthesis of peptidoglycan	*ponA, pbp F, pbpZ, pbp5, pbp A, pbpB*	Reduced susceptibility for the antibiotic
Cephalosporin		*pbp5, ponA, pbpZ*	Reduced binding affinity for the antibiotic
Glycopeptide (vancomycin)	Prevent cross-linking of peptidoglycan	*vanA, vanB, vanD, vanM, vanC, vanE, vanG, vanL, vanN, vanT, vanXY, vanT*	Reduced affinity for the antibiotic
Aminoglycosides (gentamicin, sisomicin, kanamycin, netilmicin, tobramycin)	Create pores in the cell membrane of the bacterial cell	*aac (6')-Ii, aph(2')-Ic, aph(3')- IIIa, aph(2')-Id, aph (2')-I*	Modification of the aminoglycoside structure
Oxazolidinones, linezolid	Inhibit the peptide delivery	*cfr* or *cfr(B)*	Methylation of 23S rRNA and reducing affinity to the antibiotics due to mutations
Streptogramins, macrolides, lincosamides	Dissociation of peptidyl-tRNA, preventing binding of aminoacyl-tRNA to the ribosomal and the formation of the peptide bond	*isa, msrC, eatA, msr(A), linB, mef(A), vgb(A), vat(D), vat(E)*	Efflux pump to eliminate antibiotics
Daptomycin	Alterations in the cellular membrane characteristics	*liaFSR* operon	Mutation in the specific gene to exclude the effect of antibiotics
Tetracyclines, glycylcyclines, tigecycline	Interfere with the docking of aminoacyl-tRNA in the ribosome	*tetM, tetO, tetS tetL*	Efflux mechanism, ribosomal protection overexpression of genes
Quinolones, fluoroquinolone	Disrupt DNA strand continuity as well as stop replication	*gyrA, parC, qnr*	Mutation in the specific gene, efflux pump
Rifampicin	Inhibit the process of transcription	*rpoB*	Reduced affinity due to point mutation
Trimethoprim, sulfamethoxazole	Inhibit folate biosynthesis pathway	–	Gain ability to utilize exogenous folate

18.3   Biofilm Formation in Enterococcal Infections

Biofilm is multicellular community of microbes attached on abiotic and biotic surfaces or interfaces, enclosed in a hydrated self-produced extracellular polymeric matrix (Costerton 2001). Development of biofilm is a multistep phenomenon which includes surface attachment, immobilization, cell-cell interaction, microcolony formation, confluent biofilm formation, and subsequently three-dimensional biofilm formation (O'Toole et al. 2000). Biofilms are the reservoir of many chronic infections and extremely difficult to eliminate (Mohamed and Huang 2007). As per the National Institutes of Health, about 4/5 share of all bacterial infection in the body associated with biofilm formation (Lewis 2001). Biofilm containing bacteria are phagocytosis resistant, and therefore it is an extremely challenging task to eliminate from the host or infected individual (Lewis 2001). Biofilms are reported to form in/on a broad range of medically used devices like pacemakers, catheters, orthopedic appliances, and prosthetic heart valves, which is correlated with multiple pathogenic consequences (Costerton et al. 1999).

Biofilm producing enterococci are extremely antibiotic resistant and therefore the impact of biofilm development is very crucial. Perusal of literature attested that enterococci were found to form biofilm in an array of infection like UTI, wounds, GI dysbiosis, endocarditis, etc. Though exopolymeric matrix and antibiotic resistance are the two major hurdles to eradicate enterococci, the foremost problem is the dissemination of the genetic trait of antibiotic resistance to other microbes (Ch'ng et al. 2019). As like as other biofilm-forming bacteria, adherence and biofilm formation by *E. faecalis* and *E. faecium* on diverse biomaterials and numerous medical apparatus (biliary stents, intravascular catheters, silicone gastrostomy devices, ureteral stents, etc.) have been documented (Joyanes et al. 2000; Distel et al. 2002; Dowidar et al. 1991; Sandoe et al. 2003; Dautle et al. 2003; Keane et al. 1994). Formation of enterococcal biofilm of on ocular lens has also been demonstrated (Kobayakawa et al. 2005).

18.3.1   Factors Contributing Formation of Biofilm in Enterococci

18.3.1.1   Biofilm Formation in *E. faecalis*

Development of biofilm generally consists of four phases: initial attachment, formation of microcolony, maturation of biofilm, and, finally, dispersal. There are multiple factors that influence formation of biofilm in enterococci within or outside their host condition (Dunny et al. 2014); however, the dispersion mediators have yet to be identified (Table 18.2).

Adherence to the surface is the early stage for the establishment of biofilm. Various factors like surface adhesins, glycolipids, and proteases perform significant

Table 18.2 Some nanoparticle and their effects on enterococci

Type of nanoparticles	Antibiotic resistance type/special feature	Mechanism of action	References
AgNPs	Vancomycin resistance	In combination with vancomycin causing bacterial death	Saeb et al. (2014)
	Erythromycin resistance	Cell surface damage and loss of the chain integrity	Otari et al. (2013)
	Multidrug resistance	Modification of physicochemical properties of the cell	Cavassin et al. (2015)
	Multidrug resistance	Combined effect with gentamicin and chloramphenicol	Katva et al. (2018)
Graphene oxide NPs	Multidrug resistance	UV irradiation leads to reactive oxygen species generation, multiple toxic mechanisms	Govindaraju et al. (2016)
Magnetite NPs	Biofilm forming	Effective aminoglycoside antibiotic carrier	Chifiriuc et al. (2013)
Calcium hydroxide NPs	Multidrug resistance	–	

tasks in the first step of biofilm formation. The multi-subunit (viz., A, B, and C) endocarditis and biofilm-associated pilus (Ebp) encoded by e*bpABC* facilitates surface adherence both in vivo and in vitro (Nielsen et al. 2012; Nallapareddy et al. 2011a; Singh et al. 2007). The role of Ebp in the early development of biofilm was showed by in vivo models of UTIs, catheter-associated UTI, and infective endocarditis (Nallapareddy et al. 2006, 2011a, b; Nielsen et al. 2013). Several in vivo experiments in cultured human cell also described the significance of surface adhesins in formation of biofilm (Mohamed et al. 2006; Rozdzinski et al. 2001; Sussmuth et al. 2000; Sillanpaa et al. 2010). It was also demonstrated that biofilm-associated glycolipid synthesis A influences in vitro surface adherence and subsequent biofilm development (Theilacker et al. 2009).

Initial attachment followed by formation of microcolony in which bacteria divided repeatedly and produce minute sizes of biofilm which subsequently get aggregated (Monds and O'Toole 2009). In vitro findings have clearly showed that microcolony formation is the mature stage of biofilm development, and this is significant for gut colonization. An enterococcal polysaccharide antigen gene cluster (*epaOX*) encodes a glycosyltransferase which is associated with the production of rhamnopolysaccharide associated with cell wall, and mutant *E. faecalis* for the particular trait showed a reduction in biofilm reduction (Ch'ng et al. 2019; Xu et al. 2000).

Maturation of *E. faecalis* biofilm is associated with the vigorous growth and development of extracellular matrix materials like extracellular DNA, polysaccharide, glycoprotein, modified lipid, lipoteichoic acid, etc. (Ch'ng et al. 2019; Fabretti et al. 2006). Deletion of *atl*A reduces the release of extracellular DNA, thus decreasing biofilm formation (Guiton et al. 2009). In vitro deletion of *dltABCD* operon causes inhibition of biofilm development by Gram-positive bacteria by reducing the production of D-alanine esters of lipoteichoic acid.

Biofilm formation is also contributed by population density-dependent signaling mechanism like quorum sensing and peptide pheromone signaling which upgrade expression of genes towards biofilm formation by enterococci (Krasteva et al. 2012; Camilli and Bassler 2006; Cook and Federle 2014; Li and Tian 2012; Cook et al. 2011). Recently, transfer of plasmid DNA between *E. faecalis* cells in GI tract has been documented which encourages biofilm formation (Chen et al. 2017; Hirt et al. 2018). *Eep* (Chandler and Dunny 2008), *fsrABC* (Ali et al. 2017), *bopABCD, gelE, sprE* (Dunny et al. 2014), and *AI-2* (Shao et al. 2012) are also involved in quorum sensing system of enterococcal biofilm formation.

18.3.1.2   Biofilm Formation in *E. faecium*

Multiple genes are responsible for the development of biofilm in *E. faecium* like *atlA, ebpABC, esp, fsrB, luxS, spx, acm, scm, sgrA, pilA, pilB, ecbA*, and *asrR* (Dunny et al. 2014; Lim et al. 2017; Sava et al. 2010; Hendrickx et al. 2009; Sillanpaa et al. 2008). Among these genes, *atlA, ebpABC, esp, acm*, and *asrR* are responsible to cause biofilm-associated infection in in vivo condition (Dunny et al. 2014; Sava et al. 2010). The cell surface adhesin, Esp, and EbpABC perform a crucial task in the initial attachment of *E. faecium*, followed by biofilm development in the case of UTI and infective endocarditis model (Montealegre et al. 2016a, b; Almohamad et al. 2014). Deletion of the gene *esp* and *ebpABC* operon reduced the chances of biofilm formation by the organism (Heikens et al. 2011). There are similarities in the occurrence of biofilm formation in the case of *E. faecium* and *E. faecalis* (Ch'ng et al. 2019). AtlA-dependent release of extracellular DNA plays a crucial role in biofilm formation in vitro in both the species (Paganelli et al. 2013). Several reports suggest that upregulation of gene like *ebpABC* and downregulation of genes like *fsrB, luxS*, and *spx* might regulate biofilm-forming potential of *E. faecium* (Lim et al. 2017). Moreover, deletion of asrR gene involves in growth and maturation of biofilm and also influences biofilm-associated infections (Lebreton et al. 2012).

18.4   Mechanism of Antimicrobial Drug Resistance in Enterococci

18.4.1   Mechanism of Resistance of β-Lactam Derivatives (Cell-Wall-Active Agents)

18.4.1.1   Resistance to β-Lactams

Penicillin and ampicillin are the foremost pronounced β-lactams which competitively block peptidoglycan (PPG) biosynthesis which is basic and the most common component of the bacterial cell wall. However, the lack of analogous structural component in eukaryotes excludes the lethality of these agents and makes them an ideal against bacterial infection as therapeutics. Penicillin-binding proteins (PBPs) are the flagship of the cell wall biosynthesis machinery which is broadly subdivided into two classes: class A, which exhibits bipartite enzymatic activity, namely D,D-transpeptidase and transglycosylase, and class B, which exhibits transpeptidase activity towards other enzymes.

Enterococci are inherently resistant to most β-lactams and hence less prone to restricted by the antibiotics. This is due to the expression of one kind of PBPs which have low affinity towards β-lactam antibiotics. Consequently, the minimum inhibitory concentration (MIC) of penicillin is higher in enterococci in contrast with streptococci or other Gram-positive bacteria, which do not produce chromosomally encoded low affinity PBPs. Lower MIC values of penicillin were documented for *E. faecalis* strains than *E. faecium*.

Every enterococci have at least 5 PBPs, and 6 putative PBP genes were recognized by studying the genome of *E. faecalis* and *E. faecium* (class A, *ponA*, *pbp F*, *pbpZ*; class B, *pbp5*, *pbp A*, *pbpB*) (Miller et al. 2014). Inherent tolerance against the β-lactam antibiotics is linked with the expression of species-specific *pbp5* gene (class B PBP) that minimizes binding affinity cell wall with the antibiotics. In *E. faecium*, the *pbp5* gene is a part of operon which has three genes (including *pbp5*) that take part in cell wall synthesis (*psr* and *ftsW*) (Miller et al. 2014). Enhanced resistance against β-lactam antibiotics has frequently been noticed among clinically isolated *E. faecium* but rarely noticed in the case of *E. faecalis*. High-level ampicillin resistance of *E. faecium* (MIC>128 μg/ml) has been correlated with concomitant production of Pbp5 or with specific amino acid modifications in its sequence, which minimizes affinity of the same with penicillins resulting in less vulnerable to be inhibited. The substitutions of amino acid at or near the active-site cavity (Ser-Thr-Phe-Lys, Ser-Asp-Ala, and Lys-Thr-Gly motifs) seem to be the utmost significant ones (Rybkine et al. 1998; Zorzi et al. 1996). Combinations of specific amino acid alterations in the carboxyl-terminal transpeptidase domain of PBP5 (substitution Met-485-Ala/Thr, Ala-499-Ile/Thr, Glu-629-Val and Pro-667-Ser) and the insertion of serine or aspartate after position 466 have been related to ampicillin resistance of *E. faecium* isolates (Montealegre et al. 2016a, b; Jureen et al. 2003; Poeta et al. 2007; Klibi et al. 2008; Arbeloa et al. 2004; Rice et al. 2004).

Alongside, β-lactam antibiotic resistance is also facilitated by a β-lactamase enzyme which restricts the antibiotic action by cleaving the β-lactam ring. The phenomenon was documented in both *E. faecalis* and *E. faecium* (Rice and Murray 1995; Murray 1992; Coudron et al. 1992). Selected strains of *E. faecalis* produce a plasmid-mediated β-lactamase that is similar to the enzyme produced by *Staphylococcus aureus*, encoded by the *blaZ*gene, although some polymorphisms in this gene have also been detected in some isolates (Hollenbeck and Rice 2012; Murray et al. 1992).

18.4.1.2   Resistance to Cephalosporin

As like as β-lactam antibiotic resistance, the intrinsic resistance of enterococci is correlated with a decline in the affinity of binding of cephalosporin with enterococcal PBPs, especially Pbp5 (Rice et al. 2009; Arbeloa et al. 2004). It was documented that expression of either *ponA* or *pbpF* gene in *E. faecalis* and *E. faecium* is required to exhibit cephalosporin resistance, and PbpZ alone is incapable of offering the transglycosylation property.

An array of regulatory pathways manifested by two-component system is responsible for showing cephalosporin resistance by enterococci. Downstream effector like CroRS was publicized to be imperative for the same. Besides, two-component system implicated a role in resistance also relayed by a serine/threonine kinase, namely, IreK and IreP (phosphorylated). IreB was proven as target of both the aforementioned proteins and in turn upgrade the expression of cephalosporin resistance (Comenge et al. 2003; Muller et al. 2006; Kristich et al. 2007; Hall et al. 2013). MurAA protein involved at the downstream of the IreK signaling pathway and catalyzes the first committed step in PPG biosynthesis (Miller et al. 2014).

18.4.1.3   Resistance to Glycopeptide

Vancomycin and teicoplanin belongs to glycopeptide family employed for the treatment of severe human diseases. Glycopeptides actually bind with the terminal D-alanyl-D-alanine of the pentapeptide of PPG precursors that subsequently inhibit cross-linking of PPG chains and thus restrict the bacterial cell wall synthesis. The mechanism underlying the glycopeptide resistance of enterococcal strains is the alteration of the PPG synthesis pathway. The terminus D-alanyl-D-alanine with which vancomycin binds is modified to D-alanyl-D-lactate (high resistance, MIC >64 μg/ml) or to D-alanyl-D-serine (low resistance, MIC >4–32 μg/ml). This kind of alteration in the cell wall precursors leads to reduced binding affinity of the glycopeptide with the former (Miller et al. 2014; Ahmed and Baptiste 2018; Shlaes et al. 1989; Arthur et al. 1993).

Vancomycin-resistant enterococci are formed by *van* operons, which encode the modified PPG precursors. Nine *van* operons have been recognized so far in enterococci-mediating vancomycin resistance (for D-alanyl-D-lactate modification, *van*A, *van*B, *van*D, *van*M, and for D-alanyl-D-serine modification, *van*C, *van*E, *van*G, *van*L, and *van*N) (Miller et al. 2014; Courvalin 2006; Depardieu et al. 2015). The *van*A and *van*B are the most common genotypes among VRE with acquired resistance mechanisms of humans and animals, mostly among *E. faecalis* and *E. faecium* (Ahmed and Baptiste 2018). VanC operon is the fundamental component of *E. gallinarum* and *E. casseliflavus* that helps to produce PPG precursor with terminal D-alanyl-D-serine residue reported first time (Leclercq et al. 1992; Reid et al. 2001). Apart from VanC (which is a D-alanine-D-serine ligase), the enterococcal cells encode a serine racemase (VanT), combined dipeptidase-carboxypeptidase (VanXY) and regulators encoded by *van*R and *van*S genes which encode (cytoplasmic) transcriptional regulator and membrane-bound histidine kinase, respectively (Depardieu et al. 2015; Sassi et al. 2018).

The *van*A operon is associated with the transposon Tn1546 and includes seven open reading frames (ORFs) transcribed under two different promoters. Regulation is mediated by *van*S-*van*R (sensor-kinase-response regulator) two-component system, transcribed with a common promoter. The *van*H- and *van*A-encoded protein modifies the PPG precursors, whereas *van*Y interrupt the creation of the D-alanyl-D-alanine termini of the pentapeptide by its D,D-carboxypeptidase activity. Moreover, *van*Z gene is associated with teicoplanin resistance in enterococci.

Tn1547, Tn1549, and Tn5382 are the transposons associated with *van*B operon. Among the transposons, Tn1549 is widely predominant among *van*B-type enterococci located in chromosome. *van*B has two promoters and seven ORFs. *van*B enterococci represent vancomycin resistance but susceptibility towards teicoplanin (Ahmed and Baptiste 2018; Arthur and Courvalin 1993). It was well documented that a few of van operons belong to transposable genetic element which triggers the spreading of the antibiotic resistance trait.

18.4.2 Mechanism of Resistance to Protein Synthesis Interfering Antibiotics

18.4.2.1 Resistance to Aminoglycosides

Aminoglycosides are effective bactericidal chemotherapeutic agents that interfere with the protein synthesis of the bacterial cell by binding with 30S ribosomal subunit followed by misread of genetic code. The intrinsic resistance of enterococci against aminoglycosides is imparted by inactivating the aminoglycoside through covalent modification of amino or hydroxyl groups which is carried out by enterococcal enzymes.

E. faecium express 6′-acetyltransferase enzymes [AAC (6′)-Ii] which was reported to modify tobramycin, kanamycin, sisomicin, and netilmicin. Moreover, numerous isolates from clinical samples also possess the enzyme APH(3′)-IIIa which triggers the resistance against amikacin and kanamycin owing to its phosphotransferase activity (Costa et al. 1993). Alongside, in *E. faecium*, the bypassing of the aminoglycoside action was carried out by modifying the ribosomal target through the action of rRNA methyltransferase which methylates cytidine residue at 1404 position (Galimand et al. 2011).

Gentamycin and streptomycin are the aminoglycosides that are used in clinical practice reliably because these two are not readily degraded by enterococci-produced intrinsic enzymes. APH(2′)-Ic is another gene encoding phosphotransferases reported in *E. gallinarum*, *E. faecium*, and *E. faecalis* which counteracts against gentamycin (Chow et al. 1997) and tobramycin but not in against of amikacin, whereas APH(2′)-Id, isolated from *E. casseliflavus* and *E. faecium*, confers gentamycin resistance but not against amikacin. Moreover the presence of another gene, *aph (2')-Ib*, in *E. faecium* causes amino-glycoside resistance except for amikacin and streptomycin (Eliopoulos et al. 1984; Courvalin et al. 1980).

18.4.2.2 Resistance to Oxazolidinones and Linezolid

Bacteriostatic agent linezolid binds to the 23S rRNA of Gram-positive bacteria and causes disruption in the docking of charged tRNA in ribosomal A site, followed by inhibition in the peptide delivery and elongation of the polypeptide chain subsequently (Shinabarger et al. 1997; Leach et al. 2007; Locke et al. 2009; Mendes et al. 2008). The mechanism of linezolid resistance is the gene mutation which generally encodes 23S rRNA, an important ribosomal drug-binding site (Marshall et al. 2002; Chen et al. 2013; Diaz et al. 2012, 2013). Moreover, linezolid resistance develops in enterococci through acquirement of methyltransferase gene followed by modification of A2503 in the 23S rRNA (Kehrenberg et al. 2005; Vester 2018; Wang et al. 2015). Many copy of the 23S rRNA gene present in enterococci, and as much as the gene becomes mutated, the resistance property is increased concomitantly (Boumghar-Bourtchaï et al. 2009; Bourgeois-Nicolaos et al. 2007; Toh et al. 2007).

18.4.2.3 Resistance to Streptogramins, Macrolides, and Lincosamides

Unlike *E. faecium, E. faecalis* is resistant to pristinamycin derivatives, streptomycin A and B.

In *E. faecalis* genome, *Isa* gene encodes an ATP-binding cassette (ABC) transporter protein necessary for efflux pump which eliminates the action of lincosamide and streptogramin A (Singh et al. 2002). Similar type of pumps coded by *msrC* has also been reported to act in removing the streptomycin A and B (Portillo et al. 2000). An intrinsic resistance mechanism of chromosome towards macrolides by *msr*(A) and to linosamides by *linB* in *E. faecium* has been documented (Portillo et al. 2000; Bozdogan et al. 1999). Several other genes in *Enterococcus* genus are also responsible for conferring resistance like gene *mef*(A), causing resistance to macrolides; *vgb*(A), causing resistance to lincosamide; and *vat*(D) and *vat*(E), causing resistance to streptogramins.

18.4.2.4 Resistance to Daptomycin

Daptomycin binds with cellular membrane facilitated by calcium that causes alterations in its characteristics and function. It is a cyclic lipopeptide that primarily interacts with phosphatidylglycerol and, in the presence of calcium ions, aggregates and enters into the cell membrane and reaches to the inner leaflet. This causes leakage of ions, and also formation of pores occurs on the cell membrane. It also causes lipid aggregation on the membrane surface by "lipid extraction effect." Daptomycin-resistant enterococci are reported and it is achieved by means of mutations. Report suggests that *E. faecium* repulses daptomycin from its cell surface by changing membrane phospholipids which is commonly associated with mutation in *liaFSR* operon (García-Solache and Rice 2019; Miller et al. 2016). Mutation in *liaFSR* system causes synergism between ampicillin and daptomycin in daptomycin-resistant *E. faecium* (Mishra et al. 2012).

18.4.2.5   Resistance to Tetracyclines and Glycylcyclines

Tetracyclines exhibit bacteriostatic effect by interfering with the aminoacyl-tRNA docking in the ribosome. Enterococci-acquired tetracycline resistance by ribosome shielding mechanism is facilitated by *tet(M)*, and antibiotic efflux mechanism is facilitated by *tet(L)* genes (García-Solache and Rice 2019). Several other genes like *tetO* and *tetS* confer resistance to doxycyclines, minocyclines, and tetracyclines and are transferred via the Tn*916* transposon. The encoded proteins of the above-mentioned genes hydrolyze GTP in the presence of ribosome and cause alteration of ribosomal conformation and finally displace bound tetracyclines (Rice 1998; Speer et al. 1992).

Tigecycline belongs to glycylcycline which is a broad-spectrum antibiotic used as therapeutics in severe infections in skin, soft tissues, and abdomen. It binds with the 16S rRNA and causes inhibition in the association of aminoacyl-tRNA. In tigecycline-resistant *E. faecium,* increased expressions of *tet*(M) and *tet*(L) genes were reported to confer tigecycline resistance (Fiedler et al. 2016).

18.4.3   Mechanism of Resistance to Antibiotics That Interfere in Central Dogma

18.4.3.1   Resistance to Quinolones

For the onset of cell division, starting of replication and transcription of DNA is important. Quinolones generally target two enzymes like DNA gyrase and topoisomerase IV. Those enzymes are responsible for the replication and transcription process. Administration of quinolones causes disruption of strand continuity, stopping replication process (Hawkey 2003). This antibacterial compound shows broad-spectrum effect on numerous bacteria by targeting the two said enzymes. Reduction of antibacterial activity of fluoroquinolones against *Enterococci* has also been reported (Oyamada et al. 2006). Though enterococci acquire low levels of quinolone resistance, sometimes it can also confer high-level resistance by several mechanisms (López et al. 2011; Werner et al. 2010; Yasufuku et al. 2011). Mutations in the *gyrA* and *parC genes* are responsible for the acquisition of resistance (in the case of levofloxacin and moxifloxacin) in *E. faecium* and *E. faecalis* (Tankovic et al. 1999; Kanematsu et al. 1998). EmeA and NorA like efflux pumps have also been reported for conferring the resistance of *E. faecalis* and *E. faecium* against quinolones, respectively (Hooper 2000). Another gene, *qnr*-encoded protein, is also responsible for the formation of quinolone-gyrase complex, protecting DNA gyrase, and in this way it confers resistance in *Enterobacteriaceae* (Arsène and Leclercq 2007; Tran et al. 2005).

18.4.3.2   Resistance to Rifampicin

Rifampicin binds with the β-subunit of DNA-dependent RNA polymerase and thus inhibits the process of transcription. Rifampicin-resistant *E. faecium* is developed due to substituted mutation in *rpoB* gene (H486Y) which encodes the said enzyme (Kristich and Little 2012). Moreover, *rpoB*-mutated *E. feecium* and *E. faecalis* show elevated resistance to cephalosporin (Enne et al. 2004; Rand et al. 2007).

18.4.3.3   Resistance to Trimethoprim and Sulfamethoxazole

Trimethoprim and sulfamethoxazole are the two notable antibacterial compounds that mainly target the enzymes associated with folate biosynthesis. Folate is synthesized from the *p*-amino benzoic acid and essential for synthesis of nucleic acids. The aforementioned compounds decrease the production of dihydrofolate and also blocked the conversion of tetrahydrofolate by inhibiting several enzymes in folate biosynthesis pathway. Though in vitro susceptibility is present, in vivo reports showed that these two antibiotics are ineffective against enterococci as they have gained the ability to utilize exogenous folate (Chenoweth et al. 1990; Grayson et al. 1990).

18.5   Alternative Strategies for Combating Multidrug-Resistant *Enterococcus*

The evolution of MDR enterococci has boosted interest towards alternative therapies to alleviate the disease causing potentiality of enterococci. Though virulence factors do not directly confer resistance, it will help bacteria to withstand in an unfavorable environmental condition and resist host defense mechanisms. Host biomacromolecules associated with the cell surface of *Enterococcus* and release of these molecules into the extracellular matrix inhibit the antimicrobial drugs from reaching their targeted sites (Otto 2006). Cyclic-AMP (cAMP) as an important mediator of innate immune system imparts antimicrobial activity by disturbing PPG biosynthesis and cytoplasmic membrane structure (bacterial) as well as promotes autolysins which collectively help to keep microbial populations within threshold level. However, coevolution of cAMPs and their bacterial targets is well documented (Kandaswamy et al. 2013; Gilmore et al. 2013). Exploitation of host adaptive immunity is also targeted through vaccination for the production of antibodies against enterococci. In this context, the lipoteichoic acids and diheteroglycans present over the cell walls of enterococci are marked as an epitope as they will help to induce an antibody response. This will protect the host (mouse bacteremia model) against *E. faecalis* (Theilacker et al. 2011). Application of antibodies against those specific enterococcal antigenic motifs could be a possible therapeutic to combat MDR strains in the future.

18.6 Application of Nanotechnology Against Enterococcal Infections

Development of multidrug-resistant enterococci becomes a most pressing concern in community health worldwide. The WHO (World Health Organization) and CDC (Center for Disease Control) have already expressed major concern about the gradual increase in the formation of multidrug-resistant bacteria (Baptista et al. 2018). This has boosted researchers to develop potent strategies for drug delivery and, finally, targeting bacteria. Nanostructured materials (e.g., organic, inorganic, metallic, carbon nanotubes, etc.) are being synthesized to circumvent such types of drug resistance as they easily convey antimicrobials, assist novel drug delivery, exert antimicrobial activities, and inhibit biofilm development (Baptista et al. 2018).

Several attempts were made for the synthesis of potent nanoparticles and subsequent effective delivery of the same against multidrug-resistant enterococci (Katva et al. 2018). Silver is a nontoxic, safe antimicrobial inorganic agent, and silver nanoparticles (AgNPs) have obtained much more attention as compared to other metal-based nanoparticles due to its strong antimicrobial activity. AgNPs are the utmost encouraging inorganic nanoparticles that can be applied for the alleviation of enterococcal infections. It was demonstrated that AgNPs in combination with vancomycin exhibited excellent antibacterial potential against vancomycin-resistant *E. faecalis*. Likewise, a mixture of gentamycin, chloramphenicol, and AgNPs could be promising to treat MDR *E. faecalis* infection than both the above-mentioned antibiotics separately (Katva et al. 2018). The antibacterial efficiency of AgNPs was also evaluated by Wu et al. (2014) against *E. faecalis* biofilm. Otari et al. (2013) also showed the effect of AgNPs on the erythromycin-resistant *E. faecalis*. It was suggested that AgNPs inhibit bacterial growth and proliferation by adhering on the cell wall of bacteria, leading to cell wall modification followed by penetration of AgNPs into the bacterial cell, which consequently damages the DNA leading to cell death (Aziz et al. 2015, 2016; Kumar et al. 2016; Saini et al. 2019).

Khiralla and El-Deeb (2015) developed biogenic spherical selenium nanoparticles using cell-free supernatant of *Bacillus licheniformis* which imparted paramount antimicrobial and antibiofilm potential against *E. faecalis*. Likewise, biogenic palladium nanoparticles were prepared by using flower extract of *Moringa oleifera* which showed significant antibacterial effect against the same bacteria (Anand et al. 2016).

Graphene oxide (GO) has unique physicochemical characteristics and has therefore attracted attention for antibacterial use (Hu 2010). The GO nanosheets exhibit antibacterial activity through direct interaction with bacteria and increased the reactive oxygen species (ROS) level within the cell (Akhavan and Ghaderi 2010). Govindaraju et al. (2016) demonstrated that UV-irradiated form of glucosamine-gold nanoparticle-graphene oxide composite exhibited paramount antimicrobial activity against *E. faecalis* which is better than kanamycin, and several functional groups (like carboxyl, hydroxyl, and epoxy) present in the GO-based nanomaterial are responsible for the activity. Nanocomposite of indocyanine green and GO was also reported to exhibit potential antibacterial effect against *E. faecalis* during photodynamic therapy (Akbari et al. 2017).

In order to treat vancomycin-resistant *Enterococcus*(VRE), Zhou et al. (2018) prepared Au/Ag bimetallic NPs and demonstrated that it has immense potential to be a good anti-enterococcal agent. Both in vitro (bacterial surface-enhanced Raman scattering imaging) and in vivo (mouse infection assays) results clearly revealed the effectiveness of this newly developed nanocomposite against VRE.

Chifiriuc et al. (2013) also investigated the capability of magnetic nanoparticle for a sustained and controlled release of drug which subsequently increases the effectiveness of antibiotics against resistant opportunistic pathogen, *E. faecalis*. They also suggested that magnetic nanoparticles might be a potent carrier for delivery of amino-glucoside antibiotics.

The antibacterial efficacy of calcium hydroxide nanoparticle (NCH) showed better result against *E. faecalis* in dentin block model. The MIC determination and agar diffusion test revealed that low concentration of the NCH inhibited *E. faecalis* than the native form of calcium hydroxide which is due to the enhancement of surface area due to smaller size which encourages the penetration of the NPs into the deeper layers of dentin which subsequently inhibits *E. faecalis* growth (Dianat et al. 2015).

Despite the expected potential of newly reported nanoparticles against multidrug-resistant *Enterococcus*, there are still few shortfall related to their safety when they are used in long-term basis in human. Therefore, in-depth assessment of the physical, chemical, and biological compatibility must be addressed. Experimental proof is also desirable for establishment of mechanism of action against the targeted enterococci in vivo. Moreover, the fruitful translation of the R&D work into real-life large-scale production of the newly discovered nanoparticles needs comprehensive guidelines, and effort is needed.

18.7 Conclusions and Future Challenges

Enterococcal species can colonize and survive in different biological and environmental niches. Owing to their biofilm-forming ability, multiple-drug resistance, and tendency of transfer of resistant trait to other enterococci, it became a great burden in healthcare sectors. Among the several species of enterococci, *E. faecalis* and *E. faecium* are associated with most clinical cases and hence they are marked as important nosocomial pathogens. Continuous exposure to prophylactic or metaphylactic and random application of antimicrobial agents by clinicians in human and animal hosts against enterococci contributed its ability to acquire and develop unique profiles of virulence and antimicrobial drug resistance. Moreover, expression of a wide variety of virulence characteristics promotes enterococci to colonize and also causes infections in the host body. Extensive tolerance to the antibacterial agents as well as their wondrous capacity to acquire resistance to marketed antibiotics becomes a great challenge to clinicians throughout the globe to combat with enterococcal pathogenesis. In the recent future, MDR enterococci will be immense clinical challenges to treat infections in hospitalized patients. Current trends in the epidemiology and population structure of antibiotic-resistant *Enterococcus* species clearly suggest that MDR enterococci may become the most common species isolated from patients in the upcoming eons. Nanotechnology is an emerging branch of science which could restrict the propagation of enterococci. Various attempts were already made worldwide to develop versatile nanomaterials that exhibited immense potentiality to limit enterococcal growth in in vitro and in vivo. However, with the advent and advancement of nanotechnology, more studies are extremely necessary to develop comprehensive strategies to limit the *Enterococcus*-associated infections and their large-scale implementation in upcoming eons.

Acknowledgments The authors are grateful to the Department of Science and Technology and Biotechnology, Govt. of West Bengal, India, for the financial assistance (Memo No: 532/(Sanc.)\ ST/P/S&T/2G-48/2018 dated: 27/03/2019). The first author is also thankful to the Department of Physiology, Midnapore College (Autonomous), West Bengal, India.

References

Ahmed MO, Baptiste KE (2018) Vancomycin-resistant enterococci: a review of antimicrobial resistance mechanisms and perspectives of human and animal health. Microb Drug Resist 24:590–606. https://doi.org/10.1089/mdr.2017.0147

Akbari T, Pourhajibagher M, Hosseini F et al (2017) The effect of indocyanine green loaded on a novel nano-graphene oxide for high performance of photodynamic therapy against *Enterococcus faecalis*. Photodiagn Photodyn Ther 20:148–153. https://doi.org/10.1016/j.pdpdt.2017.08.017

Akhavan O, Ghaderi E (2010) Toxicity of graphene and graphene oxide nanowalls against bacteria. ACS Nano 4:5731–5736. https://doi.org/10.1021/nn101390x

Ali L, Goraya M, Arafat Y et al (2017) Molecular mechanism of quorum-sensing in *Enterococcus faecalis*: its role in virulence and therapeutic approaches. Int J Mol Sci 18:960. https://doi.org/10.3390/ijms18050960

Almohamad S, Somarajan SR, Singh KV et al (2014) Influence of isolate origin and presence of various genes on biofilm formation by *Enterococcus faecium*. FEMS Microbiol Lett 353:151–156. https://doi.org/10.1111/1574-6968.12418

Anand K, Tiloke C, Phulukdaree A et al (2016) Biosynthesis of palladium nanoparticles by using Moringa oleifera flower extract and their catalytic and biological properties. J Photochem Photobiol B Biol 165:87–95. https://doi.org/10.1016/j.jphotobiol.2016.09.039

Anderson DJ, Murdoch DR, Sexton DJ, Reller LB (2004) Risk factors for infective endocarditis in patients with enterococcal bacteremia: a case-control study. Infection 32:72–77. https://doi.org/10.1007/s15010-004-2036-1

Arbeloa A, Segal H, Hugonnet J-E et al (2004) Role of class A penicillin-binding proteins in PBP5-mediated-lactam resistance in *Enterococcus faecalis*. J Bacteriol 186:1221–1228. https://doi.org/10.1128/JB.186.5.1221-1228.2004

Arias CA, Murray BE (2012) The rise of the Enterococcus: beyond vancomycin resistance. Nat Rev Microbiol 10:266–278. https://doi.org/10.1038/nrmicro2761

Arsène S, Leclercq R (2007) Role of a qnr-like gene in the intrinsic resistance of *Enterococcus faecalis* to fluoroquinolones. Antimicrob Agents Chemother 51:3254–3258. https://doi.org/10.1128/AAC.00274-07

Arthur M, Courvalin P (1993) Genetics and mechanisms of glycopeptide resistance in enterococci. Antimicrob Agents Chemother 37:1563–1571. https://doi.org/10.1128/AAC.37.8.1563

Arthur M, Molinas C, Depardieu F, Courvalin P (1993) Characterization of Tn*1546*, a Tn*3*-related transposon conferring glycopeptide resistance by synthesis of depsipeptide peptidoglycan precursors in *Enterococcus faecium* BM4147. J Bacteriol 175:117–127. https://doi.org/10.1128/jb.175.1.117-127.1993

Aziz N, Faraz M, Pandey R, Sakir M, Fatma T, Varma A, Barman I, Prasad R (2015) Facile algae-derived route to biogenic silver nanoparticles: Synthesis, antibacterial and photocatalytic properties. Langmuir 31:11605–11612. https://doi.org/10.1021/acs.langmuir.5b03081

Aziz N, Pandey R, Barman I, Prasad R (2016) Leveraging the attributes of Mucor hiemalis-derived silver nanoparticles for a synergistic broad-spectrum antimicrobial platform. Front Microbiol 7:1984. https://doi.org/10.3389/fmicb.2016.01984

Baddour LM, Wilson WR, Bayer AS et al (2005) Infective endocarditis: diagnosis, antimicrobial therapy, and management of complications: a statement for healthcare professionals from the Committee on Rheumatic Fever, Endocarditis, and Kawasaki Disease, Council on Cardiovascular Disease in the Young, and the Councils on Clinical Cardiology, Stroke, and Cardiovascular Surgery and Anesthesia, American Heart Association: endorsed by the Infectious Diseases Society of America. Circulation 111:e394–e434. https://doi.org/10.1161/CIRCULATIONAHA.105.165564

Baptista PV, McCusker MP, Carvalho A et al (2018) Nano-strategies to fight multidrug resistant bacteria—"A battle of the titans". Front Microbiol 9:1441. https://doi.org/10.3389/fmicb.2018.01441

Boumghar-Bourtchaï L, Dhalluin A, Malbruny B et al (2009) Influence of recombination on development of mutational resistance to linezolid in *Enterococcus faecalis* JH2-2. Antimicrob Agents Chemother 53:4007–4009. https://doi.org/10.1128/AAC.01633-08

Bourgeois-Nicolaos N, Massias L, Couson B et al (2007) Dose dependence of emergence of resistance to linezolid in *Enterococcus faecalis* in vivo. J Infect Dis 195:1480–1488. https://doi.org/10.1086/513876

Bozdogan B, Berrezouga L, Kuo MS et al (1999) A new resistance gene, linB, conferring resistance to lincosamides by nucleotidylation in *Enterococcus faecium* HM1025. Antimicrob Agents Chemother 43:925–929. https://doi.org/10.1128/AAC.43.4.925

Camilli A, Bassler BL (2006) Bacterial small-molecule signaling pathways. Science 311:1113–1116. https://doi.org/10.1126/science.1121357

Cavassin ED, de Figueiredo LF, Otoch JP et al (2015) Comparison of methods to detect the in vitro activity of silver nanoparticles (AgNP) against multidrug resistant bacteria. J Nanobiotechnol 13(1): 64. https://doi.org/10.1186/s12951-015-0120-6

Ch'ng JH, Chong KK, Lam LN et al (2019) Biofilm associated infection by enterococci. Nat Rev Microbiol 17:124–124. https://doi.org/10.1038/s41579-018-0128-7

Chandler JR, Dunny GM (2008) Characterization of the sequence specificity determinants required for processing and control of sex pheromone by the intramembrane protease Eep and the plasmid-encoded protein PrgY. J Bacteriol 190:1172–1183. https://doi.org/10.1128/JB.01327-07

Chang S, Sievert D, Hageman J et al (2003) Infection with vancomycin-resistant Staphylococcus aureus containing the vanA resistance gene. N Engl J Med 348:1342–1347. https://doi.org/10.1056/NEJMoa025025

Chen H, Wu W, Ni M et al (2013) Linezolid-resistant clinical isolates of enterococci and Staphylococcus cohnii from a multicentre study in China: molecular epidemiology and resistance mechanisms. Int J Antimicrob Agents 42:317–321. https://doi.org/10.1016/j.ijantimicag.2013.06.008

Chen Y, Bandyopadhyay A, Kozlowicz BK et al (2017) Mechanisms of peptide sex pheromone regulation of conjugation in Enterococcus faecalis. Microbiology 6:e00492. https://doi.org/10.1002/mbo3.492

Chenoweth C, Robinson K, Schaberg D (1990) Efficacy of ampicillin versus trimethoprim sulfamethoxazole in a mouse model of lethal enterococcal peritonitis. Antimicrob Agents Chemother 34:1800–1802. https://doi.org/10.1128/AAC.34.9.1800

Chifiriuc MC, Grumezescu AM, Andronescu E et al (2013) Water dispersible magnetite nanoparticles influence the efficacy of antibiotics against planktonic and biofilm embedded Enterococcus faecalis cells. Anaerobe 22:14–19. https://doi.org/10.1016/j.anaerobe.2013.04.013

Chow J, Zervos M, Lerner S et al (1997) A novel gentamicin resistance gene in Enterococcus. Antimicrob Agents Chemother 41:511–514

Comenge Y, Quintiliani R Jr, Li L et al (2003) The CroRS two-component regulatory system is required for intrinsic beta-lactam resistance in Enterococcus faecalis. J Bacteriol 185:7184–7192. https://doi.org/10.1128/jb.185.24.7184-7192.2003

Cook LC, Federle MJ (2014) Peptide pheromone signaling in Streptococcus and Enterococcus. FEMS Microbiol Rev 38(3):473–492. https://doi.org/10.1111/1574-6976.12046

Cook L, Chatterjee A, Barnes A et al (2011) Biofilm growth alters regulation of conjugation by a bacterial pheromone. Mol Microbiol 81:1499–1510. https://doi.org/10.1111/j.1365-2958.2011.07786.x

Costa Y, Galimand M, Leclercq R et al (1993) Characterization of the chromosomal aac (6′)-Ii gene specific for Enterococcus faecium. Antimicrob Agents Chemother 37:1896–1903. https://doi.org/10.1128/aac.37.9.1896

Costerton JW (2001) Cystic fibrosis pathogenesis and the role of biofilms in persistent infection. Trends Microbiol 9:50–52. https://doi.org/10.1016/S0966-842X(00)01918-1

Costerton JW, Stewart PS, Greenberg EP (1999) Bacterial biofilms: a common cause of persistent infections. Science 284:1318–1322. https://doi.org/10.1126/science.284.5418.1318

Coudron PE, Markowitz SM, Wong ES (1992) Isolation of a betalactamase-producing, aminoglycoside-resistant strain of Enterococcus faecium. Antimicrob Agents Chemother 36:1125–1126. https://doi.org/10.1128/AAC.36.5.1125

Courvalin P (2006) Vancomycin resistance in Gram-positive cocci. Clin Infect Dis 42(Supplement_1):S25–S34. https://doi.org/10.1086/491711

Courvalin P, Carlier C, Collatz E (1980) Plasmid-mediated resistance to aminocyclitol antibiotics in group D streptococci. J Bacteriol 143:541–551. https://doi.org/10.1128/aac.13.5.716

Dautle MP, Wilkinson TR, Gauderer MW (2003) Isolation and identification of biofilm microorganisms from silicone gastrostomy devices. J Pediatr Surg 38:216–220. https://doi.org/10.1053/jpsu.2003.50046

Depardieu F, Mejean V, Courvalin P (2015) Competition between VanU(G) repressor and VanR(G) activator leads to rheostatic control of vanG vancomycin resistance operon expression. PLoS Genet 11:e1005170. https://doi.org/10.1371/journal.pgen.1005170

Devriese LA, Pot B, Van Damme L, Kersters K, Haesebrouck F (1994) Identification of Enterococcus species isolated from foods of animal origin. Int J Food Microbiol 26:187–197. https://doi.org/10.1016/0168-1605(94)00119-Q

Dianat O, Saedi S, Kazem M, Alam M (2015) Antimicrobial activity of nanoparticle calcium hydroxide against Enterococcus faecalis: an in vitro study. Iran Endod J 10:39

Diaz L, Kiratisin P, Mendes R et al (2012) Transferable plasmid-mediated resistance to linezolid due to cfr in a human clinical isolate of *Enterococcus faecalis*. Antimicrob Agents Chemother 56:3917–3922. https://doi.org/10.1128/AAC.00419-12

Diaz L, Kontoyiannis DP, Panesso D et al (2013) Dissecting the mechanisms of linezolid resistance in a *Drosophila melanogaster* infection model of *Staphylococcus aureus*. J Infect Dis 208:83–91. https://doi.org/10.1093/infdis/jit138

Distel JW, Hatton JF, Gillespie MJ (2002) Biofilm formation in medicated root canals. J Endod 28:689–693. https://doi.org/10.1097/00004770-200210000-00003

Dowidar N, Moesgaard F, Matzen P (1991) Clogging and other complications of endoscopic biliary endoprostheses. Scand J Gastroenterol 26:1132–1136. https://doi.org/10.3109/00365529108998604

Dunny GM, Hancock LE, Shankar N (2014) In: Gilmore MS (ed) Enterococci: from commensals to leading causes of drug resistant infection. Massachusetts Eye and Ear Infirmary, Boston

Eliopoulos GM, Farber BF, Murray BE et al (1984) Ribosomal resistance of clinical enterococcal to streptomycin isolates. Antimicrob Agents Chemother 25:398–399. https://doi.org/10.1128/AAC.25.3.398

Emori TG, Gaynes RP (1993) An overview of nosocomial infections, including the role of the microbiology laboratory. Clin Microbiol Rev 6:428–442

Enne V, Delsol A, Roe J et al (2004) Rifampicin resistance and its fitness cost in *Enterococcus faecium*. J Antimicrob Chemother 53:203–207. https://doi.org/10.1093/jac/dkh044

Fabretti F, Theilacker C, Baldassarri L et al (2006) Alanine esters of enterococcal lipoteichoic acid play a role in biofilm formation and resistance to antimicrobial peptides. Infect Immun 74:4164–4171. https://doi.org/10.1128/IAI.00111-06

Facklam RR (1973) Comparison of several laboratory media for presumptive identification of enterococci and group D streptococci. J App Microbiol 26(2):138–145

Fiedler S, Bender JK, Klare I et al (2016) Tigecycline resistance in clinical isolates of *Enterococcus faecium* is mediated by an upregulation of plasmid-encoded tetracycline determinants *tet*(L) and *tet*(M). J Antimicrob Chemother 71:871–881. https://doi.org/10.1093/jac/dkv420

Galimand M, Schmitt E, Panvert M et al (2011) Intrinsic resistance to aminoglycosides in *Enterococcus faecium* is conferred by the 16S rRNA m5C1404-specific methyltransferase Efm M. RNA 17:251–262. https://doi.org/10.1261/rna.2233511

García-Solache M, Rice LB (2019) The enterococcus: a model of adaptability to its environment. Clin Microbiol Rev 32(2):e00058–e00018. https://doi.org/10.1128/CMR.00058-18

Gilmore M, Lebreton F, Van Tyne D (2013) Dual defensin strategy for targeting *Enterococcus faecalis*. Proc Natl Acad Sci U S A 110:19980–19981. https://doi.org/10.1073/pnas.1319939110

Govindaraju S, Samal M, Yun K (2016) Superior antibacterial activity of GlcN-AuNP-GO by ultraviolet irradiation. Mater Sci Eng C 69:366–372. https://doi.org/10.1016/j.msec.2016.06.052

Graninger W, Ragette R (1992) Nosocomial bacteremia due to *Enterococcus faecalis* without endocarditis. Clin Infect Dis 15:49–57. https://doi.org/10.1093/clinids/15.1.49

Grayson M, Thauvin-Eliopoulos C, Eliopoulos G et al (1990) Failure of trimethoprim-sulfamethoxazole therapy in experimental enterococcal endocarditis. Antimicrob Agents Chemother 34:1792–1794. https://doi.org/10.1128/AAC.34.9.1792

Guiton PS, Hung CS, Kline KA et al (2009) Contribution of autolysin and sortase a during Enterococcus faecalis DNA dependent biofilm development. Infect Immun 77:3626–3638. https://doi.org/10.1128/IAI.00219-09

Hall CL, Tschannen M, Worthey EA et al (2013) IreB, a Ser/Thr kinase substrate, influences antimicrobial resistance in *Enterococcus faecalis*. Antimicrob Agents Chemother 57:6179–6186. https://doi.org/10.1128/AAC.01472-13

Hammerum AM (2012) Enterococci of animal origin and their significance for public health. Clin Microbiol Infect 18:619–625. https://doi.org/10.1111/j.1469-0691.2012.03829.x

Hawkey P (2003) Mechanisms of quinolone action and microbial response. J Antimicrob Chemother 51:29–35. https://doi.org/10.1093/jac/dkg207

Heikens E, Singh KV, Jacques-Palaz KD et al (2011) Contribution of the enterococcal surface protein Esp to pathogenesis of *Enterococcus faecium* endocarditis. Microbes Infect 13:1185–1190. https://doi.org/10.1016/j.micinf.2011.08.006

Hendrickx AP, van Luit-Asbroek M, Schapendonk CM et al (2009) SgrA, a nidogen-binding LPXTG surface adhesin implicated in biofilm formation, and EcbA, a collagen binding MSCRAMM, are two novel adhesins of hospital-acquired *Enterococcus faecium*. Infect Immun 77:5097–5106. https://doi.org/10.1128/IAI.00275-09

Hidron AI, Edwards JR, Patel J et al (2008) NHSN annual update: antimicrobial-resistant pathogens associated with healthcare-associated infections: annual summary of data reported to the National Healthcare Safety Network at the Centers for Disease Control and Prevention, 2006–2007. Infect Control Hosp Epidemiol 29:996–1011. https://doi.org/10.1086/591861

Higuita NIA, Huycke MM (2014) Enterococcal disease, epidemiology, and implications for treatment. In: Enterococci: From Commensals to Leading Causes of Drug Resistant Infection [Internet]. Massachusetts Eye and Ear Infirmary, Boston

Hirt H, Greenwood-Quaintance KE, Karau MJ et al (2018) *Enterococcus faecalis* sex pheromone cCF10 enhances conjugative plasmid transfer in vivo. MBio 9:e00037–e00018. https://doi.org/10.1128/mBio.00037-18

Hollenbeck BL, Rice LB (2012) Intrinsic and acquired resistance mechanisms in enterococcus. Virulence 3:421–433. https://doi.org/10.4161/viru.21282

Hooper D (2000) Mechanisms of action and resistance of older and newer fluoroquinolones. Clin Infect Dis 31:S24–S28. https://doi.org/10.1086/314056

Hu W (2010) Graphene-based antibacterial paper. ACS Nano 4:4317–4323. https://doi.org/10.1021/nn101097v

Joyanes P, Pascual A, Martinez-Martinez L et al (2000) In vitro adherence of *Enterococcus faecalis* and *Enterococcus faecium* to urinary catheters. Eur J Clin Microbiol Infect Dis 19:124–127. https://doi.org/10.1111/j.1469-0691.1999.tb00160.x

Jureen R, Top J, Mohn SC et al (2003) Molecular characterization of ampicillin-resistant *Enterococcus faecium* isolates from hospitalized patients in Norway. J Clin Microbiol 41:2330–2336. https://doi.org/10.1128/JCM.41.6.2330-2336.2003

Kandaswamy K, Liew T, Wang C et al (2013) Focal targeting by human β-defensin 2 disrupts localized virulence factor assembly sites in *Enterococcus faecalis*. Proc Natl Acad Sci U S A 110:20230–20235. https://doi.org/10.1073/pnas.1319066110

Kanematsu E, Deguchi T, Yasuda M et al (1998) Alterations in the GyrA subunit of DNA gyrase and the ParC subunit of DNA topoisomerase IV associated with quinolone resistance in *Enterococcus faecalis*. Antimicrob Agents Chemother 42:433–435. https://doi.org/10.1128/AAC.42.2.433

Katva S, Das S, Moti HS et al (2018) Antibacterial synergy of silver nanoparticles with gentamicin and chloramphenicol against *Enterococcus faecalis*. Pharmacogn Mag 13:S828–S833. https://doi.org/10.4103/pm.pm_120_17

Keane PF, Bonner MC, Johnston SR et al (1994) Characterization of biofilm and encrustation on ureteric stents in vivo. Br J Urol 73:687–691. https://doi.org/10.1111/j.1464-410X.1994.tb07557.x

Kehrenberg C, Schwarz S, Jacobsen L et al (2005) A new mechanism for chloramphenicol, florfenicol and clindamycin resistance: methylation of 23S ribosomal RNA at A2503. Mol Microbiol 57:1064–1073. https://doi.org/10.1111/j.1365-2958.2005.04754.x

Khiralla GM, El-Deeb BA (2015) Antimicrobial and antibiofilm effects of selenium nanoparticles on some foodborne pathogens. LWT-Food Sci Technol 63:1001–1007. https://doi.org/10.1016/j.lwt.2015.03.086

Klein G (2003) Taxonomy, ecology and antibiotic resistance of enterococci from food and the gastro intestinal tract. Int J Food Microbiol 88:123–131. https://doi.org/10.1016/S0168-1605(03)00175-2

Klibi N, Sáenz Y, Zarazaga M et al (2008) Polymorphism in pbp5 gene detected in clinical *Enterococcus faecium* strains with different ampicillin MICs from a Tunisian hospital. J Chemother 20:436–440. https://doi.org/10.1179/joc.2008.20.4.436

Kobayakawa S, Jett BD, Gilmore MS (2005) Biofilm formation by *Enterococcus faecalis* on intraocular lens material. Curr Eye Res 30:741–745. https://doi.org/10.1080/02713680591005959

Krasteva PV, Giglio KM, Sondermann H (2012) Sensing the messenger: the diverse ways that bacteria signal through c-di-GMP. Protein Sci 21:929–948. https://doi.org/10.1002/pro.2093

Kristich C, Little J (2012) Mutations in the β subunit of RNA polymerase alter intrinsic cephalosporin resistance in Enterococci. Antimicrob Agents Chemother 56:2022–2027. https://doi.org/10.1128/AAC.06077-11

Kristich CJ, Wells CL, Dunny GM (2007) A eukaryotic-type Ser/Thr kinase in *Enterococcus faecalis* mediates antimicrobial resistance and intestinal persistence. Proc Natl Acad Sci U S A 10:3508–3508. https://doi.org/10.1073/pnas.0608742104

Kumar NI, Das SA, Jyoti AN, Kaushik SA (2016) Synergistic effect of silver nanoparticles with doxycycline against Klebsiella pneumoniae. Int J Pharm Pharm Sci 8:183–186

Leach K, Swaney S, Colca J et al (2007) The site of action of oxazolidinone antibiotics in living bacteria and in human mitochondria. Mol Cell 26:393–402. https://doi.org/10.1016/j.molcel.2007.04.005

Lebreton F, Van Schaik W, Sanguinetti M et al (2012) AsrR is an oxidative stress sensing regulator modulating *Enterococcus faecium* opportunistic traits, antimicrobial resistance, and pathogenicity. PLoS Pathog 8:e1002834. https://doi.org/10.1371/journal.ppat.1002834

Lebreton F, van Schaik W, McGuire AM et al (2013) Emergence of epidemic multidrug-resistant *Enterococcus faecium* from animal and commensal strains. MBio 4:e00534–e00513. https://doi.org/10.1128/mBio.00534-13

Lebreton F, Willems RJ, Gilmore MS (2014) Enterococcus diversity, origins in nature, and gut colonization. In: Enterococci: from commensals to leading causes of drug resistant infection [Internet]. Massachusetts Eye and Ear Infirmary, Boston

Leclercq R, Dutka-Malen S, Duval J, Courvalin P (1992) Vancomycin resistance gene *vanC* is specific to *Enterococcus gallinarum*. Antimicrob Agents Chemother 36:2005–2008. https://doi.org/10.1128/aac.36.9.2005

Lewis K (2001) Riddle of biofilm resistance. Antimicrob Agents Chemother 45:999–1007. https://doi.org/10.1128/AAC.45.4.999-1007.2001

Li YH, Tian X (2012) Quorum sensing and bacterial social interactions in biofilms. Sensors 12:2519–2538. https://doi.org/10.3390/s120302519

Lim SY, Teh CSJ, Thong KL (2017) Biofilm-related diseases and omics: global transcriptional profiling of *Enterococcus faecium* reveals different gene expression patterns in the biofilm and planktonic cells. OMICS 21:592–602. https://doi.org/10.1089/omi.2017.0119

Locke JB, Hilgers M, Shaw KJ (2009) Mutations in ribosomal protein L3 are associated with oxazolidinone resistance in staphylococci of clinical origin. Antimicrob Agents Chemother 53:5275–5278. https://doi.org/10.1128/AAC.01032-09

López M, Tenorio C, Del Campo R et al (2011) Characterization of the mechanisms of fluoroquinolone resistance in vancomycin-resistant enterococci of different origins. J Chemother 23:87–91. https://doi.org/10.1179/joc.2011.23.2.87

MacCallum WG, Hastings TW (1899) A case of acute endocarditis caused by Micrococcus zymogenes (nov. spec.), with a description of the microorganism. J Exp Med 4:521–534. https://doi.org/10.1084/jem.4.5-6.521

Maki DG, Agger WA (1988) Enterococcal bacteremia: clinical features, the risk of endocarditis, and management. Medicine 67:248–269

Malani PN, Kauffman CA, Zervos MJ (2002) Enterococcal disease, epidemiology, and treatment. In: Gilmore MS, Clewell DB, Courvalin P, Dunny GM, Murray BE, Rice LB (eds) The Enterococci: pathogenesis, molecular biology, and antibiotic resistance. ASM Press, Washington, DC, pp 385–408

Marshall S, Donskey C, Hutton-Thomas R et al (2002) Gene dosage and linezolid resistance in *Enterococcus faecium* and *Enterococcus faecalis*. Antimicrob Agents Chemother 46:3334–3336. https://doi.org/10.1128/AAC.46.10.3334-3336.2002

McDonald LC, Kuehnert MJ, Tenover FC, Jarvis WR (1997) Vancomycin-resistant enterococci outside the health-care setting: prevalence, sources, and public health implications. Emerg Infect Dis 3:311. https://doi.org/10.3201/eid0303.970307

McDonald JR, Olaison L, Anderson DJ, Hoen B, Miro JM, Eykyn S et al (2005) Enterococcal endocarditis: 107 cases from the international collaboration on endocarditis merged database. Am J Med 118:759–766. https://doi.org/10.1016/j.amjmed.2005.02.020

Mendes R, Deshpande L, Castanheira M et al (2008) First report of cfr-mediated resistance to linezolid in human staphylococcal clinical isolates recovered in the United States. Antimicrob Agents Chemother 52:2244–2246. https://doi.org/10.1128/AAC.00231-08

Miller WR, Munita JM, Arias CA (2014) Mechanisms of antibiotic resistance in enterococci. Expert Rev Anti-Infect Ther 12:1221–1236. https://doi.org/10.1586/14787210.2014.956092

Miller WR, Bayer AS, Arias CA (2016) Mechanism of action and resistance to daptomycin in *Staphylococcus aureus* and enterococci. Cold Spring Harb Perspect Med 6:a026997. https://doi.org/10.1101/cshperspect.a026997

Mishra NN, Bayer AS, Tran TT, Shamoo Y, Mileykovskaya E, Dowhan W, Guan Z, Arias CA (2012) Daptomycin resistance in enterococci is associated with distinct alterations of cell membrane phospholipid content. PLoS One 7:e43958. https://doi.org/10.1371/journal.pone.0043958

Mohamed JA, Huang DB (2007) Biofilm formation by enterococci. J Med Microbiol 56:1581–1588. https://doi.org/10.1099/jmm.0.47331-0

Mohamed JA, Teng F, Nallapareddy SR, Murray BE (2006) Pleiotrophic effects of 2 Enterococcus faecalis sagA-like genes, salA and salB, which encode proteins that are antigenic during human infection, on biofilm formation and binding to collagen type I and fibronectin. J Infect Dis 193:231–240. https://doi.org/10.1086/498871

Monds RD, O'Toole GA (2009) The developmental model of microbial biofilms: ten years of a paradigm up for review. Trends Microbiol 17:73–87. https://doi.org/10.1016/j.tim.2008.11.001

Montealegre MC, Singh KV, Somarajan SR et al (2016a) Role of the Emp pilus subunits of *Enterococcus faecium* in biofilm formation, adherence to host extracellular matrix components, and experimental infection. Infect Immun 84:1491–1500. https://doi.org/10.1128/IAI.01396-15

Montealegre MC, Roh JH, Rae M et al (2016b) Differential penicillin-binding protein 5 (PBP5) levels in the *Enterococcus faecium* clades with different levels of ampicillin resistance. Antimicrob Agents Chemother 61:e02034–e02016. https://doi.org/10.1128/AAC.02034-16

Muller C, Le Breton Y, Morin T et al (2006) The response regulator CroR modulates expression of the secreted stress-induced SalB protein in *Enterococcus faecalis*. J Bacteriol 188:2636–2645. https://doi.org/10.1128/JB.188.7.2636-2645.2006

Murray BE (1990) The life and times of the Enterococcus. Clin Microbiol Rev 3:46–65. https://doi.org/10.1128/cmr.3.1.46

Murray BE (1992) Beta-lactamase-producing enterococci. Antimicrob Agents Chemother 36:2355–2359. https://doi.org/10.1128/aac.36.11.2355

Murray BE, Lopardo HA, Rubeglio EA et al (1992) Intrahospital spread of a single gentamicin-resistant, beta-lactamase producing strain of *Enterococcus faecalis* in Argentina. Antimicrob Agents Chemother 36:230–232. https://doi.org/10.1128/AAC.36.1.230

Nallapareddy SR, Singh KV, Sillanpää J et al (2006) Endocarditis and biofilm associated pili of *Enterococcus faecalis*. J Clin Invest 116:2799–2807. https://doi.org/10.1172/JCI29021

Nallapareddy SR, Singh KV, Sillanpaa J et al (2011a) Relative contributions of Ebp Pili and the collagen adhesin ace to host extracellular matrix protein adherence and experimental urinary tract infection by *Enterococcus faecalis* OG1RF. Infect Immun 79:2901–2910. https://doi.org/10.1128/IAI.00038-11

Nallapareddy SR, Sillanpää J, Mitchell J et al (2011b) Conservation of Ebp-type pilus genes among Enterococci and demonstration of their role in adherence of *Enterococcus faecalis* to human platelets. Infect Immun 79:2911–2920. https://doi.org/10.1128/IAI.00039-11

Nielsen HV, Guiton PS, Kline KA et al (2012) The metal ion-dependent adhesion site motif of the *Enterococcus faecalis* EbpA pilin mediates pilus function in catheter-associated urinary tract infection. MBio 3:e00177–e00112. https://doi.org/10.1128/mBio.00177-12

Nielsen HV, Flores-Mireles AL, Kau AL et al (2013) Pilin and sortase residues critical for endocarditis- and biofilm-associated pilus biogenesis in *Enterococcus faecalis*. J Bacteriol 195:4484–4495. https://doi.org/10.1128/JB.00451-13

Noskin GA, Cooper I, Peterson LR (1995a) Vancomycin-resistant *Enterococcus faecium* sepsis following persistent colonization. JAMA Intern Med 155:1445–1447. https://doi.org/10.1001/archinte.1995.00430130139015

Noskin GA, Stosor V, Cooper I, Peterson LR (1995b) Recovery of vancomycin-resistant enterococci on fingertips and environmental surfaces. Infect Control Hosp Epidemiol 16:577–581. https://doi.org/10.2307/30141097

O'Toole G, Kaplan HB, Kolter R (2000) Biofilm formation as microbial development. Annu Rev Microbiol 54:49–79

Otari SV, Patil RM, Waghmare SR et al (2013) A novel microbial synthesis of catalytically active Ag–alginate biohydrogel and its antimicrobial activity. Dalt Trans 42:9966–9975. https://doi.org/10.1039/C3DT51093J

Otto M (2006) Bacterial evasion of antimicrobial peptides by biofilm formation. Curr Top Microbiol Immunol 306:251–258

Oyamada Y, Ito H, Fujimoto K et al (2006) Combination of known and unknown mechanisms confers high-level resistance to fluoroquinolones in *Enterococcus faecium*. J Med Microbiol 55:729–736. https://doi.org/10.1099/jmm.0.46303-0

Paganelli FL, Willems RJ, Jansen P et al (2013) *Enterococcus faecium* biofilm formation: identification of major autolysin AtlAEfm, associated Acm surface localization, and AtlAEfm independent extracellular DNA Release. MBio 4:e00154. https://doi.org/10.1128/mBio.00154-13

Patterson JE, Sweeney AH, Simms M et al (1995) An analysis of 100 serious enterococcal infections: epidemiology, antibiotic susceptibility, and outcome. Medicine 74:191–200. https://doi.org/10.1097/00005792-199507000-00003

Poeta P, Costa D, Igrejas G et al (2007) Polymorphisms of the pbp5 gene and correlation with ampicillin resistance in *Enterococcus faecium* isolates of animal origin. J Med Microbiol 56:236–240. https://doi.org/10.1099/jmm.0.46778-0

Portillo A, Ruiz-Larrea F, Zarazaga M et al (2000) Macrolide resistance genes in *Enterococcus* spp. Antimicrob Agents Chemother 44:967–971. https://doi.org/10.1128/aac.44.4.967-971.2000

Prabaker K, Weinstein RA (2011) Trends in antimicrobial resistance in intensive care units in the United States. Curr Opin Crit Care 17:472–479. https://doi.org/10.1097/MCC.0b013e32834a4b03

Rand K, Houck H, Silverman J et al (2007) Daptomycin-reversible rifampicin resistance in vancomycin-resistant *Enterococcus faecium*. J Antimicrob Chemother 59:1017–1020. https://doi.org/10.1093/jac/dkm045

Ray A, Pultz N, Bhalla A et al (2003) Coexistence of vancomycin-resistant enterococci and *Staphylococcus aureus* in the intestinal tracts of hospitalized patients. Clin Infect Dis 37:875–881. https://doi.org/10.1086/377451

Reid KC, Cockerill FR III, Patel R (2001) Clinical and epidemiological features of *Enterococcus casseliflavus/flavescens* and *Enterococcus gallinarum* bacteremia: a report of 20 cases. Clin Infect Dis 32:1540–1546. https://doi.org/10.1086/320542

Rice LB (1998) Tn916 family conjugative transposons and dissemination of antimicrobial resistance determinants. Antimicrob Agents Chemother 42:1871–1877. https://doi.org/10.1128/AAC.42.8.1871

Rice LB, Murray BE (1995) β-lactamase-producing enterococci. In: Brown F, Ferretti JJ (eds) Genetics of streptococci, enterococci and lactococci, Developmental and biological standards, vol 85. Karger, Basel, Switzerland

Rice LB, Calderwood SB, Eliopoulos GM et al (1991) Enterococcal endocarditis: a comparison of prosthetic and native valve disease. Rev Infect Dis 13:1–7. https://doi.org/10.1093/clinids/13.1.1

Rice LB, Bellais S, Carias LL et al (2004) Impact of specific pbp5 mutations on expression of β-lactam resistance in *Enterococcus faecium*. Antimicrob Agents Chemother 48:3028–3032. https://doi.org/10.1128/AAC.48.8.3028-3032.2004

Rice LB, Carias LL, Rudin S et al (2009) Role of class A penicillin-binding proteins in the expression of beta-lactam resistance in *Enterococcus faecium*. J Bacteriol 191:3649–3656. https://doi.org/10.1128/JB.01834-08

Robbins WC, Tompsett R (1951) Treatment of enterococcal endocarditis and bacteremia: results of combined therapy with penicillin and streptomycin. Am J Med 10:278–299. https://doi.org/10.1016/0002-9343(51)90273-2

Rozdzinski E, Marre R, Susa M et al (2001) Aggregation substance mediated adherence of *Enterococcus faecalis* to immobilized extracellular matrix proteins. Microb Pathog 30:211–220. https://doi.org/10.1006/mpat.2000.0429

Rybkine T, Mainardi J-L, Sougakoff W et al (1998) Penicillin-binding protein 5 sequence alterations in clinical isolates of *Enterococcus faecium* with different levels of β-lactam resistance. J Infect Dis 178:159–163. https://doi.org/10.1086/515605

Saeb ATM, Alshammari AS, Al-brahim H, Al-rubeaan KA (2014) Production of silver nanoparticles with strong and stable antimicrobial activity against highly pathogenic and multidrug resistant bacteria. Sci World J 2:704708. https://doi.org/10.1155/2014/704708

Saini MK, Das S, Moti HS et al (2019) Biogenic synthesis and characterization of silver nanoparticles from bacteria isolated from garden soil and its antibacterial activity against *Enterococcus faecalis*. Int J Res Pharm Sci 10:21–26. https://doi.org/10.26452/ijrps.v10i1.1774

Sandoe JA, Witherden IR, Cove JH et al (2003) Correlation between enterococcal biofilm formation in vitro and medical-device-related infection potential in vivo. J Med Microbiol 52:547–550. https://doi.org/10.1099/jmm.0.05201-0

Sassi M, Guerin F, Lesec L et al (2018) Genetic characterization of a VanG-type vancomycin-resistant *Enterococcus faecium* clinical isolate. J Antimicrob Chemother 73:852–855. https://doi.org/10.1093/jac/dkx510

Sava IG, Heikens E, Huebner J (2010) Pathogenesis and immunity in enterococcal infections. Clin Microbiol Infect 16:533–540. https://doi.org/10.1111/j.1469-0691.2010.03213.x

Shao C, Shang W, Yang Z et al (2012) LuxS-dependent AI-2 regulates versatile functions in *Enterococcus faecalis* V583. J Proteome Res 11:4465–4475. https://doi.org/10.1021/pr3002244

Shinabarger D, Marotti K, Murray R et al (1997) Mechanism of action of oxazolidinones: effects of linezolid and eperezolid on translation reactions. Antimicrob Agents Chemother 41:2132–2136. https://doi.org/10.1128/AAC.41.10.2132

Shlaes DM, Bouvet A, Devine C et al (1989) Inducible, transferable resistance to vancomycin in *Enterococcus faecalis* A256. Antimicrob Agents Chemother 33:198–203. https://doi.org/10.1128/AAC.33.2.198

Sillanpaa J, Nallapareddy SR, Prakash VP et al (2008) Identification and phenotypic characterization of a second collagen adhesin, Scm, and genome-based identification and analysis of 13 other predicted MSCRAMMs, including four distinct pilus loci, in *Enterococcus faecium*. Microbiology 154:3199–3211. https://doi.org/10.1099/mic.0.2008/017319-0

Sillanpaa J, Nallapareddy SR, Singh KV et al (2010) Characterization of the ebp(fm) pilus-encoding operon of Enterococcus faecium and its role in biofilm formation and virulence in a murine model of urinary tract infection. Virulence 1:236–246. https://doi.org/10.4161/viru.1.4.11966

Singh K, Weinstock G, Murray B (2002) An *Enterococcus faecalis* ABC homologue (Lsa) is required for the resistance of this species to clindamycin and quinupristin-dalfopristin. Antimicrob Agents Chemother 46:1845–1850. https://doi.org/10.1128/AAC.46.6.1845-1850.2002

Singh KV, Nallapareddy SR, Murray BE (2007) Importance of the ebp (endocarditis- and biofilm associated pilus) locus in the pathogenesis of *Enterococcus faecalis* ascending urinary tract infection. J Infect Dis 195:1671–1677. https://doi.org/10.1086/517524

Speer BS, Shoemaker NB, Salyers AA (1992) Bacterial resistance to tetracycline: mechanisms, transfer, and clinical significance. Clin Microbiol Rev 5:387–399. https://doi.org/10.1128/cmr.5.4.387

Sussmuth SD, Muscholl-Silberhorn A, Wirth R et al (2000) Aggregation substance promotes adherence, phagocytosis, and intracellular survival of *Enterococcus faecalis* within human macrophages and suppresses respiratory burst. Infect Immun 68:4900–4906. https://doi.org/10.1128/iai.68.9.4900-4906.2000

Tankovic J, Bachoual R, Ouabdesselam S et al (1999) In-vitro activity of moxifloxacin against fluoroquinolone-resistant strains of aerobic Gram-negative bacilli and *Enterococcus faecalis*. J Antimicrob Chemother 43:19–23. https://doi.org/10.1093/jac/43.suppl_2.19

Theilacker C, Sanchez-Carballo P, Toma I et al (2009) Glycolipids are involved in biofilm accumulation and prolonged bacteraemia in *Enterococcus faecalis*. Mol Microbiol 71:1055–1069. https://doi.org/10.1111/j.1365-2958.2009.06587.x

Theilacker C, Kaczyński Z, Kropec A et al (2011) Serodiversity of opsonic antibodies against *Enterococcus faecalis* – glycans of the cell wall revisited. PLoS One 6:e17839. https://doi.org/10.1371/journal.pone.0017839

Thiercelin ME (1899) Morphology and modes of reproduction of the enterococcus. Min Proc Soc Biol Its Affiliates 11:551–553

Toh S, Xiong L, Arias C et al (2007) Acquisition of a natural resistance gene renders a clinical strain of methicillin-resistant *Staphylococcus aureus* resistant to the synthetic antibiotic linezolid. Mol Microbiol 64:1506–1514. https://doi.org/10.1111/j.1365-2958.2007.05744.x

Tran J, Jacoby G, Hooper D (2005) Interaction of the plasmid-encoded quinolone resistance protein Qnr with *Escherichia coli* DNA gyrase. Antimicrob Agents Chemother 49:118–125. https://doi.org/10.1128/AAC.49.1.118-125.2005

van SW, Willems RJ (2010) Genome-based insights into the evolution of enterococci. Clin Microbiol Infect 16:527–532. https://doi.org/10.1111/j.1469-0691.2010.03201.x

Vester B (2018) The *cfr* and *cfr*-like multiple resistance genes. Res Microbiol 169:61–66. https://doi.org/10.1016/j.resmic.2017.12.003

Wang Y, Lv Y, Cai J et al (2015) A novel gene, *optrA*, that confers transferable resistance to oxazolidinones and phenicols and its presence in *Enterococcus faecalis* and *Enterococcus faecium* of human and animal origin. J Antimicrob Chemother 70:2182–2190. https://doi.org/10.1093/jac/dkv116

Werner G, Fleige C, Ewert B et al (2010) High-level ciprofloxacin resistance among hospital-adapted *Enterococcus faecium* (CC17). Int J Antimicrob Agents 35:119–125. https://doi.org/10.1016/j.ijantimicag.2009.10.012

Wilson WR, Wikowske CJ, Wright AJ et al (1984) Treatment of streptomycin-susceptible and streptomycin-resistant enterococcal endocarditis. Ann Intern Med 100:816–823. https://doi.org/10.7326/0003-4819-100-6-816

Wu D, Fan W, Kishen A et al (2014) Evaluation of the antibacterial efficacy of silver nanoparticles against *Enterococcus faecalis* biofilm. J Endod 40:285–290. https://doi.org/10.1016/j.joen.2013.08.022

Xu Y, Singh KV, Qin X et al (2000) Analysis of a gene cluster of *Enterococcus faecalis* involved in polysaccharide biosynthesis. Infect Immun 68:815–823. https://doi.org/10.1128/iai.68.2.815-823.2000

Yasufuku T, Shigemura K, Shirakawa T et al (2011) Mechanisms of and risk factors for fluoroquinolone resistance in clinical *Enterococcus faecalis* isolates from patients with urinary tract infections. J Clin Microbiol 49:3912–3916. https://doi.org/10.1128/JCM.05549-11

Zhou Z, Peng S, Sui M et al (2018) Multifunctional nanocomplex for surface-enhanced Raman scattering imaging and near-infrared photodynamic antimicrobial therapy of vancomycin-resistant bacteria. Colloids Surf B Biointerfaces 161:394–402. https://doi.org/10.1016/j.colsurfb.2017.11.005

Zorzi W, Zhou XY, Dardenne O et al (1996) Structure of the low-affinity penicillin-binding protein5 PBP5fm in wild-type and highly penicillin-resistant strains of *Enterococcus faecium*. J Bacteriol 178:4948–4957. https://doi.org/10.1128/jb.178.16.4948-4957.1996

Index

© Springer Nature Switzerland AG 2020
R. Prasad et al. (eds.), *Nanostructures for Antimicrobial and Antibiofilm Applications*, Nanotechnology in the Life Sciences,
https://doi.org/10.1007/978-3-030-40337-9

Printed by Printforce, the Netherlands